Solar Cells - Theory, Materials and Recent Advances

Edited by Ahmed Mourtada Elseman

Published in London, United Kingdom

IntechOpen

Supporting open minds since 2005

Solar Cells - Theory, Materials and Recent Advances
http://dx.doi.org/10.5772/intechopen.87735
Edited by Ahmed Mourtada Elseman

Contributors
Ahmed Mourtada Elseman, Alaa E. Abd El-Samad, Hager H. Zeenelabden, Radwa S. Mostafa, Menahtullah M. Mabrouk, Nasr Gad, Mostafa El-Aasser, Mohamed M. Rashad, Fridolin Tchangnwa Nya, Guy Maurel Dzifack Kenfack, Byunghong Lee, Robert Bob Chang, Ahmed M. El-Zohry, Paweł Kwasnicki, Rahmouni Salah, Nguyen Huu Hieu, Md. Mosharraf Hossain Bhuiyan, Fahmid Kabir, Md. Serajum Serajum Manir, Md. Saifur Rahaman, Md. Robiul Hossain, Prosenjit Barua, Bikram Ghosh, Fumiaki Mitsugi, Tomoaki Ikegami, Saiful Huque, Mubarak Ahmad Khan, Manoj Kumar Singh, Pratik V. Shinde, Pratap Singh, Pawan Kumar Tyagi, Selma Tchoketch Kebir, Chuan-Pei Lee, Yi-June Huang, Jhantu Kumar Saha, Animesh Dutta, Issa Faye, Ababacar Ndiaye, Elkhadji Mamado, Ghaida Salman Muhammed, QunLiang Song, Meng Wang, Sam Zhang, Johny Renoald Albert, Dishore Shunmugham Vanaja, Oleg Korotchenkov, Viktor Schlosser, Andriy Nadtochiy, Artem Podolian, Rakeshkumar Mahto, Reshma John, Harpreet Kaur Channi, Sahbi Ayachi, Rania Zaier, Badria Ibrahim, Ayah Salah Dafalla Abdalkhir, Osman Abdalrahman AL Mahdi Al Amin, Ahmed Mohamed Awadalla Mohamed, Alnazir Osman Mohammed Hamza, Fatehia Garma, Guillermo Martínez-Rodríguez, Amanda L. Fuentes-Silva

First published in London, United Kingdom, 2021 by IntechOpen
IntechOpen is the global imprint of INTECHOPEN LIMITED, registered in England and Wales, registration number: 11086078, 5 Princes Gate Court, London, SW7 2QJ, United Kingdom
Printed in Croatia

British Library Cataloguing-in-Publication Data
A catalogue record for this book is available from the British Library

Additional hard and PDF copies can be obtained from orders@intechopen.com

Solar Cells - Theory, Materials and Recent Advances
Edited by Ahmed Mourtada Elseman
p. cm.
Print ISBN 978-1-83881-016-0
Online ISBN 978-1-83881-017-7
eBook (PDF) ISBN 978-1-83881-024-5

For EU product safety concerns:
IN TECH d.o.o., Prolaz Marije Krucifikse Kozulić 3, 51000 Rijeka, Croatia, info@intechopen.com or visit our website at intechopen.com.

Meet the editor

Ahmed Mourtada Elseman obtained a BSc, MSc, and Ph.D. in Inorganic and Analytical Chemistry from the Faculty of Science, Al-Azhar University, Egypt. He obtained a Ph.D. thesis for developing perovskite solar cells in 2017. He also obtained two diplomas in Solar Energy from Inner Mongolia Institute of Scientific and Technological, Hohhot, China, in 2015, and the Institute of New Energy, Wuhan, China, in 2017, respectively. Presently, Dr. Elseman works as a research assistant professor at the Department of Electronic and Magnetic Materials, Central Metallurgical Research and Development Institute (CMRDI), Egypt. He was awarded a research fellowship under the Talent Young Scientific Program (TYSP) and received a lecturer position in the School of Materials and Energy, Southwest University, Chongqing, China (2018–2020). His current research focuses on developing scalable protocols for high-efficiency perovskite solar cells.

Contents

Preface XIII

Section 1
Theory of Solar Cells 1

Chapter 1 3
Techno Economic Feasibility Analysis of Solar PV System in Jammu: A Case Study
by Harpreet Kaur Channi

Chapter 2 19
Modeling of Photovoltaic Module
by Rakeshkumar Mahto and Reshma John

Chapter 3 31
Study of a New Hybrid Optimization-Based Method for Obtaining Parameter Values of Solar Cells
by Selma Tchoketch Kebir

Chapter 4 51
Solar Energy Assessment in Various Regions of Indian Sub-continent
by Johny Renoald Albert and Dishore Shunmugham Vanaja

Chapter 5 65
Designing Well-Organized Donor-Bridge-Acceptor Conjugated Systems Based on Cyclopentadithiophene as Donors in Bulk Heterojunction Organic Solar Cells: DFT-Based Modeling and Calculations
by Rania Zaier and Sahbi Ayachi

Chapter 6 85
Thin-Film Solar Cells Performances Optimization: Case of Cu (In, Ga) Se_2-ZnS
by Fridolin Tchangnwa Nya and Guy Maurel Dzifack Kenfack

Section 2
Crystalline Si Cells 107

Chapter 7 109
Optical Study of Porous Silicon Layers Produced Electrochemically for Photovoltaic Application
by Rahmouni Salah

Chapter 8 125
Influence of the Incidence Angle Modifier and Radiation as a Function of the Module Performance for Monocrystalline Textured Glass and No Textured in Outdoor Exposed
by Issa Faye, Ababacar Ndiaye and Elkhadji Mamadou

Chapter 9 137
Ultrasonic Processing of Si and SiGe for Photovoltaic Applications
by Andriy Nadtochiy, Artem Podolian, Oleg Korotchenkov and Viktor Schlosser

Section 3
Emerging Photovoltaics 161

Chapter 10 163
Mixed 2D-3D Halide Perovskite Solar Cells
by Alaa E. Abd El-Samad, Radwa S. Mostafa, Hager H. Zeenelabden, Menahtullah M. Mabrouk, Ahmed Mourtada Elseman, Nasr Gad, Mostafa El-Aasser and Mohamed M. Rashad

Chapter 11 185
A New Generation of Energy Harvesting Devices
by Byunghong Lee and Robert Bob Chang

Chapter 12 253
2D Organic-Inorganic Hybrid Perovskite Light-Absorbing Layer in Solar Cells
by Meng Wang, Qunliang Song and Sam Zhang

Chapter 13 265
Two-Dimensional Materials for Advanced Solar Cells
by Manoj Kumar Singh, Pratik V. Shinde, Pratap Singh and Pawan Kumar Tyagi

Chapter 14 293
Graphene-Based Material for Fabrication of Electrodes in Dye-Sensitized Solar Cells
by Nguyen Huu Hieu

Chapter 15 313
Effect of Combination of Natural Dyes and the Blocking Layer on the Performance of DSSC
by Md. Mosharraf Hossain Bhuiyan, Fahmid Kabir, Md. Serajum Manir, Md. Saifur Rahaman, Md. Robiul Hossain, Prosenjit Barua, Bikram Ghosh, Fumiaki Mitsugi, Tomoaki Ikegami, Saiful Huque and Mubarak Ahmad Khan

Chapter 16 339
IIIrd Generation Solar Cell
by Paweł Kwaśnicki

Chapter 17 351
Nanostructured Transition Metal Compounds as Highly Efficient Electrocatalysts for Dye-Sensitized Solar Cells
by Yi-June Huang and Chuan-Pei Lee

Chapter 18 367
Excited-State Dynamics of Organic Dyes in Solar Cells
by Ahmed M. El-Zohry

Chapter 19 383
Improvement of Efficiency of Dye Sensitized Solar Cells by Incorporating Carbon Nanotubes
by Md. Mosharraf Hossain Bhuiyan, Fahmid Kabir, Md. Serajum Manir, Md. Saifur Rahaman, Prosenjit Barua, Bikrom Ghosh, Fumiaki Mitsugi and Tomoaki Ikegami

Chapter 20 411
Mechanism for Flexible Solar Cells
by Ghaida Salman Muhammed

Section 4
Solar Energy Technologies 423

Chapter 21 425
Advanced Laser Processing towards Solar Cells Fabrication
by Jhantu Kumar Saha and Animesh Dutta

Chapter 22 439
Application of Solar Energy in Medical Instruments (Microscope)
by Badria Ibrahim Eisa Idris, Ahmed Mohamed, Ayah Salah, Osman Abdalrahman Almahdi Alamin, Fatehia Garma and Alnazier Osman

Chapter 23 455
Solar Energy in Industrial Processes
by Guillermo Martínez-Rodríguez and Amanda L. Fuentes-Silva

Preface

This book is about a new class of materials being developed for solar energy applications. It provides information on the fundamental scientific principles underlying these materials as well as how they are used in technological applications. Future social demands are heavily reliant on cheap and plentiful energy sources. Nonrenewable resources such as oil, coal, and natural gas provide a large portion of the world's energy needs, contributing to global warming and negatively affecting our environment. There is a pressing and rapidly growing necessity for eco-sustainable and renewable energy technologies that are CO_2 neutral. The photoelectric conversion of clean and abundant solar energy can successfully address this. Solar energy is the best solution for electricity production in rural and isolated areas due to the ability to generate electricity in situ by small (even individual) production plants, the low environmental impact of zero carbon dioxide generation, and the ease of building integration. This book outlines and discusses the best research-cell efficiencies. It also presents the state of the art of the greatest proven conversion efficiencies for research cells for a variety of solar technologies. This book covers the theory of solar cells, thin-film technologies, crystalline silicone cells, and developing photovoltaics as well as recent breakthroughs in these fields. For understanding and studying device physics, this volume uses both analytical and numerical evaluations of solar cell architectures. The appendices contain many of the details of the analytical investigations so that the growth of concepts is not hampered by the creation of equations. The book also presents the theory of solar cells, which involves light energy in photons being converted into electric current when photons strike an appropriate semiconductor device. Theoretical studies are useful because they forecast the fundamental limits of a solar cell and direct the processes that lead to solar-cell losses and efficiency.

A thin-film solar cell is a second-generation solar cell manufactured by depositing one or more thin layers or thin films (TFs) of photovoltaic material on a substrate such as glass, plastic, or metal. Cadmium telluride (CdTe), copper indium gallium diselenide (CIGS), and amorphous thin-film silicon are all commercially available thin-film solar cells (a-Si, TF-Si). The book will appeal to a broad audience and will encourage researchers, engineers, and postgraduate students to advocate for photovoltaic device technology. The goal of the book is to provide the reader with the necessary foundations to keep up with and contribute to the development of this exciting topic.

Ahmed Mourtada Elseman
Electronic and Magnetic Materials Department,
Central Metallurgical Research and Development Institute (CMRDI),
Cairo, Egypt

Section 1

Theory of Solar Cells

Chapter 1

Techno Economic Feasibility Analysis of Solar PV System in Jammu: A Case Study

Harpreet Kaur Channi

Abstract

Renewable sources of energy and related technologies are essential to the generation of energy worldwide. The photovoltaic (PV) is one of the renewable power technologies that support household electricity use. No prior research has studied the sustainability of the off-grid energy generation system in Jammu, India despite the potential of solar photovoltaics and significant amounts of global sun radiation in an area. The present work shown in the chapter is to calculate the residential load of the Patyari Kaltan situated in district Samba of Jammu by energy auditing. The NASA Surface Meteorology is used for the solar resource informa-tionof selected village. The primary sources of electricity generation are fossil fuels. Recently, the energy demand and availability deficit has worsened due to the huge population and fossil fuels cannot fulfill huge energy requirement. Meanwhile they have negative impacts on the environment as well. Therefore, renewable energy offers suitable energy way out to the residents living in remote areas and in the areas near to Borders. In this paper the main aim is to examine the feasibility of solar-battery hybrid energy system to fulfill electrical demand of a residential area in a rural region in Jammu. The research shows that the cost of construction of the project can be repaid or recovered within 1 year 6 months. To accomplish the target, 214 solar panels of 325 watt are estimated to satisfy the demand 100 percent at all times. The findings of this modeling reveal that the off-grid PV system is both technical and economically viable for power generation; they may serve as a model for the successful development of the system for practical use. Furthermore, the model can promote assistance mechanisms for players in the renewable industry to introduce a PV system in residential buildings.

Keywords: Fossil fuels, Solar PV system, Remote regions, Economic feasibility, Off grid

1. Introduction

Solar energy is the primary driving force behind natural activities on the Earth's surface. The energy expenditure on the surface of the ground which depends on the landscape is a core factor in geological, environmental and risk modeling models. The contribution of radiation is also related to the biodiversity of plants and biomass production. The Sun is an ample, infinite supply of energy available all over the world which is only minimally used [1–3]. From around beginning of the century, solar-radiation technology developments became highly political when potential solutions to fossil fuel-based traditional energy systems became recognized. Solar energy systems are widely agreed to mitigate global challenges including climate change, insecurity in the developed world and lack of stability of energy supply in most of the world's economies. As for other emerging creative developments, there is a need to make significant improvements to the way in which energy is produced, distributed and used, such as a poor understanding of technological choices, higher initial investment costs and a more conservative social environment. Country assistance policies (up to 10-20 years) should be established in order to address these obstacles, identifying suitable legal and financial instruments.

The spatial and temporal solar energy affecting the earth's surface has a seasonal dynamics (due to astronomical factors) which is modulated by stochastic weather variability. It influences solar energy systems' efficiency, reliability and economy. Photovoltaics is a fast growing technology, with a better understanding of the primary solar power resource. Enhanced expertise will lead greatly to better location and economic evaluation of new plants, to efficiency management and to energy projection. Improved understanding for solar energy systems incorporation into current energy and economic processes is also essential. The spatial dependence of renewable energy production and its distribution raise issues that often involve precise location-dependent answers in the policy process [4–6].

Much like the fossil-fuel-based energy sector relies on exploration of the energy markets and the proven reserves for discovery and economic benefit, renewable energy relies upon assessing the energy production strategy and marketing resources [7–9]. The basic resources and fuel available are solar radiation for solar-based clean energy technology such as solar thermal and photovoltaic systems. Measured data was used to determine the solar resource for this technology, where accessible. Fortunately, the uneven distribution of calculated solar data in space, and in particular over time, contributes to the use of modeled solar light as the basis for various technological and economic decisions. There are major uncertainties in the measured and modeled solar radiation. Most solar radiation models are assisted by measured data, often unknown to the uncertainty or precision of these measured data [10–13].

The energy emitted by sun and terrestrial fraction of the energy flux is marked by the Solar Constant. The solar constant is defined mainly as the measurement per unit time of solar energy flux density perpendicular to the direction of the light. Satellites outside of the earth's atmosphere are the most reliably measured. The solar constant is measured at 1367 W/m^2 at present [8, 9, 14]. This percentage ranges by 3 percent since the earth's orbit is elliptical and the length of the year is different from the Sun. There is also a little variance in the solar constant due to variations in Sun's light. This importance encompasses all forms of radiation, a large portion of which is lost as the light travels into the atmosphere.

The solar radiation is absorbed, dispersed, reflected or released as it travels through the atmosphere. The energy flux density is reduced in all these processes. In reality, the Solar Flow Density in sunny days is reduced by about 30% compared to alien radiation, which on a cloudy day is reduced by as much as 90% [15].

As a consequence the direct radiation that comes to the earth's surface (or an equipment mounted on the earth's surface) never reaches 83%. This direct radiation is known as beam radiation from the solar disc. Diffuse radiation is characterized as the dispersed, reflecting radiation that is transmitted out from all directions to the surface of the Earth (reflective of other bodies, molecules, particles, droplets etc.) [10]. Complete (or global) radiation is the sum of the beam and diffuse components. The **Table 1** shows the gadgets that are used for the measurement of different solar quantities.

Particulars	Details
Village	Patyari Kaltan
Block	Ghagwal
District	Samba
State	Jammu and Kashmir
Country	India
Time Zone	IST (UTC + 05:30)
Latitude	34.09
Longitude	74.79
Total number of Houses	35

Table 1.
Details about the village Patyari Kaltan.

Solar radiation data obtained by the instruments described above form the basis for developing any solar project. A case study of village Patyari Kaltan situated in district Samba of Jammu is explained in the next section using energy auditing.

This study seeks to reconcile demand and supply differences by investigating the feasibility of using an off-grid PV system to generate power to consume the household. The main objectives for this study are the significant global solar radiation levels in the region as well as the low home energy use. The purpose of this chapter is, via mathematical modeling, to assess the technoeconomic feasibility of an off grid PV system. This paper does not include the environmental and political aspects of using offset photovoltaic systems or other Photovoltaic hybrid systems. With respect to this investigation, solar radiation, PV peak power, inverter size, batteriesize and a charging controller are the relevant characteristics specified in this work. The remaining chapter is organized in the following subsections:

The Section 2 provides a concise background of the study while Section 3 presents the problem formulation. The detailed methodology adopted is explained in Section 4. The results and the discussion are contained in Section 5 while the conclusion of the paper is in Section 6.

2. Background

A.K.M. Sadrul Islam etc. (2012) indicated that an 8 kW PV system linked with a 15 kw gasoline generator and 25 battery counts is the most economically viable alternative (nominal power 800 Ah, nominal voltage 2 V each) [16]. Abolfazl Ghasemi et.al (2013) highlighted the potential sun rays and the lives of remote, powered, non-connected hybrid PV-diesel battery-powered communities in Iran as excellent [17]. Mohan L. Kolhe et al. described the best hybrid architecture for energy at a cost of $0.34/kWh as a 30 kW PV system, 40 kW wind, 25 kW diesel power supplies, and a bank of 222kWh batteries [18]. M. Kashif Shajzad et.al (2017) reported that the optimum solution was constructed to conduct a cost analysis of 10 kW hybrid PV panels, 8.0 kW biogas generator, 32 battery storage and 12 KW converters [19]. Simulation results for a hybrid power system of 13 kW PV modules, 14,7 kW of hydro power, 8 battery storing units, 5 kW of the diesel generator and 9 kW converters were characterized as the optimal solution with a $113201 NPC by Ali Saleh Aziz et.al (2019). [20]. Zhen-yu Zhao et.al (2019), Muhammad Ifran, discussed the cost of traditional grid power and solar PV, which are designed to assess the economic efficiency of two simulation-driven technologies. Five areas, Bhakkar, Kanewal, Multan, Bahawalnagar and Rajanpur were selected in this paper. Research has shown that Kanewal has the maximum yearly solar irradiation in this area (5.50 KWh/m^2, 22].

3. Problem formulation

Increase in energy expenditure and worldwide dependence on fossil fuels lead to power shortages and global warming. Generally the diesel generator is used in both on-grid and off-grid systems for a reliable power source [11, 21]. This is a costly and also causes environmental toxic waste. In the country there is enormous resource of renewable energy source that is not being effectively utilized. Reliable renewable hybrid systems need to be developed by using the available renewable sources These hybrid systems can be a viable option in universities, companies, hospitals, industry and rural communities to fulfill the energy needs. The construction of such systems needs detailed study of renewable energy supplies by the location, because without this the hybrid device may be massive, which raises the device expenses [12, 22].

In order to investigate the financial and technological viability of the hybrid energy systems, the feasibility of dissimilar systems configurations and energy requirements needs need to be analyzed and analyzed.

4. Methodology

4.1 Site selection

The Jammu and Kashmir shares international boundary with Pakistan & China. The Line of Control on the area of Pakistan divides the UT which turns one part as J&K'and the other part as POK'[15, 23–26]. It also shares boundary with other state like Himachal Pradesh and Punjab [23, 24, 27]. The UT has two different parts namely 1) Jammu and 2) Kashmir. The two parts of the state differ drastically from each other on the basics of climate.

Rapid population growth and hi-tech development in recent decades have led to additional energy consumption, especially in the power sector [25]. In addition, there are numerous parts of the world in rising countries which have minimal or no way to electrical energy, particularly in rural areas [28–31]. Rural electrification is also stated to be very necessary for rural development in order to achieve economic growth, elimination of deprivation, generation of jobs and improvement of village living standards [32]. According to the 2011 Indian census, out of 1.21 billion, 0.83 billion live in rural areas and about 44 per cent of the population lack grid access [33]. Electricity generation has to be improved to solve these issues. In India, fossil fuel emissions produce a large proportion of electricity [34].

Patyari Katlan is a village located in Jammu-Kashmir Samba district which falls in India. **Table 1** shows the profile of Village adopted for study. The total residential consumption of the rural community is 1083.432 kWh/day in the summer.

Load profile: Load of the selected village depends on the equipments used in the houses. Load calculated by Energy audit of the village. There are 35 houses in the village and the equipments connected in the houses are tube lights, fans, coolers, TV's, refrigerators only. The equipments connected in the village are very less because the village is just 2 km away from the Pakistan border so the village is not so well developed [16].

Energy Audit: Energy audit is defiened as "organized monitoring and review of energy usage and energy use of site, facility, system or entity with the goal of defining and reporting on energy flows and future changes in energy quality." [17, 19–21, 35].

Energy saving is extremely relevant as the demand is rising day by day in the country, looking at the situation energy auditing is being performed. This is a method of monitoring how electricity is being used, and finding places wherever pollution can be reduced if not eradicate entirely. [27, 36]

The energy use by lighting and big appliances such as fridge, fans, etc. in 24 hours is calculated by doing survey. The wattage of each equipment is represented in **Table 2**. **Figure 1** shows the hourly variation of each equipment. The highest energy is consumed by refrigerator and lowest by lighting system. The rural community's average residential consumption is 1083,432 kWh per day in the summer, and 718,952 kWh per day in the winter months. Monthly residential consumption of the village is shown in the **Table 3**. **Figure 2** depicts the graphical representation of monthly residential consumption. The **Table 3** shows that the peak load is in the July month i.e. 33586.392kWh. And the minimum energy utilization is in the February month i.e. 20130.656kWh.

4.2 Meteorology data of Patyari Kaltan (Thali)

The solar resource information used for the selected village was found from the NASA Surface Meteorology. Access was made to NASA database to assess solar irradiance in the remote rural Jammu region [13]. The most important step before using a solar photovoltaic device is to define solar power potential in a given region [3–5]. The solar irradiance records for the regions selected were

Residential load						
S.no	**Equipment**	**No. of equipment**	**Wattage (w)**	**Consumed electricity per day (kwh)**	**Consumed electricity per month (kwh)**	**Consumed electricity per year (kwh)**
1	Lighting	502	36	289.152	8674.56	104094.7
2	Fans	179	70	200.48	6014.4	36086.4
3	Coolers	41	250	164	4920	14760
4	TV	35	70	9.8	294	3528
5	Refrigerator	35	750	420	12600	151200

Table 2.
Residential load of the village.

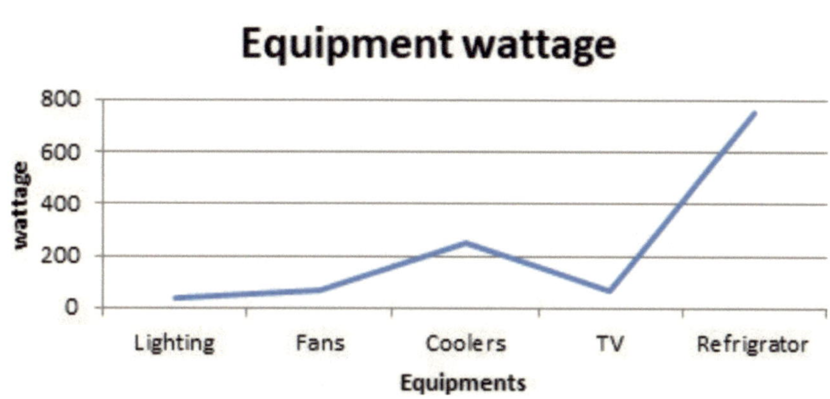

Figure 1.
Wattage of equipments connected in the houses of village.

Months	Energy consumed per month (kwh)	Months	Energy consumed per month (kwh)
January	22287.512	July	33586.392
February	20130.656	August	32502.96
March	22287.512	September	27582.96
April	27582.96	October	22287.512
May	28502.392	November	21568.56
June	32502.96	December	22287.512
Total			313109.888

Table 3.
Energy consumed per month.

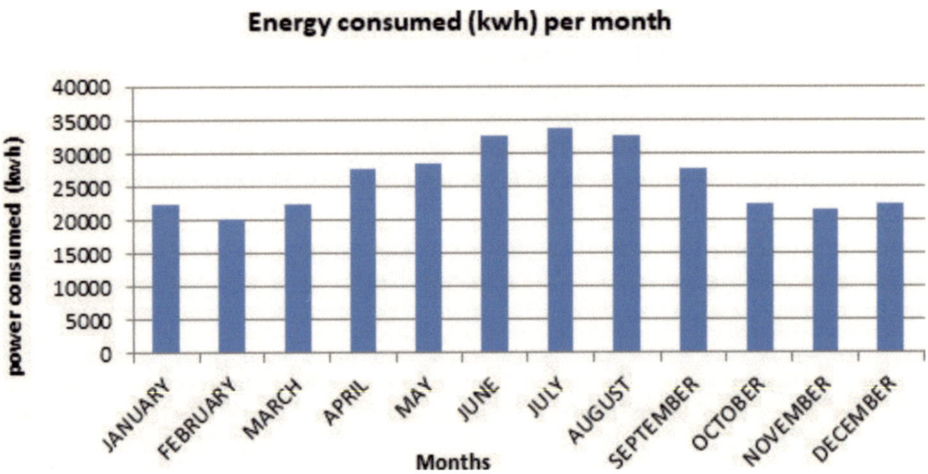

Figure 2.
Power consumed per month in village Patyari Kaltan (thali) in KWh.

provided in **Table 4**. **Figure 3** shows the daily Radiation and clearness in Patyari Katlan (Thali). Outcome of the investigation show that Patyari Kaltan received the average annual solar irradiation is 4.134 kWh / m^2 [3].

4.3 Design calculation

4.3.1 Energy balance considerations

To choose the appropriate size for the day and night time loads for the solar panel or battery storage. We describe a state of energy balance when full power from the solar system (E_{SA}) is enough to charge the battery (E_b) and the energy needed to charge electricity including system losses (E_L) as shown in **Figure 1**, i.e. without any power source from the backup.

$$E_{SA} = E_b + E_L \quad (1)$$

Assuming that the night time load is solely provided by the storage batteries with an overall efficiency factor K1 as shown in Equation (2):

$$EB = EN / K_1 \text{ where } K1 = \eta D.FU.\eta R.\eta L.\eta B \quad (2)$$

Monthly average			
Month	**Insolation clearness index**	**Horizontal surface clear sky insolation incident (kWh/ m^2/day)**	**Horizontal surface all sky insolation occurrence (kWh/m^2/day)**
January	0.424	0.929	2.213
February	0.479	0.561	3.135
March	0.447	1.065	3.788
April	0.490	0.964	4.885
May	0.468	0.829	5.199
June	0.478	0.568	5.501
July	0.510	0.000	5.754
August	0.525	0.475	5.435
September	0.639	1.777	5.707
October	0.550	1.506	3.892
November	0.310	0.532	1.744
December	0.494	1.868	2.361
Average	0.485	0.923	4.134

Table 4.
The average of monthly daily isolation incident on horizontal surface in Patyari Kaltan (thali) [3, 4].

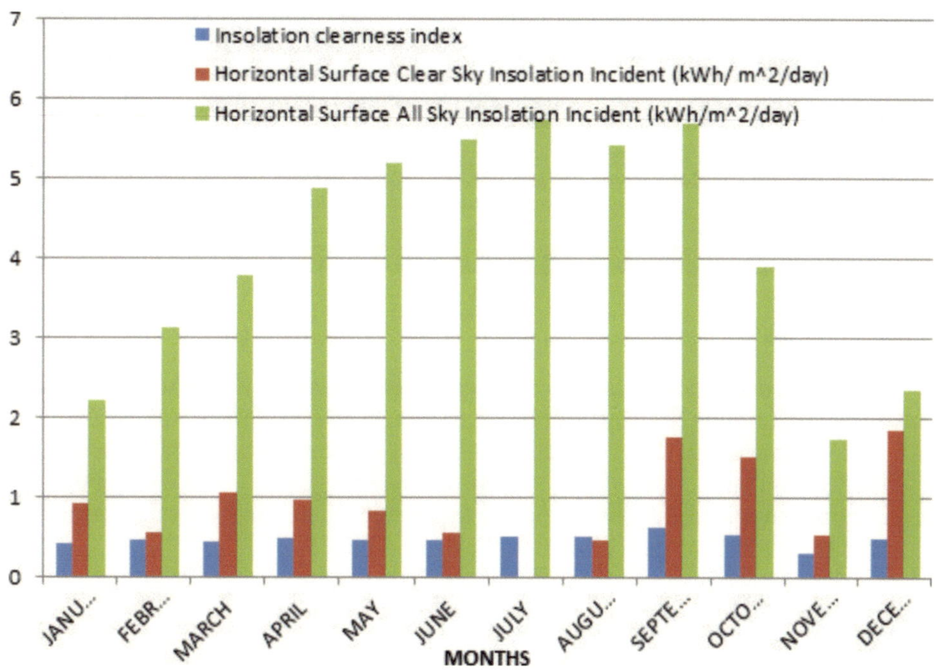

Figure 3.
Daily radiation and clearness in Patyari Katlan (thali) [3–5].

ηD being the solar array diode efficiency, FU the solar array utilization factor, ηR the regulator efficiency, ηL the line loss factor and ηB the battery WHr efficiency.

4.3.2 Balance of system components

As described above, the Power System consists of a variety of balances, apart from solar panes, of device components such as (a) mounting frames of the array module, frame support and foundations; (b) circuits, load and electricity management devices, wire interconnections, etc., etc.

There are also extra expenses for the test and inspection module, system sizing, and packaging repair installation and checkout, etc. For the construction of the PEPS, account must be taken of the amount of all these costs lumped as BOS cost. The complexity of the difference between the cost of the storage batteries and the battery power in Ahr is a significant factor in the extremely nonlinear performance of energy generation costs / peak W.

4.4 Design requirements

The module is made of solid wire or solid ribbons by attaching one cell on another. The ties may be rigid or fluid to control motion within the series, which can be caused by thermal expansions and other forces. All links should ensure the lowest resistance possible and the least possible distortion of PV output. The designer is also trying to make this relation shorter and to reduce the cross sections against increased resistance. The output from an array is connected to a manager called a bus.

4.4.1 Placing of cells

It is essential to place cells in the array and in the cell form. As the distance between the cells increases, the overall performance of the panel determined by the voltage per unit area falls. Big cells do not always improve the performance of packaging (i.e. the need for a maximum cell to panel area ratio). In building a module with desirable electric properties, cell size is an essential element. Strom from a cell varies according to the cell size, with constant voltage. Many small cells should be plugged into series for large voltage. Round cells that were halved and put in a panel in an offset pattern are used to move more of them in a unit area. This increases the cell's packaging density. Square or hexagonal cells may also be expanded. Cells are put as near or as close as possible and cannot contact so energy is cut short. There must be additional space between the cells for thermal expansion.

4.4.1.1 Array support

More is needed to create the electrical resource needs Specific solar cells can be just as delicate groups of them. The retrofitting and disassembly should be able to be held in every module. Array must be capable of resisting moderate loads, mechanical and temperature shifts pressures. The translucent cover for a module is part of the support. The cover is primarily used to shield the PV module from situations including oxygen, moisture, dust and rain.

4.4.1.2 Size of array

Solar cell size can vary from approximately 1 mm to more than 100 mm. For most standard silicone cells, the thickness range is 0.2 to 0.4 mm. For the collection of the array size we established a very basic semi empirical rule as per the Equation (3)

$$P_{ph} = \left(LH + LHd / Cr\, Bb\, x100 \right) / X \qquad (3)$$

Where Pph is the full watts array. X is the estimated annual maximum equipment time a day and is the average annual watt hour a day per poor flat hour of the PV module. L is the watts load rating, and H is the working hours a day. d is the number of storage days required. Cr is the time for charging recovery and Bb is the battery watt-hour efficiency. The value of X depends explicitly on the overall insolation of the panel on the installation site. The value of X can be calculated as shown in **Figure 4**.

$$X = EXm / 12; \text{With } Xm = \eta ov\, Im / \eta m \tag{4}$$

Where ηov is the overall device performance, I_m is the average sleeping surface insolation of the area, ηm is the efficiency of the module. Performance is the product of the efficiency of the module and the balance of device efficiency including efficiency of power conditioning, efficiency coefficients of temperature etc.

4.4.1.3 Solar panels

For terrestrial applications the majority of silicone solar cells have a round diameter of 5 cm and a diameter of 0.3-0.5 mm. The trend is to massive diameters. A 5-cm diameter cell with a surface area of approximately 20-cm has a capacity of 0.2 W with 0.45 volts during full sun and at room temperature. A variety of cells need to be mounted into a panel for higher power or higher voltage. Two cells are wired in parallel, for example, for double power at constant voltage. It can provide any amount of power at the desired voltage by joining numerous cells in parallel and series.

4.4.1.4 Battery storage

Electric storage battery is the easiest way to stored a smaller moderate scale. Solar cells generate a battery charging direct current. When needed the stored energy can then be supplied to the local load as electricity. A battery is an independent cell mixture. A cell is the elementary mixture of materials and electrolytes that form the essential energy storer electromechanical. A block box into which electrical energy is collected, electromechanically stored and then recuperated as electrical power can also be thought of as a battery. Primary batteries are non-rechargeable while secondary batteries are still able to be recharged. So secondary batteries also have a major interest in solar electricity. Sub-examples of secondary batteries include lead-acid, nickel-cadmium, iron-air, nickel-hydrogen, zinc-air etc.

Energy efficiency of a battery is defined as shown in Equation (5).

$$\eta_{energy} = \int I_1\ E_1\, dt / \int I_2\ E_2\, dt \tag{5}$$

where I_1 = battery discharging current for a period 0 to t1.
I_2 = battery charging current for a period 0 to t2.
E_1 = Battery discharging terminal voltage.
E_2 = Battery discharging terminal voltage.

Cycle life is the amount of times that the battery can be charged and unloaded, and this can differ considerably with discharge depths. Deep discharge tends to a short life cycle.

4.5 Design calculation

No. of panels required, N_S = per day demand/Rating of 1 panel (w).
= 69552w/325w.
N_S = 214.0.

20 ft.^2 = 1.858 m^2 (Area of 1 panel).
Area required = Area of 1 panel*No. of Panels.
= 1.858m^2*214.
= 397.612 m^2.

Total load per day in kwh = 1083.432 kwh.

Using 12 V, 17 Amp hour lead acid battery.

Total Capacity (C_B) = Total load kwh/Voltage of single battery.
=1083.432kwh/12 V.
=90.286 KAhr.

Number of lead acid batteries, N_B = C_B /Rating of single battery Ahr.
=90.286 KAhr / 17 Ahr.
N_B = 5310.9 or 5311 (approx).

Use charge controller of 12 V, 20 Amp is used.

Rating of charge controller in Ampere's = Total load (w)/12 V.
=69552 W/12 V.
=5796 Ampere.

12 Volt, 20 Ampere Charge Controller (N_C) needed can be measured as:

N_C = Rating of charge controller (Amp)/20 amp.

N_C = 5796 Amp/20 Amp.

N_C = 289.8 or 290 (approx).

Total load per day in watt = 69552 W.

Load per day in kw = 69552/1000.
=69.552 kw or 70 kw (approximately).

For 70 KW load, 70 KW of Inverter is needed.

Cost per watt = Rs 22.

Price of solar panel, C = overall load (watt) * price per watt.
C = 69552 W*Rs 22.
C = Rs 1530144.

Price of batteries = N_B* Cost of one battery.
=5311 * Rs1900.
=Rs 10090900.

Price of charge controller = N_C * price of one charge controller.
= 290 * Rs 798.
= Rs 231420.

Price of 70 KW inverter = overall load in KW * price per KW.
= 70* Rs 72065.
= Rs 5044550.

Total Cost = Price of solar panel + Price of charge controller + Price of Inverter + Price of battery.
=1530144 + 10090900 + 231420 + 5044550.
=Rs 16897014.

To take the cost of cabling, junction box etc. into account, 20 percent of the overall cost is applied to get the whole cost of the project.
= 20% * Rs16897014.
= 20/100 * 16897014.
=Rs 3379402.8.

Therefore the total expenditure of the project C_t,
C_t = 16897014 + 3379402.8.
C_t = Rs 20276416.8.

When we buy the electricity from the energy grid, otherwise we have to pay.
= overall demand * Price of one unit.
= 1083.432 Kwh * Rs 3.
=Rs 3250.296.

Overall cost of energy which is purchased from utility grid/year is.

D_t = Rs 3250.296*365.

D_t = Rs 1186358.04/year.

4.5.1 Pay back period

Payback period = Project costs / Annual cash inflow

$C_t - N\,D_t = 0$

Or $N = C_t/D_t$

Where C_t = Rs 20276416.8

D_t = Rs 1186358.04/year

thus

N = Rs 20276416.8/Rs 1186358.04

N = 17.09

Project costs for the project can be recovered in 17 or 18 months (1 year 6 months).

5. Results and discussion

Solar power is a huge source of electricity that can be used directly, generating other reservoirs of power: biomass, wind, hydroelectric power and wave power. While there are major differences in latitude and seasons, most Earth's area receives ample solar energy to enable low-grade heating of water and houses. Simple mirrors can focus solar energy enough at low latitudes to cook and even drive steam turbines. In certain semiconducting materials the energy of light switches electrons. This photovoltaic effect is able to produce vast amounts of electricity. However, the current low effectiveness of solar photovoltaic cells requires a great deal of energy. The only renewable way to substitute existing global electricity supplies from non-renewable sources is the immediate use of solar energy, at the cost of land areas of at least half a million km^2.

The Roof top solution is supported by the design methodology for installing solar panels in Patyari Kaltan (Thali). The incorporation of the panels into the roof of the building is the strategy used. This solution is given when it replaces the traditional roof and permits the filtering of natural sunlight. It serves as roof for structural and weather requirements with structural support, stability, protection from damage such as chemical or mechanical damage, fire-fighting protection, sun, wind and moisture protection, heat absorption and heat conservation, light diffusion control etc. It acts as a power generator in addition to those functionalities by fulfilling a portion of the building's electrical load specifications. Due to the highly flexible design of the solar cells and the storage cells, individual roof capacity can be used for specific loads – top PEPS for the same energy need as the previously described loads. The mean Horizontal insolation surface incident is 4.134 KWh / m^2 / day and the clarity index estimate has been found to be 0.485.It is closely related to the solar radiation itself, but isolation gives you a more accurate way to calculate the radiation on an energy-relevant single object, rather than just taking a sunlight measurement itself. The clearness index is a calculation of the proportion of solar radiation emitted to the Earth's surface through the atmosphere. Research shows that the payback period for the solar project of the selected village is 1 year 6 months. It shows that the cost for installing the whole project can be recovered within 18 months which means solar project can be benifical for the Patyari Kaltan village.

6. Conclusion

The study shows that the village has significant solar power capacity and is ideal for producing electricity. The cost of building the project can be recovered in 1 year 6 months. To calculate solar irradiance in the remote rural Jammu area, the NASA database was accessed, which reveals that solar irradiation obtained by Patyari Kaltan is 4,134 kWh / m^2. To meet up the demand 100% at all the time during the year, 214 solar panels of 325 watt is needed. The future work on this project is to check the feasibility and sensitivity of the PV hybrid system using HOMER software.This methodology is further extended to other parts of the country to utilize the available renewable energy resources and to meet the increasing load demand.

Author details

Harpreet Kaur Channi
Electrical Engineering Department, Chandigarh University, Mohali, India

*Address all correspondence to: harpreetchanni@yahoo.in

References

[1] kishor Verma, Jai, and Raja Sekhar Dondapati. "Techno-economic sizing analysis of solar PV system for Domestic Refrigerators." *Energy Procedia* 109 (2017): 286-292.

[2] Al Garni, Hassan, and Anjali Awasthi. "Techno-economic feasibility analysis of a solar PV grid-connected system with different tracking using HOMER software." In *2017 IEEE International Conference on Smart Energy Grid Engineering (SEGE)*, pp. 217-222. IEEE, 2017.

[3] https://power.larc.nasa.gov/text/definitions.html (accessed on 12.02.2020)

[4] https://power.larc.nasa.gov/common/php/POWER_ParametersEnergy.php (accessed on 12.02.2020)

[5] http://village.org.in/Patyari-Katlan (accessed on 12.2.2020)

[6] Pandey, Antima, Arif Iqbal, Akhilesh Kumar Mishra, and Mohammed Aslam Husain. "Economic Feasibility Analysis of Solar PV Generation at REC Ambedkar Nagar." In *2019 IEEE International Conference on Electrical, Computer and Communication Technologies (ICECCT)*, pp. 1-6. IEEE, 2019.

[7] Mughal, Shafqat, Yog Sood, and Raj Kumar Jarial. "A Proposal on Techno-Financial Design aspects of Photovoltaic System for the Twin Districts of Rajouri and Poonch (Jammu & Kashmir)." *EAI Endorsed Transactions on Energy Web* 8, no. 31 (2020): e9.

[8] Zia, Ubaid Ur Rehman, Tanzeel ur Rashid, Waqas nazir Awan, Talal Bin Ahmed, Mubeen Azhar Siddique, M. Farhan Habib, and Rana Muhammad Asid. "Technological Assessment of Bio Energy Production through Livestock Waste in Azad Jammu and Kashmir (AJK)." In *2019 International Conference on Electrical, Communication, and Computer Engineering (ICECCE)*, pp. 1-5. IEEE, 2019.

[9] Channi, Harpreet Kaur, and Inderpreet Kaur. "EROI Analysis of 2 KW PV System." In *Intelligent Computing in Engineering*, pp. 665-671. Springer, Singapore, 2020.

[10] Sagar Pandey, Harpreet Kaur Channi, Roof Top Solar Installation: A Case Study, TEST engineering and managemnt, scopus,March-April 2020,ISSN: 0193-4120 Page No. 15629 - 15633.

[11] Manhas, Rahul, and Harpreet Kaur Channi. "Analyzing the feasibility and availability of solar PV system for stand-alone PV System in (J&K)." *Solid State Technology* 64, no. 2 (2021): 2400-2406.

[12] Channi, Er Harpreet Kaur, and Sukhvir Kaur Pandher. "Study of Feasibility of Solar Panel at Chandigarh University Gharuan, Mohali."

[13] Baig, Mirza Jabbar Aziz, and Tariq Iqbal. "Design and analysis of a rooftop PV system for a University Building in Pakistan." (2020).

[14] Monteiro, L. S., K. A. Costa, E. da S. Christo, and W. K. Freitas. "Economic feasibility analysis of small hydro power projects." *International Journal of Environmental Science and Technology* (2020): 1-12.

[15] Wani, Tawseef Ahmad. "A Review of Fuzzy Logic and Artificial Neural Network Technologies Used for MPPT." *Turkish Journal of Computer and Mathematics Education (TURCOMAT)* 12, no. 2 (2021): 2912-2918.

[16] Islam, AKM Sadrul, et al. "Hybrid energy system for St. Martin Island, Bangladesh: an optimized model." Procedia Engineering 49 (2012): 179-188.

[17] Ghasemi, Abolfazl, et al. "Techno-economic analysis of stand-alone hybrid photovoltaic–diesel–battery systems for rural electrification in eastern part of Iran—A step toward sustainable rural development." Renewable and Sustainable Energy Reviews 28 (2013): 456-462.

[18] Kolhe, Mohan L., KM Iromi Udumbara Ranaweera, and AGB Sisara Gunawardana. "Techno-economic sizing of off-grid hybrid renewable energy system for rural electrification in Sri Lanka." Sustainable Energy Technologies and Assessments 11 (2015): 53-64.

[19] Shahzad, M. Kashif, et al. "Techno-economic feasibility analysis of a solar-biomass off grid system for the electrification of remote rural areas in Pakistan using HOMER software." Renewable energy 106 (2017): 264-273.

[20] Aziz, Ali Saleh, et al. "Optimization and sensitivity analysis of standalone hybrid energy systems for rural electrification: A case study of Iraq." Renewable energy 138 (2019): 775-792.

[21] Irfan, M.; Zhao, Z.-Y.; Ahmad, M.; Rehman, A. A Techno-Economic Analysis of Off-Grid Solar PV System: A Case Study for Punjab Province in Pakistan. Processes 2019, 7, 708.

[22] Mohit, Subhash Saxena, Akashdeep Singh, Varinder Singh, and Harpreet Kaur Channi. "Modeling and Designing of Solar Tracking System using Arduino." (2017).

[23] Saraswat, S. K., and K. V. S. Rao. "10 kW solar photovoltaic—Diesel hybrid energy system for different solar zones of India." In *2016 International Conference on Emerging Technological Trends (ICETT)*, pp. 1-6. IEEE, 2016.

[24] Aziz, Ali Saleh, et al. "Optimization and sensitivity analysis of standalone hybrid energy systems for rural electrification: A case study of Iraq." Renewable energy 138 (2019): 775-792.

[25] Irfan, M.; Zhao, Z.-Y.; Ahmad, M.; Rehman, A. A Techno-Economic Analysis of Off-Grid Solar PV System: A Case Study for Punjab Province in Pakistan. Processes **2019**, 7, 708.

[26] Vipasha Sharma, Harpreet Kaur, Inderpreet Kaur(2019) "Design and Implementation of Multi-Junction PV Cell for MPPT to Improve the Transformation Efficiency, International Journal of Recent Technology and Engineering (IJRTE) ISSN: 2277-3878, Volume-7, Issue-6S4, PP 248-253April.

[27] Lohan, Shiv Kumar, and Sushil Sharma. "Present status of renewable energy resources in Jammu and Kashmir State of India." *Renewable and Sustainable Energy Reviews* 16, no. 5 (2012): 3251-3258.

[28] Shah, Syed Ahsan Ali. "Feasibility study of renewable energy sources for developing the hydrogen economy in Pakistan." *International Journal of Hydrogen Energy* 45, no. 32 (2020): 15841-15854.

[29] Nouni, M. R., S. C. Mullick, and T. C. Kandpal. "Photovoltaic projects for decentralized power supply in India: A financial evaluation." *Energy policy* 34, no. 18 (2006): 3727-3738.

[30] Baba, S. H., M. H. Wani, Muzaffer Manzoor Mir, and N. A. Bazaz. "Econometric Analysis of Demand and Supply of Institutional Credit to Agriculture in Jammu and Kashmir." *Economic Affairs* 60, no. 4 (2015): 669.

[31] Tripathy, M., H. Joshi, and S. K. Panda. "Energy payback time and life-cycle cost analysis of building integrated photovoltaic thermal system influenced by adverse effect of shadow." *Applied energy* 208 (2017): 376-389.

[32] Mughal, Shafqat Nabi, Yog Raj Sood, and R. K. Jarial. "Design and techno-financial analysis of solar photovoltaic plant for school of engineering and

technology at BGSB University, Rajouri (J&K)." In *Applications of Computing, Automation and Wireless Systems in Electrical Engineering*, pp. 231-243. Springer, Singapore, 2019.

[33] Sunaina, Harpreet Kaur Channi, "Solar Power Tree Technology:A Review TEST engineering and managemnt, scopus,March-April 2020,ISSN: 0193-4120 Page No. 16597 - 1660.

[34] Kamal, Md Mustafa, and Imtiaz Ashraf. "Performance Assessment of Standalone Solar Photovoltaic System for different Load profiles in the Rural Area." *Journal of The Institution of Engineers (India): Series B* (2021): 1-20.

[35] Dutta, Aaina, and Sukanya Das. "Adoption of grid-connected solar rooftop systems in the state of Jammu and Kashmir: A stakeholder analysis." *Energy Policy* 140 (2020): 111382.

[36] Ruchira, Ruchira, R. N. Patel, and Sanjay Kumar Sinha. "Optimal Sizing of Microinverter based PV System for EWS Houses." In *2021 International Conference on Advances in Electrical, Computing, Communication and Sustainable Technologies (ICAECT)*, pp. 1-5. IEEE, 2021.

Chapter 2

Modeling of Photovoltaic Module

Rakeshkumar Mahto and Reshma John

Abstract

A Photovoltaic (PV) cell is a device that converts sunlight or incident light into direct current (DC) based electricity. Among other forms of renewable energy, PV-based power sources are considered a cleaner form of energy generation. Due to lower prices and increased efficiency, they have become much more popular than any other renewable energy source. In a PV module, PV cells are connected in a series and parallel configuration, depending on the voltage and current rating, respectively. Hence, PV modules tend to have a fixed topology. However, in the case of partial shading, mismatching or failure of a single PV cell can lead to many anomalies in a PV module's functioning. If proper attention is not given, it can lead to the forward biasing of healthy PV cells in the module, causing them to consume the electricity instead of producing it, hence reducing the PV module's overall efficiency. Hence, to further the PV module research, it is essential to have an approximate way to model them. Doing so allows for understanding the design's pros and cons before deploying the PV module-based power system in the field. In the last decade, many mathematical models for PV cell simulation and modeling techniques have been proposed. The most popular among all the techniques are diode based PV modeling. In this book chapter, the author will present a double diode based PV cell modeling. Later, the PV module modeling will be presented using these techniques that incorporate mismatch, partial shading, and open/short fault. The partial shading and mismatch are reduced by incorporating a bypass diode along with a group of four PV cells. The mathematical model for showing the effectiveness of bypass diode with PV cells in reducing partial shading effect will also be presented. Additionally, in recent times besides fixed topology of series–parallel, Total Cross-Tied (TCT), Bridge Link (BL), and Honey-Comb (H-C) have shown a better capability in dealing with partial shading and mismatch. The book chapter will also cover PV module modeling using TCT, BL, and H-C in detail.

Keywords: photovoltaics, complementary metal oxide semiconductor (CMOS), bypass diode, partial shading, metal oxide semiconductor field effect transistor (MOSFET)

1. Introduction

A Photovoltaic (PV) cell is a device that by the principle of photovoltaics effect converts solar energy into electricity [1, 2]. In a PV module, PV cells are connected in a series and parallel configuration, depending on the voltage and current rating, respectively [1]. In recent times PV based energy is gaining prominence due to the advances in the PV cells [3, 4], lowering PV cells cost [5], and government incentives [6–8]. Compared to any other renewable energy-based power source, PV is considered portable and easy to use [5]. Hence, the PV based power source is used in a wide range of applications that include residential and commercial building, drones, vehicle, satellites, embedded systems, sensors and many others [9].

The PV-based power source is not ideal and performance can cause many anomalies [10, 11]. One of the most significant issues that affect PV modules performance is the shading caused due to clouds, physical objects, and living beings [10, 11]. Generally, there are two types of shading, complete shading and partial shading [12]. The complete shading occurs when the whole PV module is under the shade. If only a few of the PV cells are under the shade, it results in partial shading conditions. Both of these shading types reduce the power efficiency of PV modules. However, the partial shading condition can have much more severe after effects [13]. The current flow in a row of PV cells connected in series is governed by the PV cells that are affected by the shade [13, 14]. This phenomenon can lead to forward biasing of unshaded PV cells, leading to them consuming the power instead of generating it [13, 14].

Additionally, the partial shading condition can cause hotspot generation in the panel's neighboring PV cells [15–17]. This hotspot can even instigate a fire hazard [18–20]. Mismatch in the PV cells in the PV module can also create abnormalities like the partial shading conditions [14]. Hence, to prevent such a phenomenon from happening, the PV modules are equipped with bypass and blocking diode [2, 9, 21]. The bypass diode causes bypassing of the shaded or damage PV cells in the panel [22]. Simultaneously, the blocking diode prevents the current from flowing in the reverse direction in case of a mismatch in the output voltage, which can lead to forward basing of the PV cells [21]. Similarly, faults such as open and short circuiting of PV cells in the module degenerate the solar panel's performance [12].

Due to all these abnormalities that reduce the PV panel's performance, it is desirable to model a PV module that can emulate its electrical characteristics to derive a better way to tackle them. Also, equivalent modeling helps to better understand the PV panel characteristics before they are being deployed for real-world applications.

2. SPICE based PV cell and module modeling

2.1 Equivalent circuit model of a PV cell

To model the PV cell, a SPICE based 2-diode based equivalent circuit is used as shown in **Figure 1** [23]. All the parameters shown in **Figure 1**, are presented in **Table 1** [23]. Two diode-based PV cell modeling techniques are selected over single diodes since they are considered more accurate [24]. The resistance Rs and Rp as shown in **Figure 1** are the internal resistance of the PV Cell.

The current i_{PV} *ge*nerated by the PV cell, as shown in **Figure 1**, can be computed by:

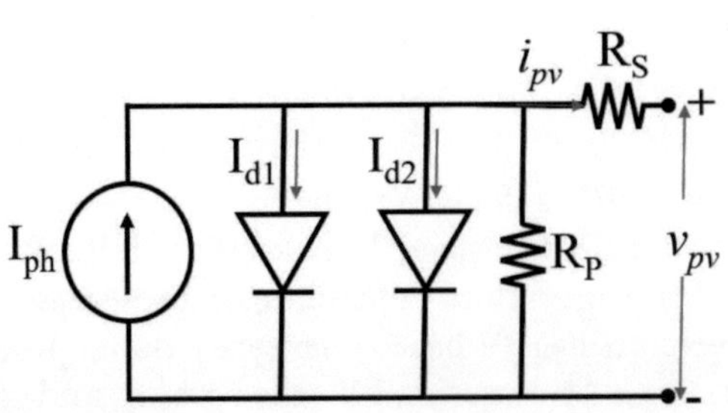

Figure 1.
2-diode based equivalent PV cell model.

Symbol	Description	Value and units
V_{PV}	Output voltage across the PV module	Volts
I_{PV}	Output current across the PV module	Amperes
v_{pv}	Output voltage across a PV cell	Volts
i_{pv}	Output current generated from the PV cell	Amperes
V_{OV}	Open load voltage across PV module	Volts
v_{ov}	Open load voltage across PV cell	0.55 Volts
I_{PH}	Total photon current of the PV module	2.17 Amperes
R_P	Shunt resistor	1 mΩ
R_S	Series resistor	0.1 MΩ
A_{PV}	Area of PV cell	126.6 cm^2
J_{SC}	Short-circuit current density	34.3 mA/cm^2
V_T	Thermal voltage = KT / q	26 mV
A	Diode ideality constant	1.2
I_S	Diode saturation current	I_{S1} = 10pA and I_{S2} = 1 nA
G	Solar irradiation data	1000 W/m^2
G_{Sh}	Shaded solar irradiation data	500 W/m^2
N_S	Number of PV cells in series	
N_P	Number of PV cells in parallel	
N	Number of PV cells in a PV panel = N_S x N_P	

Table 1.
The parameters descriptions for modeling PV cells and module, their assumed constant value, and units.

$$i_{PV} = I_{ph} - I_{d1} - I_{d2} - \frac{v_{PV} + i_{pv} R_S}{R_P} \tag{1}$$

Where the I_{d1} and I_{d2} can be computed by:

$$I_d = I_s \left(e^{\left(\frac{v_{PV} + i_{pv} R_S}{A \times V_T} \right)} - 1 \right) \tag{2}$$

The total photon current of a PV cell is dependent on the area of the PV cell, short-circuit current density of the PV cell, and the solar irradiance. The following equation can compute the total photon current for a given PV cell:

$$I_{Ph} = A_{PV} \times J_{SC} \times \frac{G}{G_{Sh}} \tag{3}$$

The current vs. voltage (I-V) and power vs. voltage (P-V) characteristics obtained for a single PV cell using SPICE-based equivalent PV cell are shown in **Figure 2**. The SPICE simulation uses Eq. (1), Eq. (2), and Eq. (3).

2.2 Equivalent modeling of PV module

A series–parallel topology is used to model a PV module using the equivalent PV cell shown in **Figure 1**. The total number of PV cells in the PV panel is equal to N. The number of PV cells connected in series is equal to Ns. The Ns number of series connected PV cells are then tied together to form a PV panel or module. The total number of PV cells connected in parallel is equal to Np. Thus, the number of PV cells in a PV module is given by:

$$N = N_S \times N_P \tag{4}$$

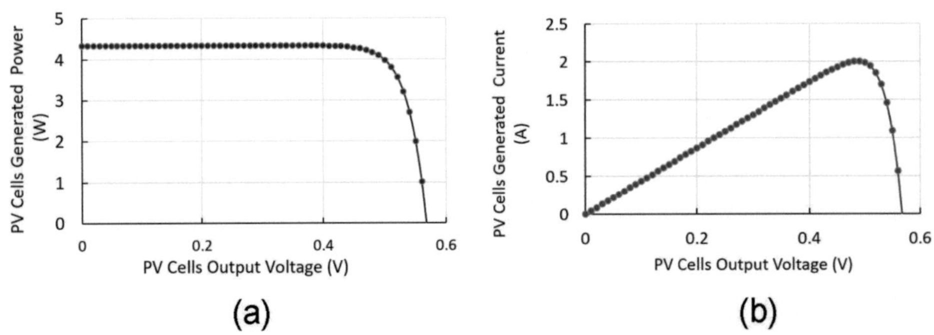

Figure 2.
a) Current vs. voltage (I-V) characteristics of PV cell b) power vs. voltage (P-V) characteristics of PV cell.

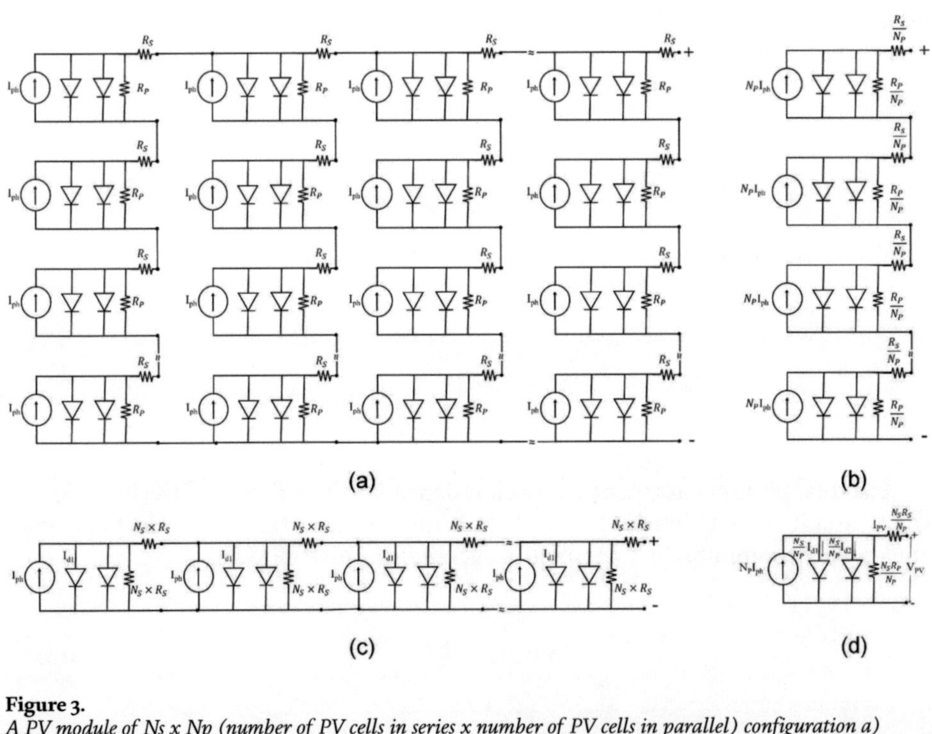

Figure 3.
A PV module of Ns x Np (number of PV cells in series x number of PV cells in parallel) configuration a) double-diode based equivalent PV module b) column decomposition c) row decomposition d) approximate equivalent model.

If all the PV cells in a PV module are homogenous and are receiving an identical solar irradiance, it can be modeled in **Figure 3** [25]. The total current generated from the PV module with Ns number of PV cells in series and Np number of PV cells in parallel can be modeled by using Eq. (1) and the equivalent double diode based model shown in **Figure 3d**).

The photon current shown in **Figure 3** is given by the summation of photon current across each column in the PV module. Thus, the total photon current of a PV module is equal to the Np times the single PV cell's photon current. In the partial shading in the PV module, the current across the series connected PV cell is determined by the current generated from the shaded PV cells. Thus, in this case, the PV module's total photon current is not equal to Np times the photon current. Instead, it is given by Eq. (5).

$$Total\ I_{PH} = \sum_{i=1}^{N_P} min\left(I_{Ph}\right) \tag{5}$$

3. Different types of topologies for creating a PV module and effect of shading

3.1 Fixed topology

Typically, the topology used for creating PV modules from a single PV cell is mostly fixed. To reduce the partial shading and mismatch impact, bypass diode is connected across PV cells in the panel to reduce its impact on power generation [22]. To reduce the effect of partial shading condition, different techniques of placing bypass diode were analyzed [26]. However, they make the PV module much more complicated for mass production. There are different topologies of connecting PV cells in the module and these are important for understanding the effectiveness of different topologies for mitigating the ailing effect of partial shading condition, the shading pattern shown in **Figure 4** is used [27]. For simplifying the simulation, the number of series connected PV cells is equal to 4, and the number of parallel connected PV cells is 5, i.e., N_S is equal to 4 and N_P is equal to 5. Thus, the total number of PV cells in the PV module is considered 20, as shown in **Figure 4**.

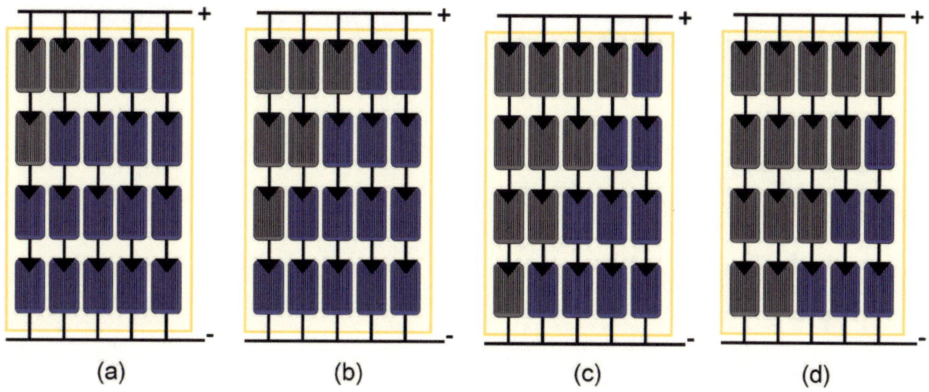

Figure 4.
Different shading pattern a) 10% b) 25% c) 50% d) 75%.

3.1.1 Series–parallel (SP)

The SP PV module configuration is the easiest and simplest way of creating a PV module. As discussed earlier, it is created by connecting solar cells in series. The total number of PV cells connected in series is dependent on the total voltage ratings of the PV module. Later, a similar configuration of series connected PV cells is connected in parallel based on the PV module's current and power rating. A bypass diode is connected across each PV cell in the module (**Figure 5a**). The I-V and P-V characteristics of the PV module using the SPICE based PV is shown in **Figure 5a)** and **b)**, respectively.

3.1.2 Total crossed-tied (TCT)

The TCT topology for the PV module is like the SP. The only difference is that each PV cell is connected in parallel with the neighboring PV cells in that column [28–31]. The neighboring PV cells in the module are shorted together as per **Figure 6a)** for modeling this configuration. The I-V and P-V characteristics for different shading condition is shown in **Figure 6b)** and **c)**.

3.1.3 Bridge linked (BL)

Figure 7a) shows the 20 PV cells-based module connected in BL configuration. The I-V and P-V characteristics obtained through the SPICE PV module is shown in **Figure 7b)** and **c)**. This configuration got the name since the module's PV cells are connected like a bridge rectifier [28–31]. In this configuration, two or more PV cells

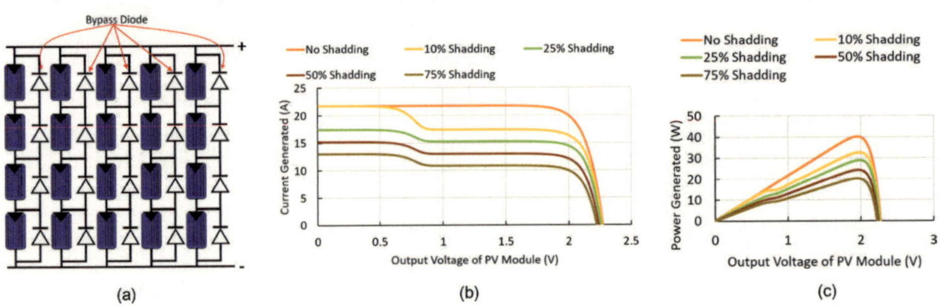

Figure 5.
The PV module connected in SP topology. a) the configuration of PV module is 4 × 5 b) I-V characteristics of PV module c) P-V characteristics of PV module under different shading condition.

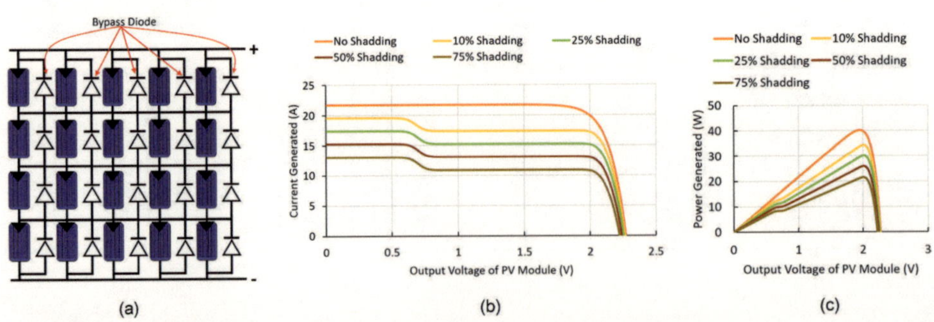

Figure 6.
The PV module connected in TCT topology. a) the configuration of PV module is 4 × 5 b) I-V characteristics of PV module c) P-V characteristics of PV module under different shading condition.

that are connected in series are tied with neighboring two or more PV cells in parallel. For simplicity a bypass diode is connected across each PV cells in the module.

3.1.4 Honeycomb (HC)

PV cells are connected in a hexagonal shape in this configuration, creating a honey bee house like structure [28–31]. Hence, six PV cells are connected together to form a hexagon as shown in **Figure 8a)**. The I-V and P-V characteristics obtained using the SPICE-based PV cell are shown in **Figure 8b)** and **c)**.

3.2 Reconfigurable PV module

It is shown in different research work the ways different fixed PV cell topologies mitigate the effect of partial shading and mismatch [28–31]. The usage of bypass diode achieves the effectiveness of fixed topologies. However, a fault in the bypass diodes can make various topologies for creating PV modules ineffective. Compared to using bypass diode, a novel complementary metal-oxide-semiconductor (CMOS) switch embedded PV module is proposed [32]. Thus having CMOS based switches the PV modules configuration, i.e., the number of PV cells in series vs. the number of PV cells in parallel, can be changed in real time [12, 32] in case of a fault in PV cells or partial shading condition. A CMOS embedded PV module is shown in **Figure 9**. The circuit diagram of the switches used is presented in detailed in [33].

In [27], it is presented that reconfigurable PV modules are much better in tackling the effect partial shading condition. However, it is also shown on the resistance of metal oxide semiconductor field-effect transistor (MOSFET) can reduce the

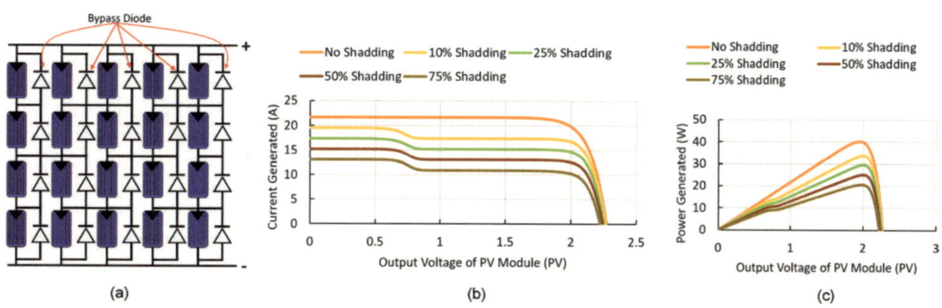

Figure 7.
The PV module connected in BL topology. a) the configuration of PV module is 4 × 5 b) I-V characteristics of PV module c) P-V characteristics of PV module under different shading condition.

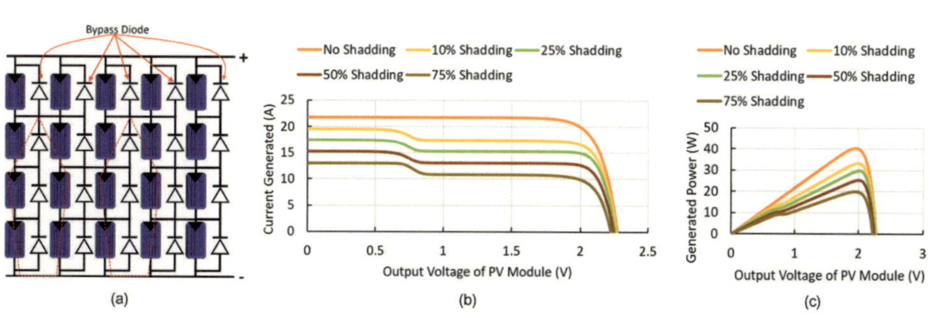

Figure 8.
The PV module connected in HC topology. a) the configuration of PV module is 4 × 5 b) I-V characteristics of PV module c) P-V characteristics of PV module under different shading condition.

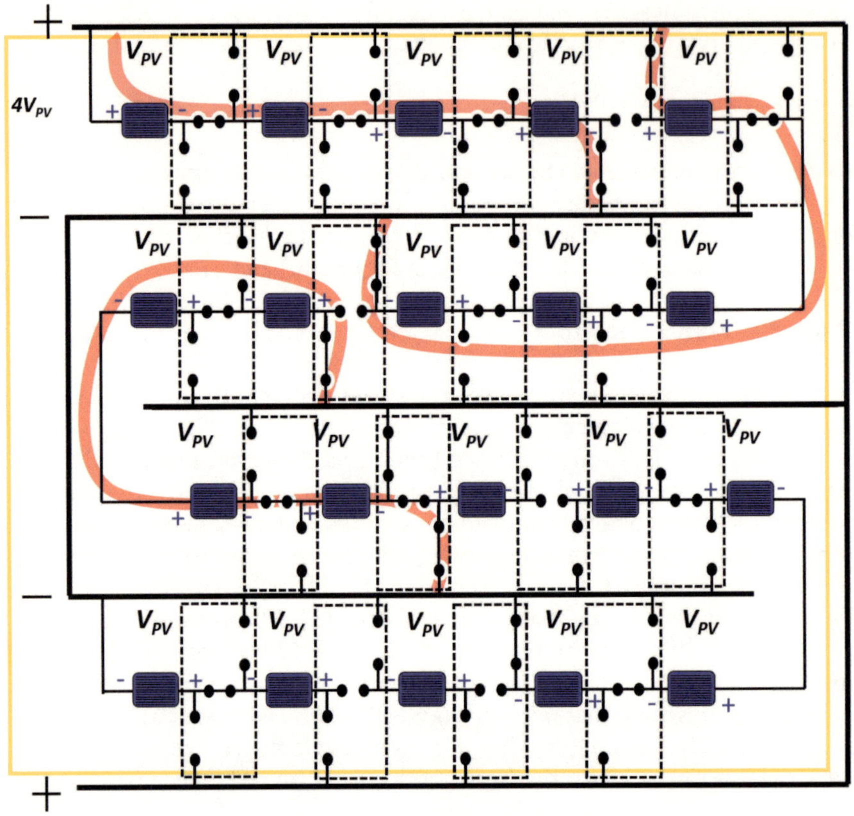

Figure 9.
CMOS switch embedded PV module. A group of four transistors are connected in series by turning ON and OFF CMOS based transistors.

effectiveness of the CMOS embedded PV module. Hence, it is a necessity to develop a modeling technique for reconfigurable PV modules for further development and deployment of their usage in various applications. Three different types of MOSFETs are used for modeling the reconfigurable, which is presented in **Table 2** [12].

The array decomposition for modeling the CMOS embedded PV module is shown in **Figure 10**. For simplicity, the two diodes are combined into a single diode shown in **Figure 10**. The total current generated by the reconfigurable PV module in configuration Ns × Np (number of PV cells in series x number of PV cells in parallel) is given by [33]:

$$I_{PV} \approx N_P\left[I_{ph} - I_{d1} - I_{d2} - \frac{B}{R_P}\right] \tag{6}$$

$$I_d = I_s\left(e^{\left(\frac{B}{A\times V_T}\right)} - 1\right) \tag{7}$$

$$B = \frac{v_{pv}}{N_S} + i_{pv}\left(\frac{R_S}{N_P} + \frac{(N_S-1)\cdot R_T}{N_P\cdot N_S} + \frac{2\cdot R_{ns}}{(N_P-1)\cdot N_S}\right) \tag{8}$$

In Eq. (8), the R_{ns} is equal to R_{PS}. It is possible to achieve it by sizing the N-type and P-type MOSFET transistor's width based on the process transconductance. The

Symbol	Description	Value and units
R_{Ps}	ON resistance of P-type MOSFET transistor	1mΩ
R_T	ON resistance of transmission gate transistor	1mΩ
R_{ns}	ON resistance of N-type MOSFET transistor	1mΩ

Table 2.
CMOS transistors resistance variable name and values.

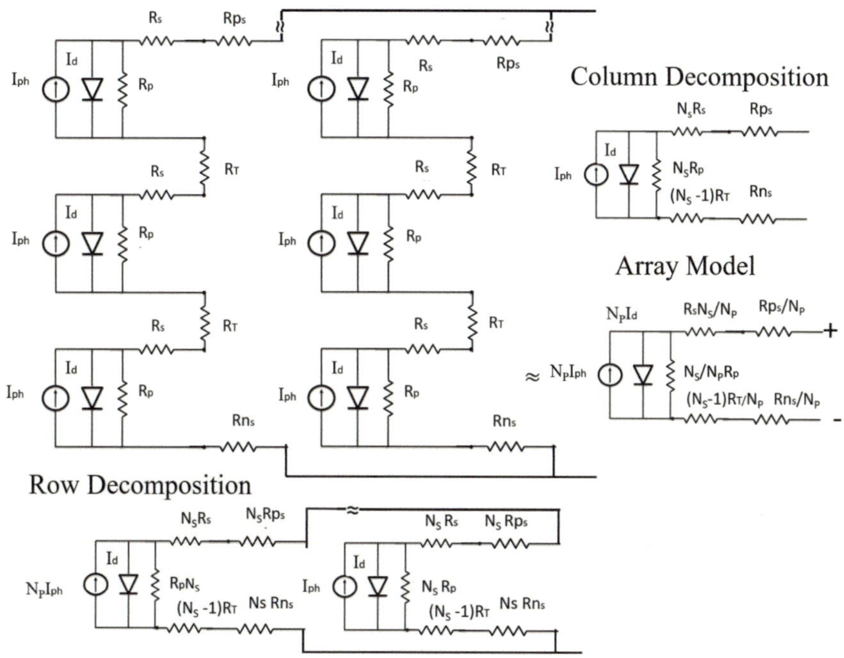

Figure 10.
PV cell decomposition technique for modeling CMOS embedded PV module.

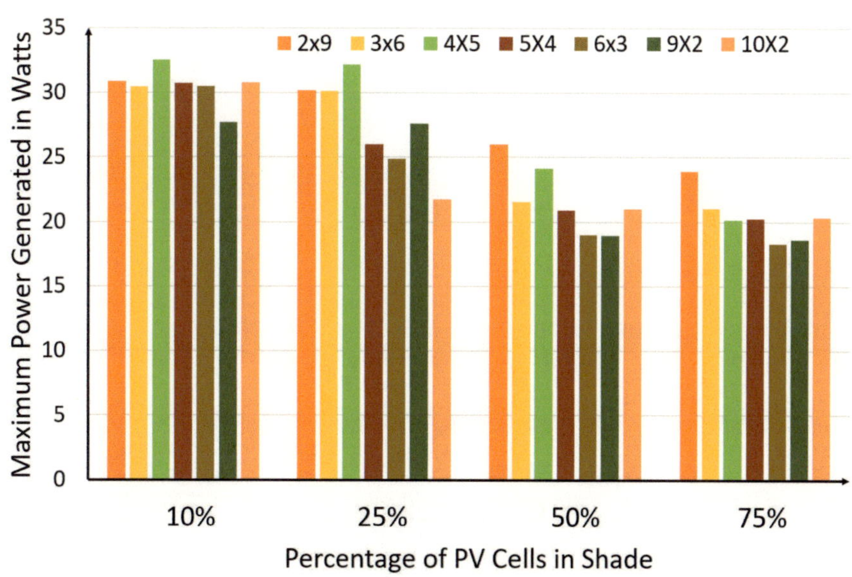

Figure 11.
The maximum power under different shading patterns over PV module for different configurations (number of PV cells in series x number of PV cells in parallel).

maximum power generated by the CMOS embedded PV module using SPICE based simulation is shown in **Figure 11**.

4. Conclusions

The authors have presented SPICE based modeling of PV cells and modules in detail. The I-V and P-V characteristics of popular fixed topologies for creating PV modules using SPICE modeling are presented. The contribution presented in this work will be significant in creating a smarter PV module that can better able to reduce the impact of partial shading, mismatch, and open/short fault.

In recent times many research works are done on creating a reconfigurable PV module. The modeling technique shown in the chapter presented the modeling technique of the reconfigurable PV module. These silicon switches embedded PV modules have the capability to create futuristic smart PV panels. Hence, for studying their performance in different operating conditions, the modeling technique presented in this chapter will be significant for their acceptance in different applications.

Conflict of interest

The authors declare no conflict of interest.

Author details

Rakeshkumar Mahto* and Reshma John
Computer Engineering Program, California State University, Fullerton, CA, USA

*Address all correspondence to: ramahto@fullerton.edu

References

[1] Reinders A, Sark W van, Verlinden P. Introduction. Photovoltaic Solar Energy. John Wiley & Sons, Ltd; 2016. pp. 1-12. doi:10.1002/9781118927496.ch1

[2] PVEducation. [cited 17 Feb 2021]. Available: https://www.pveducation.org/

[3] Shubbak MH. Advances in solar photovoltaics: Technology review and patent trends. Renew Sustain Energy Rev. 2019;115: 109383. doi:10.1016/j.rser.2019.109383

[4] Garnett EC, Ehrler B, Polman A, Alarcon-Llado E. Photonics for Photovoltaics: Advances and Opportunities. ACS Photonics. 2021;8: 61-70. doi:10.1021/acsphotonics.0c01045

[5] Green MA. How Did Solar Cells Get So Cheap? Joule. 2019;3: 631-633. doi:10.1016/j.joule.2019.02.010

[6] Ding H, Zhou DQ, Liu GQ, Zhou P. Cost reduction or electricity penetration: Government R&D-induced PV development and future policy schemes. Renew Sustain Energy Rev. 2020;124: 109752.

[7] Singh R, Tripathi P, Yatendra K. Impact of Solar Photovoltaic Penetration In Distribution Network. 2019 3rd International Conference on Recent Developments in Control, Automation Power Engineering (RDCAPE). 2019. pp. 551-556. doi:10.1109/RDCAPE47089.2019.8979014

[8] Dong CG, Wiser R, Rai V. Incentive pass-through for residential solar systems in California. 2014.

[9] Fraas LM, Partain LD. Solar Cells and Their Applications. John Wiley & Sons; 2010.

[10] Dewi T, Risma P, Oktarina Y. A Review of Factors Affecting the Efficiency and Output of a PV System Applied in Tropical Climate. 2018; 12.

[11] Meral ME, Dinçer F. A review of the factors affecting operation and efficiency of photovoltaic based electricity generation systems. Renew Sustain Energy Rev. 2011;15: 2176-2184. doi:10.1016/j.rser.2011.01.010

[12] Mahto R. Fault resilient and reconfigurable power management using photovoltaic integrated with CMOS switches. 2016.

[13] Bai J, Cao Y, Hao Y, Zhang Z, Liu S, Cao F. Characteristic output of PV systems under partial shading or mismatch conditions. Sol Energy. 2015;112: 41-54.

[14] Pendem SR, Mikkili S. Modelling and performance assessment of PV array topologies under partial shading conditions to mitigate the mismatching power losses. Sol Energy. 2018;160: 303-321.

[15] Wendlandt S, Drobisch A, Buseth T, Krauter S, Grunow P. Hot spot risk analysis on silicon cell modules. 25th European Photovoltaic Solar Energy Conference and Exhibition. Valencia, Spain; 2010. pp. 4002-4006.

[16] Molenbroek E, Waddington DW, Emery KA. Hot spot susceptibility and testing of PV modules. Photovoltaic Specialists Conference. Las Vegas; 1991. pp. 547-552.

[17] Skomedal \AAsmund F., Aarseth BL, Haug H, Selj J, Marstein ES. How much power is lost in a hot-spot? A case study quantifying the effect of thermal anomalies in two utility scale PV power plants. Sol Energy. 2020;211: 1255-1262.

[18] Dhere NG, Shiradkar NS. Fire hazard and other safety concerns of

photovoltaic systems. J Photonics Energy. 2012;2: 022006.

[19] Pandian A, Bansal K, Thiruvadigal DJ, Sakthivel S. Fire hazards and overheating caused by shading faults on photo voltaic solar panel. Fire Technol. 2016;52: 349-364.

[20] Laukamp H, Bopp G, Grab R, Wittwer C, Häberlin H, van Heeckeren B, et al. PV fire hazard-analysis and assessment of fire incidents. 26th EUPVSEC. 2013.

[21] Kato K, Koizumi H. A study on effect of blocking and bypass diodes on partial shaded PV string with compensating circuit using voltage equalizer. 2015 IEEE international symposium on circuits and systems (ISCAS). IEEE; 2015. pp. 241-244.

[22] Silvestre S, Boronat A, Chouder A. Study of bypass diodes configuration on PV modules. Appl Energy. 2009;86: 1632-1640.

[23] Castaner L, Silvestre S. Modelling photovoltaic systems using PSpice. John Wiley and Sons; 2002.

[24] Bana S, Saini RP. A mathematical modeling framework to evaluate the performance of single diode and double diode based SPV systems. Energy Rep. 2016;2: 171-187. doi:10.1016/j.egyr.2016.06.004

[25] Mahto R, Zarkesh-Ha P, Lavrova O. MOSFET-based modeling and simulation of photovoltaics module. 2016 IEEE 43rd Photovoltaic Specialists Conference (PVSC). 2016. pp. 3078-3081. doi:10.1109/PVSC.2016.7750231

[26] Silvestre S, Boronat A, Chouder A. Study of bypass diodes configuration on PV modules. Appl Energy. 2009;86: 1632-1640. doi:10.1016/j.apenergy.2009.01.020

[27] Mahto RV, Sharma DK, Xavier DX, Raghavan R. Improving performance of photovoltaic panel by reconfigurability in partial shading condition. J Photonics Energy. 2020;10: 042004.

[28] Yadav AS, Pachauri RK, Chauhan YK, Choudhury S, Singh R. Performance enhancement of partially shaded PV array using novel shade dispersion effect on magic-square puzzle configuration. Sol Energy. 2017;144: 780-797. doi:10.1016/j.solener.2017.01.011

[29] Wang Y-J, Hsu P-C. An investigation on partial shading of PV modules with different connection configurations of PV cells. Energy. 2011;36: 3069-3078. doi:10.1016/j.energy.2011.02.052

[30] Bingöl O, Özkaya B. Analysis and comparison of different PV array configurations under partial shading conditions. Sol Energy. 2018;160: 336-343. doi:10.1016/j.solener.2017.12.004

[31] Jazayeri M, Uysal S, Jazayeri K. A comparative study on different photovoltaic array topologies under partial shading conditions. 2014 IEEE PES T D Conference and Exposition. 2014. pp. 1-5. doi:10.1109/TDC.2014.6863384

[32] Mahto R, Lavrova O, Zarkesh-Ha P, Lester L. Reconfigurable and Programmable Photovoltaic Power for Micro Autonomous Systems. ReSpaceMAPLD 2010. 2010.

[33] Mahto R, Zarkesh-Ha P, Lavrova O. Reconfigurable photovoltaic integrated with CMOS for a fault tolerant system. 2016 IEEE 43rd Photovoltaic Specialists Conference (PVSC). IEEE; 2016. pp. 2578-2581.

Chapter 3

Study of a New Hybrid Optimization-Based Method for Obtaining Parameter Values of Solar Cells

Selma Tchoketch Kebir

Abstract

This chapter presents a comprehensive study of a new hybrid method developed for obtaining the electrical unknown parameters of solar cells. The combination of a traditional method and a recent smart swarm-based optimization method is done, with a big focus on the application of the topic of artificial intelligence algorithms into solar photovoltaic production. The combined approach was done between the traditional method, which is the noniterative Levenberg-Marquardt technic and between the recent meta-heuristic optimization technic, called Grey Wolf optimizer algorithm. For comparison purposes, some other classical solar cell parameter determination optimization-based methods are carried out, such as the numerical (iterative, noniterative) methods, the meta-heuristics (evolution, human, physic, and swarm) methods, and other hybrid methods. The final obtained results show that the used hybrid method outperforms the above-mentioned classical methods, under this study.

Keywords: solar cell, identification, optimization, meta-heuristics, swarm-based intelligence

1. Introduction

Solar photovoltaic energy is becoming the most popular renewable energy used in the world, at many caring of installations. Modeling and characterization are important topics that necessitate the determination of the exact solar photovoltaic (PV) cell's unknown parameters values and thus optimizing the PV power generated. Solar PV generator's performances are affected by many factors, where some of them are external, related to the environmental conditions like the weather's variations (irradiation and temperature), shading phenomenon, hotspot [1], dust, cell damage, wind velocity, and soiling [2]. Others are internal, related to the electrical, physical, and mathematical modeling. After the modeling step of any PV generator, their identified parameters values are used in the established model. Therefore, it is necessary to find the accurate values of the unknown electrical PV parameters by an appropriate approach. Besides, accurate parameter values of PV cells are essential for the development of good controlling techniques for Maximum Power Point Tracking (MPPT) based power electronic converters [3]. As shown in the **Figure 1** the importance of PV parameters' obtaining accurate values for a whole PV system.

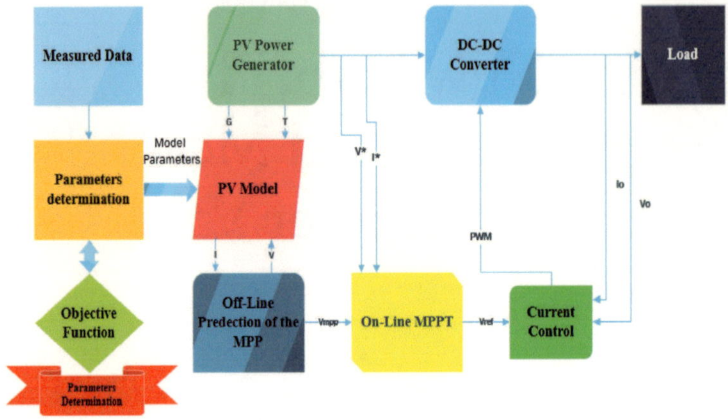

Figure 1.
Importance of determination of accurate PV parameter values in PV systems [3].

With the complexity of the nonlinearity founded from the current-voltage relationship through the model used to represent the physical behavior of PV cells (Single Diode, Double Diode, Three Diode, and so on) [4]. The parameters to be found become more numerous, as the complexity of models increases. The PV parameters of a Single Diode Model (SDM), which are the most determined in literature, and which are the light and the diode saturation currents, the diode ideality factor, the series, and the shunt resistances. A challenge is to obtain the values of all the PV model's parameters value while keeping a reasonable compromise of some criteria, such as the fast speed of convergence, low implementation complexity, and so on.

Artificial intelligent (AI) algorithms have attracted attention lately, by the scientific community to be used for resolving many topic's problems. Among them solar photovoltaics' problems, such as modeling, identification, prediction, optimization, sizing, control, and many others [5]. The hybrid optimization-based methods have attracted more attention to achieve more efficiency and precision. For this reason, this paper presents a combination of the traditional Levenberg-Marquardt (LM) with the recent meta-heuristic Grey Wolf optimizer (GWO) method. This hybrid LMGWO method has seemed to be the most performing, which we finally have chosen to be used in this work.

The remaining of this chapter is structured as follow. After the introduction given in this Section 1. Section 2 presents a classification of the optimization-based methods used in the literature to estimate the PV parameters values. In the Section 3, models are presented of the PV cell (single diode), and the problem of PV parameters obtaining values is presented. Section 4 gives some details about the hybrid used method to estimate the SDM parameters values. Moreover, this section provides the basic concept of the LM and the GWO. Finally, Section 5 shows some results comparing one method of each type proposed in the classification and the LMGWO. Some conclusions are given in Section 6.

2. Classification and literature review

Earliest, numerous research workings have been developed only for obtaining parasitic resistances (series and shunt) values [6] by the cause of their high influences on the PV's performances. Then, it has been observed some influences of all PV's electrical parameters on the PV's performances [7], which leads the researchers, for doing a large number of studies for obtaining their accurate values.

In literature, different approaches, that allow the evaluation of the PV cell's electrical parameters values, exist [8–11]. There are some analytical approaches [12–15], and those based on the optimization process. By the cause of limits of the analytical method to achieve with high precision the PV parameters values, our interest is gone for the optimization-based methods. This latter can be classified as in **Figure 2**.

The optimization algorithms are classified into traditional, heuristic/meta-heuristic, and hybrid groups. More details about each group of optimization-based are given at the following subsections.

2.1 Numeric traditional

The numeric traditional optimization-based methods are used to find the optimum of a function using gradient or hessian. These numeric-traditional methods applied for PV parameters obtaining values, are based on the reduction of the number of parameters to be evaluated, such as Kashif's one [16]. In this subsection's methods, the traditional iterative Newton-Raphson (NR) approach [17], iterative curve-fitting [18], can also be used. It necessitates an iterative process with good initialization guess of PV parameters values, to converge to the best solutions. Others build a set of nonlinear transcendental equations (based on short-circuit, open-circuit, MPP, derivatives of the I-V curve) and execute an optimization problem instead of solving by numerical methods [17]. For the noniterative method, the Levenberg-Marquardt [19] can be cited.

Even though with their effectiveness to get a good local search, they still have other limitations, such as the need of a convex, continuous, and differentiability of the objective function. Besides, good guessing of initial parameters values is necessary for a good converging process. Also, as the complexity of the modeling process increase, as the optimizer loses the ability for obtaining better results.

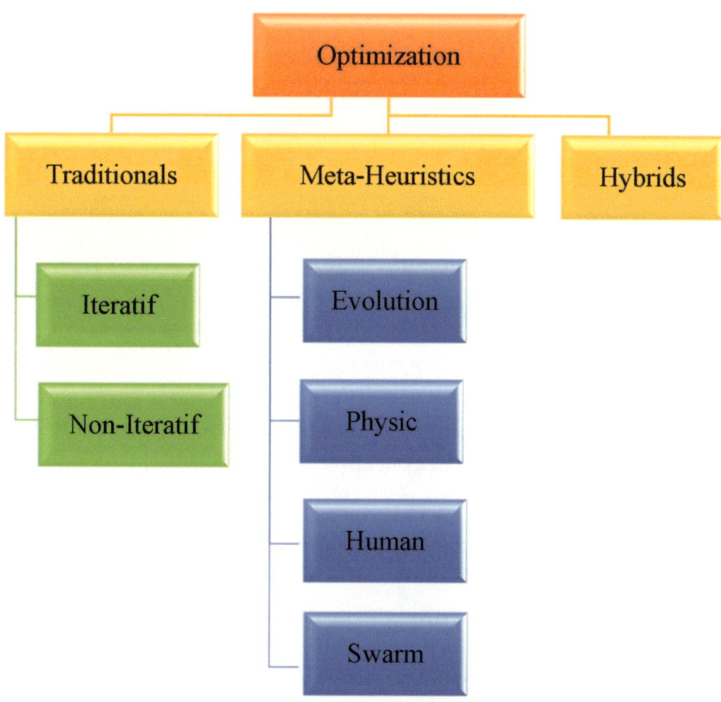

Figure 2.
Classification of PV cell's electrical parameter determination optimization-based methods.

2.2 Meta-heuristics

In recent times, meta-heuristic optimization-based methods, using Artificial-Intelligence (AI) inspired algorithms, have attracted the care of researchers to obtain with good precision, the unknown PV parameters values.

The metaheuristic methods use bio-inspired algorithms in the search process to identify the PV parameters values at real-time, using the errors between the real experimental data and the simulated data. These approaches are based on an experimental process and are known as identification methods [20]. These approaches are graphically based on curve characteristics fitting.

Meta-heuristics are categorized into four main sets such as evolution-based [21], physic-based [22], immune-human-based [23] and swarm-based intelligence methods [24]. Some of each category is used for obtaining PV parameters values as presented on what follow.

2.2.1 Evolution-based

Evolutionary Algorithm (EA) [25], Differential Evolutionary (DE) [23], Genetic Algorithms (GAs) [25], Pattern Search (PS) [21], Simulated Annealing (SA) [26], Improved Shuffled Complex Evolution (ISCE) [27], Repaired Adaptive Differential Evolution (Rcr-IJADE) [28].

2.2.2 Physic-based

Electromagnetic Field Optimization (EFO), Gravitational Search Algorithm (GSA), Electromagnetism-Like Algorithm (EMA), Weighted Superposition Attraction (WSA) [29].

2.2.3 Human-based

Harmony Search (HS) [30], Bacterial Foraging Algorithm (BFA) [31], Simplified Teaching-Learning-Based Optimization (STLBO) [32], Discrete Symbiosis Organism Search (DSOS) [33], Artificial Immune system (AIS) [34].

2.2.4 Swarm-based

The swarm-based, Particle Swarm Optimization (PSO) [3, 35, 36], Bird Mating Optimization (BMO) [37], Artificial Bee Swarm Optimization (ABSO) [38]. Grey Wolf Optimizer (GWO) [39], Chaotic Whale Optimization Algorithm (CWOA) [40], Cat Swarm Optimization (CSO) [41], and Cluster Analysis (CA) [3].

The metaheuristics are more attractive than the deterministic traditional methods in terms of accuracy and robustness, by the cause of their good global research achieving. Besides, they do not require a gradient or differentiable of the objective function. Besides, the initial guess of parameters values is not a necessity, but it necessitates the upper and lower limits of an interval of research.

2.3 Hybrids

The hybrid method combines different approaches. These methods make a mix of other methods, i.e. analytical and numeric-traditional methods [15]; analytical and meta-heuristics, numeric-traditional and meta-heuristics optimization; a combination of two different meta-heuristics, etc. [38]. We can site, hybrid adaptive Nelder-Mead simplex algorithm based on eagle strategy (EHA-NMS) [41], Nelder-Mead simplex algorithm based on eagle strategy (EHA-NMS) [41], Nelder-Mead and Modified

Particle Swarm Optimization (NM-MPSO) [42], Artificial Bee Colony-Differential Evolution (ABC-DE) [43], Trust-Region Reflective deterministic algorithm with the Artificial Bee Colony (ABC-TRR) [43], Teaching–learning–based Artificial Bee Colony (TLABC) [43]. Our proposed Levenberg-Marquardt with Grey Wolf optimizer (LM-GWO), and so on. Those methods, which are called hybrid, have excellent performances because they restrict the universe in the search process without losing precision (without losing the optimum). They achieve outstanding results with a smaller number of iterative steps when compared with pure optimization methods.

3. Modeling and problem formulation

There are several electrical models, used by researchers, to describe the physical behaviors of PV cells. The Single Diode Model, containing the five unknown parameters, used in this paper is represented in **Figure 3**. By the cause of compromise between accuracy and simplicity, the SDM is selected herein.

The mathematical expressions related to the current-voltage, (I-V) relationship of the PV cell is as follow.

$$I = I_L - I_D - I_{sh} \tag{1}$$

$$I = I_L - I_{ds}\left(e^{\left(\frac{V+R_s.I}{n.V_t}\right)} - 1\right) - \frac{V+R_s.I}{R_{sh}} \tag{2}$$

The overhead mathematical equation is in a nonlinear form and has a set of five unknown parameters (I_L, I_{ds}, n, R_s, R_{sh}). The main challenge is to get the accurate values of all the PV model's parameters values while keeping a reasonable computational effort.

Several approaches permit the formulation of the optimal nonlinear PV parameters determination problem, using the error (between real and simulated data) [10].

Our focus is to estimate the PV parameters values of the SDM model using RTC France data at the conditions of irradiance about 1000 W/m^2 and of temperature about 300°C. We do not review the identification process as detailed on our previous work [20]; our focus is restricted on the third part of identification process, which is the estimation of PV parameters values. The big focus is to optimize the damping factor of LM through GWO. The characteristics of RTC France Silicon-cell data from datasheet are presented on the following **Table 1**.

The real experimental data used of RTC France are presented on the following **Table 2**.

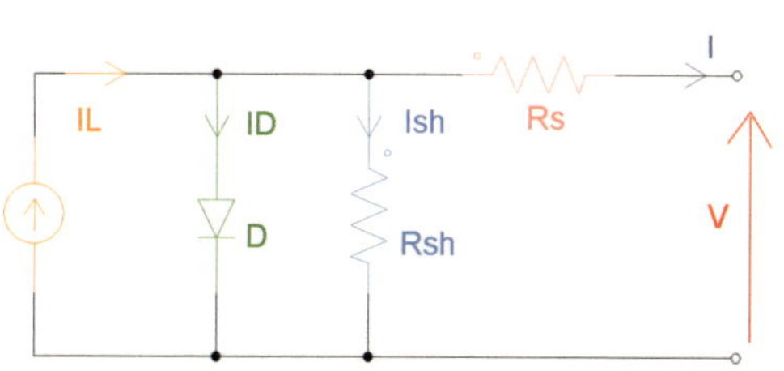

Figure 3.
PV cell's electrical equivalent circuit (SDM) [12].

Characteristic data	R.T.C France
I_{sc} (A)	0.7603
V_{oc} (V)	0.5728
V_{mpp} (V)	0.4507
I_{mpp} (A)	0.6894
P_{mpp}(W)	0.311
R_{sho} (Ω)	246.80*
R_{so} (Ω)	0.0907*
T (K)	306.15
N	1
K_v	71.44
K_i	0.035

Table 1.
Characteristic data from R.T.C. France (Si solar cell).

Measurement	V (Volts)	I (Ampere)
1	−0.2057	0.764
2	−0.1291	0.762
3	−0.058	0.7605
4	0.00057	0.7605
5	0.06460	0.76
6	0.1185	0.759
7	0.1678	0.757
8	0.2132	0.757
9	0.2545	0.7555
10	0.2924	0.754
11	0.3269	0.7505
12	0.3585	0.7465
13	0.3873	0.7385
14	0.4137	0.728
15	0.4373	0.7065
16	0.459	0.6755
17	0.4787	0.632
18	0.496	0.753
19	0.5119	0.499
20	0.5265	0.413
21	0.5398	0.3165
22	0.5521	0.2120
23	0.5633	0.1035
24	0.5736	−0.0100
25	0.5833	−0.1230
26	0.5900	−0.2100

Table 2.
Real data from RTC [38].

4. Hybrid optimization-based method

Hybrid optimization-based algorithms have become the modern choice for resolving challenging problems [41–43]. A compromise is gotten in this work, from a combination of a traditional numeric optimization-based with a metaheuristic swarm-based method.

The estimation/identification process can be gotten in three major steps, such as the initial step of prediction through the use of least-squares mean (LSM), the getting of optimal PV parameters values through Levenberg-Marquardt (LM), and the optimization of a dominant factor through GWO as detailed below.

4.1 Least squares mean (initial phase of prediction)

Prediction of initial PV parameters values using LSM [44, 45] for the two parts of the introduced real experimental points of I-V curve characteristics as described below.

- For the linear part:

The prediction in the linear part [46, 47] of the model can be obtained simply through the use of the following expressions.

$$I_{Model}(i) = a * V_{Model}(i) + b \tag{3}$$

$$Error(i) = I_{Real}(i) - I_{Model}(i) \tag{4}$$

$$J(i) = J(i-1) + error(i)^2 \tag{5}$$

where a and b are constants depending on a determinant and others constants introduced by user.

- For the nonlinear part:

The prediction in the nonlinear part [19, 48] of the model can be obtained with a logarithmic way through the use of the following logarithmic expression.

$$I_{Model}(i) = C_0 + C_1 * I_{Model}(i) + C_2 * \log\left(1 - \frac{I_{Real}(i)}{b}\right) \tag{6}$$

$$Error(i) = I_{Real}(i) - I_{Model}(i) \tag{7}$$

$$J(i+1) = J(i) + error(i)^2 \tag{8}$$

where C_0, C_1, C_2 and b are constants depending on a determinant, on the hessian and other constants introduced by the user.

Once obtaining initial values of PV parameters values, we introduce them on the LM in order to optimize their values, as explained in the following subsection.

4.2 Levenberg Marquardt (get of optimal PV parameter values)

The traditional Levenberg-Marquardt approach is a gradient order from Steepest-Descent (SD) in its first step and from Gauss-Newton (GN) in its second step [48–50]. It is mainly based on an optimization of the error between real data and data from the model through the following expression.

$$Ecart - Quad = \sum_{i=1}^{N} Error(i)^2 \tag{9}$$

where N is the number of measured I-V data.

$$Error = I_{Real}(i) - I_{Model}(i) \tag{10}$$

The real and simulated data are denoted by I_{Real} and I_{Model}, respectively. While I_{Model} is the objective function given as Eq. (2),

$$I_{Model}(i) = f(I, V, \theta) \tag{11}$$

Evaluate the objective function $f(\Theta)|_{\Theta = \Theta k}$. Here, Θ is considered as the PV parameters vector.

$$\theta = \{I_L, I_{ds}, n, R_s, R_{sh}\} \tag{12}$$

Calculus of Jacobian of $f(I,V,\Theta)$ for Θ_k, as the derivative calculation of I (Eq. (2)) with respect to parameters:

$$J = -\left[\frac{\partial f(\theta)}{\partial \theta}\right]_{\theta=\theta_k} \tag{13}$$

For (damping optimized) update Θ_k. The PV parameters to be found are updated at each iteration by the use of the expression below.

$$\theta_{k+1} = \theta_k - \left[\frac{J' * \varepsilon}{J' * J + \lambda_k * I}\right]_{\theta=\theta_k} \tag{14}$$

The dominant factor λ is considered as responsible parameters for switching from SD to GN in the LM process [19].

For this reason, it is important to get an optimal value of this damping factor by the use of another optimization-based method, our choice was for the recent swarm-based method called GWO, through the following idea:

$$Ecart - Quad(I, V, \theta, \lambda) \rightarrow Ecart - Quad(\lambda)\big|_{\theta=\theta_k} \tag{15}$$

In addition, it is mentioned that at each iteration of the LM process that the damping factor must be found and is considered as crucial factor for the convergence process of the algorithm. Therefore, its value must be optimized by the use of another approach such as the GWO approach.

4.3 Grey Wolf optimizer (optimization of damping factor's value)

In this subsection, our focus is on the evolution of the function$f(I,V,\Theta,\lambda)$ indicated by$f(\lambda)$ for Θ fixed at Θ_k, as regards with various varied values of the damping factor, at each iteration of the LM. As it is observed that at each iteration different local minimums values of$f(\lambda)$ exist. So, for obtaining the global minimum of$f(\lambda)$, which correspond to the best minimal value of the objective function$f(I,V,\Theta)$, we suggest using the swarm-based meta-heuristic GWO method.

The meta-heuristic methods are known for their simplicity, flexibility, derivation free process and the ability to find the global optimal solution. They are also appropriate for a diversity of problems without changing on their main structure. These methods can be based on a single solution or on population of solutions. The basic concepts can be obtained through exploration (exploring all of the search space and thus avoiding local optimum) and exploitation (investigating process in detail of the promising search space area).

Swarm-based intelligence (SI) methods, which derive from meta-heuristics, are based on the smart collective behavior of decentralized and self-organized swarms to ensure some biological needing such as food or security. A detailed discussion about the recent smart swarm-based algorithm, known as GWO is presented as follow.

Grey Wolf optimizer (GWO) algorithm, developed by Mirjalili in 2014, is a recent smart swarm-based meta-heuristic approach [50–52]. This algorithm mimics the leadership hierarchy and hunting process of Grey wolves in the wildlife. The following points represent the hierarchy in a wolf's group, which is about 5 to 12 members.

1. The alphas wolves (α): are the leading wolves that are responsible for managing and making decisions. These are the first level of the wolves' social hierarchical structure. This later is presented in **Figure 4**.

2. The betas wolves (β): represent the second level. Their main job is to help and support alpha's decisions.

3. The deltas wolves (δ): represent the third level in the pack and are called subordinates. They use to follow alpha and beta wolves. The delta wolves can divide their tasks into five categories as follows:

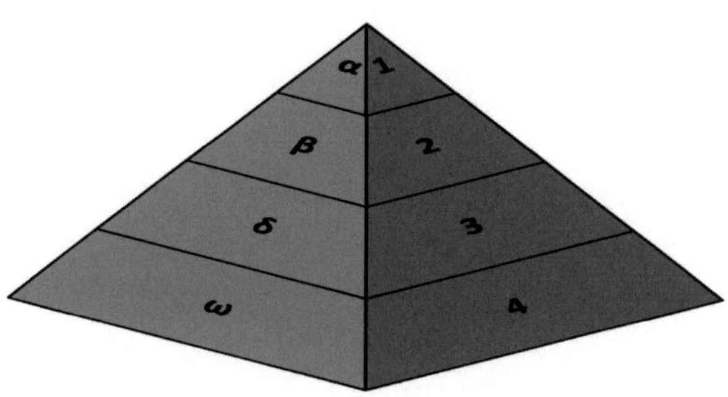

Figure 4.
The social hierarchical structure of Grey wolves (dominance decreases from the top-down) [51].

- Scouts: used to control the boundaries of the territory and alert the pack in case of danger.

- Sentinels: protect and guarantee the safety of the pack.

- Elders: among these strong and mature wolves, some of them become either alpha or beta.

- Hunters: help alpha and beta in the hunting prey, providing food to the pack.

- Caretakers: responsible for caring the ill, wounded and weak wolves.

4. The omegas wolves (ω): represent the lowest level. They have to follow alpha, beta and delta wolves.

When a pack of wolves sees a prey such as (gazelle, rabbit or a buffalo) they attack it in three steps and do not recede, **Figure 5**.

These three steps of the hunting process can be mentioned as follows.

- Encircling, tracking, chasing, and approaching the prey (**Figure 5**: A, B).

- Pursuing, encircling, and harassing the prey until it stops moving (**Figure 5**: C).

- Attacking the prey (**Figure 5**: D, E).

Figure 5.
The process of hunting prey by a group of wolves [51].

The mentioned above social hierarchy and hunting process of Grey wolves have been mathematically modeled in GWO, as follows [51, 52]:

- The first, second and third best solutions are considered as α, β and δ wolves, respectively.
- The rest of the candidate solutions are considered as ω.

The following equations are used to model the encircling first step of Grey wolves hunting process:

$$\vec{D} = \left|\vec{C} * \overrightarrow{X_P}(i) - \vec{X}(i)\right| \tag{16}$$

$$\vec{X}(i+1) = \overrightarrow{X_p}(i) - \vec{A} * \vec{D} \tag{17}$$

where i represents the current iteration. X and X_p represent the position vectors of the wolves and the prey, respectively. A and C are the coefficients and are calculated as follows:

$$\vec{A} = 2 * \vec{a} * \overrightarrow{r_1} - \vec{a} \tag{18}$$

$$\vec{C} = 2 * \overrightarrow{r_2} \tag{19}$$

where a is linearly decreasing from 2 to 0 throughout iterations, and r_1, r_2 are random values in an interval from 0 to 1. In GWO, decreasing the values of A, from 2 to 0 during the optimization process, simulates the prey approach and provides the exploration ability of the algorithm. Besides, the exploitation ability of the GWO comes from the random value of C.

To mathematically simulate the second step of the Grey wolves hunting process, we suppose that the alpha (best candidate solution), beta and delta have a better knowledge about the potential location of the prey [53]. Therefore, the first three best solutions obtained so far are saved and oblige the other search agents (including the omegas) to update their positions according to the position of the best search agents. In this regard, the following formulas are used.

$$\overrightarrow{D_{\alpha,\beta,\delta}} = \left|\overrightarrow{C_{1,2,3}} * \overrightarrow{X_{\alpha,\beta,\delta}} - \vec{X}\right| \tag{20}$$

$$\overrightarrow{X_{1,2,3}} = \overrightarrow{X_{\alpha,\beta,\delta}} - \overrightarrow{A_{1,2,3}} * \overrightarrow{D_{\alpha,\beta,\delta}} \tag{21}$$

$$\vec{X}(i+1) = \frac{\vec{X}_1 + \vec{X}_2 + \vec{X}_3}{3} \tag{22}$$

The final third step is the hunting process as attacking the prey as soon as it stops moving.

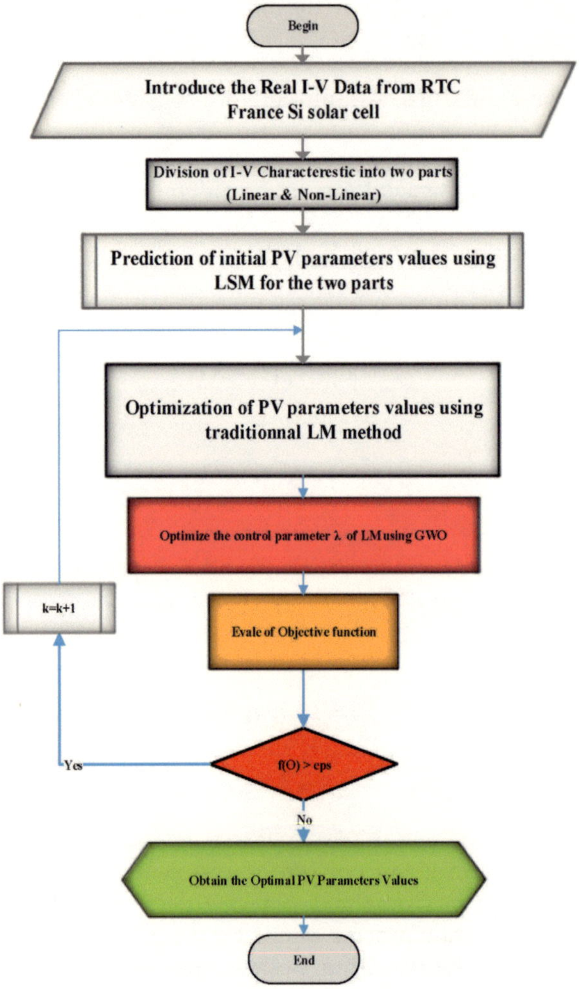

Figure 6.
PV parameters identification steps using the hybrid LM approach with GWO approach.

4.4 LMGWO

The main steps of the used hybrid LMGWO method applied for the PV parameters obtaining values are presented in **Figure 6**.

5. Results and discussions

The following **Table 3** presents PV parameters results for the all classified optimization-based method discussed in Section 2.

From the above **Table 3** it is clear that for the traditional methods, the LM is more accurate than Newton's method, which in turn outperforms Kashif's method. Then, for the metaheuristic methods for each of their category as follow.

- Evolution-based:

  It is observed that ISCE, Rcr-IJADE, and PCE outperform PS, which in turn is better than GA and SA.

Methods	Parameters	I_L (A)	$I_{ds}(\mu A)$	n	$R_s(\Omega)$	$R_{sh}(\Omega)$	RMSE
Traditional	Kashif [16]	0.760300	2.624738e-09	1.200000	0.014000	19.000032	7.090000e-02
	LM [19]	0.760782	3.166611e-07	1.479182	0.036461	53.271523	9.8680e- 4
	Newton [17]	0.7608	0.3223	1.4837	0.0364	53.7634	9.70E−03
Meta-heuristics	**Evolution-based**						
	GA [25]	0.7619	0.8087	1.5751	0.0299	42.3729	0.019
	SA [26]	0.762	0.4798	1.5172	0.0345	43.1034	0.019
	PS [21]	0.7617	0.998	1.6	0.0313	64.1026	0.0149
	ISCE [27]	0.760776	0.32302	1.48118	0.03638	53.7185	9.8602E−04
	Rcr-IJADE [28]	0.760776	0.32302	1.48118	0.03638	53.7185	9.8602E−04
	PCE	0.760776	0.323021	1.481074	0.03638	53.7185	9.8602E−04
	Swarm-based						
	PSO [35, 36]	0.76077	0.32454	1.48165	0.03636	53.8550	9.8606E−04
	ABSO [38]	0.7608	0.30623	1.47583	0.03659	52.2903	9.9124E−04
	BMO [37]	0.76077	0.32479	1.48173	0.03636	53.8716	9.8608E−04
	CSO [41]	0.76078	0.323	1.48118	0.03638	53.7185	9.8602E−04
	CWOA [40]	0.76077	0.3239	1.4812	0.03636	53.7987	9.8602E−04
	Human-based						
	BFA [31]	0.7602	0.8000	1.6951	0.0325	50.8691	0.029
	HS [30]	0.7607	0.305	1.4754	0.0366	53.5946	9.95E−04
	STLBO [32]	0.76078	0.32302	1.48114	0.03638	53.7187	9.8602E−04
	Physic-based						
	EFO [29]	0.760776	0.323022	1.481184	0.036377	53.718646	9.860219E-04
	GSA [29]	0.760977	0.847206	1.585214	0.032130	82.871489	2.166195E-03
	EMA [29]	0.760590	0.329155	1.483019	0.036365	57.025188	9.972880E-04
	WSA [29]	0.754454	1.000000	1.607072	0.027957	97.854073	7.702232E-03
Hybrid	LM-GWO	0.760776	0.32306	1.48118	0.03637	53.7222	9.8601E-04
	LMSA [19]	0.7608	0.3185	1.4798	0.0364	53.3264	9.86E−04
	EHA-NMS [27]	0.760776	0.32302	1.48118	0.03638	53.7185	9.8602E−04
	ABC-TRR [43]	0.760776	0.32302	1.48118	0.03638	53.7185	9.8602E−04
	ABC-DE [43]	0.76077	0.32302	1.47986	0.03637	53.7185	9.8602E−04
	NM-MPSO	0.76078	0.32306	1.4812	0.03638	53.7222	9.8602E−04
	TLABC [43]	0.76078	0.32302	1.48118	0.03638	53.7164	9.8602E−04

Table 3.
Parameter extraction results for 57-mm diameter R.T.C. France commercial silicon solar cell using the single diode model.

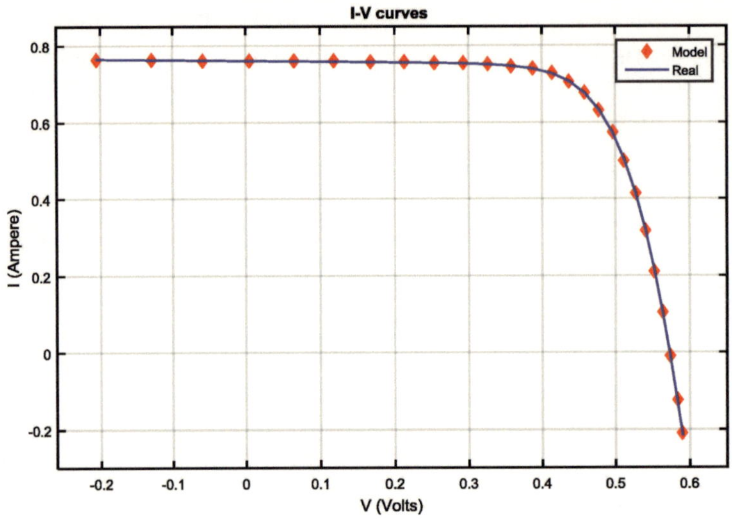

Figure 7.
Fitted I-V curve characteristic for the RTC silicon solar cell, using the hybrid LM-GWO method.

- Swarm-based:

  All the swarm-based used outperform ABSO in terms of precision.

- Human-based:

  CSO and STBLO outperform HS, which is better than BFA.

- Physics-based

  EFO is more accurate than EMA, which is more accurate than WSA, which is more accurate then GSA.

It is mentioned that the swarm-based got the best results compared to the other metaheuristic's category.

Finally, for the hybrid methods, it is clear that all of them have achieved the highest best optimized (minimum) values for RMSE, until now with the value of (9.8601E−04).

In addition, the hybrid methods outperform the metaheuristics, which in turn outperform the traditional methods.

The fitting obtained curves of real and simulated data, using the proposed LMGWO are illustrated in **Figure 7**.

The best approximation gotten from the fitted curves in **Figure 7** has proved the effectiveness of our hybrid LMGWO method.

6. Conclusion

Since nonlinear and multi-parameter PV models are used to represent a PV cell's physical behavior, classical methods are incapable of evaluating the exact parameters' values of these models. For these reasons, the present paper presents a proposed hybrid method of obtaining the unknown electrical parameters of solar photovoltaic cells. To do so, we applied our hybrid method, the LM combined with GWO method, after having initial guess using least squares mean, and then compared it with other previous optimization-based methods. The application of LMGWO has shown high precision for the obtained solutions' values. The LMGWO outperforms the other tested algorithms in many aspects. It is simple and accurate and converges rapidly to the optimum in every test. In addition, it has fewer parameters to set then it is easily implemented. The obtained results demonstrate the efficiency of the hybrid LMGWO approach compared to the other meta-heuristics and some of the other traditional methods.

Conflict of interest

The authors declare no conflict of interest.

Author details

Selma Tchoketch Kebir[1,2]

1 Laboratoire de Dispositifs de Communications et de Conversion Photovoltaique, Electronic Department, Ecole Nationale Polytechnique, 16000 El-harrach, Algiers, Algeria

2 Unité de Développement des Equipements Solaires, UDES/Centre de Développement des Energies Renouvelables, CDER, Bou-Ismail, 42415 W, Tipaza, Algeria

*Address all correspondence to: selma.tchoketch_kebir@g.enp.edu.dz

References

[1] Guerriero P, Daliento S. Toward a hot spot free PV module. IEEE Journal of Photovoltaics. May 2019;**9**(3):796-802

[2] Coello M, Boyle L. Simple model for predicting time series soiling of photovoltaic panels. IEEE Journal of Photovoltaics. September 2019;**9**(5):1382-1387

[3] Jieming M. Optimization approaches for parameter estimation and maximum power point tracking (MPPT) of photovoltaic systems [thesis]. Liverpool: University of Liverpool for the degree of Doctor in Philosophy; 2014

[4] Saha C, Agbu N, Jinks R, Nazmul Huda M. Review article of the solar PV parameters estimation using evolutionary algorithms. MOJ Solar and Photoenergy Systems. 2018;**2**(2):66-78

[5] Leva S, Ogliari E. Computational intelligence in photovoltaic systems. Applied Sciences. 2019;**9**:1826. DOI: 10.3390/app9091826

[6] Bashahu M, Habyarimana A. Review and Test of Methods for Determination of the Solar Cell Series Resistance. Pergamon: Renewable Energy; 1995

[7] Kenneth KL. Analysis of performance degradation in CdS solar cells. IEEE Transactions on Aerospace and Electronic Systems. 1969;**AES-5**(6):912-917

[8] Tchoketch-Kebir S, Haddadi M, Ait-Cheikh MS. An overview of solar cells parameters extraction methods. In: IEEE Proceedings in the 3rd International Conference on Control, Engineering & Information Technology CEIT, Telmcen, Algeria, 2015

[9] Humada Ali M, Hojabri M, Mekhilef S, Hamada Hussein M. Solar cell parameters extraction based on single and double-diode models: A review. Renewable and Sustainable Energy Reviews. 2016;**56**:494-509

[10] Pillai Dhanup S, Rajasekar N. Metaheuristic algorithms for PV parameter identification: A comprehensive review with an application to threshold setting for fault detection in PV systems. Renewable and Sustainable Energy Reviews. [Accepted: 28 October 2017]

[11] Abbassia R, Abbassic A, Jemlic M, Chebbi S. Identification of unknown parameters of solar cell models: A comprehensive overview of available approaches. Renewable and Sustainable Energy Reviews. 2018;**90**:453-474

[12] Shongwe S, Hanif M. Comparative analysis of different single-diode PV modeling methods. IEEE Journal of Photovoltaics. 2015;**5**(3)

[13] Mehta Hitesh K, Warke H, Kukadiya K, Panchal Ashish K. Accurate expressions for single-diode-model solar cell parameterization. IEEE Journal of Photovoltaics. 2019;**9**(3)

[14] Chitti Babu B, Gurjar S. A novel simplified two-diode model of photovoltaic (PV) module. IEEE Journal of Photovoltaics. 2014;**4**(4):1156-1161

[15] Brano VL, Ciulla G. An efficient analytical approach for obtaining a five parameters model of photovoltaic modules using only reference data. Applied Energy. 2013;**111**:894-903

[16] Kashif I, Zainal S, Hamed T. Simple, fast and accurate two-diode model for photovoltaic modules. Solar Energy Materials & Solar Cells. 2011;**95**:586-594

[17] Cabestany J, Castanier L. Evaluation of solar cell parameters by nonlinear algorithms. Journal of Physics D: Applied Physics. 1983;**16**:2547

[18] Easwarakhanthan T, Bottin J, Bouhouch I, Boutrit C. Nonlinear minimization algorithm for determining the solar cell parameters with microcomputers. International Journal of Solar Energy. 1986;**4**(1):1-12

[19] Dkhichi F, Oukarfi B. Levenberg-Marquardt algorithm for parameter identification of solar cell model. In: Conference Paper June. 288 014. pp. 781-788

[20] Tchoketch_Kebir S, Ait_Cheikh MS, Haddadi M. A detailed step-by-step electrical parameters identification method for photovoltaic generators using a combination of two approaches. ASTES Journal. 2018;**3**(4):44-52

[21] AlHajri MF, El-Naggar KM, AlRashidi MR, Al-Othman AK. Optimal extraction of solar cell parameters using pattern search. Renewable Energy. 2012;**44**:238-245

[22] Rashedi E, Nezamabadi-Pour H, Saryazdi S. GSA: A gravitational search algorithm. Information Sciences. 2009;**179**:2232-2248

[23] Chena M-H, Chang P-C, Wu J-L. A population-based incremental learning approach with artificial immune system for network intrusion detection. Engineering Applications of Artificial Intelligence. 2016

[24] Beni G, Wang J. Swarm intelligence in cellular robotic systems. In: Robots and Biological Systems: Towards a New Bionics. Riverside, California 92521, USA: Springer; 1993. pp. 703-712

[25] Gao X, Cui Y, Hu J, Xu G, Wang Z, Qu J. Parameter extraction of solar cell models using improved shuffled complex evolution algorithm. Energy Conversion and Management. 2018;**157**:460-479

[26] El-Naggar KM, AlRashidi MR, AlHajri MF, Al-Othman AK. Simulated annealing algorithm for photovoltaic parameters identification. Solar Energy. 2012;**86**:266-274

[27] Chen Z, Wu L, Lin P, Wu Y, Cheng S. Parameters identification of photovoltaic models using hybrid adaptive Nelder-Mead simplex algorithm based on eagle strategy. Applied Energy. 2016;**182**:47-57

[28] Gong W, Zhihua C. Parameter extraction of solar cell models using repaired adaptive differential evolution. Solar Energy. 2013;**94**

[29] Yurtkuran A, Küçükoğlu İ. Comparative study of physics-inspired meta-heuristic algorithms for the solar cell parameter identification problem. In: 16th International Conference on Clean Energy (ICCE-2018)

[30] Askarzadeh A, Rezazadeh A. Parameter identification for solar cell models using harmony search-based algorithms. Solar Energy. 2012;**86**:3241-3249

[31] Rajasekar Neeraja N, Kumar K, Venugopala R. Bacterial foraging algorithm based solar PV parameter estimation. Solar Energy. 2013;**97**:255-265

[32] Niu Q, Zhang H, Li K. An improved TLBO with elite strategy for parameters identification of PEM fuel cell and solar cell models. International Journal of Hydrogen Energy. 2014

[33] Bouali C, Schulte H, Mami A. A high performance optimizing method for modeling photovoltaic cells and modules array based on discrete symbiosis organism search. Energies. 2019;**12**:2246. DOI: 10.3390/en12122246

[34] Jacob B, Balasubramanian K, Sudhakar Babu T, Azharuddin SM, Rajasekar N. Solar PV modelling and parameter extraction using artificial immune system. Energy Procedia. 2015;**75**:331-336

[35] Wang X, Xu Y, Ye M. Parameter extraction of solar cells using particle swarm optimization. Journal of Applied Physics. 2009;**105**:094502

[36] Soon JJ, Low KS. Photovoltaic model identification using particle swarm optimization with inverse barrier constraint. IEEE Transactions on Power Electronics. 2012;**27**:3975-3983

[37] Askarzadeh A, Rezazadeh A. Extraction of maximum power point in solar cells using bird mating optimizer-based parameters identification approach. Solar Energy. 2013;**90**:123-133

[38] Chen X, Xu B, Mei C, Ding Y, Li K. Teaching–learning–based artificial bee colony for solar photovoltaic parameter estimation. Applied Energy. Elsevier. 2018;**212**:1578-1588

[39] Robandi DI. Photovoltaic parameter estimation using Grey Wolf optimization. In: 3rd International Conference on Control, Automation and Robotics, 2017

[40] Oliva D, Abd El Aziz M, Ella Hassanien A. Parameter estimation of photovoltaic cells using an improved chaotic whale optimization algorithm. Applied Energy. 2017;**200**:141-154

[41] Guo L, Meng Z, Sun Y, Wang L. Parameter identification and sensitivity analysis of solar cell models with cat swarm optimization algorithm. Energy Conversion and Management. 2016;**108**:520-528

[42] Hamid NFA, Rahim NA, Selvaraj J. Solar cell parameters identification using hybrid Nelder-Mead and modified particle swarm optimization. Journal of Renewable and Sustainable Energy. 2016

[43] Hachana O, Hemsas KE, Tina GM, Ventura C. Comparison of different metaheuristic algorithms for parameter identification of photovoltaic cell/ module. Journal of Renewable and Sustainable Energy. 2013;**5**

[44] Miller SJ. The Method of Least Squares. Providence, RI: Mathematics Department Brown University. p. 02912

[45] Lawson C, Hanson R. Solving Least Squares Problems. Society for Industrial and Applied Mathematics, Classics in applied mathematics, 15; SIAM e-books. Philadelphia, Pa, Englewood Cliffs, NJ: Prentice-Hall; 1974

[46] Kong M, Li D, Zhang D. Research on the application of improved least square method in linear fitting. IOP Conference Series: Earth and Environmental Science. 2019;**252**:052158. DOI: 10.1088/1755-1315/252/5/052158

[47] Whittle P, Sargent T. Least-square approximation. In: Prediction and Regulation by Linear Least-Square Methods. 2nd ed. Minneapolis: University of Minnesota Press; 1983. pp. 46-55

[48] Walling DD. Numerical methods for non-linear least squares curve fitting [thesis]. Iowa State University; 1963

[49] Gavin HP. The Levenberg-Marquardt Method for Nonlinear Least Squares Curve-Fitting Problems. Department of Civil and Environmental Engineering, Duke University; 2016

[50] Manolis I, Lourakis A. A Brief Description of the Levenberg-Marquardt Algorithm Implemened by levmar. Crete, Greece: Institute of Computer Science, Foundation for Research and Technology - Hellas (FORTH); 2005

[51] Mirjalili S, Mirjalili SM, Lewis A. Grey Wolf optimizer. Advances in Engineering Software. 2014;**69**:46-61

[52] Mirjalilia S, Saremia BS, Mirjalilic BSM, Coelhod LDS. Multi-objective grey wolf optimizer: A

novel algorithm for multi-criterion optimization. Expert Systems with Applications. 2016;**47**:106-119

[53] Lakshminarayanan S. Nature inspired Grey Wolf optimizer algorithm for minimizing operating cost in green smart home [thesis]. The University of Toledo; 2015

Chapter 4

Solar Energy Assessment in Various Regions of Indian Sub-continent

Johny Renoald Albert and Dishore Shunmugham Vanaja

Abstract

The demand for sustainable energy has increased significantly over the years due to the rapid depletion of fossil fuels. The solar photovoltaic system has been the advantage of converting solar irradiation directly to electricity, and it is suitable for most of the regions. But in the case of solar energy conversion, the voltage evolved from the solar photovoltaic cells is not adequate to meet the energy demand. Therefore, the converters and inverters with energy storage systems are used to fulfill the energy demand. These conversion architectures create new challenges for effective management of the grid. Due to the evaluation of power generation, load in a particular region or area, let us simplify with the help of the duck curve. The study is focused on the energy auditing, assessment, and measurement of solar irradiation from PV system design software. This graphical representation is implemented with a typical electricity load pattern at any region.

Keywords: solar-PV system, solar irradiation, modular scale battery energy storage system, multilevel converters, duck curves

1. Introduction

The global solar power demand capacity will grow by 9% every year between 2018 and 2050. According to that a new strategy has been released by the International Renewable Energy Agency (IRENA). The study of future Photovoltaic energy states the global solar capacity will be increase from 480 GW in 2018 to over 8000 Giga Watt by 2050 [1, 2]. The solar energy can be predicted to some degree from analysis level of climate conditions at the project site, but for the basic explanation that the atmosphere cannot be tracked. But, the solar panels can supply power on demand [3]. The evaluation of power generation load in a certain region shortens by using for duck curve [4, 5]. In the case of solar energy consumers often know in advance that their investments can produce energy only during the daytime. A solar system has a constant power output, the system only needs to be sized larger and it's needed for excess energy storage system [6–8]. There are 3 emerging storage technologies that is viable energy solution for renewables such as solar or wind in recent scenario such as 1. Smart batteries, 2. Heat based energy storage, 3. Hydrogen fuel-cells. This approach is not suitable in reality, because the energy loss during charge/discharging duty cycle is up to 10–15%, and the large scale energy storage is currently much expensive [9, 10]. This practical problem as renewable energy has become more widespread, and to get consent power output from a solar system.

To implement the intermittency challenges for effective Modular Scale Battery Energy Storage System (MSBESS), and it's established as a necessary component for solar integrated micro grid system [11–13]. India has set itself an admiral target of 175 GW of Renewable energy by 2022. The target will principally comprise of 40 GW Rooftop and 60 GW through large and medium scale grid connected solar power projects in India [14–16]. In a span of 3 hrs in the evening the conventional sources need to ramp up production by almost 10 GW. A framework was improved to the utilization of localized solar irradiation, and availability of Indian sub-continent region with associating with the open source archive database [17–22]. The results and analysis are presented for detailed study for various region of solar resource potential evaluation. The measurements and demonstration of a simplified with PV software system design latest version 7.0.2 under the simulation of regional solar energy requirements of real-time basis level. Accordingly, one of the main restriction of solar energy, there is no control over when the PV system will be producing the power. In this situation can be solved by converters to interconnection in between the load, and as well as PV storage management system.

2. Major solar irradiation level in India region

India has a high potential for solar power Generation on about 300 direct sunshine days per year. The regular solar incident in India varies with an annual sunlight of 4 to 7 kWh/m^2, which is about 1500 to 2000 hours above the irradiation level gross energy consumption. The renewable energy generated by India in 2020 amounted to grown up 9.46%. **Table 1** shows the major irradiation level solar hot spot evaluated in India. Total solar power capacity was installed in India 35,739 MW. **Figure 1** show the annual average insolation solar hotspot map.

Monthly global average insolation data is collected the entire topography of India with in longitudes 67° to 97°E and 9° to 39°N. The various region of global insolation like as solar power generation identified hotspot in India based on surface measurements obtained from solar radiation station. The average global insolation map is employed to produce for using global information system [28, 29]. The insolation solar direct global is given by the Eq. (1),

State	Total area (1500 ha)	Total solar potential energy	High irradiation insolation level
Karnataka [2]	19,050	24.7 GW	3.5–4.0 kWh/m^2/day
Rajasthan [17]	34,270	4.8 GW	5–7 kWh/m^2/day
Maharashtra [17]	30,758	64.32 GW	3–4 kWh/m^2/day
Himachal Pradesh [23]	4548	33.8 GW	3–4 kWh/m^2/day
Jammu Kashmir [23]	3781	111.05 GW	3–4 kWh/m^2/day
Andhra Pradesh [24]	27,505	38.4 GW	3–4.5 kWh/m^2/day
Gujarat [24]	18,866	35.7 GW	4–4.7 kWh/m^2/day
Odisha [25]	15,043	25.7 GW	4–4.7 kWh/m^2/day
Madhya Pradesh [26]	30,756	61.6 GW	2.9–4.0 kWh/m^2/day
Uttar Pradesh [26]	24,170	22.8 GW	2.9–3.9 kWh/m^2/day
Haryana [27]	18,096	73.2 MW	3.5–4.5 kWh/m^2/day

Table 1.
Major solar irradiation hot spot.

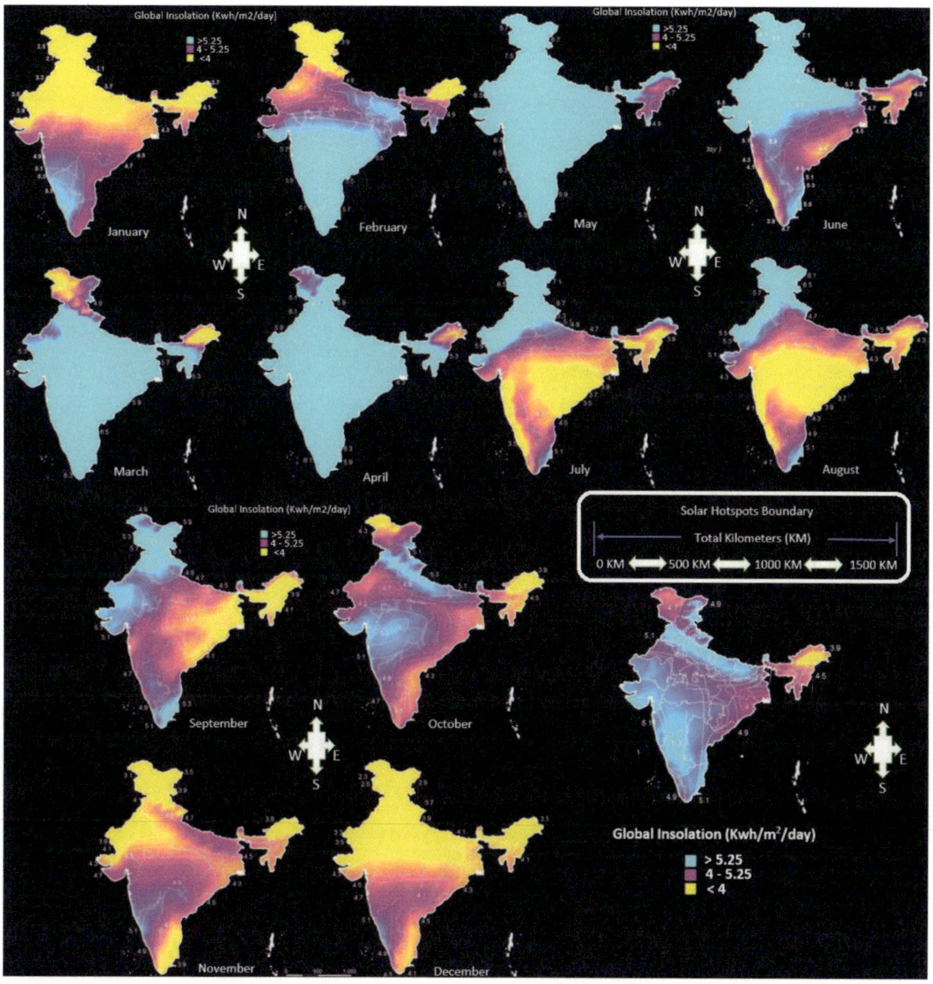

Figure 1.
Annual average insolation solar hotspot map.

$$I = (G\text{-}D) / \sin\phi \quad (1)$$

Where, G is solar direct global insolation, D is the diffuse component, and Φ is the sun elevation angle. The installation by five regional solar power generation as given in **Table 2**. There are three main strategy of maximum power generated from solar in India: 1. ground mounted system, 2. rooftop, and 3. off-grid (stand-alone mode). The global radiation is the sum of the total horizontal radiation at any location calculated by radiation directly (I_d) and radiation diffusely (I_b) given in the Eq. (2). Most of the individual researchers have proposed numerous models for estimating the meteorological application of global radiation parameters include such as cloudiness, air temperature etc. The global radiation level is calculated by daily solar radiation level in horizontal surface [30, 31]. That is average of the hourly global radiation on the surface as given in Eq. (3). Firz et al. (2004) have suggested change of solar variation depends on the effect of sun earth distance and also the extraterrestrial global radiation on horizontal surface at a location is given by the Eq. (4).

Month	Indian Regional solar power generation (GW)					Total (GW)
	North	West	South	East	North-East	
Jan 2020	945.6	904.8	2712.8	48.3	1.04	4612.7
Feb 2020	1151.8	979.1	2906.1	51.9	1.5	5090.6
Mar 2020	1218.1	1091.0	3253.8	68.6	1.59	5633.3
Apr 2020	839.9	903.7	2358.8	64.6	1.41	4168.6
May 2020	942.8	926.4	2402.7	53.9	1.3	4327.4
June 2020	932.4	787.4	2136.1	61.1	1.02	3918.1
July 2020	785.6	702.8	1889.8	48.4	1.2	3428.0
Aug 2020	796.6	630.7	2111.3	36.0	0.97	3575.7
Sep 2020	885.5	585.1	2054.6	38.8	0.93	3565.1
Oct 2020	988.5	763.8	2074.8	54.2	0.9	3882.41
Nov 2020	807.4	776.9	2305.0	46.2	1.07	3936.8

Table 2.
Installation of five regional solar power generation.

$$I_H = I_d + I_b \tag{2}$$

$$H_H = H_b + H_d \tag{3}$$

$$H_o = \frac{24 \times 3.6}{\Pi} I_{sc} E_0 \left[\left(\frac{\Pi}{180} \right) w_s \left(\sin\delta \sin\varphi \right) + \left(\cos\delta \cos\varphi \sin w_s \right) \right] \tag{4}$$

Where, H_0 - daily extraterrestrial global radiation ($KJm^{-2} \cdot day^{-1}$), E_0 - the eccentricity correction factor, W_s - Sunrise hour angle, δ - Declination angle, and φ - latitude.

3. Solar energy assessment in various region using duck curves

The duck curve is a graph of power production in electricity generation on a daily basis which shows a timeline disparity between peak demand and the production of renewable energy. As an example of solar production is increasing the net load curve is taking the shape of a duck's belly. When the sun goes down, the energy demand from conventional power plants needs to quickly ramp up. **Figure 2** shows the evolution of duck curve from 2012 to 2020. In 2013, National Renewable Energy Laboratory (NREL) initially used the phrase of duck curve. NREL issued a graph containing the projected power load less, and its anticipated grid integrated solar power supply. The capacity of photovoltaic systems are highest generation of solar power during the day at 10 a.m. towards 5 p.m. it poses a danger that the grid will destabilize over generation [32]. Similarly, after 5 p.m., solar power generation falls quickly, leading to increases the electricity consumption from other sources which need to accelerate their production shortly. This raises the possibility that generation costs and blackouts will be increased if demand is not met.

The peak demand occurs after sunset in many energy markets when solar power is not available anymore. In areas where there has been a significant amount of solar

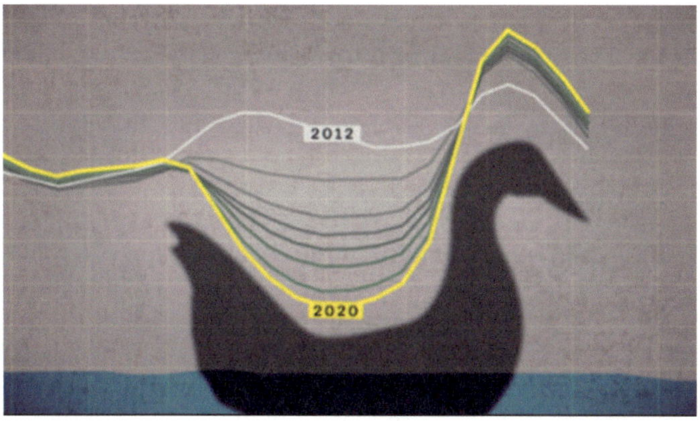

Figure 2.
Evolution of the duck curve.

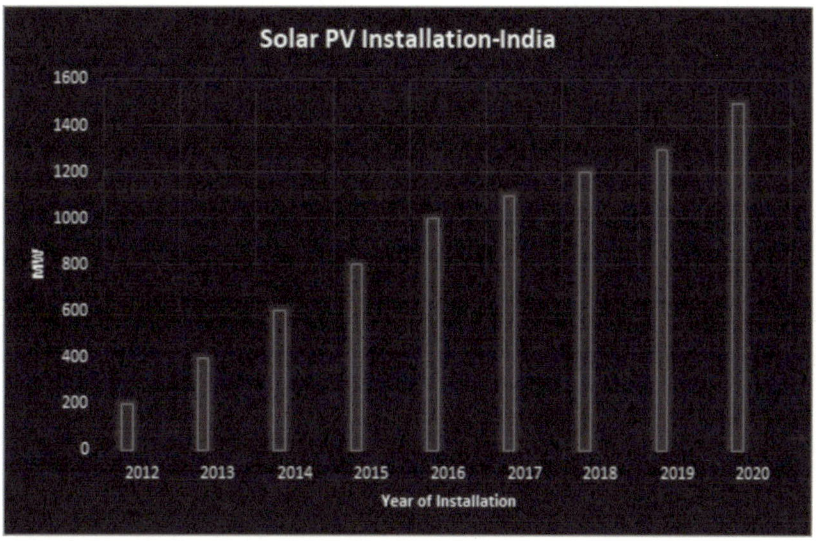

Figure 3.
PV installation in India.

power, the amount of power to be generated from sources other than solar or wind shows that around the sunset and peaks in the middle of the night, a chart similar to the silhouette of a duck is created. Unless energy storage is available in some way, after high solar generation companies have to rapidly increase their power production around sunset to repay the loss of solar generation, which is a major concern for grid operators, where photovoltaic expand rapidly [33]. Fly wheeled batteries were found to provide excellent frequency control [34]. Short-term usage of batteries, large enough in use, can help flatten the curve of the duck and avoid fluctuation by generators, which can help maintain the voltage profile. The issue of a duck curve is mainly in India with because of high solar integration, while there is no integration of the grid in other major solar energy producing states such as China and United states [35]. **Figure 3** shows the PV plant installations in India from 2012 to 2020. The analysis has been taken from 2012 since the duck curve came into existence from the year 2012. This article explains the promising solution of the Duck curve as to implement the battery storage systems.

Figure 4 shows the duck curve analysis of Rajasthan since it has a PV power plant installed in an area of 34,270 ha (hectare), and the net output evolved from the PV station is about 4.8 GW [36]. The highest irradiation insolation level is 5–7 kWh/m^2/day.

Since that desert is located in Rajasthan (state of Indian sub-continent) the amount of solar radiation received in that geographical area is greater than other states but If we consider the duck curve in **Figure 4** shows that from 2012 to 2020 the ducks belly goes on increasing during the hours 4 am to 8 pm. A blue color line in **Figure 4** indicates that the best solution to overcome the duck curve is to implement storage system (SS). If we observe the graph with SS the ducks belly has decreased whereas if SS is not used then the Ducks belly is increased which is indicated by red color. The ramp reduction without SS is only about 12% whereas the ramp reduction with SS is about 57%.

Figure 5 shows the duck curve analysis of Haryana (state of Indian sub-continent)since it has a PV power plant installed in an area of 18,096 ha, and the net output evolved from the PV station is about 73.2 MW [37]. The highest irradiation insolation level is 3.5–4.5 kWh/m^2/day. Since Haryana is located in the northern part of India and has a very cold climate the amount of solar radiation received in that geographical area is lesser than other states but If we consider the duck curve in **Figure 5** shows that from 2012 to 2020 the ducks belly goes on increasing during the hours 12 pm to 6 pm. A blue color line in **Figure 5** indicates that the best solution to overcome the duck curve is to implement storage system (SS). If we observe

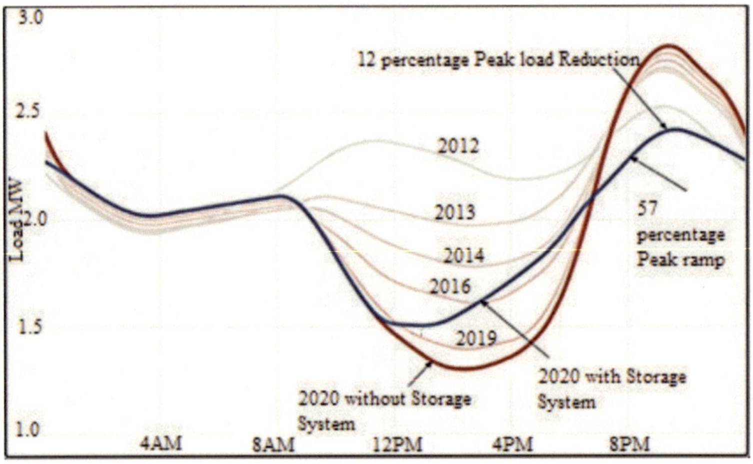

Figure 4.
Graphical analysis of duck curve in Rajasthan.

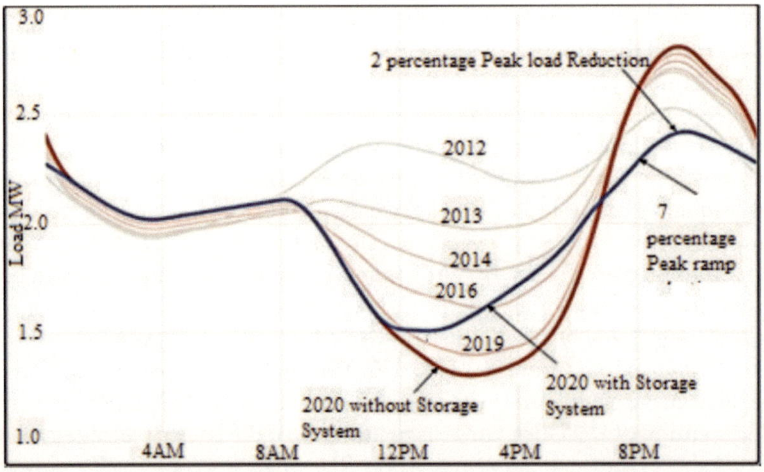

Figure 5.
Graphical analysis of duck curve in Haryana.

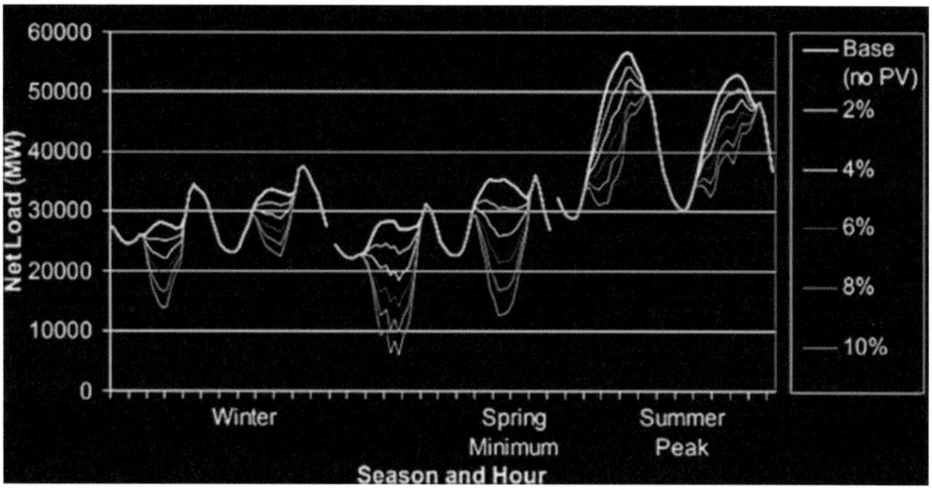

Figure 6.
Graphical analysis of duck curve in Rajasthan during winter spring and summer.

the graph with SS the ducks belly has decreased whereas if SS is not used then the Ducks belly is increased which is indicated by red color. The ramp reduction without SS is only about 2% whereas the ramp reduction with SS is about 7%.

Figure 6 shows the Duck curve analysis from 2014 to 2020 of the Rajasthan state with respect to the seasons. Three seasons taken into consideration such as the winter, spring and summer, the amount of solar radiation received in the geographical area of Rajasthan during winter is very less since less solar radiation. So the use of conventional energy sources like hydro or thermal power has demand. But later when seasons change the graph shows changes in the irradiation level and energy utilization.

4. Modular scale-battery energy storage system

In India, the fast moving towards a renewable energy future with solar battery system. In order to continuous growing electricity demand with minimize the fossil fuel and environmental pollution. The maximum solar energy is wasted without energy storage devices i.e., battery or capacitor bank during in daytime [38], because in India sub-continental is the largest country in square 3,287,263 kilometers, and Solar-PV (S-PV) system can contribute in most of the production region in India. Due to extensive development of renewable resources are used to interconnect with micro-grid/smart-grid approach [39]. Modular Scale -Battery Energy Storage System (MS-BESS) is enable power system operators, and it can interconnect utility provider with stored energy for lateral uses. The purpose of MS-BESS connected to a solar system could also work with protecting storage and reducing peak demand [40]. In existing solar batteries are manufactured with some limitations. It can be used for Li-Ion and Li-Po model batteries, that is incorporate the roles of cell balance, charging, discharge, cell display and defense. These tasks are done autonomously charging and discharging with 10–15% losses, the internal battery supplies the electricity for the analysis to be carried out, and extruded batteries with active batteries should be used. However, to charge normally MS-BMSS batteries must be attached in multilevel converters [41, 42]. The battery charging line is attached in parallel to solve this problem, while charging is achieved at low voltage using by parallel charging [43], and in India maximum rooftop [44], standalone models are fixed with residential, domestic purpose.

The battery storage system in India is proposed periodically power absorption to the grid during without peak load time. In this strategy selection is achieved 3 times per day to better match the consumption peak load of domestic user. Which occurs early in morning time slot 1: Starts at midnight and ends at 6.00 A.M, next time slot: 2 in between 6.00 A.M to 6.00 P.M, third time slot start at 6.00 P.M, and it can finish at mid-night. Finally satisfied by PV production is especially in winter time consumption level. In fact at 6.00 P.M the PV module system is absorbed weather forecasts for the next 24 hours, so the PV production period is almost getting over. The calculation of update storage battery management strategy and provisional energy balanced conditions are accurately find the quality of stored energy. In this time the battery do not charged and its supply transfer to the grid will not to be considered. The first model is shown in **Figure 7**. It can present the S-PV system and peak load for day one: in this first day the sky is clear at 6.00 P.M (Total discharging time is 12 hours). The battery storage and load will be mainly supplied with the help of PV module. The S-PV system in day two and three: the PV will minimum production, due to cloudy weather conditions. Thus, the MS-BMSS will maintain and manage the discharge of storage until in morning of the day, a total discharging time is 36 hours (discharging time in between 18 hours to 54 hours) as shown in **Figure 8**.

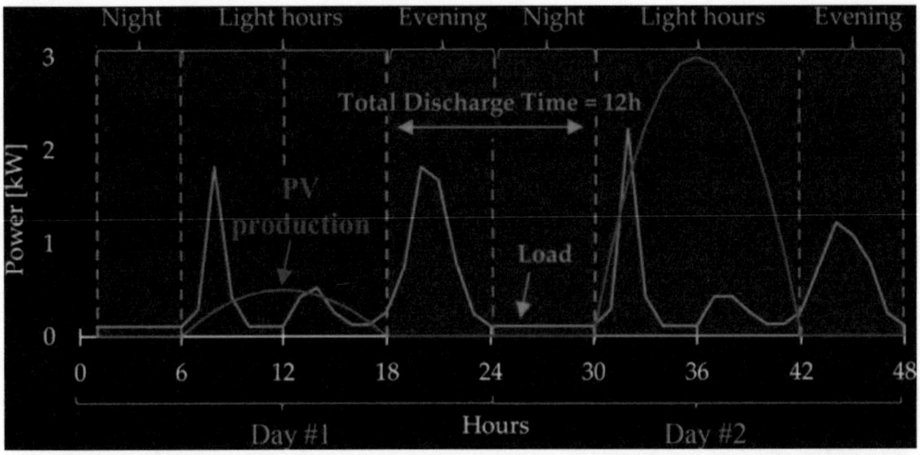

Figure 7.
PV module and load profile for two day (12 hours discharging time).

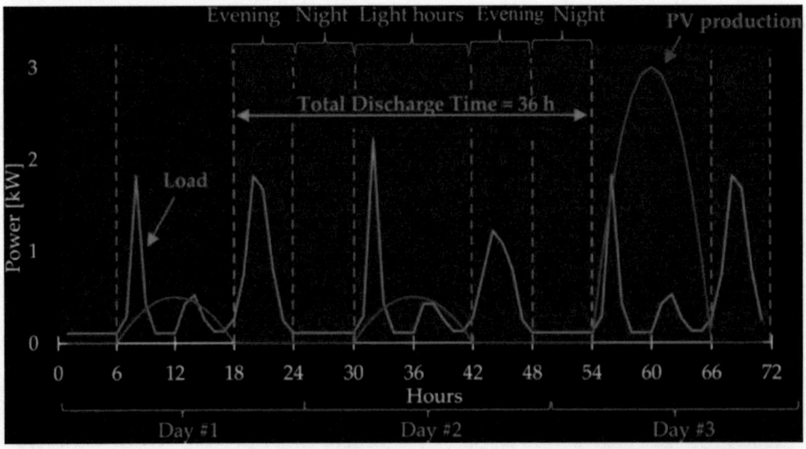

Figure 8.
PV module and load profile for three days (36 hours discharging time).

5. Measurements of solar grid system pre-sizing

The solar-grid connected system presizing in India as of July 2020. Predominantly, the measurement of irradiation level and system grid interaction in south India. S-PV system module type is standard, polycrystalline cells, and mounting method is tilt roof. Especially, the geographical site latitude are 11.41°N, longitude 76.70°E (Ooty), altitude 0 m, and time zone UT + 5.5. The PV field nominal standard test condition power is 30 kWp. The PV software system 7.0.2 Installation of solar collector plane

Month	Global horizontal plane kWh/m^2/day	Global in tilted plane kWh/m^2/day	System Output kWh/day	System Output kWh
Jan	6.15	7.46	188.3	5836
Feb	6.71	7.58	191.3	5357
Mar	6.56	6.85	173.2	5368
Apr	6.11	5.87	148.3	4450
May	5.89	5.10	129.2	4004
Jun	4.07	3.60	91.07	2732
July	4.01	3.58	90.45	2804
Aug	4.38	4.12	103.9	3221
Sep	4.24	4.28	107.8	3234
Oct	4.76	5.14	129.7	4021
Nov	4.75	5.57	140.4	4212
Dec	5.87	7.23	182.3	5651
Year	2.28	5.53	139.4	50,890

Table 3.
System output graphical site in Ooty.

Month	Global horizontal plane kWh/m^2/day	Global in tilted plane kWh/m^2/day	System Output kWh/day	System Output kWh
Jan	4.82	5.55	139.8	4335
Feb	5.87	6.51	164.1	4594
Mar	6.33	6.62	166.9	5175
Apr	6.47	6.35	160.0	4800
May	6.21	5.82	146.7	4547
Jun	5.64	5.18	130.5	3914
July	5.20	4.82	121.6	3770
Aug	5.24	5.03	126.9	3933
Sep	5.35	5.44	137.1	4113
Oct	4.43	4.70	118.4	3671
Nov	3.84	4.25	107.2	3217
Dec	3.98	4.56	115.0	3565
Year	5.28	5.39	136.0	49,634

Table 4.
System output graphical site in Chennai.

orientation tilt 15°, Azimuth 0°, and It will be produced maximum system output 139.4 kWh/day as shown in **Table 3**. In this zone solar sun radiation is very low due to wet season is overcast, and the dry season is cloudy weather zone. The temperature typically varies from 46°F to 74°F, and it is rarely below 40°F or above 80°F.

Another, geographical site latitude are 13.09°N, longitude 80.28°E (Chennai), time zone UT + 5.5, and altitude 0 m. The PV module and system installation should be the same characteristics pre-sizing evaluation, and it will be produced maximum system output 136.0 kWh/day as shown in **Table 4**. But, in this zone sun radiation is high. This report delivered the solar output does not depend on the sun temperature. Solar panel is increases heat with the effect of sun temperature, automatically the output current increases. However, the output voltage is linearly decreases. It's directly by changing the rate of solar heat delivered from atmosphere, and earth.

Year	Total capacity (MW)	Installation per year (MW)
2010	161	—
2011	461	300
2012	1205	744
2013	2319	1114
2014	2632	313
2015	3744	1112
2016	6763	3019
2017	12,289	5526
2018	21,651	9362
2019	28,181	6530
2020	34627	6446

Table 5.
Statistical information on PV installations in India from 2010-2020.

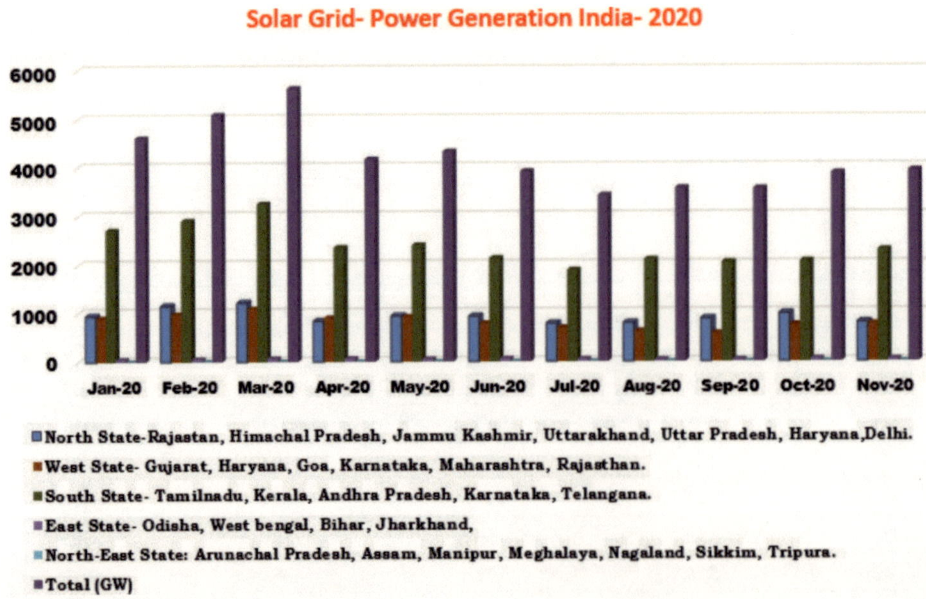

Figure 9.
Solar grid connected generation in India 2020.

It can achieve indirectly by changing the cloud forming process. The maximum annual energy is 49.6 MW using for back ventilation property with free air circulation. **Table 5** shows the total PV installation as per year wise in India. The maximum energy is consumed by solar power grid connected module, and PV irradiations level are measurement in maximum production ranges in Rajasthan (North-state), Karnataka (South-state), and Telangana (South-state). **Figure 9** shows the solar power installed generation in Indian sub continental with different state.

6. Conclusion

The results has reveal that maximum region in Indian sub-continental an enormous solar power potential located in an different region, where solar radiations are regularly available in all overt the year (expect rainy or cloud season). Furthermore, PV energy is easy –maintain, cost saving, and high durable. So, in this type of PV energy is more suitable for Indian region with identified maximum solar irradiation under state-wise, assessment of PV power spectrum analysis, PV grid system pre-sizing, and battery management system. Finally, the statistical summary that solar insolation values ranges of 2.9–4.0 $kWh/m^2/day$ to 5–7 $kWh/m^2/day$ with an average value of 5.9 kWh/m^2. The important role have been recommended for solar power in order to fulfill the country energy demand. Its way to a future need of sustainable development play major role in Indian sub-continental.

Author details

Johny Renoald Albert[1*] and Dishore Shunmugham Vanaja[2]

1 Vivekanandha College of Engineering for Women, Affiliated by Anna University, Namakkal, India

2 Narayanaguru College of Engineering, Affiliated by Anna University, Kanyakumari, India

*Address all correspondence to: jorenoeee@gmail.com

References

[1] Irfan M, Zhao Z-Y, Ikram M, Gilal NG, Li H, Rehman A. *Assessment of India's energy dynamics: Prospects of solar energy*. Journal of Renewable and Sustainable Energy. 2020;*12(5):053701*. DOI: 10.1063/1.5140236

[2] Kumar D. Satellite-based solar energy potential analysis for southern states of India. Energy Reports. 2020;**6**:1487-1500. DOI: 10.1016/j.egyr.2020.05.028

[3] Raina G, Sinha S. Outlook on the Indian scenario of solar energy strategies: Policies and challenges. Energy Strategy Reviews. 2019;**24**:331-341. DOI: 10.1016/j.esr.2019.04.005

[4] Wang, Chang, Bai, Liu, Dai, & Tang. (2019). Mitigation strategy for duck curve in high photovoltaic penetration power system using concentrating solar Power Station. Energies, 12(18), 3521. doi:10.3390/en12183521.

[5] Jain Anil, Chakraborty Arindam, Gupta Sanjay Kumar, Kandpal Krishna Kumar, Shukla Nitu, Prashant (2019), Evaluating pumped hydro storage technology in the era of renewable generation and ancillary markets. Water and Energy International, 62(1), 33-38.

[6] Nwaigwe KN, Mutabilwa P, Dintwa E. An overview of solar power (PV systems) integration into electricity grids. Materials Science for Energy Technologies. 2019. DOI: 10.1016/j.mset.2019.07.002

[7] Blanco H, Faaij A. A review at the role of storage in energy systems with a focus on power to gas and long-term storage. Renewable and Sustainable Energy Reviews. 2018;**81**:1049-1086. DOI: 10.1016/j.rser.2017.07.062

[8] Fathima H, Palanisamy K. Optimized sizing, selection, and economic analysis of battery energy storage for grid-connected wind-PV hybrid system. Modelling and Simulation in Engineering. 2015;**2015**:1-16. DOI: 10.1155/2015/713530

[9] May GJ, Davidson A, Monahov B. Lead batteries for utility energy storage: A review. Journal of Energy Storage. 2018;**15**:145-157. DOI: 10.1016/j.est.2017.11.008

[10] Chen H, Cong TN, Yang W, Tan C, Li Y, Ding Y. Progress in electrical energy storage system: A critical review. Progress in Natural Science. 2009;**19**(3):291-312. DOI: 10.1016/j.pnsc.2008.07.014

[11] Hirsch A, Parag Y, Guerrero J. Microgrids: A review of technologies, key drivers, and outstanding issues. Renewable and Sustainable Energy Reviews. 2018;**90**:402-411. DOI: 10.1016/j.rser.2018.03.040

[12] Papageorgiou A, Ashok A, Hashemi Farzad T, Sundberg C. Climate change impact of integrating a solar microgrid system into the Swedish electricity grid. Applied Energy. 2020;**268**:114981. DOI: 10.1016/j.apenergy.2020.114981

[13] Wang F, Zhu Y, Yan J. Performance of solar PV micro-grid systems: A comparison study. Energy Procedia. 2018;**145**:570-575. DOI: 10.1016/j.egypro.2018.04.083

[14] 18thElectric Power Survey; Central Electricity Authority, Report of the expert group on 175GW renewable energy by 2020, National institution of transforming India. http://www.niti.gov.in/

[15] Das P, Mathuria P, Bhakar R, Mathur J, Kanudia A, Singh A. Flexibility requirement for large-scale renewable energy integration in Indian power system: Technology, policy and modeling options. Energy Strategy Reviews. 2020;**29**:100482. DOI: 10.1016/j.esr.2020.100482

[16] Dondariya C et al. Performance simulation of grid-connected rooftop solar PV system for small households: A case study of Ujjain, India. Energy Reports. 2018;**4**:546-553 http://dx.doi.org/10.1016/j.egyr.2018.08.002

[17] Jin Q, Wang C. The greening of northwest Indian subcontinent and reduction of dust abundance resulting from Indian summer monsoon revival. Scientific Reports. 2018;**8**:4573. DOI: https://doi.org/10.1038/s41598-018-23055-5

[18] Kathayat G, Cheng H, Sinha A, Yi L, Li X, Zhang H, et al. The Indian monsoon variability and civilization changes in the Indian subcontinent. Science Advances. 2017;**3**(12):e1701296. DOI: 10.1126/sciadv.1701296

[19] Varikoden H, Ravadekar JV. On the extreme rainfall events during the southwest monsoon season in northeast regions of the Indian subcontinent. Meteorological Applications. 2019. DOI: 10.1002/met.1822

[20] Liepert BG, Giannini A. Global warming, the atmospheric brown cloud, and the changing Indian summer monsoon. Bulletin of the Atomic Scientists. 2015;**71**(4):23-30. DOI: 10.1177/0096340215590802

[21] Toon OB, Bardeen CG, Robock A, Xia L, Kristensen H, McKinzie M, et al. *Rapidly expanding nuclear arsenals in Pakistan and India portend regional and global catastrophe*. Science Advances. 2019;*5(10):eaay5478*. DOI: 10.1126/sciadv.aay5478

[22] Bergin MH, Ghoroi C, Dixit D, Schauer JJ, Shindell DT. *Large reductions in solar energy production due to dust and particulate air pollution*. Environmental Science & Technology Letters. 2017;*4(8):339-344*. DOI: 10.1021/acs.estlett.7b00197

[23] T. V. Ramachandra, Gautham Krishnadas, Rishabh Jain (2012). Solar potential in the Himalayan landscape, ISRN renewable energy, 10(2), 1-15. https://doi.org/10.5402/2012/203149.

[24] Kumar D. Mapping solar energy potential of southern India through geospatial technology. Geocarto International. 2019;**34**(13):1477-1495 http://dx.doi.org/10.1080/10106049.2018.1494759

[25] Patil KN, Garg SN, Kaushik SC. Luminous efficacy model validation and computation of solar illuminance for different climates of India. Journal of Renewable and Sustainable Energy. 2013;**5**(6):063120. DOI: 10.1063/1.4841195

[26] Shukla KN, Rangnekar S, Sudhakar K. *Mathematical modelling of solar radiation incident on tilted surface for photovoltaic application at Bhopal, M.P., India*. International Journal of Ambient Energy. 2015;*37(6):579-588*. DOI: 10.1080/01430750.2015.1023834

[27] Bhattacharjee S, Bhattacharjee R. Comprehensive solar energy resource characterization for an intricate Indian province. International Journal of Ambient Energy. 2018:1-34. DOI: 10.1080/01430750.2018.1531257

[28] Masoom A, Kosmopoulos P, Bansal A, Kazadzis S. *Solar energy estimations in India using remote sensing technologies and validation with sun photometers in urban areas*. Remote Sensing. 2020;*12(2):254*. DOI: 10.3390/rs12020254

[29] Choi, Suh, & Kim. (2019). *GIS-Based Solar Radiation Mapping, Site Evaluation, and Potential Assessment: A Review. Applied Sciences, 9(9), 1960.* doi:10.3390/app9091960.

[30] Mousavi Maleki S, Hizam H, Gomes C. *Estimation of hourly, daily and monthly global solar radiation on*

inclined surfaces: Models Re-visited. Energies. 2017;*10*(*1*):*134*. DOI: 10.3390/en10010134

[31] Okundamiya MS, Nzeako AN. *Empirical model for estimating global solar radiation on horizontal surfaces for selected cities in the six geopolitical zones in Nigeria*. Journal of Control Science and Engineering. 2011;*2011:1-7*. DOI: 10.1155/2011/356405

[32] Sultan V, Hilton B. Electric grid reliability research. Energy Inform. 2019;**2**:3. DOI: https://doi.org/10.1186/s42162-019-0069-z

[33] Harinarayana T, Jaya Kashyap K. Solar energy generation potential estimation in India and Gujarat, Andhra, Telangana states. Smart Grid Renew. Energy. 2014;**05**(11):275-289

[34] Khan N, Dilshad S, Khalid R, Kalair AR, Abas N. Review of Energy Storage and Transportation of Energy. Energy Storage. 2019:*e49*. DOI: 10.1002/est2.49

[35] Hou Q, Zhang N, Du E, Miao M, Peng F, Kang C. *Probabilistic duck curve in high PV penetration power system: Concept, modeling, and empirical analysis in China*. Applied Energy. 2019;*242:205-215*. DOI: 10.1016/j.apenergy.2019.03.067

[36] Gupta SK, Anand RS. *Development of solar electricity supply system in India: An overview*. Journal of Solar Energy. 2013;*2013:1-10*. DOI: 10.1155/2013/632364

[37] Khare V, Nema S, Baredar P. *Status of solar wind renewable energy in India*. Renewable and Sustainable Energy Reviews. 2013;*27:1-10*. DOI: 10.1016/j.rser.2013.06.018

[38] Soni VK, Pandithurai G, Pai DS. *Evaluation of long-term changes of solar radiation in India*. International Journal of Climatology. 2011;*32*(*4*):*540-551*. DOI: 10.1002/joc.2294

[39] Johnson SC, Papageorgiou DJ, Mallapragada DS, Deetjen TA, Rhodes JD, Webber ME. Evaluating rotational inertia as a component of grid reliability with high penetrations of variable renewable energy. Energy. 2019. DOI: 10.1016/j.energy.2019.04.216

[40] Bangash KN, Farrag MEA, Osman AH. Investigation of energy storage batteries in stability enforcement of low inertia active distribution network. Technol Econ Smart Grids Sustain Energy. 2019;**4**:1. DOI: https://doi.org/10.1007/s40866-018-0059-4

[41] A, J. R., & STONIER, A. (2020). *Design and Development of Symmetrical Super-Lift DC-AC Converter using Firefly Algorithm for Solar-Photovoltaic Applications. IET Circuits, Devices & Systems.* doi:10.1049/iet-cds.2018.5292.

[42] [inverter]Kumar NM, Dasari S, Reddy JB. Availability factor of a PV power plant: Evaluation based on generation and inverter running periods. Energy Procedia. 2018;**147**:71-77 http://dx.doi.org/10.1016/j.egypro.2018.07.035

[43] Bruen T, Marco J. *Modelling and experimental evaluation of parallel connected lithium ion cells for an electric vehicle battery system*. Journal of Power Sources. 2016;*310:91-101*. DOI: 10.1016/j.jpowsour.2016.01.001

[44] Goel M. *Solar rooftop in India: Policies, challenges and outlook*. Green Energy & Environment. 2016;*1*(*2*):*129-137*. DOI: 10.1016/j.gee.2016.08.003

Chapter 5

Designing Well-Organized Donor-Bridge-Acceptor Conjugated Systems Based on Cyclopentadithiophene as Donors in Bulk Heterojunction Organic Solar Cells: DFT-Based Modeling and Calculations

Rania Zaier and Sahbi Ayachi

Abstract

Two host materials based on CPDT as donors in bulk heterojunction organic solar cells were designed and investigated by means of DFT calculations. The first one (P-CPDTBT3) is a copolymer with D-A configuration and the second one (SM-CPDTDPP) is a D-π-A-π-D type small molecule. The investigated materials exhibited interesting structural properties with high planarity and rigidity originated from intra-molecular non-covalent interactions between the different building blocks. Thanks to their narrow band gaps, the optical absorption spectra have covered the main part of solar spectrum of interest. In addition, some general transport properties have been established. The transition density matrix (TDM) was used to get insight into the interaction of hole–electron localization and the electronic excitation processes. The photovoltaic parameters (FF, Voc) were calculated. The obtained results have been attempted to provide novel structure–property relationships for the rational design strategies of high-performance photovoltaic materials with power conversion efficiency of nearly 10%.

Keywords: cyclopentadithiophene, benzothiadiazole, diketopyrrolopyrrole, DFT, bulk heterojunction organic solar cells, TDM

1. Introduction

The organic photovoltaic solar cells have gained most attention compared to the inorganic counterparts thanks to their exclusive characters such as the flexibility, the light weight, the transparency and the low-cost of fabrication [1–6]. Bulk heterojunction organic solar cells (BHJ-OSCs), have been largely emerged regarding the several advantages especially their impressive photo-physical properties. Introducing a high performance material remains a challenge for researchers [7–11].

Recently, polymers and small molecules organic semiconductors have received great attention to be used in BHJ-OSCs, because of their well-defined molecular structure, simple synthesis, high mobility and the structure could be easily modified [12–14].

Particularly, π-conjugated systems incorporating donor-acceptor (D-A) and donor-π-acceptor-π-donor (D-π-A-π-A) configurations have been emerged as promising category of materials for photovoltaic applications. In fact, thanks to the high electron delocalization and the intra-molecular charge transfer (ICT) that takes place within the conjugated skeleton, D-A materials have shown interesting optoelectronic properties and high charge carrier mobility [15, 16]. These particular characteristics of these kinds of π-conjugated systems leaded to further improve the conjugated arrangement for more increasing the device performance of materials-based BHJ-OSCs.

Polymers and small molecules based on cyclopentadithiophene (CPDT) were widely used in organic electronic applications thanks tothe high rigidity, planarity and charge transfer ability delivered by CPDT group [17–23]. Further, Benzothiadiazole (BT) [24–27] and Diketopyrrolopyrrole (DPP) [28–30] have been extensively utilized as electron deficient acceptor patterns in conjugated systems. These acceptor groups offer the advantages of the high electron withdrawing ability, the excellent electro-optical properties and the important carrier mobility within conjugated materials. Hence, a rational design incorporating donor and acceptor moieties may improves the optoelectronic properties to assure high performance BHJ-OSCs [31, 32].

This study aims to investigate two donor materials as illustrated in **Figure 1**, the first one is a copolymer based on CPDT and BT with donor-acceptor (D-A) configuration, namely P-CPDTBT3. The choice of the conjugated chain length of this copolymer with n = 3 is based on the simulated results in our previous work [33]. The second one is a small molecule based on CPDT and DPP with D-π-A-π-D configuration, namely SM-CPDTDPP, in which thiophene was used as a potential π-spacer building block regarding its high electron abilities [34–36]. These compounds are desired to be blended with fullerene-based acceptor material to form

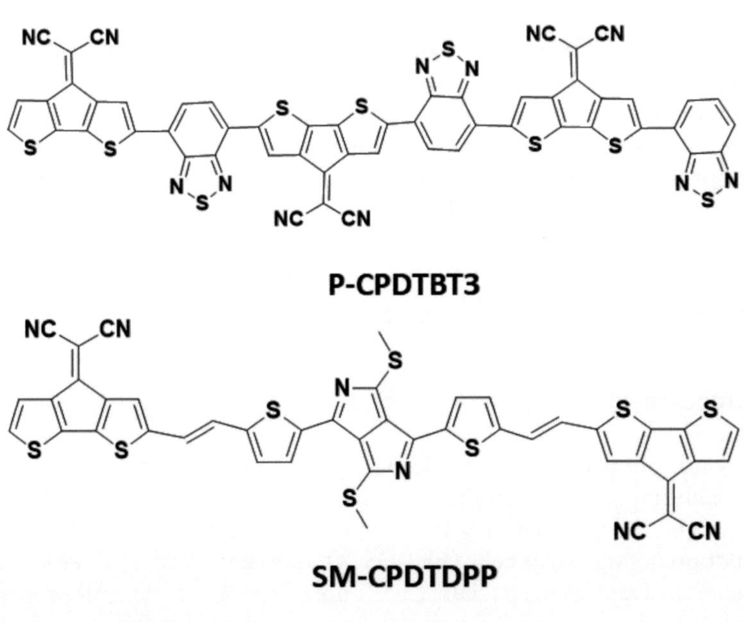

Figure 1.
Molecular structures of investigated compounds.

active layer within BHJ-OSC architecture. A computational investigation applying the DFT approach based on the effects of tuning the acceptor building block together with the molecular configuration on the optoelectronic properties has been reported.

2. Computational methods

All the calculations have been performed by means of GAUSSIAN 09 software package [37]. The ground state geometric optimizations were carried out using density functional theory (DFT) with the B3LYP hybrid functional method at 6-311 g(d,p) basis set in gaseous phase. The frontier molecular orbitals (FMOs) properties (HOMO, LUMO and Eg) were determined based on the optimized ground state geometries. The electronic parameters including the ionization potential (IP) and electron affinity (EA) were calculated from neutral, cation and anion optimized structures. The optical absorption characteristics were investigated using time dependent DFT (TD-DFT) method at B3LYP/6-311 g(d,p) level of theory [38, 39]. Charge transfer properties were investigated based on the reorganization energies of hole and electron within the studied molecules. Transition density matrix (TDM) plots were carried out using Multiwfn [40] to understand the electron–hole coherence correlation and the exciton dissociation at the first excited state. Finally, photovoltaic parameters were computed and power conversion efficiencies (PCE) were estimated using Scharber diagram.

3. Results and discussions

3.1 Ground-state geometry optimizations

The studied conjugated molecules are constructed based on CPDT units as donors with BT and DPP as acceptor units. Hence, these compounds are of "push-pull" type conjugated molecules [41, 42]. Both P-CPDTBT3 and SM-CPDTDPP were optimized in the ground state using DFT//B3LYP/6-311 g(d,p) method. This study aims to examine the effect of the conjugated molecular design on the optoelectronic and photovoltaic properties. Here, we have maintained the CPDT donor building block and we have tuned the acceptor moieties based on BT and DPP units. Besides, we are looking to reveal the difference of behavior between polymer and small molecule.

As it can be seen from **Figure 2**, both compounds exhibit a high planar optimized geometry. The dihedral angles are almost 0°, as observed from the side view of these molecules. These planar configurations are arising from the intramolecular non-covalent interactions of S---H, N---H and S---N types that take place within the conjugated framework [43]. These non-covalent bonds are found smaller than the sum of Van der Waals radii of the considered atoms. The planar backbone structure is one of the key factors to enhance the conjugation degree and accordingly increasing the π-staking for more charge transfer capability.

The bridge bonds are described as the bonds that link between the distinct building blocks such as electron donating units, electron acceptor units and π-spacer within the conjugated backbone. The interest of examining the bridge bond length is to get an idea about the interactions among the different building blocks. Where, the shorter bridge bond length leads to stronger intra-molecular interactions and higher charge transfer [44, 45]. For the studied compounds, the bridge bond defines the bond C-C between the CPDT donor and BT or DPP

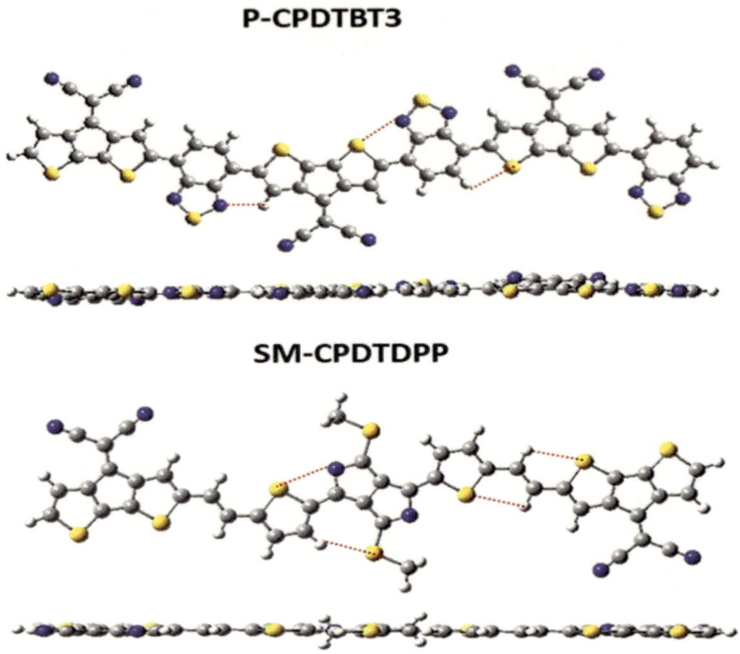

Figure 2.
Ground state optimized structures of P-CPDTBT3 and SM-CPDTDPP at DFT/B3LYP/6-311 g(d,p) level of theory.

acceptor units. The bridge bond lengths are found around 1.45 Å for P-CPDTBT3 and 1.43 Å for SM-CPDTDPP. The obtained values are higher than the regular C=C bond length (1.34 Å) and smaller than the regular C-C bond length (1.54 Å) which indicates that these bonds are still found to have double-bond character.

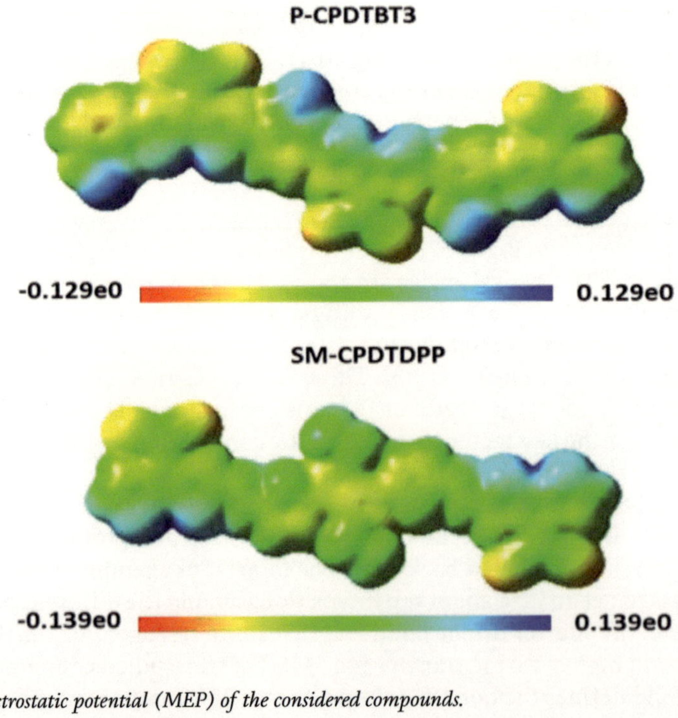

Figure 3.
Molecular electrostatic potential (MEP) of the considered compounds.

Based on the optimized ground state geometries, we deduce that important π-electron delocalization, within the conjugated frameworks; can induce intramolecular charge transfer (ICT) characteristics in push-pull donor materials.

Further, Molecular Electrostatic Potential Surfaces (MEPs) were simulated to identify the electronic properties and molecular stability. The MEP is a helpful tool for specifying the reactive sites as it is related to the topology of molecular electron density [46]. The colors displayed in the MEP represent the different electrostatic potential values and charge distributions within the molecules. From **Figure 3**, the electron rich-regions (red color), usually have negative potentials, are mainly located over the dicyanomethylene bridge groups whereas the blue color depicts regions of more positive electrostatic potentials (electron-deficient) color are concentrated over H and S atoms. The MEP plots have shown the dominance of the zero potential which is presents green color. This observation revel the high stability of the considered compounds.

3.2 Frontier molecular orbitals (FMOs) analysis

The analysis of Frontier molecular orbitals (FMOs) gives a description of the electron delocalization as well as the electron transport capacities within the conjugated skeleton. The highest occupied molecular orbital (HOMO) and the lowest unoccupied molecular orbital (LUMO) strongly determine the optoelectronic properties of conjugated compounds, pointedly on the photovoltaic properties of donor materials. Largely, donor compound should tend to have a deep HOMO level to assure a high open circuit voltage V_{OC} and a suitable LUMO energy level with respect to that of the acceptor unit [47–49].

The FMOs of the considered materials are carried out based on DFT/B3LYP method at 6-311 g(d,p) and listed in **Table 1**. The FMOs contour plots are illustrated in **Figure 4**.

Compound	E_{H-1} (eV)	E_H (eV)	E_L (eV)	E_{L+1} (eV)	E_{gap} (eV)	IP (eV)	EA (eV)
P-CPDTBT3	−5.90	−5.49	−3.87	−7.74	1.62	6.10	3.20
SM-CPDTDPP	−5.83	−5.20	−3.78	−3.72	1.42	5.90	3.09

Table 1.
Electronic properties for studied materials obtained at DFT/B3LYP 6-311 g(d,p) level of theory.

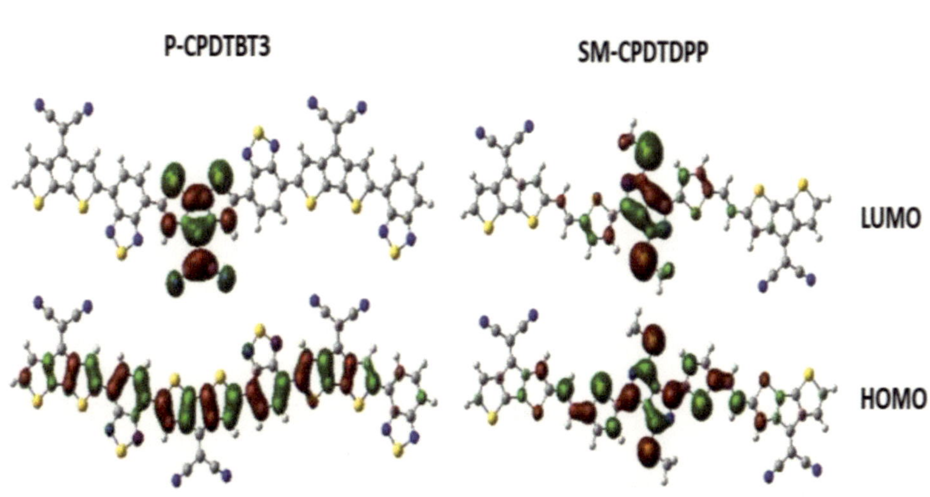

Figure 4.
FMOs contour plots at the optimized ground state of the considered materials.

The simultaneous interactions of donor and acceptor groups are the responsible of the electron delocalization and thus producing the electronic charge distribution within the HOMOs and LUMOs. As it can be seen from **Figure 4**, there is considerable discrepancy of molecular orbital distributions resulting from the particular molecular configurations of P-CPDTBT3 and SM-CPDTDPP.

The spatial distribution of the HOMO orbital of P-CPDTBT3 is dominantly localized over the main conjugated backbone. While, that of SM-CPDTDPP is mainly located on the central part of the conjugated framework. The LUMOof P-CPDTBT3 is dispersed over the central CPDT unit indicating a high steric hindrance rising from the strong electron withdrawing group effect of dicyanomethylene group [50]. In the case of SM-CPDTDPP, the LUMO is centered over the DPP substituted group and the thiophene π-spacer units. These distributions may increase the $\pi \rightarrow \pi^*$ electronic transitions and reinforce the ICT ability. Besides, these materials dispose narrow band gap energies (1.62 eV) for P-CPDTBT3 and 1.42 eV for SM-CPDTDPP) that lead to improve the electron transition and light harvesting. The 2D molecular electrostatic maps of studied materials have been simulated to better understand the intra-molecular interactions (See **Figure 5**). As revealed from **Figure 5**, the central part is the most conjecturable zone into the conjugated framework of the studied molecules that is in good agreement with the FMOs analysis.

Ionization potential (IP) and Electron Affinity (EA) were calculated from the neutral, cation and anion optimized structures. IP and EA describe the barrier injection energies of electron and hole, respectively. The application of the considered materials in OCSs requires relevant IP and EA in order to promote the electron injection and hole transport. Thus, it is revealed from the FMOs analysis the significant effect of building blocks on the electronic properties that are related to the charge delocalization.

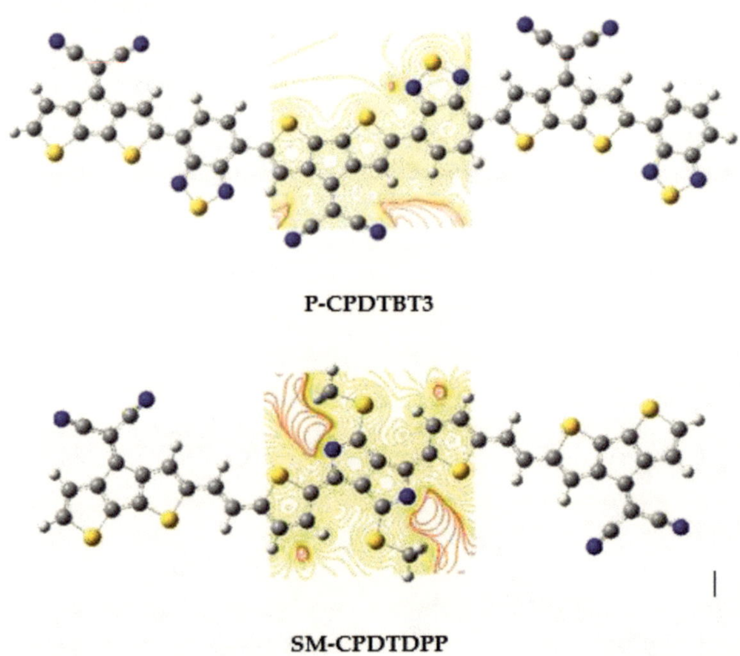

Figure 5.
2D molecular electrostatic maps of studied materials.

3.3 Optical properties

The optical absorption spectrum in the solar spectral zone also its intensity are the main factors that influence the value of short-circuit current density (J_{SC}) of OSCs [51]. Fundamentally, the J_{SC} is a function of the external quantum efficiency (EQE) with the photon number $S(\lambda)$ coveringall the frequencies providedfrom the solar spectrum, as above [52]:

$$J_{SC} = q\int EQE.S(\lambda)d\lambda \tag{1}$$

Where, EQE presents the product of light harvesting efficiency (η_λ), exciton diffusion efficiency (η_{ED}), charge separation efficiency (η_{CS}), and charge collection efficiency (η_{CC}). As revealed from the following expression, the donor material absorption capability remains a crucial parameter for increasing the organic solar cell efficiency. The light harvesting efficiency (η_λ) is related to the oscillator strength (f) of the maximum optical absorption wavelength as expressed above [53]:

$$\eta_\lambda = 1 - 10^{-f} \tag{2}$$

In order to explore the photo-physical properties of the considered compounds, the optical absorption spectra were simulated using TD-DFT approach as cost-effective method [54, 55].

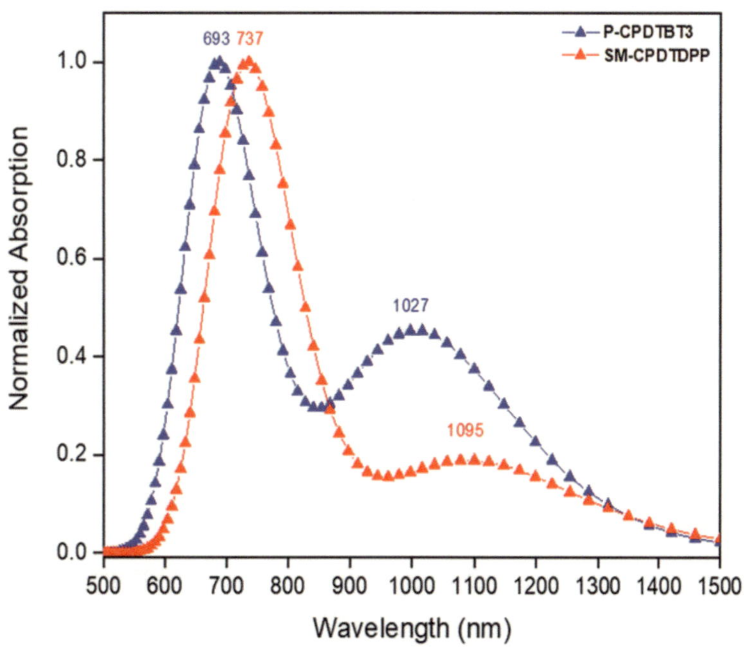

Figure 6.
Optical absorption spectra of P-CPDTBT3 and SM-CPDTDPP simulated at TD-DFT//B3LYP/6-311 g(d,p) level of theory.

Compound	E^{ex}	λ_{max}	f	Major configuration	η_λ
P-CPDTBT3	1.7887	693	1.0184	H → L + 3 (50%)	0.9041
	1.2065	1027	0.5411	H → L (82%)	
SM-CPDTDPP	1.6803	737	1.1021	H → L + 2 (60%)	0.9209
	1.1318	1095	0.3962	H → L (88%)	

Table 2.
Calculated electronic transition energy E^{ex} (eV), maximum absorption wavelengths, λ_{max} (nm), oscillator strength (f) and major configuration at TD-DT//B3LYP/6-311 g(d,p) level.

TD-DFT simulations were performed at the optimized ground state (S_0) geometries in gaseous phase (See **Figure 6**) and the related optical parameters are listed in **Table 2**.

As we can see from **Figure 6**, these materials exhibit, as expected, abroad absorption bands in the wavelength range from 550 nm to 900 nm which covers a relevant part of the solar spectrum, where the maximum optical absorption within the solar spectrum is at about 700 nm [56]. The broad absorption in the visible and near infrared region, displayed by the considered materials, leads to reinforce BHJ-OCSs performances. The maximum absorption peaks were found at 693 nm and 737 nm for P-CPDTBT3 and SM-CPDTDPP, respectively.

In fact, these maximum wavelengths are generated mainly from HOMOs to LUMOs electronic transitions of ground to first excited state ($S_0 \rightarrow S_1$) of electrons associated with high oscillator strength (f) values. Where, the pronounced absorption peaks are generated by $\pi \rightarrow \pi^*$ electronic transitions from the electron donating CPDT moieties to theelectron acceptor BT or DPP moieties [57]. The simulated absorption spectrum of P-CPDTBT3 is in convenient agreement with the experimental results reported in ref. [58], that confirms the accuracy of TD-DFT approach in reproducing the experimental data.

SM-CPDTDPP absorption spectrum was found red shifted by 44 nm compared to that of P-CPDTBT3. The slight red shift detected can be explained by the present of the thiophene π-spacer that may enhance the electron delocalization within the main conjugated framework.

The large absorption band ranging from 900 nm to 1500 nm is attributed to the ICT generated from the sulfur rich electron to the electron withdrawing dicyanomethylene group within the CPDT units [59]. A promising organic donor material should exhibit a large light harvesting efficiency (η_λ) in order to reach high photocurrent signal [60, 61]. From **Table 2**, we reveal that these materials exhibit high η_λ values close to one leading to an important light harvesting.

Overall, the molecules under investigation have shown interesting absorption properties by covering the amount of the visible and the near infrared regions which leads to potential photo-physical properties and J_{SC} improvement.

3.4 Charge transfer properties

Efficient BHJ-OSCs dispose high charge carrier's mobility. The free chargers generated from the exciton dissociation/separation will be diffused/transported within the compound. Thus, efficient donor material should exhibit high hole transport ability to improve the photo-generation of charge carriers, and then the J_{SC}.

The charge hopping process is selected to arrange the hole mobility into the compound at room temperature. This process is commonly described as the

Compound	λ_{hole}	t_{hole}	k_{hole}
P-CPDTBT3	0.255	0.205	$1.66 \times 10^{+14}$
SM-CPDTDPP	0.224	0.315	$4.99 \times 10^{+14}$

Table 3.
Reorganization energies for hole transport (λ_{hole}), hole integral transfer (t_{hole}) and hole transport rate (k_{hole}) of the considered molecules. All these parameters are given in eV.

self-exchange and charge transport between two adjacent molecules. The hole transport rate (k_{hole}) is approximated based on Marcus theory, as above [62]:

$$k_{hole} = \frac{2\pi t_{hole}^2}{\text{h}} \left(\frac{\pi}{\lambda_{hole} k_B T} \right)^{\frac{1}{2}} \exp\left(\frac{-\lambda_{hole}}{4k_B T} \right) \quad (3)$$

Where, h is Planck's constant, k_B is Boltzmann's constant and T is the temperature (298 K).

From Eq. (3), hole transfer integral (t_{hole}) and reorganization energy for hole transport (λ_{hole}) arecrucial parameters to precisely evaluate the charge transport abilities. The hole transfer integral is influenced by the intra-molecular staking of conjugated molecules, as expressed bellow [63]:

$$t_{hole} = \frac{1}{2}(\text{E}_{\text{HOMO}} - \text{E}_{\text{HOMO}-1}) \quad (4)$$

Where, E_{HOMO} and E_{HOMO-1} define the energies of HOMO and HOMO-1 at neutral state, respectively. This expression defines the electron coupling strength of two adjacent segments of the molecule. The hole reorganization energy (λ_{hole}) is determined fromenergy system's variation between neutral and charge states. The charge transport properties (λ_{hole} , t_{hole} and k_{hole}) of the studied materials are listed in **Table 3**.

The λ_{hole} value of SM-CPDTDPP is lower than that of P-CPDTBT3 that could be explained by the electronegative discrepancy within the over conjugated framework. As well, t_{hole} value found for P-CPDTBT3 is lower than that found for SM-CPDTDPP, which shows the higher energy levels overlap within the polymer than the small molecule. Mostly, the investigated materials possess interesting hole mobility capability that will improve the electrical properties of BHJ-OSC devices.

3.5 Transition density matrix (TDM) analysis

Transition density matrix (TDM) analysis provides an insight into the interactions of donor and acceptor fragments at the first excited state (S_1), the electron excitation process and the electron–hole coherence. TDM is a helpful tool to estimate the exciton escape possibility from the Coulomb attraction [64, 65]. The efficient separation of created exciton improves the charge transfer ability within the BHJ-OSC.

TDM plots simulated upon $S_0 \rightarrow S_1$ excitation configuration are shown in **Figure 7**. It is observed from the **Figure 7** that the electron–hole coherences are primarily concentrated upon the diagonal box (D-D, A-A) and the off-diagonal (D-A) for photo-excitation. The wide distribution in the diagonal box (D-D, A-A) validates the high $\pi \rightarrow \pi^*$ transitions within donor and acceptor regions.

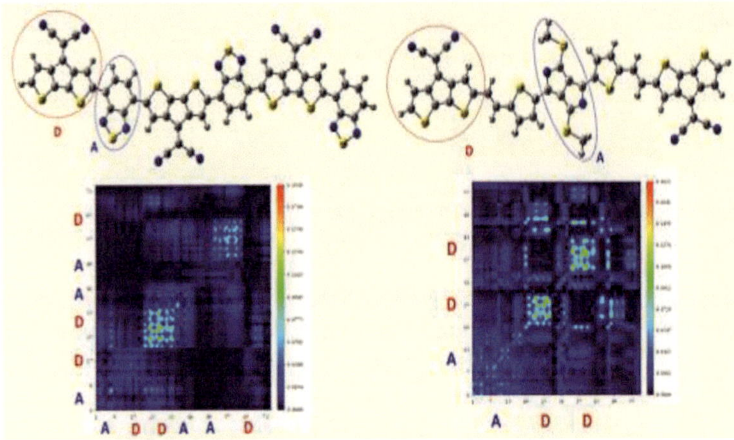

Figure 7.
TDM plots at the first excited state (S_1) of the investigated materials.

The weaker coupling of electron and holes makes easier the dissociation of exciton. The contour plots of TDM show also the exciton dissociation in the studied molecules may be easy regarding the weak electron–hole correlation that involves the charge transfer from main CPDT units to the dicyanomethylene bridge group [66]. The coefficients correlation of D-A within P-CPDTBT3 are slightly higher than those of SM-CPDTNDPP. Hence, the exciton dissociation is expected to be comparatively easier in the case of SM-CPDTNDPP than that in the case of P-CPDTBT3. The TDM analysis demonstrates the efficiency of charge separation within these molecules which leads to a considerable improvement of the Jsc.

3.6 Photovoltaic properties

The advanced performance photovoltaic devices exhibit fundamentally significant power conversion efficiency (PCE). For high PCE, the photovoltaic devices should possess high fill factor (FF) and large open-voltage circuit (V_{OC}). In fact, these conditions dispose a challenge for narrow band gap materials to cover as much of the solar spectrum as possible.

BHJ-OSCs contain principally an electron donor material blinded with an electron acceptor fullerene derivative named (6,6)-Phenyl-C71 Butyric Acid Methyl Ester ([70] PCBM). Accordingly, we have proposed a schematic energy diagram of BHJ-OSCs (P-CPDTBT3/SM-CPDTDPP: [70] PCBM), as shown in **Figure 8**. The experimental [70] PCBM energy level values of were recorded in ref. [67].

FF is one of the crucial factors that influence the PCE and can be estimated using the expression above [68]:

$$FF = \frac{v_{oc} - \ln(v_{oc} + 0.72)}{v_{oc} + 1} \tag{5}$$

Where, v_{oc} is the dimensionless voltage:

$$v_{oc} = \frac{eV_{oc}}{k_B T} \tag{6}$$

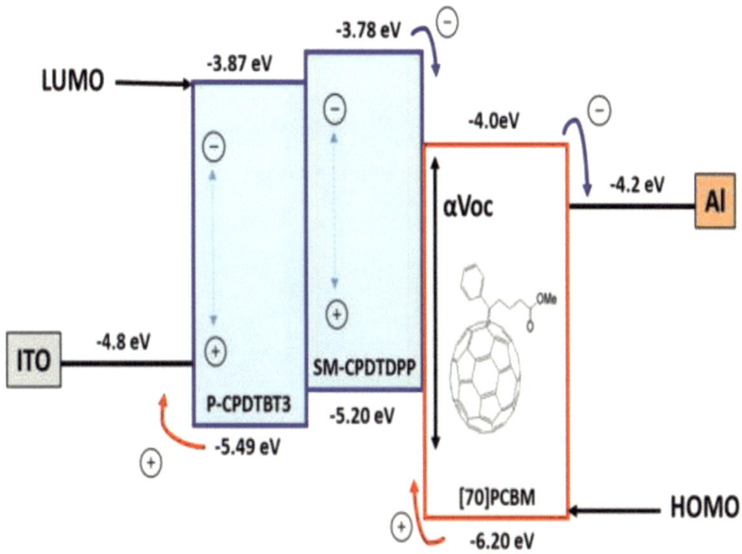

Figure 8.
Schematic energy diagram of the proposed (P-CPDTBT3/SM-CPDTDPP: [67] PCBM) BHJ OSCs.

Here, e, k_B, T and V_{OC} are the elementary charge, Boltzmann's constant, temperature and open circuit voltage, respectively.

The V_{OC} can be approximated as [69]:

$$V_{oc} = \frac{1}{e}\left(\left|H^D\right| - \left|L^A\right|\right) - 0.3 \tag{7}$$

Where, e, H^D and L^A are the elementary charge, HOMO of donor and LUMO of acceptor, respectively.

The calculated photovoltaic parameters are listed in **Table 4**. As we can see from the table, there is a growth tendency of the photovoltaic parameters from SM-CPDTDPP to P-CPDTBT3. The P-CPDTBT3 copolymer exhibits larger V_{oc} which is expected as this latter dispose a deeper HOMO energy level value.

Further, Scharber diagram was used to estimate the power conversion efficiency (PCE) of BHJ-OSCs [70]. From **Figure 9**, the predicted PCE of P-CPDTBT3 and SM-CPDTBT materials are found to be 9.5% and 8.2%, respectively. Thus, we can reveal from these results the fruitful molecular design of the investigated compounds to ensure a promising PCE for developing efficient materials for BHJ-OSCs.

Compound	V_{oc}	v_{oc}	*FF*
P-CPDTBT3	1.11	43.22	0.89
SM-CPDTDPP	0.86	33.48	0.86

Table 4.
Photovoltaic properties calculated at DFT/B3LYP/6-311 g(d,p) level.

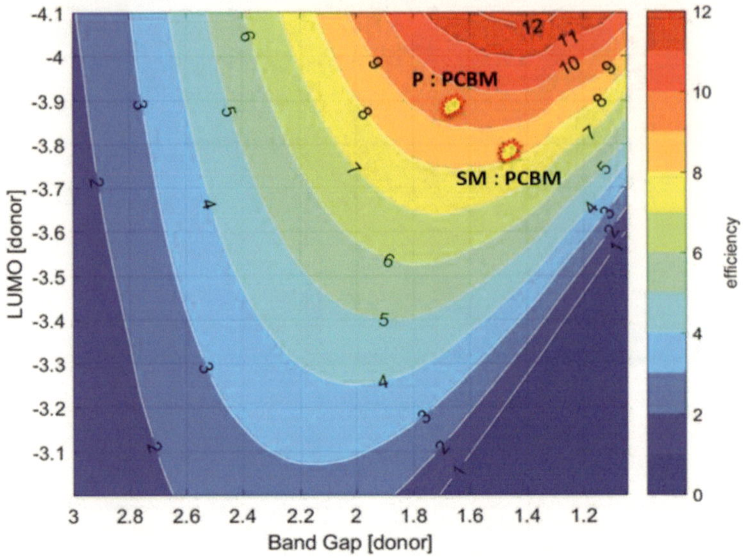

Figure 9.
Scharber diagram for estimating the power conversion (PCE) efficiency of the studied compounds.

4. Conclusion

We have reported in this work a theoretical investigation on a D-A type copolymer and a D-π-A-π-D type small molecule based on CPDT derivatives as electron donor with BT or DPP as electron acceptor moiety.

The studied materials exhibited narrow band-gap energies with high planar structures governed by intra-molecular non-covalent interactions. The particular conjugated arrangement added to the dicyanomethylene Bridge groups has leaded to considerable intra-molecular charge transfer (ICT) within these molecules. The optical absorption spectra covered an interesting part of the solar spectrum which indicated that the studied molecules exhibit high light harvesting efficiency. The deep HOMO energy levels of these molecule encourage their application as donor materials in a BHJ-OSCs with the [70] PCBM as acceptor. The TDM analysis showed the ease dissociation of excitons within the considered materials at the first exited state that leads to enhance the photovoltaic properties. The PCEs, estimated using Scharber diagram, close to 10% were reached.

This study revealed that the reliability of the designed donor materials for developing efficient materials for BHJ OSCs. A molecular tuning based on these designed conjugated materials may enhance the donor materials efficiency in photovoltaic applications.

Acknowledgements

This research was supported by the Ministry of Higher Education and Scientific Research, Tunisia.

Author details

Rania Zaier* and Sahbi Ayachi*
Laboratory of Physico-Chemistry of Materials (LR01ES19), Faculty of Sciences of Monastir (FSM), Avenue of the Environment 5019 Monastir, University of Monastir, Tunisia

*Address all correspondence to: rania.zaier02@gmail.com; ayachi_sahbi@yahoo.fr

References

[1] Peumans P, Yakimov A, Forrest S.R, Small molecular weight organic thin-film photodetectors and solar cells. J. Appl. Phy. 2003;93:3693-3723. https://doi.org/10.1063/1.1534621

[2] Zhang Q, Kan B, Liu F, Long G, Wan X, Chen X, Zuo Y, Ni W, Zhang H, Li M, Hu Z, Huang F, Cao Y, Liang Z, Zhang M, Russell T.P, Chen Y, Small-molecule solar cells with efficiency over 9%. Nat. Phot. 2015;9:35-41. https://doi.org/10.1038/nphoton.2014.269

[3] Cowan S.R, Roy A, Heeger A.J, Recombination in polymer-fullerene bulk heterojunction solar cells, Phys. Rev. B. 2010;82:245207(1-10). https://doi.org/10.1103/PhysRevB.82.245207

[4] Zhao W, Li S, Yao H, Zhang S, Zhang Y, Yang B, Hou J, Molecular Optimization Enables over 13% Efficiency in Organic Solar Cells. J. Am. Chem. Sci. 2017;139(21):7148-7151. DOI: 10.1021/jacs.7b02677

[5] Gao J, Gao W, Ma X, Hu Z, Xu C, Wang X, An Q, Yang C, Zhang X, Zhang F, Over 14.5% efficiency and 71.6% fill factor of ternary organic solar cells with 300 nm thick active layers, Energy Environ. Sci. 13;2020:958-967. https://doi.org/10.1039/C9EE04020J

[6] Ma R, Liu T, Luo Z, Guo Q, Xiao Y, Chen Y, Li X, Luo S, Lu X, Zhang M, Li Y, Yan H, Improving open-circuit voltage by a chlorinated polymer donor endows binary organic solar cells efficiencies over 17%. Sci China Chem, 2020;63:325-330. https://doi.org/10.1007/s11426-019-9669-3

[7] Lu G, Usta H, Risko C, Wang L, Facchetti A, Ratner M.A, Marks T.J, Synthesis, Characterization, and Transistor Response of Semiconducting Silole Polymers with Substantial Hole Mobility and Air Stability. Experiment and Theory. J. Am. Chem. Soc. 2008;130(24):7670-7685. DOI: 10.1021/ja800424m

[8] Wang G, Melkonyan F.S, Facchetti A, Marks T.J, All-Polymer Solar Cells: Recent Progress, Challenges, and Prospects, Angew. Chem. Int .Ed. 2019;58(13);4129-4142, https://doi.org/10.1002/anie.201808976

[9] Facchetti A, Polymer donor–polymer acceptor (all-polymer) solar cells, Mater. Today. 2013;16:123-132. https://doi.org/10.1016/j.mattod.2013.04.005

[10] Mohajeri A, Omidvar A, Fullerene-based materials for solar cell applications: design of novel acceptors for efficient polymer solar cells-a DFT study, Phys. Chem. Chem. Phys. 2015;17:22367-22376. https://doi.org/10.1039/C5CP02453F

[11] Ayoub S.A, Lagowski J.B, Optimizing the Performance of the Bulk Heterojunction Organic Solar Cells Based on DFT Simulations of Their Interfacial Properties. Mater. Des. 2018;156:558-569. https://doi.org/10.1016/j.matdes.2018.07.016

[12] Hachi, M, Slimi A, Fitri A, ElKhattabi S, Benjelloun A.T, Benzakour M, Mcharfi M, New small organic molecules based on thieno[2,3-b] indole for efficient bulk heterojunction organic solar cells: a computational study, Mol. Phys.2019;118:1-14. doi:10.1080/00268976.2019.1662956

[13] Zhang S, Qin Y, Zhu J, Hou J, Over 14% Efficiency in Polymer Solar Cells Enabled by a Chlorinated Polymer Donor, Adv. Mat. 2018;30(20):1800868 (1-7). https://doi.org/10.1002/adma.201800868

[14] Sun C, Pan F, Bin H, Zhang J, Xue L, Qiu B, Wei Z, Zhang Z.G, Li Y, A low cost and high performance polymer donor material for polymer solar cells,

Nat. Comm. 2018;9:743(1-10). doi: 10.1038/s41467-018-03207-x

[15] Akkuratov A, Prudnov F, Mukhacheva O, Luchkin S, Sagdullina D, Obrezkov F, Kuznetsov P, Volyniuk D, Grazulevichus J.V, Troshin P, New cyclopentadithiophene-based (X-DAD'AD)n conjugated polymers for organic solar cells, Sol. Energy Mater. Sol. Cel. 2019;193:66-72. DOI: 10.1016/j.solmat.2018.12.035

[16] Gao K, Kan Y, Chen X, Liu F, Kan B, Nian L, Wan X, Chen Y, Peng X, Russell T.P, Cao Y, Jen A.K.Y, Low-Bandgap Porphyrins for Highly Efficient Organic Solar Cells: Materials, Morphology, and Applications, Adv. Mat. 2020;32(32):1906129(1-19). DOI: 10.1002/adma.201906129

[17] Burrezo P.M, Domínguez R, Zafra J.L, Pappenfus T.M, de la Cruz P, Welte L, Janzen D.E, Navarrete J.T.L, Langa F, Casado J, Oligomers of cyclopentadithiophene-vinylene in aromatic and quinoidal versions and redox species with intermediate forms, Chem. Sci. Chem. Sci. 2017:8;8106-8114. DOI: 10.1039/C7SC02756G

[18] Zaier R, Hajaji S, Kozaki M, Ayachi S, DFT and TD-DFT studies on the electronic and optical properties of linear π-conjugated cyclopentadithiophene (CPDT) dimer for efficient blue OLED, Opt. Mat. 2019;91:108-114. https://doi.org/10.1016/j.optmat.2019.03.013

[19] Lee J, Hernandez J.L, Pelse I, Reynolds J.R, Yang C, Semi-transparent low-donor content organic solar cells employing cyclopentadithiophene-based conjugated molecules, J. Mater. Chem. C. 2018;6:10532-10537. https://doi.org/10.1039/C8TC02863J

[20] Guijarro F.G, Malhotra P, Gupta G, Caballero R, de la Cruz P, Singhal R, Sharma G.D, Langa F, The influence of the terminal acceptor and oligomer length on the photovoltaic properties of A-D-A small molecule donors, J. Mat. Chem. C. 2020;8:4763-4770. https://doi.org/10.1039/D0TC00154F

[21] Zhang X.F, Liu H, Chen Y, Cheng J.Z, Wang L.H, Ye D.N, Chen L, Liu S.Y, One-pot synthesis of cyclopentadithiophene-isoindigo based low bandgap long-chain π-conjugated oligomers, Mater Today Commun. 2020;22:100850(1-8). https://doi.org/10.1016/j.mtcomm.2019.100850

[22] Zaier R, Hajaji S, Kozaki M, Ayachi S, Designing New Small Molecules from Cyclopentadithiophene (CPDT) Derviatives for Highly Efficient Blue Emitters in OLEDs: DFT Computational Modeling, J. Mater. Environ. Sci. 2019;10(3):195-207. ISSN(:(2028;2508

[23] Zaier R, Hajaji S, Kozaki M, Ayachi S. DFT Computational Modeling and Design of New Cyclopentadithiophene (CPDT) Derivatives for Highly Efficient Blue Emitters in OLEDs (Chap.3,). In Polymers for Light-Emitting Devices and Displays (eds Inamuddin, R. Boddula, M.I. Ahamed and A.M. Asiri) Scrivener Publishing; 2020. p. 51-76. doi:10.1002/9781119654643.ch3.

[24] Lu J, Liang F, Drolet N, Ding J, Tao Y, Movileanu R, Crystalline low band-gap alternating indolocarbazole and benzothiadiazole-cored oligothiophene copolymer for organic solar cell applications, Chem. Commun. 2008;42: 5315-5317. https://doi.org/10.1039/B811031J

[25] Alqurashy B.A, Iraqi A, Zhang Y, Lidzey D.G, Pyrene-benzo[1,2,5]thiadiazole based conjugated polymers for application in BHJ solar cells, J. Saudi Chem. Soc. 2020;24(6):484-491. https://doi.org/10.1016/j.jscs.2020.04.004

[26] Cameron J, Abed M.M, Chapman S.J, Findlay N.J, Skabara P.J,

Horton P.N, Coles S.J, Investigating the effect of heteroatom substitution in 2,1,3-benzoxadiazole and 2,1,3-benzothiadiazole compounds for organic photovoltaics, J. Mater. Chem. C. 2018;6:3709-3714. https://doi.org/10.1039/C7TC05075E

[27] Ayachi S, Mabrouk A, Bouachrine M, Alimi K, Photophysical Properties of Two New Donor-Acceptor Conjugated Copolymers and Their Model Compounds: Applications in Polymer Light Emitting Diodes (PLEDs) and Polymer Photovoltaic Cells (PPCs). In Organic Light Emitting Diode, edited by Jai Singh. IntechOpen, 2012. p. 105-142. DOI: 10.5772/54254

[28] Tang A, Zhan C, Yao J, Zhou E, Design of Diketopyrrolopyrrole (DPP)-Based Small Molecules for Organic-Solar-Cell Applications, Adv. Mat. 2017;29:1600013(1-39). https://doi.org/10.1002/adma.201600013

[29] Hendsbee A.D, Sun J.P, Rutledge L.R, Hill I.G, Welch G.C, Electron deficient diketopyrrolopyrrole dyes for organic electronics: synthesis by direct arylation, optoelectronic characterization, and charge carrier mobility, J. Mater. Chem. A. 2014;2:4198-4207. https://doi.org/10.1039/C3TA14414C

[30] Maglione C, Carella A, Centore R, Chávez P, Lévêque P, Fall S, Leclerc N, Novel low bandgap phenothiazine functionalized DPP derivatives prepared by direct heteroarylation: Application in bulk heterojunction organic solar cells, Deyes Pigm. 2017;141:169-178. https://doi.org/10.1016/j.dyepig.2017.02.012

[31] Chen Y, Yan Y, Du Z, Bao X, Liu Q, Roy V, Sun M, Yang R, Lee C.S, Two-dimensional benzodithiophene and benzothiadiazole based solution-processed small molecular organic field-effect transistors & solar cells, J. Mat. Chem. C. 2014;2:3921-3927. https://doi.org/10.1039/C4TC00086B

[32] Luo Z, Liu T, Wang Y, Zhang G, Sun R, Chen Z, Zhong C, Wu J, Chen Y, Zhang M, Zou Y, Ma W, Yan H, Min J, Li Y, Yang C, Reduced Energy Loss Enabled by a Chlorinated Thiophene-Fused Ending-Group Small Molecular Acceptor for Efficient Nonfullerene Organic Solar Cells with 13.6% Efficiency, Adv. Energy Mat. 2019;9:1900041(1-9). https://doi.org/10.1002/aenm.201900041

[33] Zaier R, De La Cruz M.P, De La Puente F.L, Ayachi S, Optoelectronic properties of cyclopentadithiophene-based donor–acceptor copolymers as donors in bulk heterojunction organic solar cells: A theoretical study, J. Phys. Chem . Sol. 2020;145:109532(1-7). https://doi.org/10.1016/j.jpcs.2020.109532

[34] Yang Z, Bao C, Zhang G, Liu Z, Zhu W, Cui S, Peng Q, Liu Y, Low-cost donors based on a dicarboxylic ester side-chain substituted thieno[3,2b] thiophene unit for efficient polymer solar cells, Deyes Pigm. 2020;182:108698(1-10). https://doi.org/10.1016/j.dyepig.2020.108698

[35] Bi X, Zhang T, An C, Bi P, Ma K, Li S, Xian K, Lv Q, Zhang S, Yao H, Xu B, Zhang J, Cao S, Hou J, Enhanced photovoltaic effect from naphtho[2,3-c] thiophene-4,9-dione-based polymers through alkyl side chain induced backbone distortion, J. Mat. Chem. A. 2020;8:14706-14712. https://doi.org/10.1039/D0TA02859B

[36] Hu L, Qiao W, Gu W, Zhu X, Wang C, Wang Z.Y, Broadband polymer photodetectors with a good trade-off between broad response and high detectivity by using combined electron-deficient moieties, J. Mater. Chem. C. 2020;8:3431-3437. https://doi.org/10.1039/C9TC05224K

[37] Gaussian 09, Revision A.1, Frisch M.J. Trucks G.W, Schlegel H.B, Scuseria G.E,

Robb M.A, Cheeseman J.R, Scalmani G, Barone V, Mennucci B, Petersson G.A, Nakatsuji H, Caricato M, Li X, Hratchian H.P, Izmaylov A.F, Bloino J, Zheng G, Sonnenberg J.L, Hada M, Ehara M, Toyota K, Fukuda R, Hasegawa J, Ishida M, Nakajima T, Honda Y, Kitao O, Nakai H, Vreven T, Montgomery J.A, Peralta Jr.J.E, Ogliaro F. Bearpark M, Heyd J.J, Brothers E, Kudin K.N, Staroverov V.N, Kobayashi R, Normand J, Raghavachari K, Rendell A, Burant J.C, Iyengar S.S, Tomasi J, Cossi M, Rega N, Millam J.M, Klene M, Knox J.E, Cross J.B, Bakken V, Adamo C, Jaramillo J, Gomperts R, Stratmann R.E, Yazyev O, Austin A.J, Cammi R, Pomelli C, Ochterski J.W, Martin R.L, Morokuma K, Zakrzewski V.G, Voth G.A, Salvador P, Dannenberg J.J, Dapprich S, Daniels A.D, Farkas O, Foresman J.B, Ortiz J.V, Cioslowski J, Fox D.J, Gaussian, Inc., Wallingford CT, 2009.

[38] Hwang H, Ko H, Park S, Suranagi S.R, Sin D.H, Cho K, Fluorine-functionalization of an isoindoline-1,3-dione-based conjugated polymer for organic solar cells, Org. Elect. 2018;59:247-252. https://doi.org/10.1016/j.orgel.2018.05.009

[39] Chou S.H, Chen H.C, Wang C.K, Chung C.L, Hung C.M, Hsu J.C., Wong K.T, Synthesis and characterization of new asymmetric thieno[3,4-b]pyrazine-based D-π-A-A type small molecular donors with near-infrared absorption and their photovoltaic applications, Org. Elect. 2019;68:159-167. https://doi.org/10.1016/j.orgel.2019.02.013

[40] Lu T, Chen F, Multiwfn: A multifunctional wavefunction analyzer, J. Comp. Chem. 2012;33:580-592. https://doi.org/10.1002/jcc.22885

[41] Ku J, Gim Y, Lansac Y, Jang Y.H, N-Alkylthienopyrroledione versus benzothiadiazole pulling units in push-pull copolymers used for photovoltaic applications: density functional theory study, Phys. Chem. Chem. Phys. 2016;18:1017-1024. https://doi.org/10.1039/C5CP06075C

[42] de Gier, H.D, Rietberg B.J, Broer, R, Havenith R.W.A, Influence of push-pull group substitution patterns on excited state properties of donor-acceptor co-monomers and their trimers. Comp. Theo. Chem. 2014;1040-1041: 202-211. https://doi.org/10.1016/j.comptc.2014.03.002

[43] Liu D, Kan B, Ke X, Zheng N, Xie Z, Lu D, Liu Y, Extended Conjugation Length of Nonfullerene Acceptors with Improved Planarity via Noncovalent Interactions for High-Performance Organic Solar Cells, Adv. Ener. Mat. 2018;8:1801618(1-8). https://doi.org/10.1002/aenm.201801618

[44] Bhattacharya L, Sahoo S.R, Sharma S, Sahu S, Effect of electron-withdrawing groups on photovoltaic performance of thiophene-vinyl-thiophene derivative and benzochalcogenadiazole based copolymers: A computational study, Inter. J. Quant. Chem. 2019;119:e25982(1-14). https://doi.org/10.1002/qua.25982

[45] Zhang L, Shen W, He R, Liu X, Tang X, Yang Y, Li M, Fine structural tuning of diketopyrrolopyrrole-cored donor materials for small molecule-fullerene organic solar cells: A theoretical study, Org. Elect. 2016;32:134-144. https://doi.org/10.1016/j.orgel.2016.01.023

[46] Yao H, Cui Y, Qian D, Ponseca Jr C.S, Honarfar A, Xu Y, Xin J, Chen Z, Hong L, Gao B, 14.7% Efficiency Organic Photovoltaic Cells Enabled by Active Materials with a Large Electrostatic Potential Difference, J. Am. Chem. Soc. 2019;141(19):7743-7750. DOI: 10.1021/jacs.8b12937

[47] Kim B, Ma B, Donuru V.R, Liu H, Fréchet J.M.J, Bodipy-backboned

polymers as electron donor in bulk heterojunction solar cells, Chem. Commun., 2010;46:4148-4150. https://doi.org/10.1039/B927350F

[48] Tsuji M, Saeki A, Koizumi Y, Matsuyama N, Vijayakumar C, Seki S, Benzobisthiazole as Weak Donor for Improved Photovoltaic Performance: Microwave Conductivity Technique Assisted Molecular Engineering, Adv. Funct. Mat. 2014;24:28-36. https://doi.org/10.1002/adfm.201301371

[49] Kularatne R.S, Magurudeniya H.D, Sista P, Biewer M.C, Stefan M.C, Donor-Acceptor Semiconducting Polymers for Organic Solar Cells, J. Polym. Sci, Part A: Polym. Chem. 2013;51:743-768. DOI: 10.1002/pola.26425

[50] Tarkuc S, Eelkema R, Grozema F, The relationship between molecular structure and electronic properties in dicyanovinyl substituted acceptor-donor-acceptor chromophores, Tetra. Lett. 2017;73:4994-5004. https://doi.org/10.1016/j.tet.2017.04.037

[51] Bartelt J.A, Lam D, Burke T.M, Sweetnam S.M, McGehee M.D, Charge-Carrier Mobility Requirements for Bulk Heterojunction Solar Cells with High Fill Factor and External Quantum Effi ciency >90% , Adv. Ener. Mat. 2015;5:1500577(1-10). DOI: 10.1002/aenm.201500577

[52] Bérubé N, Gosselin V, Gaudreau J, Côté M, Designing Polymers for Photovoltaic Applications Using ab Initio Calculations, J. Phys. Chem. C. 2013;117(16):7964-7972. DOI: 10.1021/jp309800f

[53] Ostovan A, Mahdavifar Z, Bamdad M, Evaluation of photovoltaic properties and effective conjugated length of DTTTD-based polymers as donor in BHJ solar cells; quantum chemical approach, Polym. 2017;126:162-176. https://doi.org/10.1016/j.polymer.2017.08.044

[54] Adamo C, Jacquemin D, The calculations of excited-state properties with Time-Dependent Density Functional Theory, Chem. Soc. Rev. 2013;42:845-856. https://doi.org/10.1039/C2CS35394F

[55] Chung H.Y, Oh J, Park J.H, Cho I, Yoon W.S, Kwon J.E, Kim D, Park S.Y, Spectroscopic Studies on Intramolecular Charge-Transfer Characteristics in Small-Molecule Organic Solar Cell Donors: A Case Study on ADA and DAD Triad Donors, J. Phys. Chem. C. 2020;124(34):18502-18512. DOI: 10.1021/acs.jpcc.0c06741

[56] Azazi A, Mabrouk A, Chemek M, Kreher D, Alimi K, DFT modeling of conjugated copolymers photophysical properties: Towards organic solar cell application, Synt. Met. 2014;198:314-322. https://doi.org/10.1016/j.synthmet.2014.10.015

[57] Domínguez R, Schulz G.L, de la Cruz P, Bäuerle P, Langa F.J.D, Pigments, Cyclopentadithiophene-based co-oligomers for solution-processed organic solar cells, Deyes Pigm. 2017;143:112-122. https://doi.org/10.1016/j.dyepig.2017.04.029

[58] You J, Dou L, Yoshimura K, Kato T, Ohya K, Moriarty T, Emery K, Chen C.C, Gao J, Li G, A polymer tandem solar cell with 10.6% power conversion efficiency. Nat. Commun. 2013;4:1446(1-10). doi: 10.1038/ncomms2411

[59] Wang Z, Putta A, Mottishaw J.D, Wei Q, Wang H, Sun H, Molecular Origin of Isomerization Effects on Solid State Structures and Optoelectronic Properties: A Comparative Case Study of Isomerically Pure Dicyanomethylene Substituted Fused Dithiophenes, J. Phys. Chem. C. 2013;117(33):16759-16768. DOI: 10.1021/jp4033029.

[60] Hedström S, Henriksson P, Wang E, Andersson M.R, Persson P,

Light-harvesting capabilities of low band gap donor-acceptor polymers, Phys. Chem. Chem. Phys. 2014;16:24853-24865. https://doi.org/10.1039/C4CP03191A

[61] Fu Z. Shen W, He R, Liu X, Sun H, Yin W, Li M, Theoretical studies on the effect of a bithiophene bridge with different substituent groups (R = H, CH_3, OCH_3 and CN) in donor-π-acceptor copolymers for organic solar cell applications, Phys. Chem. Chem. Phys. 2015;17:, 2043-2053. https://doi.org/10.1039/C4CP04103H

[62] Wang L, Nan G, Yang X, Peng Q, Li Q, Shuai Z, Computational methods for design of organic materials with high charge mobility, Chem. Soc. Rev. 2010;39:423-434. https://doi.org/10.1039/B816406C

[63] Liu H, Kang S, Lee J.Y, Electronic Structures and Charge Transport of Stacked Annelated β-Trithiophenes, J. Phys. Chem. B. 2011;115(18):5113-5120. https://doi.org/10.1021/jp1045595

[64] Ans M, Ayub K, Xiao X, Iqbal J, Tuning opto-electronic properties of alkoxy-induced based electron acceptors in infrared region for high performance organic solar cells, J. Mol. Liq. 2019;298:111963. https://doi.org/10.1016/j.molliq.2019.111963

[65] Taouali W, Casida M.E, Darghouth A.A.M, Alimi K, Theoretical design of new small molecules with a low band-gap for organic solar cell applications: DFT and TD-DFT study, Comp. Mater. Sci. 2018;150:54-61. https://doi.org/10.1016/j.commatsci.2018.03.038

[66] Li S.B, Duan Y.A, Geng Y, Gao H.Z, Qiu Y.Q, Su Z.M, Theoretical design and characterization of pyridalthiadiazole-based chromophores with fast charge transfer at donor/acceptor interface toward small molecule organic photovoltaics, RSC Adv. 2015;5:29401-29411. https://doi.org/10.1039/C5RA00785B

[67] Pal S.K, Kesti T, Maiti M, Zhang F, Inganäs O, Hellström S, Andersson M.R, Oswald F, Langa F, Österman T, Pascher T, Yartsev A, Sundström V, Geminate Charge Recombination in Polymer/Fullerene Bulk Heterojunction Films and Implications for Solar Cell Function, J. Am. Chem. Sc. 2010;132(35):12440-12451. DOI: 10.1021/ja104786x

[68] Guo X, Zhou N, Lou S.J, Smith J, Tice D.B, Hennek J.W, Ortiz R.P, Navarrete J.T.L, Li S, Strzalka J, Chen L.X, Chang R.P.H, Facchetti A, Marks T.J, Polymer solar cells with enhanced fill factors. Nat. Phot.2013;7(10):825-833. https://doi.org/10.1038/nphoton.2013.207

[69] Liu X, Shen W, He R, Luo Y, Li M, Strategy to Modulate the Electron-Rich Units in Donor-Acceptor Copolymers for Improvements of Organic Photovoltaics, J. Phys. Chem. C. 2014;118(31):17266-17278. https://doi.org/10.1021/jp503248a

[70] Scharber M.C, Mühlbacher D, Koppe M, Denk P, Waldauf C, Heeger A.J, Brabec C.J, Design Rules for Donors in Bulk-Heterojunction Solar Cells-Towards 10% Energy-Conversion Efficiency, Adv. Mat. 2006;18:789-794. https://doi.org/10.1002/adma.200501717

Chapter 6

Thin-Film Solar Cells Performances Optimization: Case of Cu (In, Ga) Se_2-ZnS

Fridolin Tchangnwa Nya and Guy Maurel Dzifack Kenfack

Abstract

In this chapter, we investigate a way of improving solar cells performances. By focusing studies on optimizing the structural, the opto-electrical and electronic properties of materials that constitute the layers and interfaces of a solar device, such as electrical susceptibility, doping concentration, mobility of charge carriers and crystallographic structure, it is possible to improve the output parameters of a solar cell. Working on a CIGSe-based second-generation ultra-thin solar cell model, and using Zinc Sulfide (ZnS) as a window layer, and based on recent studies, vital information are found on the optimal values of these properties that may enhance the efficiency of the cell. A correct modeling of the device with a trusted software such as SCAPS and an appropriate set of the exact conditions and parameters of simulation allow to obtain very promising results. In particular, for nanoscale and microscale thicknesses of buffer and absorber layers materials respectively, and with an appropriate choice of other materials properties such as intrinsic doping concentration, electrons and holes mobilities, it is possible to record efficiencies and fill factors of more than 26% and 85% respectively. These values are very promising for solar energy harvesting technologies development through CIGSe – ZnS based solar devices.

Keywords: thin-film, solar cell, CIGSe, ZnS, efficiency optimization, SCAPS numerical simulation, structural properties, Opto-electrical properties

1. Introduction

These two last decades have been marked by a dizzying rise in the field of solar energy harvesting thanks to photovoltaic technologies. Among the three well-known fields of solar device technologies, thin-film solar cells are nowadays gaining more interest among researchers and industrial applications due to the reduction in the quantities of materials, their flexibility, their low environmental impact, reduction in time and new opportunities for higher yields goals. The achievement of a record yields of 22.6% [1] for CIGSe-based thin-film solar cells at the end of 2016 marked a decisive turning point towards the quest for higher yields. On the basis of recent studies, the efficiency of CIGSe-based ultra-thin solar cells can be further considerably improved by directing studies on obtaining the best structural and opto-electrical properties of the different materials layers that constitute the cell structure. For example, among the alternative materials for the buffer layer, scientific community recognizes the Zinc Sulfide ZnS material as the one which stands

IntechOpen

out as the most promising, although the highest yield obtained with it to date does not exceed 22%. In the process of reducing the thickness of the layers, numerous works have been identified. In 2015 a yield of about 20.75% was achieved with CIGSe device using only 1 μm of absorber layer thickness and ZnS as buffer layer [2]. A couple of years later, a yield of 22.62% is recorded using nanoscale thicknesses for the different layers but using the controversy CdS material as buffer layer [3]. By going into continuation of the above-mentioned works, a correct modeling with the trusted numerical simulation software SCAPS allow to investigate the influence of the values of certain properties of layers and interfaces mainly, their electrical susceptibility, their crystallographic structure, their intrinsic doping level and the mobilities of charge carriers, on the performances of CIGSe-ZnS based solar cell structures. In this chapter, we carry out investigations and we highlight key properties values that allow to record very promising results. A good knowledge of these values is a basis to better control the material design and layers deposition processes, to be closed to these theoretical results [4, 5]. To achieve this goal a good understanding of the micro-activities which occur at the level of the layer interfaces which are the recombination mechanisms of the photo-generated electron-hole pairs, and the effects of the defect states which inevitably appear with thicknesses reduction, is required to limit their harmful effects on the cell performances [6, 7].

This chapter will be structured as follow, in the first section we are presenting the state of the art on CIGSe-based second-generation thin film solar cells. The different properties of materials are highlighted in the second section, and we recall the mathematical relationships that support the microscopic phenomena within layers and interfaces, and which constitute the basis set of our solar cell modeling. In the last section, investigations are carried out to understand the influence of certain material properties on the overall performance of our device and depending on the results obtained, we highlight those which lead to better yields.

2. State of the art on solar cells - introduction on CIGSe-based thin films

2.1 State of the art in thin film industry

The photovoltaic market has experienced rapid development since 2003, with a growth rate of 40% until 2009 and 135% in 2010, reaching an installed capacity of 40 GW. This market is dominated by technologies based on crystalline silicon. Developed for 60 years, the silicon industry witnessed many advances. In 2012, it represented 90% of the photovoltaic market share [8]. We can then distinguish two large families of cells depending on the nature of the silicon wafer, monocrystalline (m-Si) or poly crystalline (p-Si), for which the record yields are 26.8% (homo-junction), 24.9% (heterojunction) and 21.7% (homojunction), 20.3% (heterojunction) respectively [9]. Research in this field has been extremely active in recent years and advances have been rapid, which has allowed the advent of photovoltaic technologies that consume less energy and only require a few microns in thickness (compared to around 200 microns for silicon) and costs of smaller productions. This is the so-called second generation thin-film cell industry. Their main advantage comes from the small amount of materials needed to make a cell. This generation experienced a 39% increase in production between 2010 and 2014 and more than 50% between 2015 and 2019. There are mainly three (03) sectors in that field:

- **Amorphous silicon (a-Si) thin films**: they have record efficiencies of around 11% for single junction amorphous silicon cells and 12.6% for tandem double junction amorphous silicon/micro-amorphous silicon cells.

- **Cadmium Telluride (CdTe) thin films**: the record yield is 22.1%. It is the leading thin film technology in the photovoltaic industry. However, a promising future is not guaranteed because of the use of cadmium (Cd) which is a highly harmful material for environment purpose.

- **Thin films based on Indium Gallium Copper DiSelenide Cu (In, Ga) Se_2**, known as CIGSe based solar cells: which represent the best yields with a conversion yield of 22.6% in 2016 [1].

The third generation includes all the new approaches proposed and developed in recent years, this generation is to reduce manufacturing costs (Organic solar cells, Gräetzel[1] cells, etc.). It seeks to overcome the current limits of yields by resorting to original concepts such as multi-junction cells, intermediate gap cells or using hot carriers. The majority of third generation systems are currently under development and target more or less long-term industrial applications. Gräetzel cells for example can offer a yield of up to 11.5% [1].

2.2 CIGSe-based cells: evolution

The scientific advances which have enabled the realization of very high efficiency thin film solar cells have taken place by successive technological leaps, the study of these different key stages is essential to understanding the complexity of the structure of a solar cell based on standard CIGSe and the problematic of this work.

Initially intended for the manufacture of photo-detectors, the first solar cells consisted of single crystals of $CuInSe_2$ (CISe) evaporated on an alumina/molybdenum substrate. Interest in photovoltaic applications grew very quickly in regards to the good yields of around 9% obtained by BOEING in 1981. Since the 1980s, four main developments have made it possible to obtain current yields.

- ***Modification of the structure***

Modification of the buffer layer by the Cadmium Selenide (CdS) layer and the introduction of the ZnO: Al window layer enhanced absorption of the solar spectrum at short wavelengths. As of today, many materials have already been used as a buffer layer in particular (In_2S_3, ZnS, ZnSe, Zn (S, OH)) and as a window layer we will see for example zinc oxide doped with Boron (ZnO: B).

- ***Introduction of Gallium***

Beginning in 1987, Chen et al. [9] attempted to incorporate Gallium atoms into the CISe structure. The partial substitution of indium by gallium has improved the electrical performance of the solar cell.

- **Influence of sodium**

In the 1990s, Hedson et al., working on the substitution of the initial substrate to soda glass in order to reduce costs, found that the performance of solar cells was

[1] It was in 1991 that Gräetzel's cell was discovered by Michael GRÄETZEL, a Swiss chemist and professor at the Swiss Federal Institute of Technology in Lausanne.

greatly improved. They attributed this observed beneficial effect to the influence of sodium from the glass on the doping of the CIGSe layer.

- **Filing procedures**

The importance of the deposition processes of the absorbent layer was studied in the Boeing laboratory, going from one to two, then to three stages, the deposition processes have made it possible to improve the performance of solar cells by various phenomena. That are the burial of the P-N junction, the Gallium gradient and recrystallization.

2.3 Challenges to take up

Advances in the field have enabled CIGSe based technology to exceed the 20% yield recorded in 2010 and yields close to 23% in 2016. If this technology is actually offering very good yields for thin films structures, it is also facing many economic, environmental and competitive challenges strongly linked to this last decade requirements.

- **The first challenge: Reducing production costs.**

The European Union has classified Indium as a critical material since 2010 due to the drastic lowering of its reserves following its multi-field use where demand is increasingly growing in addition to the photovoltaic market (ITO screen, LED, etc.) and the instability of its cost is already hindering the development of CIGSe technology. Among the distinct avenues of research aiming at reducing the use of Indium, the reduction of the thickness of the CIGSe layer is the subject of this work.

- **The second challenge: Reducing the environmental impact, replacing the buffer layer with a less harmful layer.**

Known CIGSe-based cell models use Cadmium Sulfide (CdS) as the buffer layer material. The latter has so far provided the best returns. However, the cadmium present in the material is a highly carcinogenic element representing a potential danger during the manufacturing and recycling phase at the end of the lifetime.

- **The third challenge: Obtain better performance and compete with other technologies.**

So far, the record performance achieved by CIGSe-based thin-film photovoltaic technology is 22.6% [1], close to the highest yield of 24.9% recorded by silicon technology. The main goal of the present development, is to undertake scientific investigations to go beyond the above-mentioned values.

2.4 Thin film issue

In principle, a thin layer of a given material is an element of this material, one of its dimensions, called the thickness, has been greatly reduced so that it is expressed in nanoscale and that this small distance between the two boundary surfaces (almost two-dimensionality) leads to a disturbance of the majority of the physical properties.

The second essential characteristic of a thin film is such that whatever the procedure used for its manufacture; a thin film is almost part of the support on which it is built. Consequently, the support has a very strong influence on the structural properties of the layer deposited thereon.

Much work already exists in the context of optimizing the performance of CIGSe cells with various materials as a buffer layer. It is in this tertiary perspective that this work is also part of, that is to say to seek the optimized properties of the material used as a buffer layer and to control the micro-activities behind performance losses.

3. Material choosing, methods and general principle

3.1 Material choosing and structure of our solar cell

3.1.1 Reason for choosing ZnS

Zinc Sulfide (ZnS) is a semiconductor formed by the association of an element atom from column II with another element atom from column VI of the periodic table of chemical elements. It has intrinsic properties which make it a material of choice in the search for good performance in the CIGSe-based thin film chain. It is recognized as having the following properties: It is non-toxic to the environment and its constituents are abundant in nature (Zinc and Sulfur) (**Table 1**).

3.1.2 ZnS material global properties

3.1.2.1 ZnS crystallographic properties

The crystallography of compounds II-VI6 to which Zinc Sulfide belongs poses some problems because of the polymorphism of these compounds. They can have crystallographic structures of two main types: the cubic structure of the sphalerite type (Zinc Blende), and the hexagonal structure of the Wurtzite type. The cubic structure (Zinc Blende) is stable at room temperature (27°C), while the hexagonal structure is more stable at very high-top temperatures of around 1020°C [10]. Indeed, with a lattice parameter $a_{CIGS} = 0.58\ nm$ for the CIGSe chalcopyrite structure and $a_{0\_ZnS} = 0.541\ nm$ respectively$a_{0_ZnS} = b_{0_ZnS} = 0.3811nm$ for ZnS in its cubic and hexagonal structure respectively, CIGSe forms a better lattice agreement with ZnS in its sphalerite structure which is furthermore its stable structure [11].

A deposition technique by ALD [12] or by laser sputtering [11] would make it possible to obtain a layer of ZnS having a crystallographic orientation preferentially sphalerite.

Column I A-B	Column II A-B	Column III B	Column IV B	Column V B	Column VI B	Column VII B
Li_3	Be_4	B_5	C_6	N_7	O_8	F_9
Na_{11}	Mg_{12}	Al_{13}	Si_{14}	P_{15}	$[S_{16}]$	Cl_{17}
Cu_{29}	$[Zn_{30}]$	Ga_{31}	Ge_{32}	As_{33}	Se_{34}	Br_{35}
Ag_{47}	Cd_{48}	In_{49}	Sn_{50}	Sb_{51}	Te_{52}	I_{53}
Au_{79}	Hg_{80}	Ti_{81}	Pb_{82}	Bi_{83}	Po_{84}	At_{85}

Table 1.
Chemical elements of the Mendeleev table of columns II and VI [13].

3.1.2.2 ZnS Opto-electrical properties

The spectrum of white light extends from the ultraviolet characterized by short wavelengths (λ < 380 nm) to the infrared characterized by long wavelengths (λ > 780 nm) through the visible spectrum whose wavelengths are between 380 nm < λ < 780 nm. The absorption spectrum of ZnS is an important element that characterizes its absorption power when subjected to illumination. It is all the more significant as it extends over a frequency band that is difficult to absorb by other materials. The important element to characterize it is the value of the band gap of the material. The energy value of the band gap, E_g of ZnS can be determined by an optical approach through its absorption spectrum. The formula that describes light absorption ability of a material given his band gap is well highlighted by Rayan, Elseman and co-workers [14–16] and is given by:

$$(\alpha h\nu)^2 = A\left(h\nu - E_g\right) \tag{1}$$

This formula is only valid for authorized direct transitions. α is the absorption coefficient, A is a constant to be determined, ﬁ is Planck's constant and ν is the frequency of the incident photon.

Literature tells us that ZnS is a semiconductor with a wide band gap (> 3.5 eV). This wide band gap gives it high optical transparency in the visible and infrared regions of the solar spectrum and a high absorption coefficient in the ultraviolet region. Its transmission spectrum is recorded between 400 and 850 nm [17], which allows the transmission of photons of higher energy, thus increasing the absorption of the light spectrum in the absorber layer.

It is also a direct gap semiconductor. Indeed, the maximum of the valence band and the minimum of the conduction band are found at the centre of the Brillouin zone (where $\vec{K} = 0$). The transitions are made from band to band without the intervention of phonons, therefore without loss or dissipation of energy in thermal form [18].

The refractive index of ZnS is 2.41 to 0.5 μm and 2.29 to 1.1 μm in depth [19].

Remember that the electronic structures of Sulfur and Zinc are:

$$Zn : [Ar]4d^{10}5s^2 \; S : [Ne]3d^2 3p^4$$

The $3p$ states of Sulfur form the valence band, the $5\,s$ states of Zinc constitute the conduction band. This gives ZnS a wide forbidden band. This wide band gap makes it a very promising material for optoelectronic and solar applications. Its forbidden band is between 3.68 eV and 3.9 eV depending on whether we are in a cubic or hexagonal structure. This band gap value may vary depending on the preparation method and the doping rate [19]. We can also find from the literature that it has a relatively high exciton binding energy (34 meV); its structure exhibits a better lattice matching with absorbers having energy bands in the range (1,2–1,5 eV) [13] and finally it has a high electrons mobility (165 $cm^2/V.\,s$).

3.1.3 Structure of our solar cell

In its most common structure, a CIGSe-based cell is formed by a stack of several thin film materials deposited successively on a substrate. Let us consider the following structure:

(Ni/Al)/MgF2/ZnO: B/i-ZnO/ZnS/CuInGaSe_2/Mo/SLG (**Figure 1**).

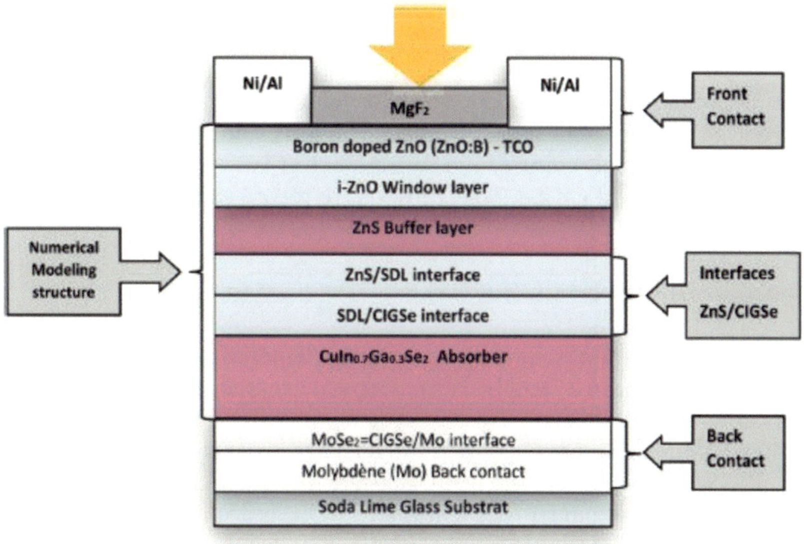

Figure 1.
One-dimensional structure of the cell.

- **Second window layer: i-ZnO**

Zinc oxide is a semiconductor belonging to group II-VI with a number of properties that make it a material widely used in several fields, these are its piezoelectricity, its wide band gap and its intrinsic doping type N. It can exhibit an N-type doping process, for example by adding the atoms of Al, Ga and B and it also exhibits an N-type conductivity due to structural defects. Its forbidden band is 3.3 eV and can vary by adding Mg, Cd or S atoms. It can crystallize according to the Zinc Blende, wurtzite or diamond structures. Only the wurtzite structure is stable under ambient conditions [20].

- **First window layer: ZnO: B**

Most CIGSe-based solar cells use Aluminum-doped Zinc oxide (ZnO: Al) as the Transparent Conductive Oxide (TCO). Within the framework of this work we opt for ZnO doped with boron (ZnO: B) as OCT. Boron doping would be more beneficial for solar cells. Its standard thickness varies between 450 and 1400 nm [18].

- **The buffer layer: ZnS**

Its standard thickness varies between 40 and 60 nm. It has an N-type conductivity and its gap is greater than that of the absorber layer. Three roles are mainly attributed to it:

- **An electrical role**: it adapts the width of the forbidden band between the absorber and the window layer and limits the recombination of carriers at the interface;

- **An optical role**: due to its wide forbidden band (E_g = 3.68 eV), it makes it possible to absorb the maximum of the light spectrum in the region not absorbed by the active layer and thus minimizes optical losses;

- **A role of protective layer**: it protects the surface of the absorber during the deposition of the ZnO layer, which can cause defects on the surface of the CIGSe.

From Won Song et al. work [19], only the values of properties recorded for the cubic structure will be taken into account during this study, since that structure is more stable at room temperature, and forms a better lattice matching with the chalcopyrite structure of CIGSe. That allows us to record as dielectric constant value $\varepsilon_r = 8.3$.

- **The absorbent layer: Cu (In, Ga) Se_2**

It is a semiconductor obtained by combining elements from groups I-III-VI2 of the periodic table (**Table 2**) and has a chalcopyrite crystal structure. Its standard thickness is between 1.5 - 3 μm. This tetragonal structure can be described as a stack of two Zinc Blende structures in which the tetrahedral sites are occupied by atoms of group VI (Se) (anions) and the other sites are occupied in an orderly manner by atoms of groups I (Cu) and III (In) (cations). CIGSe is a solid solution of the semiconductor materials $CuInSe_2$ and $CuGaSe_2$ which have direct gaps of 1.06 eV and 1.7 eV respectively.

$$\text{The ratio}: x = \frac{[Ga]}{[Ga] + [In]} \tag{2}$$

determine the rate of Gallium atoms that replace Indium atoms in the structure. The value of the band gap and the electrical susceptibility of the material vary as a function of x between the values of pure CIS and pure CGS according to the following empirical laws [21]:

$$E_g = 1.06 + 0.39238x + 0.24762x^2 \tag{3}$$

Material properties	ZnS
Melting point (K)	2038 (WZ, 150 atm)
Band gap E_g at 300 K (eV) (ZB/WZ)	3.68/3.911
Electrical susceptibility (eV)	3.9-4.5
Lattice parameter (ZB) a0 at 300 K (nm)	0.541
ZB structure density at 300 K (g/cm-3)	4.11
Lattice parameters (WZ) at 300 K (nm) a0 = b0 c0 c0/a0	0.3811 0.6234 1.602
WZ structure density at 300 K (g/cm-3)	3.98
Heat capacity Cp (Cal/mol K)	11
Relative dielectric constant ε_r	8.3
Refractive index (ZB/WZ)	2.368/2.378
Absorption coefficient	≤0.15
Electron effective mass (m*/m$_0$)	−0.4
Electrons mobility at 300 K (cm2/V. s)	165
Holes mobility at 300 K (cm2/V. s)	5

Table 2.
Properties of zinc sulfide ZnS [22].

$$\chi_e = 4.6 - 1.15667x + 0.03333x^2 \tag{4}$$

The choice of the value of E_g (and therefore of χ_e) depends on several factors.

The best yields are obtained with a value of the band gap E_g of about 1.2 eV. This corresponds to a Gallium concentration level close to [Ga] = 30%. This is the value that will be considered in this study.

- **Rear contact: Molybdenum (Mo)**

The back contact here is a thin layer of Molybdenum (Mo) which is 300 nm thick (the standard thickness is between 0.3-1 micron). It has the ability to form ohmic contact with CIGSe [23]. Indeed, Mo can react with selenium (Se) during the deposition of CIGSe to form $MoSe_2$. Consequently, the CIGSe/Mo structure then becomes *CIGSe/$MoSe_2$/Mo* with a thickness of $MoSe_2$ of about 10 nm. $MoSe_2$ is a semiconductor with a gap of 1.41 eV and its existence has the effect of giving an ohmic behavior to the CIGSe/Mo hetero-contact, while reducing recombination at the interface.

- **Substrate: Soda-lime glass**

The standard substrate used to make CIGSe cells is soda lime glass. It's benefit effects were described above in §2.2. Its thickness varies from 1 to 3 mm.

3.1.4 Investigations on the ZnS – Cu (In, Ga) Se_2 interface: Highlighting of the surface defect layer (SDL)

Between the ZnS and the absorber, a layer called OVC (Ordered Vacancy Compound) has been identified. Investigations carried out within recent works identify his properties and found that they were similar to that of a Surface Defects Layer (SDL). Ouédraogo et al. [24] and Tchangnwa et al. [25, 26], highlighted the beneficial effect related to the existence of the SDL on cell performance. In fact, the presence of an Indium-enrich micro-layer which exhibits an N-type conductivity (N-SDL), at the top of the P-type conductivity CIGSe material, is responsible of the existence of that defect state layer. That leads to a discontinuity at the band gap level at the interface. It has a positive effect since it enhance the transport of the charge carriers through the junction. Another interpretation is given for the existence of that defect state layer from other authors [27], that is, it results from a copper-poor film at the top of the CIGSe material. Both interpretations complete each other, since the N-SDL is almost located within the CIGSe layer, but exhibits different opto-electrical, electronic and structural properties. As a consequence, at the P-N interface between the N-type ZnS and the P-type CIGSe materials, an homojunction is formed. That is, the N-type SDL is almost fully integrated within the CIGSe structure. That explains why in our structure, we will model our P-N heterojunction (the ZnS-CIGSe junction) as two different interfaces: the ZnS/SDL interface and SDL/CIGSe interface, each of them exhibiting different properties.

3.2 General principle

3.2.1 Mechanism of photovoltaic conversion

The photovoltaic effect is based on the properties of semiconductor materials. Indeed, the latter are capable of absorbing photons of frequency ν whose energy is:

$$\left[E_{photon} = h\nu - E_g\right] \geq \left[E_g = E_C - E_V\right] \tag{5}$$

Where E_g is the value of the forbidden band, $E_C$and E_V are respectively the energy values of the conduction band and the valence band of the illuminated material.

A semiconductor material alone cannot generate electric current. In order to generate the electric current, one must assemble two semiconductors of different types, thus creating a P-N junction [21]. In this chapter, let consider the denomination heterojunction because the constituent materials of our two semiconductors are different. The P-doped zone is that containing the CIGSe, and the N-doped zone groups together the buffer layer (ZnS) and the window layers (i-ZnO, ZnO: B). During contact between the P and N zones, the majority carriers of each diffuse through the contact surface, at this time a depletion zone is created, positively charged on the side of the N-type semiconductor and a negatively charged zone on the P-type semiconductor side. This transition zone is called the Space Charge Zone (ZCE). The Fermi levels of the two zones equalize, causing the band diagram to bend, introducing a potential barrier V_e at the interface.

The concentration gradient of the majority carriers induces the presence of a permanent electric field in this ZCE at equilibrium. The electric field thus created leads each type of carrier towards the zone where it is the majority carrier (the electrons towards the N zone and the holes towards the P zone). It follows the mechanism of the collection of each majority carrier.

3.2.2 Densities of states and concentrations of charge carriers

Electrons and holes obey the Fermi-Dirac statistic, the probability that an energy level E is occupied by a charge carrier is given by:

$$f_{FD}(E) = \frac{1}{1 + \exp\left(\frac{E - E_f}{K_b T}\right)} \tag{6}$$

Assuming we are in non-degenerate states of energies we have$(E - E_f)/K_b T \gg 1$ and$f_{FD}(E) \to \exp - \left(\frac{E - E_f}{K_b T}\right)$.

The densities of electrons and holes in the conduction and valence band respectively are given by the following integrals [28].

$$n = \int_{E_C}^{\infty} N(E) f(E) dE = f_{FD}(E_c) = \mathrm{Nc}\ \exp - \left(\frac{E_c - E_f}{K_b T}\right) \tag{7}$$

$$p = \int_{-\infty}^{E_V} N(E)\left(1 - f(E) dE = f_{FD}(E_V) = \mathrm{Nv}\ \exp\left(\frac{E_V - E_f}{K_b T}\right)\right. \tag{8}$$

With:

$$N_C = \frac{2\left(2\pi m_e^* KT\right)^{3/2}}{\hbar^3} \tag{9}$$

and

$$N_V = \frac{2\left(2\pi m_h^* KT\right)^{3/2}}{\hbar^{3.}} \tag{10}$$

The fermi level energy is given by:

$$E_F = \frac{E_C + E_V}{2} + \frac{KT}{2} \ln\left(\frac{N_V}{N_C}\right) = \frac{E_C + E_V}{2} + \frac{KT}{2} \ln\left(\frac{m_h^*}{m_e^*}\right)^{\frac{3}{2}} \tag{11}$$

with: $m_e^* = 0.4m_e$ and $m_h^* = 1.7m_e$

3.2.3 Current generation

These are the Poisson equation in the presence of an electric potential φ, and the continuity equations of electrons and holes with well-specified boundary conditions [23].

The Poisson equation for semiconductors is:

$$\frac{d}{dx}\left(\varepsilon \frac{d\varphi}{dx}\right) = -q\left(p - n + N_D^+ - N_A^-\right) \tag{12}$$

The continuity equations for electrons and holes are:

$$\frac{d}{dx}(J_n) = q(R - G) + q\frac{\partial n}{\partial t}; \tag{13}$$

$$\frac{d}{dx}\left(J_p\right) = -q(R - G) + q\frac{\partial p}{\partial t} \tag{14}$$

ε is the dielectric constant of the material, φ is the electrostatic potential, n and p are respectively the concentration of free carriers for electrons and holes, N_D^+ and N_A^-- are the densities of ionized donors and acceptors, J_n and J_p are the current densities due to electrons and holes. R and G are the rates of recombination and generation of electron-hole pairs, respectively.

$$J_n = q\mu_e n \nabla\varphi + qD_e \nabla n; \tag{15}$$

$$J_p = q\mu_p p \nabla\varphi + qD_p \nabla p \tag{16}$$

The current density in the cell can be written in the following form: (17)

$$J = J_{Gen\acute{e}ration} - J_{Recombinaison} = -q \int_{-d}^{W+L} G_L(\lambda, x)dx - J_{ir} + q \int_{0}^{W+L} R(x)dx \tag{17}$$

Where G_L is the generation function, J_{ir} is the recombination current at the interface, R is the recombination function in the volume of the absorbent layer. d, W and L are the widths of the buffer layer, ZCE and ZQN respectively [23].

The rate of generation of electron-hole pairs at one dimension of the surface of the semiconductor is given by:

$$G_L(\lambda, x) = \alpha_1(\lambda)F(\lambda)(1 - R(\lambda)) \exp(-\alpha_1 x) \tag{18}$$

α_1 (λ) is the number of incident photons per cm^2 per s per unit wavelength. $R(\lambda)$ is the fraction of photons reflected from the surface, $\alpha_1$is the absorption coefficient in the semiconductor.

3.3 Study of the performance reduction mechanisms within the device

Some micro-activities took place within the junction when the two materials (ZnS and CIGSe) are gathered together and are submitted to an external electric field and light. These activities which tend to reduce the global performances of the device, are the recombination mechanisms of the photo-generated electron-hole pairs. To recall what we said in **Section 1**, the first challenge of the current researches in the field of solar energy harvesting is the reduction of the quantities of materials while keeping the device performance or enhancing them. This is a great challenge since, the effects of the defect states, which inevitably appear with the reduction of thicknesses, have very detrimental effects on the performance of the cell.

3.3.1 Charge carrier's recombination mechanisms

These mechanisms are characterized by their rates R which describes the number of recombination per unit time and per unit volume of the material and by the lifetime of the charge carriers. There are three of them, but let us just consider the radiative and the Auger recombination.

- **Radiative recombination**

It takes place by the direct transition of an electron from the conduction band to the valence band. The energy of the transition is released as a photon. If the trap levels created by defect states are close to the middle of the forbidden band, the rate of radiative recombination and the lifetime of electron can be given by [21]:

$$R_R \approx \frac{n - n_0}{\tau_{R,n}} \tag{19}$$

with

$$\tau_{R,n} = \frac{1}{p_0 C_r} \tag{20}$$

and,
$\tau_{Rad,n}$ is the lifetime of the electrons, C_r is the radiative recombination coefficient, p_0 the hole density at equilibrium.

However, this mechanism is not always so harmful on the performance of the cell because in fact, it is possible that the photon released during this mechanism is reabsorbed and thus forming another electron-hole pair. Indeed, the energy value of the emitted photon is close to that of the band gap.

- **Auger recombination**

These are direct carrier band-to-band transfers. It could be an electron or a hole. However, instead of being emitted as a photon, the energy is transferred to another carrier of the same type as thermal energy. The latter will return to its initial state by interacting with the crystal lattice, it will therefore emit a phonon [21]. The rate of radiative recombination can be given by:

$$R_A \approx \frac{n - n_0}{\tau_{A,n}} \tag{21}$$

with

$$\tau_{A,n} \approx \frac{1}{p_0^2 C_p} \tag{22}$$

$\tau_{A,n}$ is the lifetime of the electrons, p_0 the hole density at equilibrium, C_p is the Auger recombination coefficient.

We recall these mathematical expressions to highlight the influence of doping process on recombination mechanism rate at a certain point. We will more explain it a little further in the third section.

3.3.2 Investigations on the effects of defect states on cell performances

Most of the time, defect states are a result of the way the processes of preparation or deposition of layer sheets are performed. They act on material properties, mainly on material electrical properties the same way as for adding impurities. Generally, the main goal is to improve charge carriers transport within the material or to reduce recombination mechanism rate. However, as we will see a little further, they have detrimental effects on the cell performances. Recalling what has been said in **§3.1.6**, and because of the highlighted SDL, at the heterojunction level, we will investigate these effects on two interfaces: ZnS/SDL and /CIGSe.

- **Investigation on the ZnS/SDL sub-heterojunction.**

In the **§3.1.5**, we highlighted one of the numerous positive roles of the ZnS on enhancing the global cell performance, that is "it protects the surface of the absorber during the deposition of the ZnO layer, which can cause defects on the surface of the CIGSe". Therefore, a good choice of the deposition process can significantly reduce these defects density at ZnS/SDL sub-heterojunction. If we consider an ideal case, where the preparation and the deposition processes are perfect, that means with no defect reported within the sub-heterojunction, then we may have this situation:

- The negative charges concentration located in the Space Charge Region (SCR), that is almost localized within the top film part of the CIGSe layer is compensated with the positive charges concentration on the other side, resulting from top to bottom of ZnO: B, i-ZnO and ZnS materials.

That is, at equilibrium, it can be described by the following equation:

$$Q_n + qt_{i-ZnO}N_w + qt_{ZnS}N_b = qN_a t_{SCR} \tag{23}$$

where Q_n is the surface charge in the depletion zone of the boron-doped ZnO window layer, q is the elementary charge, N_w, N_b and N_a are the doping concentrations in the i-ZnO, ZnS and CIGSe top film part layers with respective thicknesses t_{i-ZnO}; t_{ZnS} and t_{SCR}.

Introducing a defect state within a material crystallographic structure, is generally materialized by adding a negative charge. Let say, this situation happens at the ZnS/SDL sub-heterojunction section, to compensate the negative charge added due to defect states, we should reduce the width of the SCR in the CIGSe top film part. That action we will increase the recombination rate at the ZnS-SDL level and negatively affect the cell output parameters. This interpretation is confirmed by previous work reports [27].

Worker reference	Subject of study	Reports
[27]	Excess defect at the CdS/CIGS interface solar cells.	Considering a density of defect of $10^{12}cm^2$, and an electron-hole capture section through the interface of $10^{-12}cm^{-2}$, the yield decreases from 18.9-14%.
[29]	Effects of defect states on the performance of $CuInGaSe_2$ solar cells.	For a defect state density of less than $10^{14}cm^{-3}$, in the CIGSe material and not more than $10^{18}cm^{-3}$, in the CdS material, the detrimental effects of defect states are not really perceived.

Table 3.
Reports of recent studies on the effect of defect states on the cell performances.

- **Investigation on the SDL/CIGSe sub-heterojunction.**

Recent studies were carried out on the effects of the density of defect states on the global performance of CIGSe/CdS-based solar cells and their conclusions are summarized within the **Table 3** above [27, 29].

3.4 Modeling and calculation tools

We have modeled our solar cell device by using the version 3.3 of the one-dimensional numerical simulation software SCAPS-1 D[2] [30].

4. Results and discussion

4.1 Optimal value of ZnS electronic affinity

Considering the electrical susceptibility of Zinc Sulfide (χ_e), within the range of 3.9 [31] and 4.5 [32], we investigate the effects of his value on the performance of our modeled device. Observations are mainly carried out on Fill Factor (FF) and Efficiency (η) and for a thickness of the absorber layer of 0.5 μm. The results obtained are plotted in **Figure 2** above.

It shows that the optimum value of the ZnS electrical susceptibility for recording the best performance of our solar cell is defined for χ_e $[4.1; 4.5eV]$. This result is in very good agreement with those reported in literature [17, 19]. The maximum of the parameters is obtained for $\chi_e = 4.3eV$. This value can be registered as the optimum value of the electrical susceptibility of ZnS for this structure of solar cell.

4.2 Influence of the doping concentration of the CIGSe layer on the performance of the cell

Recent works with the CdS as buffer layer [3] has shown that, beyond a concentration of holes $N_A = 0.8\ 10^{14}cm^{-3}$, the overall parameters of the cell degrade considerably. Ouédraogo et al. [19] have shown in their work that beyond $N_A = 10^{16}cm^{-3}$ the voltage reaches saturation independently of the thickness of the absorber layer. The resulted obtained are plotted on the **Figure 3** below. From these plots, we can see that, varying N_A influences the performances of the cell. For a fixe value of t_{CIGSe}, V_{OC}, increases significantly with the increase of N_A. If we decide as well to increase t_{CIGSe},

[2] SCAPS-1D: Solar Cell Capacitance Simulator in One Dimension. Free software developed by M. Burgelman, Nollet and Degrave from the University of Gent in Belgium in 2000.

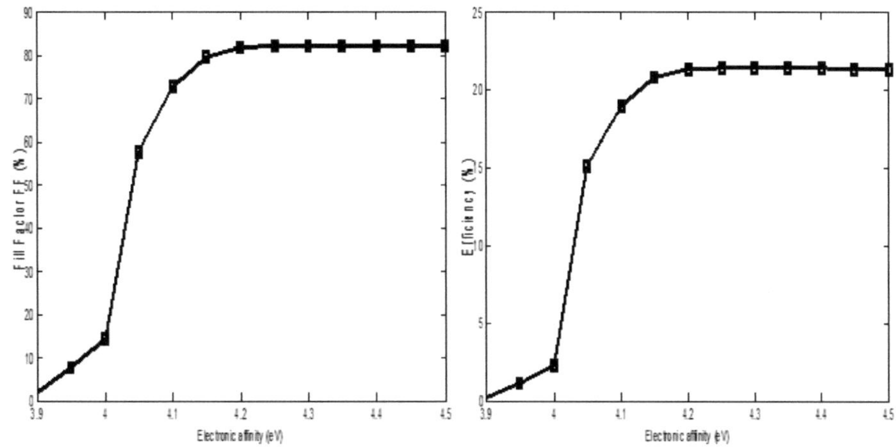

Figure 2.
Influence of ZnS χ_e, on the fill factor (FF) and efficiency.

then for the same value of N_A, V_{OC} will increase. This observation is the same with efficiency (**Figure 3(d)**). Conversely, J_{SC} (**Figure 3(b)**) and the fill factor (FF) (**Figure 3(c)**) decrease significantly with the increase of N_A. By setting the value of N_A, and by varying t_{ZnS}, the output parameters of the cell are almost constant. When the thickness of the absorber layer is less than 250 nm, the performances of the cell are not influenced with variation of N_A in particular FF and η.

But a significant increase on the output results is observed for a doping level $N_A = 10^{15} cm^{-3}$ and with a thickness of the absorber layer no more than 0.5 μm. For $N_A \in [10^{14} : 10^{15} cm^{-3}]$, and for $t_{CIGSe} < 2500$ nm, the parameters of the cell are globally interesting because $V_{OC} < V_{SAT}$ (**Figure 3(a)**), whereV_{SAT} is the saturation voltage of our device. In addition, the short-circuit current is at its maximum value $J_{SC} = J_{(SC)\,max}$ (**Figure 3(b)**) and the values of the fill factor (**Figure 3(c)**) $FF \approx 85\%$. This observation is important because it will allow to circumscribe the optimal value of N_A. For$N_A > 10^{15}$, the overall performance of the cell increases significantly. Believing that the best performances can be obtained with a high level of intrinsic doping recorded in the absorber layer is an utopia; two factors limit that way of thinking: The quality factor in **Figure 5(c)** decreases significantly and characterizes the poor quality and instability of the solar device in question; For $N_A = 10^{16} cm^{-3}$ and considering $t_{CIGSe} > 500$ nm, saturation is automatically reached. Thus, for $t_{CIGSe} = 500nm$, we have $V_{(OC)SAT} = 0.8V; J_{SC} = 34.39\ mA/cm^2, \eta = 22.27\%$, and $FF = 79.47\%$. let us recall the mathematical expressions given in §3.3.1, those are Eq. (20) and Eq. (22). According to Eq. (20), when we add the holes density *P*, the lifetime of negative charge carriers decreases since the latter is inversely proportional to hole concentration and to the square root of hole concentration (Eq. (22)). As a conclusion, the benefits of higher doping for P-type conductivity materials are limited by the Auger and radiative mechanisms. Moreover, for $N_A \in [10^{15} : 4.10^{15} cm^{-3}]$ the performances of the cell are globally very interesting. This would probably justify why Daouda et al. [23] obtained a good yield (18.6%) by working with$N_A = 7.10^{15} cm^{-3}$. However, they quickly reached saturation as soon as$t_{CIGSe} > 1000$ nm. The analysis of the different graphs shows that the optimal intrinsic doping level of CIGSe is $N_A = 10^{15} cm^{-3}$.

$$(V_{oc,ZCE})_{dom} = \frac{E_g}{q} - \frac{nkT}{q} \ln\left(\frac{1}{J_{ph}} * \frac{qD_nN_CN_V}{L_nN_A}\right) \tag{24}$$

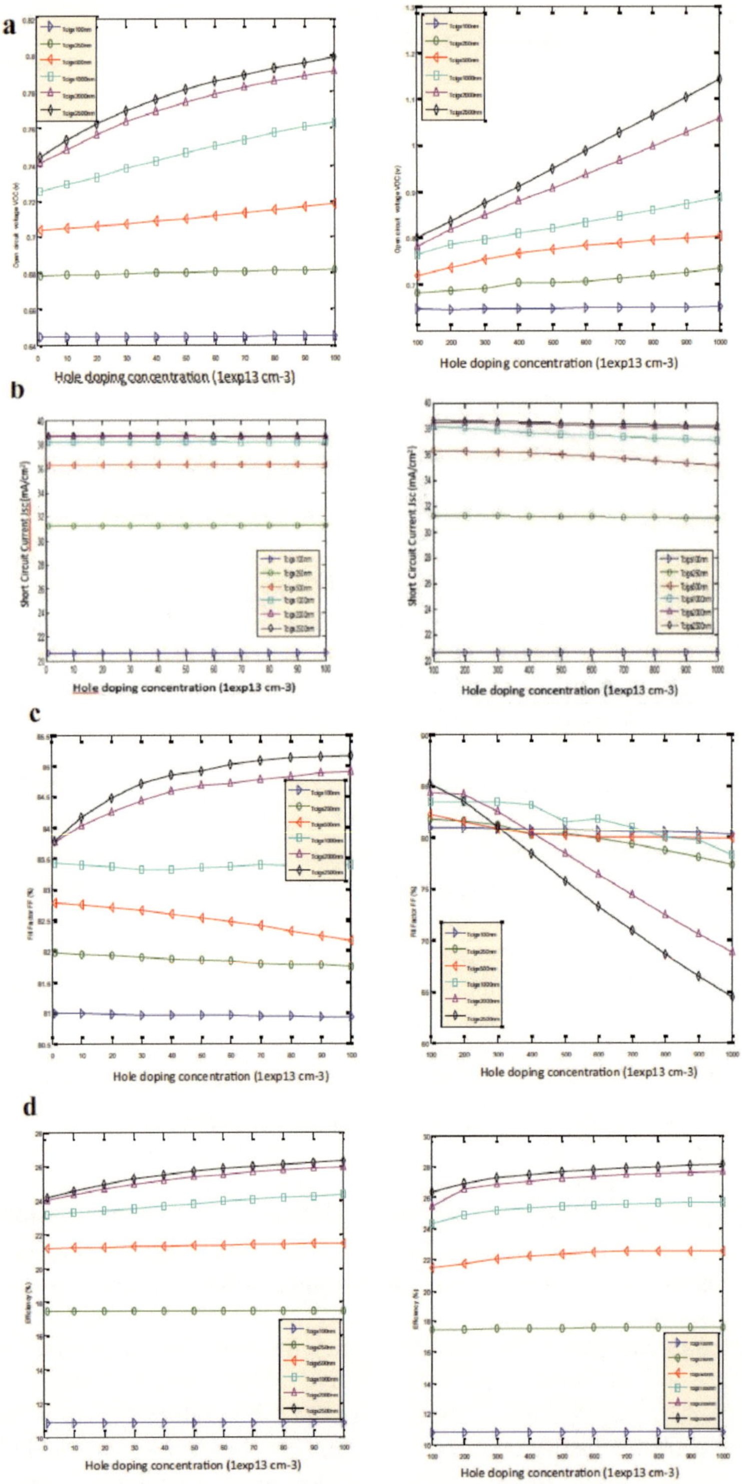

Figure 3.
Influence of the acceptor density N_A on the cell parameters for different CIGSe thickness values. (a) Open circuit voltage (V_{OC}), (b) short-circuit current density (J_{SC}), (c) fill factor (FF), and (d) efficiency (η).

where D_n is the electron scattering coefficient, N_C and N_V the state densities in the conductance and valence bands, L_n the electron scattering length, N_A the density of acceptor states in the CIGSe layer. From this relation when N_A

increases, the voltage $(V_{oc,ZCE})_{dom}$ tends to reach critical value materialized as saturation.

The results obtained are in good agreement with those reported by Ouédraogo and co-workers [23] who were investigating on a CIGSE-based cell structure with CdS as buffer layer. We bring out from these results that both the choice of the buffer layer material and its thickness are not limiting input parameters for hole doping level within the absorber layer. Th high density of recombination mechanisms and a weak collection mechanism of free charge carriers can explain very well the decreasing of short-circuit current (J_{SC}). For a doping level of more than $10^{15}\ cm^{-3}$, within the CIGSe material, the output parameters of our device are not enhanced.

4.3 Effects of the thickness of CIGSe on cell performance

In this subsection, we are investigating how the global output parameters of the device is affected when the thickness of the active layer is a variable. The simulations are carried considering the optimized properties of the other layers. From literature, the commonly used thickness for absorber layer is within the range [2000: 3000 nm]. Since we are on the way of reducing materials quantities, we will observe the performances of our cell with CIGSe thickness ranged from 100 nm to 3000 nm.

There are two main areas of interpretation,

- the first for $100\ nm < t_{CIGSe} < 500\ nm$, and
- the second for $500\ nm < t_{CIGSe} < 2500\ nm$.

Vital information came out from the results obtained. The global performances of the cell (J_{SC}, V_{OC}, FF, η) are improved significantly when the thickness of the absorber layer t_{CIGSe} is gradually increased.J_{SC} is the most sensible parameter, the values of $20.65mA/Cm^2$ and $30.31mA/Cm^2$ are recorded for t_{CIGSe} values of 100 nm and 500 nm respectively. That is a jump of about $16mA/Cm^2$ (**Figure 4(b)**). In fact, at the absorber and Molybdenum junction, the rate of backward recombination mechanisms are reduced significantly because of the Space Charge Region which is completely localized within the active material layer. A similar situation is observed with the plot of the open-circuit voltage, which value increases from 0.65 V for 100 nm of absorber thickness, to 0.76 V for 1000 nm of thickness respectively. That is a gain of 0.13 V! (**Figure 4(a)**). That is not all, the efficiency increased from 10.78% to 24.31%, a jump of almost 14% (**Figure 4(d)**). A value of 83.4% was recorded for the Fill Factor (**Figure 4(c)**). Those results are just impressive. Let us remember that we are undertaking our calculations based on optimized values of properties of the other materials that constitute our structure. These are reported in the **Table 4** below.

For a thickness of the absorber layer between 500 nm and 2500 nm, we record very good value of our cell output performance. Of course, for a thicker CIGSe layer, the probability of absorbing a wide range of light Spectrum is higher and thus, and since the materials properties have been optimized, the quantum efficiency will be higher too. From **Figure 4(a)**, we can easily notice how remarkably V_{OC} increases with absorber layer thickness.

Pogrebjak and co-workers [7, 32] who worked on the influence of temperature on CIGSe-based ultra-thin solar cells and on nano-scale technologies, obtained results which are in good agreement with those obtain during our simulations. The slight difference in results can be explained by global condition of calculation such

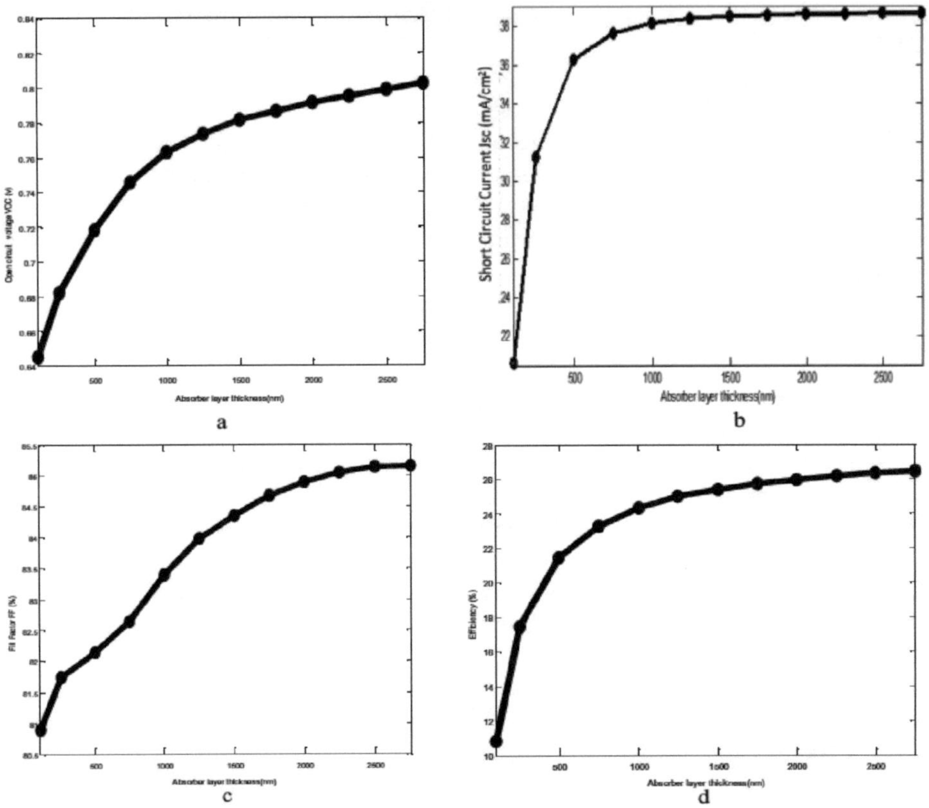

Figure 4.
Influence of the thickness of the absorber layer on the output parameters of the cell for $t_{ZnS} = 5$ nm $N_A = 10^{15} cm^{-3}$. (a) Open circuit voltage (V_{OC}), (b) short-circuit current (J_{SC}), (c) fill factor (FF), and (d) efficiency (η).

Material/layer	Optimized properties considering each material			
	Crystallographic structure	Electrical susceptibility	Dielectric constant	Thickness
ZnS	**Zinc Blende**	**4.3**	**8.3**	**5 nm**
		Intrinsic doping level		
Cu (In, Ga) Se_2		**$10^{15} cm^{-3}$**		
		Assumption on SDL		
Surface defect layer (SDL)	Modeling of SDL properties and splitting it in two sub-interfaces: ZnS/SDL and SDL/CIGSe with different properties			

Table 4.
Optimized value of some layers/interface properties used for simulation.

Absorber layer thickness (nm)	V_{OC} (V)	JSC (mA/cm^{-2})	FF	η
500	0.7180	36.31	82.16%	21.42%
1000	0.7633	38.18	83.39%	24.31%
2500	0.7989	38.66	85.15%	26.30%

Table 5.
Output parameters of our photovoltaic device.

as external temperature, incident light power, and defect states density input values.

Running several calculations allowed us to detect the critical values of some properties values beyond which the device is no more stable, even if the efficiency is higher. This is for example 2750 nm for absorber layer thickness or an intrinsic doping level of more than 10^{15}. Since we manufacture our solar cell device based on the model proposed in this work and watching out not to have these highlighted critical values, our device will definitely work in good condition. The **Table 5** above reports the best performance of our cell obtained during all our calculations.

5. Conclusion

This chapter focused on enhancing efficiencies of solar cell devices working on a CIGSe-based second – generation ultra-thin model, and using Zinc Sulfide (ZnS) as a window layer. Vital information is found when investigating the influence of layers and interfaces properties on output parameters of the device. The challenging part is not the use of the promising ZnS material itself, but it is to find through literature and recent works, the key values of the ZnS properties in a preferential crystallographic orientation, that allow to obtain better performances and also the good choice of materials that make up the other layers. Starting on that point, the following cell *(Ni/Al)/MgF2/ZnO: B/i-ZnO/ZnS/CuInGaSe$_2$/Mo/Substrate* has been modeled and simulations were ran from version 3.3 of the SCAPS-1D software. The benefits associated with the existence of the Surface Defects Layer (SDL) on the device stability have been highlighted. The Blende structure of Zinc Sulfide material (ZnS) forms a more stable lattice matching with CIGSe absorber layer chalcopyrite structure. That is why most of the key values of its intrinsic properties are obtained from that orientation, especially its band gap $E_{g\_ZnS} = 3.68eV$, its electrical susceptibility $\chi_e = 4.3eV$, its dielectric constant $\varepsilon_r = 8.3$ according to simulation results. After running numerous simulations, very promising performances are recorded, a conversion efficiency of 26.30% and a fill factor of 85.14%. Going further in research, some may obtain even more interesting results by directing the work towards implementation of additional manufacturing technologies, including the use of antireflective coatings and the texturization of the inner back layers.

Author details

Fridolin Tchangnwa Nya* and Guy Maurel Dzifack Kenfack
Department of Physics, Faculty of Science, University of Maroua, Maroua, Cameroon

*Address all correspondence to: nyafridolin@yahoo.fr

References

[1] Martin Green A., Keith Emery, Yoshihiro Hishikawa, Wilhelm Warta, Ewan Dunlop D., Dean Levi H. and Anita Ho-Baillie W. Y. Mint: Solar cell efficiency tables (Version 49). Progress in Photovoltaics and applications. 2020. DOI:10.1002/pip.2855.

[2] Mostefaoui M., Mazari H., Khelifi S., Bouraiou A., Dabou R. Mint: Simulation of high efficiency CIGS solar cells with SCAPS-1D software. Elsevier, Energy Procedia 74, 736–744. 2015. https://doi.org/10.1016/j.egypro.2015.07.809.

[3] Tchangnwa Nya F., Touogam Touolak B., Ejuh W. G., Ouédraogo S. and Njaka, J. M. Energy and Environment Focus Vol. 6, pp. 1-9. 2017.

[4] Powalla M., Cemernjak M., Eberhardt J., Kessler F., Kniese R., Mohring H.D., Dimmler B. Mint: Large-area CIGS modules: pilot line production and new developments. Sol. Energy Mater. Sol. 2006. Cells 90, 3158.

[5] Liao D., Rockett A. Mint: Cd doping at the CuInSe2/CdS heterojunction. J. Appl. Phys. 2003. 93, 9380.

[6] Li Z., Yu-Ming X., Chuan-Ming X., Qing H., Fang L.F., Chang-Jian L., Yun, S. Mint: Microstructural characterization of Cu-poor Cu (In,Ga) Se2 surface layer. Thin Solid Films. 2012. 520, 2873.

[7] Pogrebjak A.D., Muhammed A.K.M. Mint: Simulation study of effect operating temperature and layer thickness on thin film CIGS solar cell performance nano and electronic. Physics 3 (3), 2011, 2012. 051–058.

[8] Duchatelet Aurelien. Synthèse de couches minces de Cu (In,Ga)Se2 pour les cellules solaires par électro-dépôt d'oxydes mixtes de cuivre-indium-gallium, Molécules et Matière condensée. [thesis]. Université des Sciences et Technologies de Lille. France; 2013. http://doc.univ-lille1.fr

[9] W. Chen et al. Dans Proc. 19th IEEE Photovoltaic Specialist Conf. 1987.

[10] Eman N. M. Mint: Surface morphology and structural properties of ZnS and ZnS:Al thin films. International journal of Innovative Research in Sciences, Engineering and Technology, Vol. 3, ISSN: 2319-8753. Issued 1, January 2014.

[11] Won Song, Bo-Ra Koo, Seok Eui Choi, Yong-Taeg Oh, and Dong-Chan Shin. Mint: Optical and structural properties of a ZnS buffer layer fabricated with deposition temperature of RF magnetron sputtering system. International Journal of Chemical, Molecular, Nuclear, Materials and Metallurgical Engineering Vol:6, No: 12. 2012.

[12] Wang Cai-Feng, LI Qing-Shan, LV Lei, ZHANG Li-Chun, QI Hong-Xia, CHEN Hou. Mint: Structural, optical and electrical properties of ZnS/porous silicon heterostructures. CHIN.PHYS. LETT, Vol. 24, No. 3 (2007) 825.

[13] Benghabrit S. Elaboration et caractérisation de couches minces CdS par bain chimique CBD pour application photovoltaïque. [thesis]. Université des Sciences et de la Technologie d'Oran. 2015.

[14] Rayan, D.A., Elseman, A.M. & Rashad, M.M. Mint: Remarkable impact of Ni2+ ion on the structural, optical, and magnetic properties of hexagonal wurtzite ZnS nanopowders. Appl. Phys. A 124, 659 (2018). https://doi.org/10.1007/s00339-018-2084-5

[15] Appl. Phys. A 117, 877–890 (2014).

[16] ACS Sustainable Chem. Eng. 2016, 4, 9, 4875–4886.

[17] Pawan Kumar, Aravind Kumar, Dixit P.N. and Sharma T.P. Mint: Optical, structural and electrical properties of ZnS thin film. Indian Journal of pure & applied physics, Vol 44, pp.690-693. September 2006. IPC Code: H01L 27/00.

[18] Zenebe Assefa Tsegaye. Density functional theory studies of electronic and optical properties of ZnS alloyed with Mn and Cr, NTNU-Throndheim, Condensed Matter Physics, Norwegian University of Sciences and Technology. 2012.

[19] Habibah, B. L'effet de la température du substrat et de la molarité sur les propriétés des couches minces de sulfure de zinc déposées par spray ultrasonique. Université Mohamed Kheider-Biskra. 2013.

[20] Platzer-Bjorkman C. Band Alignment Between ZnO-Based and Cu (In,Ga)Se2 Thin Films for High Efficiency Solar Cells. Acta Universitatis Upsaliensis. Digital Comprehensive Summaries of Uppsala Dissertations from the faculty of Sciences and technology 136. 80 pp. Uppsala. 2006. ISBN 91-554-6435-1.

[21] Charles Roger. Développement de cellules photovoltaïques à base de CIGS sur substrats métalliques. HAL archives-ouvertes.fr. [thesis]. Université de Grenoble Alpes. Français. NNT: 2013GRENI050. Tel-00965592. 2013. https://tel.archives-ouvertes.fr/tel-00965592

[22] Kassap S., Capper P. Mint: Springer Handbook of Electronic and Photonic Material. 2007. Url: http://www.springer.com/978-0-387-26059-4.

[23] Daouda O., Bawindsome Kebre M., Zougmoré F., Njomo D., Ouattara F. Mint: Numerical simulation of Cu (In, Ga)Se2 solar cells performances. Jounal of Energy and Power Engineering 9 1047-1055. 2015. doi : 10.17265/1934-8975/2015.12.002.

[24] Ouédraogo S., Zougmoré F. and Njaka J.M. Mint: Numerical Analysis of Copper-Indium-Gallium-Diselenide-based Solar Cells by SCAPS – 1D. Indawi Publishing Corporation, International Journal of Photoenergy, Article ID 421076, 9 pages. 2013. http://dx.doi.org/10.1155/2013/421076

[25] Tchangnwa Nya F., Dzifack Kenfack G. M., Geh Ejuh W., Touogam Touolak B., Ndjaka J. M. Mint: Highlighting some layers properties in performances optimization of CIGSe based solar cells: Case of Cu (In, Ga) Se–ZnS. Journal of King Saud University – Science 31 (2019) 1404–1413 https://doi.org/10.1016/j.jksus.2018.03.026

[26] Touafek, N., Mahamdi, R. Excess defect at the CdS/CIGS interface solar cells. Chalcogenide Lett. 11 (11), 589–596. 2014.

[27] Farhan Rana. Semiconductor Optoelectronics, Chapter 2: Semiconductor Heterostructures. Cornell University. Accessed online March 04, 2017.

[28] Rüdiger Memming Wildbret. Semiconductor Electrochemistry, Second Edition, Edited by, 25462 Rellingen Germany. Published by WILEY-VCH, Germany. 2015.

[29] Wan Fucheng, Tang Fuli, Xue Hongtao, Lu Wenjiang, Feng Yudong, Rui Zhiyuan. Mint: Effects of defect states on the performance of CuInGaSe2 solar cells. J. Semiconductors 35 (2). 2014. DOI:10.1088/1674-4926/35/2/024011

[30] Burgelman M., Nollet P., Degrave S. Mint: Modelling polycrystalline semiconductor solar cells. Thin Solid Films. 2000. 361–362, 527–532.

[31] Al-Ani S.K., Ba-Yashoot A.Kh., Makadsi M.N., Ai-Sharbaty A.M. Photovoltaic properties of n-(ZnS) x (CdTe)1-x/p-Si. 2006.

[32] Ramli H., Kamal S., Abd Rahim T., Muhaimin M. Optimization of Zinc Sulfide (ZnS) electron affinity in Copper Indium Sulfide (CIS) based photovoltaic cell. Chalcogenide Lett. 10 (6), 189–195. 2013

Section 2

Crystalline Si Cells

Chapter 7

Optical Study of Porous Silicon Layers Produced Electrochemically for Photovoltaic Application

Rahmouni Salah

Abstract

In previous years, porous silicon is rapidly attracting increasing interest in various fields and has received a great deal of attention from researchers because of its potential use in a variety of industrial applications such as photovoltaic device applications. The present study conclusively suggested that in order to prepare porous silicon samples, we need to determine the optimal conditions that lead to the increase of the optical efficiency. Porous silicon layers were elaborated by the electrochemical etching method using doped *p*-type ⟨100⟩-oriented silicon substrate. The photoluminescence (PL) and the spectroscopic ellipsometry (SE) measurements were used to calculate the physical and optical parameters (porosity, thickness) (refractive index and extinction coefficient). This study can give a very important interest in the photovoltaic field.

Keywords: porous silicon, antireflective coating, electrochemical anodization, photoluminescence, spectroscopic ellipsometry

1. Introduction

Porous silicon has undergone many developments; it has a very wide field of application and has received a great deal of attention from researchers because of its potential use in a variety of industrial applications such as photovoltaic device applications [1–4], chemical and gas sensors [5–12], biosensors [13, 14], biomedical applications [15], micromachining [16–18], templates for micro- and nano-fabrication [19–21], and solar cells and photoluminescence [1, 22, 23]. However, a very limited data of optoelectronic uses in this field are available [24, 25]. It is reported that the photoluminescence of porous silicon (PS) has achieved a large-scale investigation, giving an explanation of the photoluminescence phenomenon with obtaining the optical properties of porous silicon, as well as determining its refractive index and the gap energy, which can be determined directly by the absorption measurement, or by a non-destructive technique called spectroscopic ellipsometry (SE).

However, the physical and optical properties could be studied. In addition, some of the physical phenomena are still poorly understood because of the strong relationship between the PS nanostructure and the elaboration conditions (HF concentration,

IntechOpen

current density, and anodization time) [26]. The present study conclusively suggested that in order to prepare porous silicon samples, we need to determine the optimal conditions that lead to increase the optical efficiency. Herein, we need to study the correlation between the results extracted from the PL analysis and those obtained by ellipsometry. The study of the evolution of the intensities of the emission spectra obtained by the measurement of PL as a function of the porosity and the thickness determined by the ellipsometry of the layers for silicon substrates oriented P-100 of low resistivity is made to precisely clarify the evolution of optical parameters.

2. Photoluminescence study

The PL measurements were carried out by a solid laser 447 nm and detected through a Jobin Yvon 250-mm HR mono-chromator, with a GaAs photomultiplier associated to standard lock-in technique. Of note, the laser power of 7.66 mW was applied on the surface of the sample.

The optical characterization used in our work is based on "the radiation-matter interaction"; it tells us about the optical properties of the material. Photoluminescence spectroscopy is widely used to study the electronic structure of materials and the processes of radiative recombinations. It is a nondestructive optical characterization technique for nanomaterials.

In this part, using photoluminescence (PL) spectroscopy, we will determine the optical properties of porous silicon samples produced by electrochemical anodization. First, we analyze the PL spectra of SiP at room temperature. Second, we study the variation of the intensity, and the integrated intensity of PL in the temperature range [10–300 K].

2.1 Experimental details

Serial of porous silicon samples were prepared of *p*-type (100)-oriented silicon substrate with a resistivity of 0.001–0.02 Ω cm.

The experimental set-up consists first of all of a pulsed laser, the emission wavelength of which can be changed. The laser we have allows us to take measurements with an excitation of 447 nm. The laser beam is then conveyed to the sample by sets of mirrors. The sample is on a sample holder that can be placed in a cryostat (for low temperature studies) (**Figure 1**).

Figure 1.
Experimental device used for the measurement of photoluminescence.

2.2 Photoluminescence of porous silicon layers

2.2.1 Evolution of photoluminescence spectra of P-type porous silicon prepared at different etching time

All of the PS samples showed a visible PL at a room temperature. **Figure 2** illustrates the PL spectra of samples of p-type elaborated at different etching times (60, 120, 180, and 240 s) as well as an etching current density of 15 mA/cm^2.

Figure 2 shows that the observed large band, that ranges from 550 to 780 nm, decreases in time with a peak of 668 nm, and a decreased full width at half maximum (FWHM) from 226 to 112 is obtained along with an increase of PL intensity from 0.07 to 21.06.

The width of this PL band is attributed to the wide size distribution of the silicon nanocrystallites which constitute the porous layer. The intense spectrum is characterized by a maximum at the energy of 1.86 eV and a width at mid-height of the order of 112 meV.

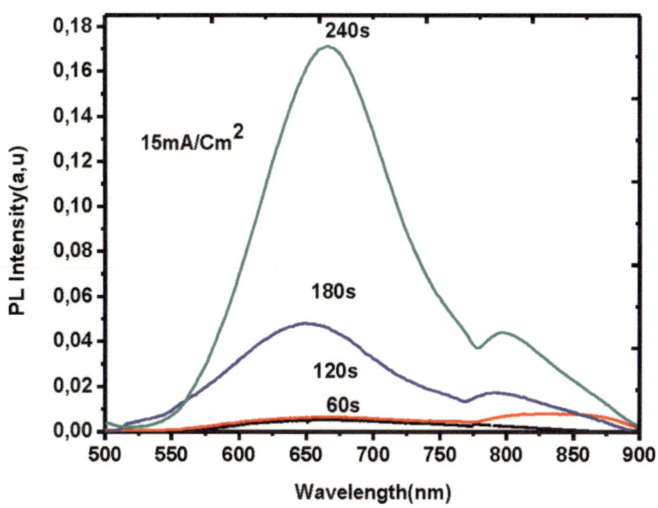

Figure 2.
Evolution of photoluminescence spectra of P-type porous silicon prepared at different etching times.

In addition, we noticed a second less intense band compared to the first. This may be due to an oxide layer located in a band between 780 and 900 nm (**Table 1**).

Samples	Etching time (s)	λPic (nm)	FWHM (meV)	IPLMax (u,a)	Eg (eV)
S1	60	628	226	0.07	1.96
S2	120	666	189	0.89	1.86
S3	180	652	122	6.81	1.91
S4	240	668	112	21.06	1.86

Table 1.
Optimized fitting parameters corresponding to the theoretical curves of porous silicon samples prepared at different etching time and current density of 15 mA/cm^2.

Figure 3 shows the variation of the PL intensity and the width at half height (FWHM) as a function of the anodization time.

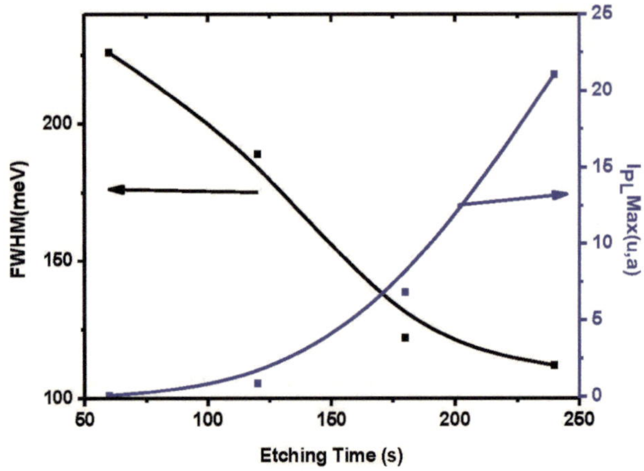

Figure 3.
The variation of the intensity, PL, and the width at half height (FWHM) as a function of the anodization time.

From **Figure 3**, we notice that the width at half height of the PL spectra decreases as a function of the anodization time. As known the thickness of the porous layer increases as a function of the anodization time, then deduces that the width at half height decreases, and the intensity of PL increases as a function of the thickness of the porous layer. This is due to the decrease in the sizes of nano-crystallites. However, note that the intensity of PL increases as a function of the anodization time.

2.2.2 Evolution of photoluminescence spectra as a function of the anodization current density

Figure 4 shows the PL spectra of p-type samples that were obtained from different etching current densities, 5, 10, 15 and 20 mA/cm^2, and etching time of 180 s. The presented serial in **Figure 1** indicates near similar variations of the PL

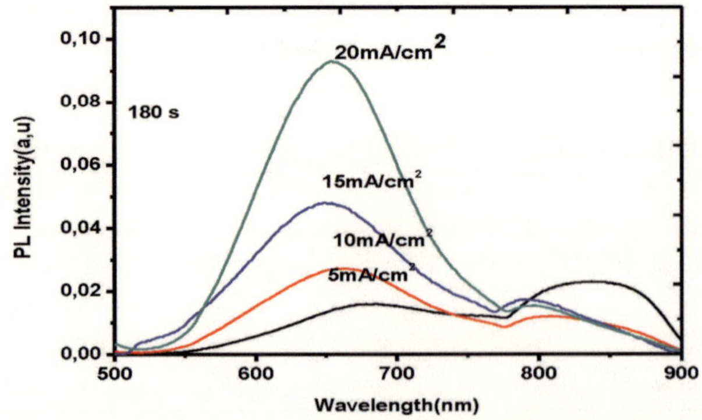

Figure 4.
Evolution of photoluminescence spectra of P-type porous layers prepared at different etching current density.

intensity. However, a remarkable increase of the PL intensity ranges from 2.59 to 11.72 was noticed (**Table 2**).

The two spectra show an improvement in the PL intensity as a function of etching current density and etching time. On the other hand, we noticed that the PL intensity reached the maximum value of JCM, that equals 20 mA/cm^2, with tm = 240 s. The slight blue-shift energy of PL band allow us to attribute a confinement of more and more wells and wire that leads to diminution of nanocrystallite size [27, 28].

The porous layer obtained from p-type silicon wafer is presented as cylindrical and spherical crystallites [29, 30]. Therefore, PS is defined as a mixture of quantum wells (QWs) and quantum wire following different concentrations and sizes. In the case of the increased etching time, the thickness of porous silicon layer increases from 106.3 to 1027.2 nm (**Table 3**).

Figure 5 shows the variation of the PL intensity and the width at half height (FWHM) as a function of the anodization current density.

The width at half height (FWHM) decreases according to the current density; this is due to the increase in porosity, therefore the increase in the intensity of PL.

Figure 6 depicts that the thickness and the integral intensity of the PL increase as function of etching time, meanwhile the porosity and the integral intensity of PL increase as a function of the current density **Figure** 7. On the other hand, the PL intensity of the porous layer increases as a function of etching time and current density.

Samples	Current density (mA/Cm2)	λPeak (nm)	FWHM (meV)	IPLMax (u,a)	Eg (eV)
S5	5	682	249	2.59	1.81
S6	10	665	210	4.12	1.87
S7	15	652	122	6.81	1.91
S8	20	655	111	11.72	1.90

Table 2.
Optimized fitting parameters corresponding to the theoretical curves of porous silicon samples prepared at different etching current densities and etching time of 180 s.

Samples	Etching time (s)	Current density (mA/Cm^2)	Thickness (μm) (d)	Porosity (%) (P)	n	k
S1	60	15	0.1063	36.02	/	/
S2	120	15	0.7061	58.84	/	/
S3	180	15	0.9058	70.20	/	/
S4	240	15	1.0272	76.89	/	/
S5	180	5	0.0985	34.53	1.77	0.0035
S6	180	10	0.7221	61.45	1.43	0.0022
S7	180	15	0.9058	70.20	1.33	0.0018
S8	180	20	1.0311	78.23	1.22	0.0014

Table 3.
Different parameters evaluated from the optical model of samples obtained at different etching times and different current densities [31].

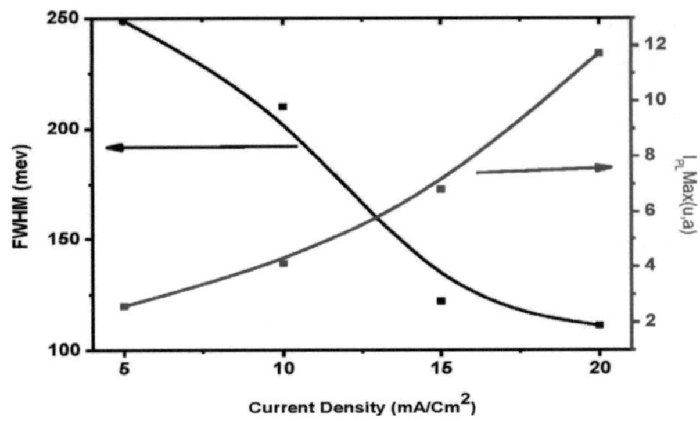

Figure 5.
The variation of the PL intensity and the width at half-height (FWHM) as a function of the anodization current density.

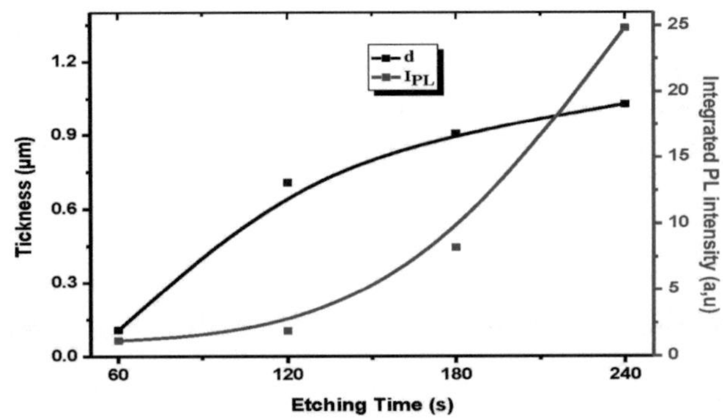

Figure 6.
Variation of thickness and integrated PL intensity according to etching time [31].

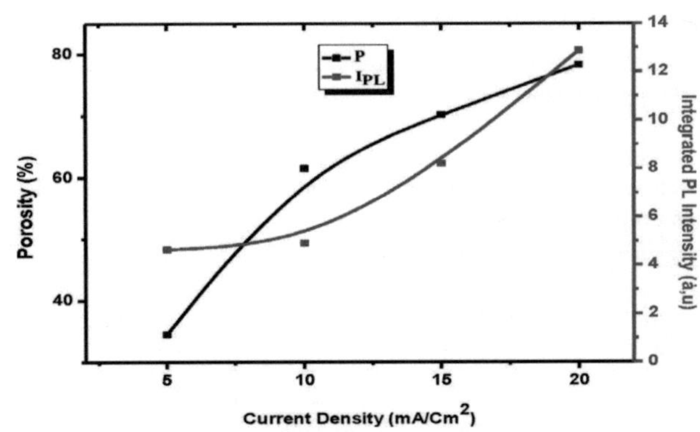

Figure 7.
Variation of porosity and integrated PL intensity according to etching current [31].

2.3 Study of low temperature photoluminescence

This part is devoted to the study of the evolution of the PL band as a function of temperature in order to identify the nature of the energy levels which are at the origin of this PL and to understand the mechanisms of radiative recombinations which participate in this PL. We have studied the variation of intensity and the integrated intensity all as a function of temperature in the range [10–300 K].

The sample studied is of type N (100), produced under these conditions: [HF] = 16%, t = 3 min and j = 20 mA/cm^2.

2.3.1 Evolution of PL intensities as a function of temperature

i. At low temperatures (T < 50 K), the decrease in IPL as a function of the increase in temperature is attributed to the transfer of excitons to states of lower energies in a non-radiative manner by thermal activation (**Figure 8**).

ii. While increasing the intensity of PL in the temperature range of (50 K ≤ T ≤ 80 K) indicates that the excitons are trapped in the lower localized states are thermally activated toward the higher states and then recombine radiatively and generate an increase in IPL intensity [32].

iii. At higher temperatures (T ≥ 80 K), thermal activation becomes more dominant and localized excitons become free and can diffuse in a non-radiative manner in the structure leading to a decrease in the intensity of PL [32].

iv. In the same figure, we notice the appearance of an intensity peak of PL at a characteristic temperature T_M = 80 K.

2.3.2 The intensity of PL as a function of the inverse of temperature

According to Zhao et al. and Weng et al., there are two types of transfer processes:

i. The first is a tunnel transfer [33, 34].

ii. The second is transfer by thermal dissociation [34] which is negligible at low temperatures.

But we have shown that the variation in the intensity of PL with temperature is due to the presence of these two exciton injection processes.

Ten et al. have shown that the increase in temperature could increase the tunneling process [35]. However, Hua et al. confirm that thermal dissociation of excitons increases with temperature, which favors the leakage of excitons from Q_{Ds} to Q_{Ws} at lower energy levels [36]. In **Figure 9**, we represent the integrated

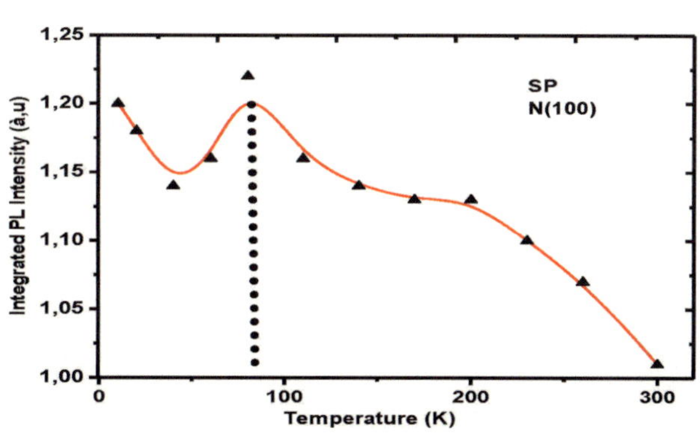

Figure 8.
Evolution of PL intensities as a function of the temperature of a SiP layer produced on an N-type substrate (100).

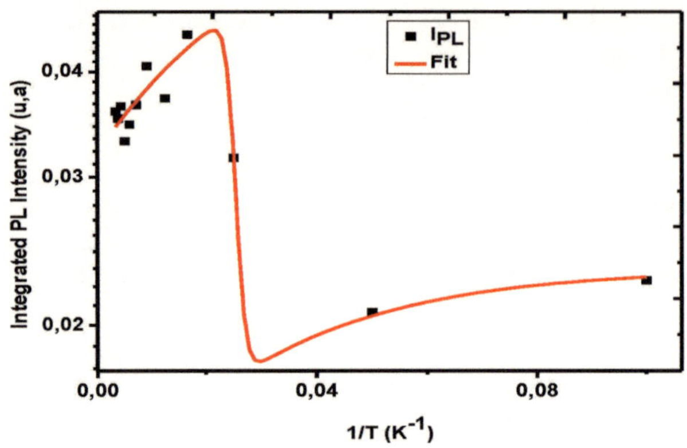

Figure 9.
Variation of the integrated intensity of PL as a function of the inverse of the temperature.

IPL	0.0232
a1	0.4046
e1	36.6044*8.33E-5 = 3.05 meV
A	1.7293
e2	1205.8422*8.33E-5 = 100.44 meV
a2	1.1545E13

Table 4.
The values obtained from the parameters of the equation.

intensity of PL as a function of the inverse of the temperature. Taking into account these two transfer processes, we fitted the experimental curve using the two-energy model, this empirical model is presented in Eq. (1) below [37]:

$$I_{PL}(T)=\frac{I_{PL}(0)}{\left[1+a_1\exp\left(-\frac{e_1}{kT}\right)\right]^2}\times\left(1+\frac{A}{\left[1+\frac{1}{a_2}\times\exp\left(\frac{e_2}{kT}\right)\right]}\right) \tag{1}$$

where e_1: first thermal activation energy; e_2: second thermal activation energy; A, a_1 and a_2: the fixing parameters; and I_{PL}: the intensity PL.

The different values found are presented in **Table 4**.

From the results, it can be seen that the thermal activation energy of one energy level is different from that of another level. This phenomenon can be explained by the increase in the Tunnel process with the increase in temperature [35]. We can also know the thermal activation energy of phonons in porous silicon.

3. Ellipsometric spectroscopy measurement

Spectroscopic ellipsometry is a very sensitive optical surface analysis method that allows the physical and morphological properties of a flat sample to be probed at different scales and at different energies. It has experienced significant growth over the past 100 years [38] and particularly over the past 20 years with modern computing. The technique makes it possible to obtain information on the surface of a massive sample, on the volume of a thin film or even on the interfaces. Ellipsometry has the advantage of being very simple and quick to implement, nondestructive, of allowing in-situ and real-time monitoring, and being applicable to a very wide range of samples.

3.1 Expérimental details

Ellipsometric measurements were performed on the porous silicon substrate for an angle of incidence of 78° in the spectral range 250–2000 nm using the GES5 Sopra made rotating polarizer spectroscopic ellipsometer was used for the (SE) spectroscopic ellipsometry measurement, controlled by the WinElli-II software.

To extract the thickness of the PS layer from measuring of SE, an optical model must be assumed, as well as the calculated data have to follow experimental spectra. In applying EMA software, we utilized a model of a multi-layer model, whereas the first mixture of the layer (void/SiO_2), while the second layer is (Sic/void), As indicated in **Figure 10**, a high improvement in the fit quality was observed [31].

3.2 Experimental device used

Figure 11 shows the "GES5" spectroscopic ellipsometer used in our work; this ellipsometer is assisted by a computer and controlled by the WinElli-II software.

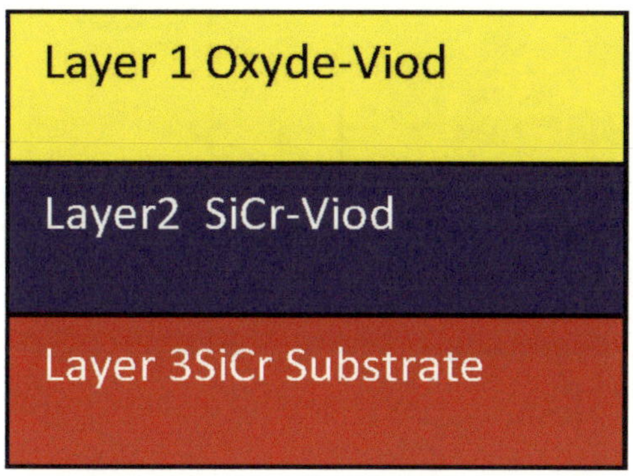

Figure 10.
The multilayer model of the PSL [31].

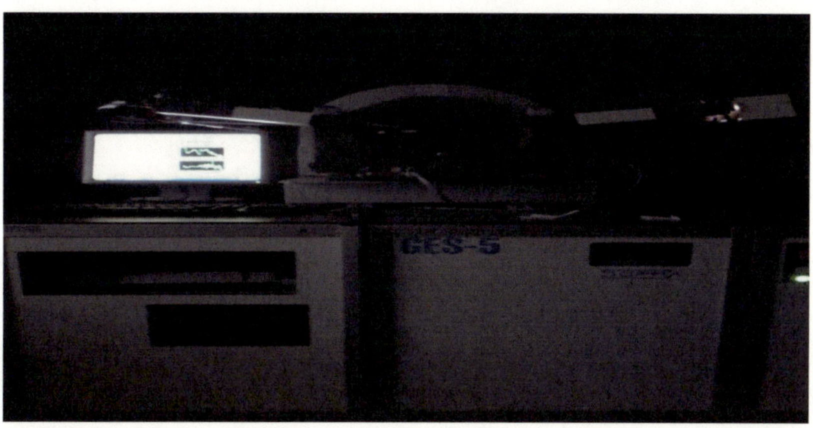

Figure 11.
Real photo of the GES 5 Sopra Ellipsometry used in our work.

This work was carried out at the techno pole photovoltaic laboratory of bordj çedria in Tunisia.

3.3 Thickness measurement

We measured the thicknesses of the series A samples (see **Table 1**).
Figures 12 and **13**, respectively, represent the ellipsometric spectra of samples S2 and S3.

The measure and the shape of the ellipsometrical spectra are presented in **Figures 12** and **13**.

The first serial allows us to determine ellipsometry. The adjustment parameters for the thickness of separated layers, using each sub-layer of the optical model are summarized in **Table 3**.

Figure 6 shows an agreement between the PL measurement presented by the integral intensity of PL along with the SE measurement presented by the thickness of porous layers obtained through varying etching time.

3.4 Porosity measurement

The second serial of samples was obtained following an etching time of 180 s and a current density varying from 5 to 20 mA/cm^2. **Figures 14** and **15** show the measurement and the shape of ellipsomety spectra.

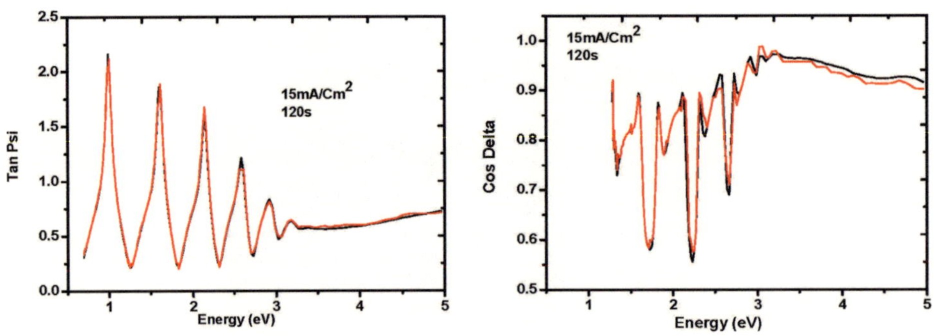

Figure 12.
SE measurements of the PSL obtained by using 15 mA/cm², 120 s, and the calculated spectra based on the best-fitted parameters [31].

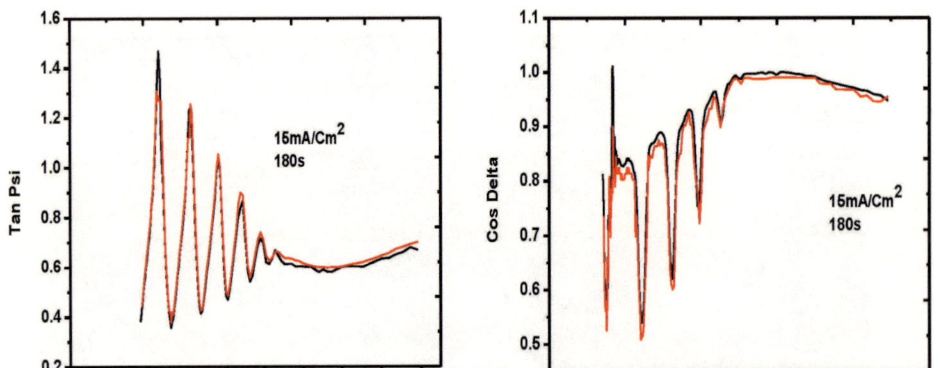

Figure 13.
SE measurements of the PSL obtained by using 15 mA/cm², 180 s, and the calculated spectra based on the best-fitted parameters [31].

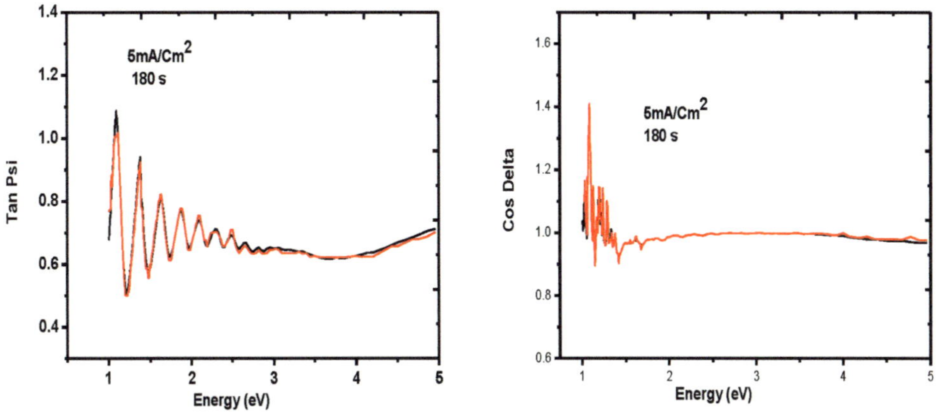

Figure 14.
SE measurements of the PSL obtained in using 5 mA/cm², 180 s, and the calculated spectra based on the best-fitted parameters [31].

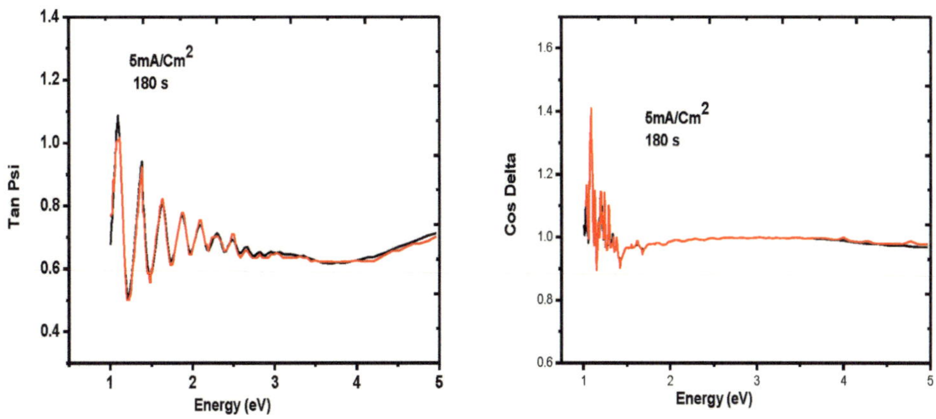

Figure 15.
SE measurements of the PSL obtained in using 10 mA/cm², 180 s, and the calculated spectra based on the best-fitted parameters [31].

The second serial allows us to determine the porosity of separated layers, using ellipsometrical method. The adjustment parameters for each sub-layer of an optical model are summarized in **Table 3**.

Figures 14 and **15**, respectively, show the measurement and shape of the spectra obtained by ellipsometry, the porous layers (S5, S6) obtained by J = 5 mA/cm2 and J = 10 mA/cm2, and an anodization time of 180 s.

Figure 7 shows an agreement between the PL measurement presented by the integral intensity of PL and SE measurements presented by the porosity of porous layers obtained following varying current density.

The two produced curves using values obtained by ellipsometry show that the thickness is as an increased function following etching time; meanwhile the porosity is as an increased function following the current density [39].

In this case, the PL comportment can be explained by the absence of the laser interference in the cleaned layer, as previously indicated. The contrary case was noticed in the thickness and the porous layer of silicon variations.

Adjusting the parameters of the proposed model to the experimental measurements taken on the sample, allowed us to obtain the values presented in **Table 3**.

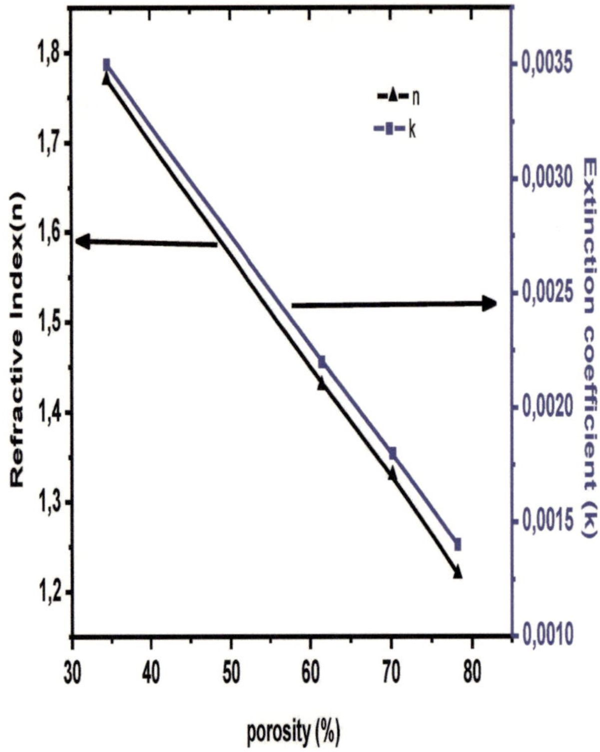

Figure 16.
Variation of refractive index and extinction coefficient as a function of porosity [31].

3.5 Calculation of the refractive index (n) and the extinction coefficient (k)

The results illustrated in **Table 3** and **Figure 16** show that the refractive index and the extinction coefficient are as a decreased function along with the porosity.

In case of a porosity of 34%, we get n = 1.77 and k = 0.0035, while for porosity of 78%, we get n = 1.22 and k = 0.0014 [31].

This result shows that the remained porous layer is more proper, but less thick and gives us a best PL intensity. Hence, the laser diffuses in wells by confinement effect. This confinement means to confine the incident laser radiation in crystallites seals and therefore the laser reflection is reduced following the big values of thickness and the porous layers.

4. Conclusion

In this study, we developed porous silicon layers by electrochemical anodization, the optical characterization made by spectroscopic ellipsometry (SE), and photoluminescence (PL); this characterization enabled us to calculate the physical and optical parameters (porosity, thickness; refractive index, extinction coefficient). To determine the effect of the etching parameters, we have suggested that for a good elaboration of the porous layers, it would be necessary to know the optimal conditions of anodization, taking into account the great role of the oxide layer on the surface. We note that the results obtained demonstrate that it excites a correlation between photoluminescence characterization (PL) and measurement of spectroscopic ellipsometry (SE), and these results are very important in the photovoltaic field.

Author details

Rahmouni Salah[1,2]

1 Higher School for Professor of Technological Education, ENSET, Skikda, Algeria

2 Laboratory of Physical Chemistry and Biology of Materials (LPCBM), ENSET, Skikda, Algeria

*Address all correspondence to: rahmouni.eln@gmail.com

References

[1] Ramirez A, Aziz WJ, Hassan Z, Omar K, Ibrahim K. Improved performance of solar cell based on porous silicon surfaces. Optik-International Journal for Light and Electron Optics. 2011;**122**(23):2075-2077

[2] Mohamed BA, Anouar H, Brahim B. Improvement of multicrystalline silicon solar cell performance via chemical vapor etching method-based porous silicon nanostructures. Solar Energy. 2012;**86**(5):1411-1415

[3] Sanchez de la Morena S, Recio-Sanchez G, Torres-Costa V, Martin-Palma RJ. Hybrid gold/porous silicon thin films for plasmonic solar cells. Scripta Materialia. 2014;**74**:33

[4] Atyaoui M, Dimassi W, Atyaoui A, Elyagoubi J, Ouertani R, Ezzaouia H. Improvement in photovoltaic properties of silicon solar cells with a doped porous silicon layer with rare earth (Ce, La) as antireflection coatings. Journal of Luminescence. 2013;**141**:1-5

[5] Harraz FA, Ismail AA, Bouzid H, Al-Sayari SA, Al-Hajry A, Al-Assiri MS. A capacitive chemical sensor based on porous silicon for detection of polar and non-polar organic solvents. Applied Surface Science. 2014;**307**:704-711

[6] Bardaoui A, Chtourou R, Amlouk M. Hand book, porous silicon multilayers: Synthesis and applications. Nova Science Publishers; 2012

[7] Salem MS, Sailor MJ, Harraz FA, Sakka T, Ogata YH. Electrochemical stabilization of porous silicon multilayers for sensing various chemical compounds. Journal of Applied Physics. 2006;**100**(8):083520

[8] Karthik TVK, Martinez L, Agarwal V. Porous silicon ZnO/SnO_2 structures for CO_2 detection. Journal of Alloys and Compounds. 2018;**731**:853

[9] Salem MS, Sailor MJ, Harraz FA, Sakka I, Ogata YH. Sensing of chemical vapor using a porous multilayer prepared from lightly doped silicon. Physica Status Solidi C. 2007;**4**(6):2073-2077

[10] Li M, Hu M, Zeng P, Ma S, Yan W, Qin Y. Effect of etching current density on microstructure and NH_3-sensing properties of porous silicon with intermediate-sized pores. Electrochimica Acta. 2013;**108**:167-174

[11] Dwivedi P, Das SD. Synthesized MoS_2/Porous Silicon Nanostructures for Efficient and Selective Ethanol Sensing at Room Temperature. ACS Applied Materials & Interfaces. 2017;**9**(24):21017-21024

[12] Harraz FA, Ismail AA, Bouzid H, Al-Sayari SA, Al-Hajry A, Al-Assiri MS. Mesoporous silicon layer as a highly sensitive ethanol sensor. International Journal of Electrochemical Science. 2014;**9**:2149-2157

[13] Myndrul V, Viter R, Savchuk M, Shpyrka N, Erts D, Jevdokimovs D, et al. Porous silicon based photoluminescence immunosensor for rapid and highly-sensitive detection of Ochratoxin A. Biosensors and Bioelectronics. 2018;**102**:661-667

[14] Harrez FA. Porous silicon chemical sensors and biosensors: A review. Sensors and Actuators B: Chemical. 2014;**202**:897-912

[15] Sarkar T, Basu D, Mukherjee N, Das J. Comparison of glucose sensitivity of nano and macro porous silicon. Materials Today - ScienceDirect. 2018;**5**(3) Part 3:9798

[16] Sun X, Parish G, Keating A. Micromachined microbeams made from porous silicon for dynamic and static mode sensing. Sensors and Actuators, A: Physical. 2018;**269**:91-98

[17] Wallace RA, Sepaniak MJ, Lavrik NV, Datskos PG. Evaluation of porous silicon oxide on silicon microcantilevers for sensitive detection of gaseous HF. Analytical Chemistry. 2017;**89**(11):6272-6276

[18] Lysenko V, Perichon S, Remaki B, Barbier D. Thermal isolation in microsystems with porous silicon. Sensors and Actuators A. 2002;**99**:13-24

[19] Jia C, He S, Song J, Jin Y, Zhu W, Cheng Q. Fabrication of micro/mesoporous silica tubes templated by electrospun cellulose acetate fibers. Cellulose Chemistry and Technology. 2017;**51**(7-8):693-701

[20] Bandarenka HV, Girel V, Zavatski SA, Panarin A, Terekhov SN. Progress in the development of SERS-active substrates based on metal-coated porous silicon. Materials. 2018;**11**(5):852

[21] Harraz FA. Impregnation of porous silicon with conducting polymers. Physica Status Solidi C. 2011;**8**(6):1883-1887

[22] Lenshin AS, Seredin PV, Kashkarov VM, Minakov DA. Origins of photoluminescence degradation in porous silicon under irradiation and the way of its elimination. Materials Science in Semiconductor Processing. 2017;**64**:71-76

[23] Omar K, Salman KA. Effects of electrochemical etching time on the performance of porous silicon solar cells on crystalline n-type (100) and (111). Journal of Nanoparticle Research. 2017;**46**:45-46

[24] Mazzoleni C, Pavesi L. Application to optical components of dielectric porous silicon multilayers. Applied Physics Letters. 1995;**67**(20):2983-2985

[25] Hadi HA, Ismail RA, Habubi NF. Optoelectronic properties of porous silicon heterojunction photodetector. Indian Journal of Physics. 2014;**88**(1):59-63

[26] Ji J-M, He X-X, Duan Q, Wang Z-X. Preparation of porous silicon substrate for protein microarray fabrication by double-cell electrochemical etching method. Chinese Journal of Analytical Chemistry. 2013;**41**(5):698-703

[27] Ee DTJ, Sheng CK. MIN Isa. photoluminescence of porous silicon prepared by chemical etching method. The Malaysian Journal of Analytical Sciences. 2011;**15**(2):227-231

[28] Gelloz B. Possible explanation of the contradictory results on the porous silicon photoluminescence evolution after low temperature treatments. Applied Surface Science. 1997;**108**(4):449-454

[29] Cullis AG, Canham LT, Dosser OD. The structure of porous silicon revealed by electron microscopy materials. Research Society Symposium Proceedings. 1992;**256**(7)

[30] Berbezier I, Halimaoui A. A microstructural study of porous silicon. Journal of Applied Physics. 1993;**74**(9):5421

[31] Rahmouni S, Zighed L, Chaguetmi S, Daoudi M, Khelifa M, Karyaoui M, et al. Correlation between photoluminescence and ellipsometric measurements of porous silicon layers. Optoelectronics and Advanced Materials. 2018;**12**(9-10):5535-5558

[32] Mabrouk A et al. Correlation between optical properties surface morphology of porous silicon electrodeposited by Fe^{3+} ion. Superlattices and Microstructures, Elsevier. 2015;**77**:219-231

[33] Weng GE, Zhang BP, Liang MM, Lv XQ, Zhang JY, Ying LY, et al. Optical properties and carrier dynamics in asymetric coupled InGasN multiple

quantum wells. Materials Letters. 2013;**6**:2

[34] Zhao D, Li B, Wu C, Lu Y, Shen D, Zhang J, et al. Temperature dependence of carrier transfer and exciton localization in ZnO/MgZnO heterostructure. Journal of Luminescence. 2006;**119-120**:304-308

[35] Ten S, Henneberger F, Rab M, Peyghambarian N. Femtosecond study of exciton tunneling in (Zn,Cd)Se/ZnSe asymmetric double quantum wells. Physical Review B. 1996;**53**:12637

[36] Hua J, Li-Gong Z, Zhu-hong Z, Anli-Nan, You-Ming L, Ji-Ying Z, et al. Temperature-Dependent Photoluminescence in Coupling Structures of CdSe Quantum Dots and a ZnCdSe Quantum Well. Chinese Physics Letters. 2005;**22**:1518-1521

[37] Saint-Girons G, Sagnes I. Photoluminescence quenching of a low-pressure metal-organic vapor-phase-epitaxy grown quantum dots array with bimodal inhomogeneous broadening. Journal of Applied Physics. 2002;**91**:10115

[38] Hall AC. A century of ellipsometry. Surface Science. 1969;**16**:1-13

[39] Rahmouni S, Zighed L, Tifouti I, Hadnine S, Aida MS. Experimental study of porous silicon films prepared on N and P type monocrystalline silicon wafers. Optoelectronics and Advanced Materials. 2017;**11**(1-2):105-108

Chapter 8

Influence of the Incidence Angle Modifier and Radiation as a Function of the Module Performance for Monocrystalline Textured Glass and No Textured in Outdoor Exposed

Issa Faye, Ababacar Ndiaye and Elkhadji Mamadou

Abstract

The variation of the incidence angle over the year is an important parameter determined the performance of the module. The standard orientation of the module or a PV system, the perpendicular positioning of the sun to the module's surface occurs twice a year. In outdoor exposed, angular losses of the module decrease the output of the PV or the system of PV. Although these losses are not always negligible, they are commonly not taken into account when correcting the electrical characteristics of the PV module or estimating the energy production of PV systems. This chapter is focused on the measurement of the angular response and spectral radiation (global and direct radiation) of solar cells based on two different silicon technologies, monocrystalline textured (m-Si) and non textured (mc-Si). The analysis of the source of deviation from the theoretical response, especially those due to the surface reflectance. As main contributions, the effects of glass encapsulation on the angular response of the modules are investigated by comparing the electrical parameter of the textured module to no textured and quantify electrical angular losses in this measurement area.

Keywords: textured, cell center, incidence angle

1. Introduction

The conversion of solar energy in solar modules is subject to electrical and optical losses [1, 2]. Optical losses are substantially depending on light incidence angle relative to the module plane. Manufactures information of photovoltaic panels typically provide electrical parameters at only one operating condition. Photovoltaic panels operate over a large range of conditions so the manufacturer information is not sufficient to determine overall performance. The electrical power output from a photovoltaic panel depends on the solar incidence irradiation, the cell temperature, the solar incidence angle [1, 3, 4]. To minimize reflection losses and thus maximize

IntechOpen

the electric yield, the PV industry introduced several different concepts and materials, such as antireflective coatings or structured glass with inverted pyramids [5]. To measure the nominal power, the incidence angle at normal incidence in standard test conditions is allowed. In real conditions, modules are exposed in different environments conditions. In some location, the main losses mechanism performance of PV or a system of PV is that the angle dependence. The PV module characteristics can help the company to predict accurately the PV performance. A new standard for performance testing and energy rating is under development [6]. The angle dependence losse of a PV module or a cell can be measured in several manner such the angle dependence reflectance. While reflectance measurements do not account for absorption losses, common I – V curve measurements can suffer from incorrect injection dependence and mismatch corrections.

Precise characterization methods and measurement systems are needed to assess angular dependent module performance. The incidence angle is measure of deviation from the direct solar radiation to the PV panel surface. The incidence angle is directly involved in the determination of the radiation incident angle affects the amount of solar radiation transmitted through the projective cover and converted to electricity by the cell [3, 7]. Significant effects of inclination occur at incidence angles greater than 65 degree [8]. The main of this paper is to evaluate and a comparative study of two crystalline silicone technologies in reel outdoor pyramidal textured and non-textured for different incidence angles and spectral radiation. The major difference between outdoor operation in natural sunlight and the laboratory test conditions is the existence of diffuse light, which is dependent on the climatic conditions on the location.

2. Incident angle distribution

The incident angle of the light on a PV module depends on tree parameter such as the module orientation, the time of the year and the geographical location [9]. However, due to cosine losses in those operation times with higher AOI, this only corresponds to 29% of the energy share in energy share in module plane [5]. To effectively reduce angular of incidence loss mechanisms and to utilize effects which boost the performance at lower angles of incidence it is crucial to effectively separate the different mechanisms which play a role when the AOI is varied. The **Figure 1**, shows the situation in which oblique incidence plays an important role for the performance.

Advances in solar glass production.

In the past few years, glass–glass module received a significant increased in attention, triggered by number of reasons.

Light trapping glass.

A major requirement for front cover glass in solar modules is high transmission in the wavelength range of the semiconductor material. One option to boost transmission is texturing the front surface in a similar manner to crystalline silicon solar cells.

Reduction of reflection at oblique light incidence.

As show in **Figure 2,** the light reflection on a mountain lake. In the front, the observer can see through the water surface: the reflection is low. With increasing angles, the reflection constantly increases and in the background only the mountains can be seen on the water surface.

Hence, the reflection is higher than the transmission at oblique light incidence. The reflection for equally (uniformly) polarized light can be described by the following equation.

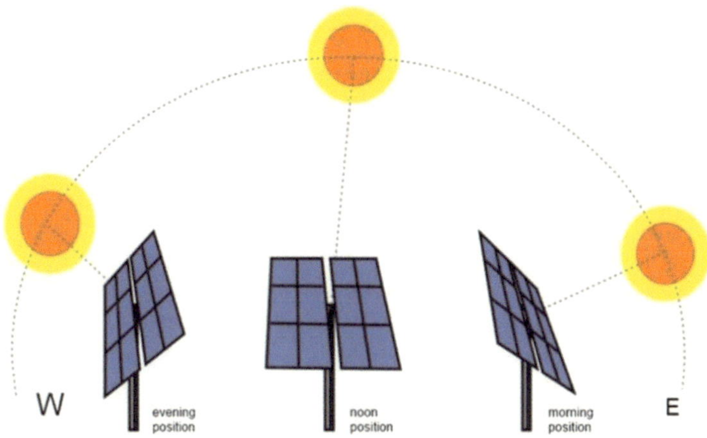

Figure 1.
Situations in which oblique light incidence plays an important role for the performance of the PV-module. The important part of light is normal to the module surface. This enormous part of light derive under diffuse light. Thus, understanding the oblique light performance of solar modules is crucial to understand its diffuse light performance.

Figure 2.
Light reflection as function of observed angle [10].

3. Methods of solar tracking

To track a maximum power point (MPPT) of a PV module, we always use a solar tracker. Different solar tracker are used as follows.

3.1 Passive trackers

The principle behind passive trackers is to make use of the solar heat to cause an imbalance, which leads to a movement in the tracker. They work on thermal expansion and commonly employ a low boiling point compressed gas fluid or shape memory alloys. For the high precision of the concentration of solar Power, the passive tracker are not always used. However, they can be employed for common flat PV systems. The passive trackers is more useful than the active tracker, however, the efficient in low temperature is less. The SMA actuator can easily be deformed even at relatively low temperatures (by tracker actuators below 70°C). It produces mechanical work by returning back to its original shape when heated above transformation temperature. The study found that the tracker worked very well in the short term field tests and the SMA actuators provided an efficiency of approximately 2% [11].

3.2 Active trackers

The mechanism of a active solar tracker is to use a motor to enable control the mobility of the tracker. These motors are usually fed by a (**Figure 3**).

The motor executed the command from a signal which have a main aims to provide magnitude and direction and incidence angle. The incidence angle is enable to change from 0° to 90°. We highlight that the active tractors made use performed accurately however, they consume more energy. They are more efficiently than the passive trackers.

3.3 Active trackers with single-axis system

To compare to a two axis systems that provide two degrees freedom, a single axis provide for only bone. Hence, a single axis consumed less energy than a two axis system. The 1A-3P sun tracker was designed to operate at only 3 different angles as shown in **Figure 4**. A simple designed tracker is combined with a DC motor to turn the system The tracker rotation is enable by a timer IC which provides the time signal to trigger the motor to turn at the turning angle. The measuring functions for tracker motion, PV generation and all the control algorithms are implemented using microcontroller PIC18F452 [12] A standalone single axis active solar tracker and presented the modeling and simulation of the photovoltaic system under a constant load using MATLAB/Simulink was designed by [13]. The different components

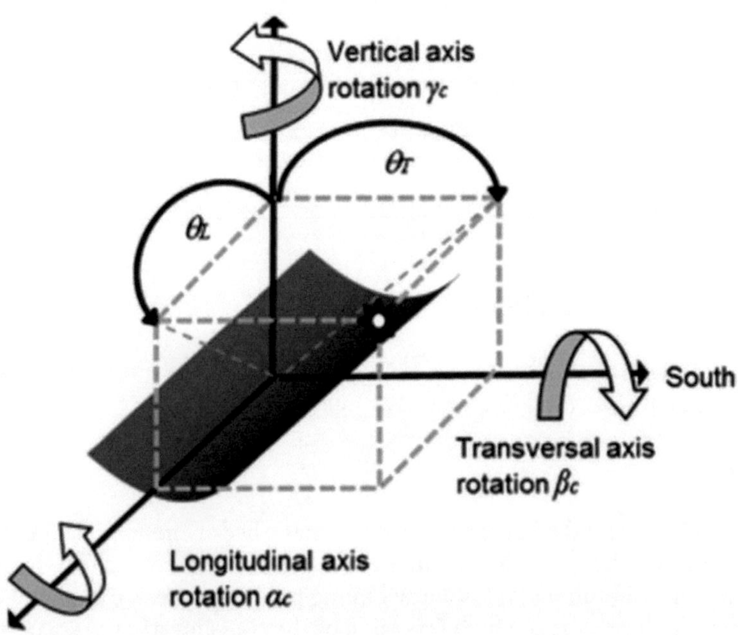

Figure 3.
Scheme of the rotation angles [11].

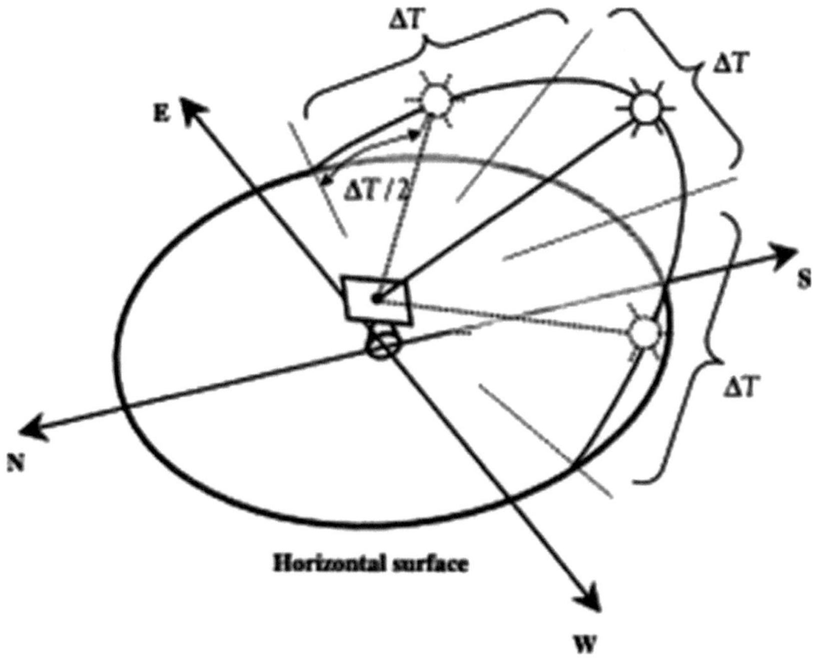

Figure 4.
Schematic diagram of azimuth three step tracking [11].

such as a servermotor, a batetery, a charger, two LDR sensors and external load and microcontroller provide the output. The aims of the tracker is to rotate in a single axis, then it's designer take into accunt of the number of axis rotation and activate the motor to had a single-axis freedom. The sunlight intensity was sensed using the LDR sensors, which would then send a signal to the microcontroller to rotate the panel using the servo motor. To control the system, a Lead Acid battery polarize the components which is controlling via a charge controller. Konar et al. designed a single axis microcontroller based automatic, position control scheme. The system is controlled by two axis, the first one is a reflector that tilted optimally across it; the second one is controlled by the tracker by changing the azimuth angle. The tracking system was designed to search for the maximum solar irradiance in the whole azimuth angle of 360° during the locking cycle, and hence the system was not constrained by the geographical location of installation. The system also employed step tracking scheme instead of continuous tracking which keeps the motors idle for a longer time to save energy [11] reported a single-axis solar tracker on a small size Parabolic-Trough Collector (PTC). To locate the azimuth of the sun, a algorithm that classified as an an open-loop is elaborate. The angular tracking error was accurately characterized using a digital inclinometer. The transversal Incidence Angle Modifier.

(IAM) curve is determined by ray tracking simulation for all longitudinal incidence angles as well as the transversal incidence plane is showing in **Figure 4**. The proposed procedure gives a better accuracy for the tracking error than the theoretical acceptance angle [14]. derived formulae to evaluate the daily and hourly radiation incident on azimuth three step tracking system, hour angle three step tracking systems and compared the results with the radiation received by a horizontal surface. A yield tilted surface performed an a optimal angle gain 30,2% higher radiation than a horizontal surface. In comparison, a two axis azimuthal three step tracking performed better with a 72% higher radiation [15]. As shown in **Figure 5** a global illustration of the schematic digram. A theoretical study to analyze the performance

Figure 5.
Solar tracking system at the Cologne University of Applied Sciences.

of an east–west oriented single-axis tracking panel was done by [16] claimed the performance of a single axis tracking, the system consist of a panel fixed in a horizontal axis, in different time of the year. They compared the gain in an est-west to north-sud. The results show that the yearly gains obtained in est-west is less than the north-sud. The results show that, the correlation between the irradiation and the latitude is hight, it's 12% at the equator and 14,3% in the Arctic. To increase the efficiencies of the PV plant, a ray tracking is performed; besides, the change of the angle may have an impact of the output of the system. The mathematical formula can accurately investigate the optical performance, As make a use by [11] with a single axis in South–North, they estimate the annual radiation of the panels composed the system when using a single axis. Some of study haves been done in china, they compare single axis and dual axis and conclude that the dual axis performed 96–97% higher than the single axis. The results illustrate that in some areas, at low resources, the tracking is unsuitable. The sun tracking may perform better than the traditional fixed if both are compared over the year, the gain performance were estimated to higher than 30% with high solar irradiation, however it is less than 20% in low irradiation areas [17–19].

4. Experimental work

The outdoor system are developed and investigated. It's installed on the roof of the Technical university of Cologne. The angle of incidence starting position with orientation to the sun gradually changing the angle of inclination at a constant azimuth angle up to an angle of incidence of 90 selection of the distances between the approached angles of incidence helicopter measurement horizontal module storage (inclination = 90°) gradual change in solar tracker azimuth orientation over a total angle of 180 choice of the distance between the approached angles. The angular incidence effect is measured with the short circuit current which is assumed to be proportional to the light reaching the solar cell and thus the photon generation.

4.1 Isotextured solar cell

The incident angle modifier is varied for 0° to 90°. The measurement of the short circuit current of the two modules is done when the incidence angle is changing. A textured surface retains more light than a plane surface in a active area of a PV or cell. However, the advantage of the rough surface can be offset by the mechanism of the recombination at the surface of the PV module or cell textured. At the comparably low angles of incidence, the reflection is reduced to 4% [20]. A textured such as the Alberino glass will drastically lower the reflection for angles of incidence ≥50° and thus will keep the incidence angle midifier closer to unity. The texture of Alberino P is the best described as inverted and rounded pyramid. The **Table 1** shows the electrical characteristics of the Two PV plants in monocrystalline silicone.

Within the framework of this work a solar tracking device was constructed at the Cologne University of Applied Sciences (CUAS) which is capable of measuring the IAM of solar cells and whole modules, see **Figure 5**.

4.2 Global and direct irradiation

Global solar radiation on horizontal surfaces can be measured with a pyrometer or reference cell which is an instrument that measures global solar radiation from all directions. To calculatee the global solar radiation we use the formulate as follows.

Diffuse solar radiation (G_b).

Direct beam solar radiation (G_d).

Solar radiation on a horizontal surface is the sum of the horizontal direct and diffuse radiation.

$$G_H = G_d + G_b$$

The solar module is irradiated by permanently changing solar spectrum. To nevertheless be able to compare the performance of different solar cells technology, a standard spectrum with a relative air mass (AM) of 1.5 was defined:

$$\text{Irradiance} = \frac{Avarage\ value\left[W/m\text{^}2\right]*115{,}3\,mV}{1000\left[W/m\text{^}2\right]} \tag{1}$$

Power (Pmax)	240
Open circuit voltage (Voc)	36,72
Short circuit current (Isc)	8,74
Voltage at maximum power (Vmp)	28,89
Current at maximum power (Imp)	8,31
Permissible system voltage	1000VDC
Maximum reverse current	
Application class	A

Table 1.
Electrical characteristics of both PV modules.

$$Isc,stc = Isc,mes + Isc0 * \left(\frac{G0}{G} - 1\right) + \alpha * (T2 - T1) \quad (2)$$

Where, $I_{sc,stc}$ represents the measured current which is standardized, I_{scmes} is the measured current and I_{sc0} is the short circuit current measured.

5. Results and discussion

The results show that, (see **Figure 6**) and **Figure 7**. For direct irradiance, reflectance losses will also be higher than for orthogonale incidence most of time, if stationary modules are considered. There is a decrease short circuit current I_{sc} in direct irradiation from 6.95 to 2.84 for monotextured and from 7.93 to 3.12 in the incident angle from 0° to 60°. This is due to the effect of variation in the minority carrier concentration as said by [21].

The textured module shows a good performance at low irradiance there is still a high share of direct light with oblique incidence angles. So the performance of the textured module is strongly affected by the factor $K_{\tau\alpha}$ as said by (20) Nils Reihners (2018). The short circuit current for the global irradiation decrease from 9.91 to 5.80 for the monotextured and from 11.09 to 6.68 for no textured when the incidence angle increase from 0° to 60°. On top of the cosine effect, additional losses occur, particularly at incidence angles beyong 60° as showed in **Figure 7**. They provide correction factors for device short circuit current Isc for incidence angle β from 0° to 90° with respect to normal incidence of 0°. The short circuit current dependence losse angle is called incidence angle modifier. It is defined as the ratio of the short circuit current measured at an angle of incidence and the short circuit current at perpendicular incidence. The value of corrected by cosines to eliminate the cosine effect and to keep all other angular effects

$$K_{\alpha}(\beta) = \frac{I_{sc}(\beta)}{I_{sc}(0°).\cos(\beta)}$$

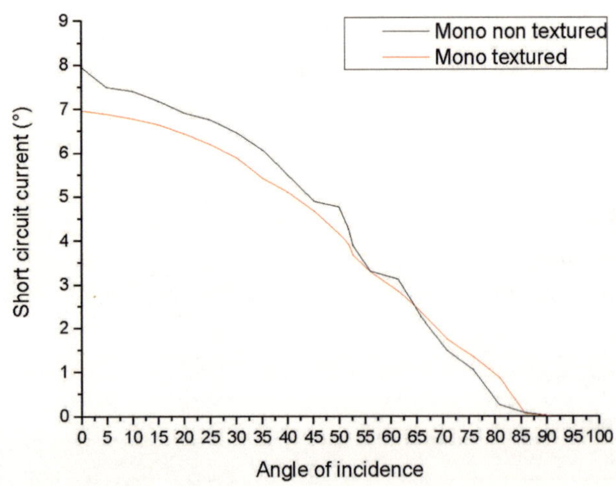

Figure 6.
Curve of direct irradiation in dependence on the angle incidence.

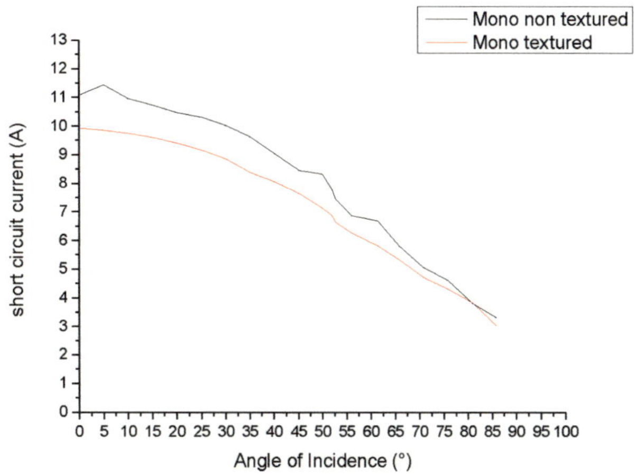

Figure 7.
Curve of global irradiation in dependence on the angle incidence.

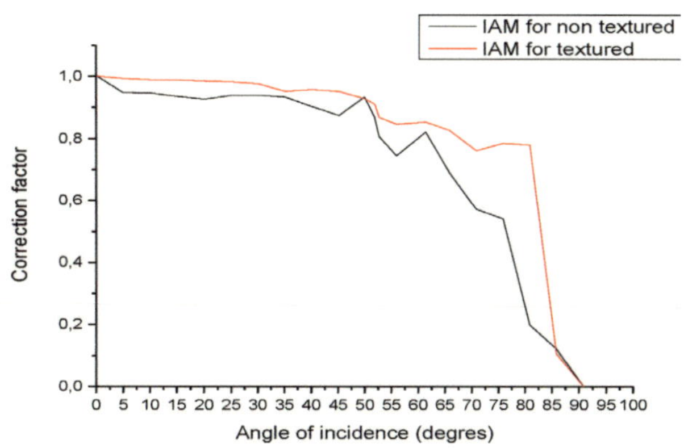

Figure 8.
IAM with direct radiance comparison between textured to no texture. Correction factor $K_{\alpha\tau}$ for flat glass and a pyramidal textured surface.

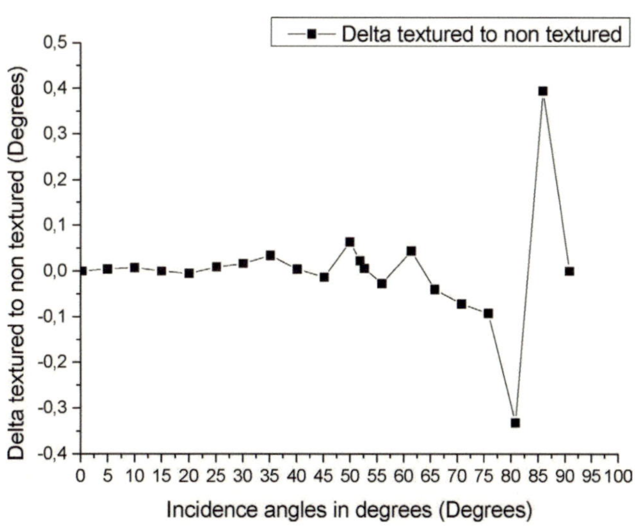

Figure 9.
IAM between delta textured to non-textured in global irradiation.

To include the influence of a changing solar spectrum, it is common to derive a correction factor, which depends on the air mass that the light needs to traverse before hitting the module.

The **Figure 8** shows the IAM comparison between textured and no textured module.

To better understand the IAM of the textured and non-textured module, we introduce the difference between the textured and no textured as showed in **Figure 9.**

6. Conclusion

Angle dependence can be a major loss mechanism in photovoltaic. Our approach enables the measurement the solar irradiation and the angle dependence for modules spectrally resolved. We evaluated the measurement uncertainty of our setup according to calibration standards and measured textured and no textured PV modules. This effort is needed to quantify in reel outdoor exposed the effects of cell textured and the incidence angle modifier. The results show that, the textured PV module performs better than the flat PV module in global irradiance and diffuse irradiance.

Author details

Issa Faye[1*], Ababacar Ndiaye[1] and Elkhadji Mamadou[2]

1 Assane Seck University, Ziguinchor, Sénégal

2 Keita Cheikh Anta Diop University, Dakar, Senegal

*Address all correspondence to: issafaye1211@gmail.com

References

[1] M. A. Green et S. P. Bremner, « Energy conversion approaches and materials for high-efficiency photovoltaics », *Nat. Mater.*, vol. 16, n° 1, p. 23, 2017.

[2] A. Pospischil, M. M. Furchi, et T. Mueller, « Solar-energy conversion and light emission in an atomic monolayer pn diode », *Nat. Nanotechnol.*, vol. 9, n° 4, p. 257-261, 2014.

[3] Ö. Ayvazoğluyüksel et Ü. B. Filik, « Estimation methods of global solar radiation, cell temperature and solar power forecasting: A review and case study in Eskişehir », *Renew. Sustain. Energy Rev.*, vol. 91, p. 639-653, 2018.

[4] J. J. Michael, S. M. Iqbal, S. Iniyan, et R. Goic, « Enhanced electrical performance in a solar photovoltaic module using V-trough concentrators », *Energy*, vol. 148, p. 605-613, 2018.

[5] M. Ebert, H. Stascheit, I. Hädrich, et U. Eitner, « The impact of angular dependent loss measurement on PV module energy yield prediction », *29th Eur. PV Sol. Energy Coference Exhib. Amst. Neth.*, 2014.

[6] R. P. Kenny, E. D. Dunlop, H. A. Ossenbrink, et H. Müllejans, « A practical method for the energy rating of c-Si photovoltaic modules based on standard tests », *Prog. Photovolt. Res. Appl.*, vol. 14, n° 2, p. 155-166, 2006.

[7] E. Lorenzo, « Energy collected and delivered by PV modules », *Handb. Photovolt. Sci. Eng. Chapitre*, vol. 22, p. 984-1042, 2011.

[8] W. De Soto, S. A. Klein, et W. A. Beckman, « Improvement and validation of a model for photovoltaic array performance », *Sol. Energy*, vol. 80, n° 1, p. 78-88, 2006.

[9] S. Dey, M. K. Lakshmanan, et B. Pesala, « Optimal solar tree design for increased flexibility in seasonal energy extraction », *Renew. Energy*, vol. 125, p. 1038-1048, 2018.

[10] U. Blieske et G. Stollwerck, « Glass and Other Encapsulation Materials », in *Semiconductors and Semimetals*, vol. 89, Elsevier, 2013, p. 199-258.

[11] V. Sumathi, R. Jayapragash, A. Bakshi, et P. K. Akella, « Solar tracking methods to maximize PV system output–A review of the methods adopted in recent decade », *Renew. Sustain. Energy Rev.*, vol. 74, p. 130-138, 2017.

[12] H. Mousazadeh, A. Keyhani, A. Javadi, H. Mobli, K. Abrinia, et A. Sharifi, « A review of principle and sun-tracking methods for maximizing solar systems output », *Renew. Sustain. Energy Rev.*, vol. 13, n° 8, p. 1800-1818, 2009.

[13] C. S. Chin, A. Babu, et W. McBride, « Design, modeling and testing of a standalone single axis active solar tracker using MATLAB/Simulink », *Renew. Energy*, vol. 36, n° 11, p. 3075-3090, 2011.

[14] F. Sallaberry, R. Pujol-Nadal, M. Larcher, et M. H. Rittmann-Frank, « Direct tracking error characterization on a single-axis solar tracker », *Energy Convers. Manag.*, vol. 105, p. 1281-1290, 2015.

[15] R. G. Vieira, F. Guerra, M. Vale, et M. M. Araújo, « Comparative performance analysis between static solar panels and single-axis tracking system on a hot climate region near to the equator », *Renew. Sustain. Energy Rev.*, vol. 64, p. 672-681, 2016.

[16] S. Abdallah et S. Nijmeh, « Two axes sun tracking system with PLC control », *Energy Convers. Manag.*, vol. 45, n° 11-12, p. 1931-1939, 2004.

[17] Z. Li, X. Liu, et R. Tang, « Optical performance of inclined south-north single-axis tracked solar panels », *Energy*, vol. 35, n° 6, p. 2511-2516, 2010.

[18] J. Jean, P. R. Brown, R. L. Jaffe, T. Buonassisi, et V. Bulović, « Pathways for solar photovoltaics », *Energy Environ. Sci.*, vol. 8, n° 4, p. 1200-1219, 2015.

[19] G. Makrides, B. Zinsser, M. Norton, G. E. Georghiou, M. Schubert, et J. H. Werner, « Potential of photovoltaic systems in countries with high solar irradiation », *Renew. Sustain. Energy Rev.*, vol. 14, n° 2, p. 754-762, 2010.

[20] G. Dennler *et al.*, « Angle dependence of external and internal quantum efficiencies in bulk-heterojunction organic solar cells », *J. Appl. Phys.*, vol. 102, n° 5, p. 054516, 2007.

[21] J.-A. Jiang, J.-C. Wang, K.-C. Kuo, Y.-L. Su, et J.-C. Shieh, « On evaluating the effects of the incident angle on the energy harvesting performance and MPP estimation of PV modules », *Int. J. Energy Res.*, vol. 38, n° 10, p. 1304-1317, 2014.

Chapter 9

Ultrasonic Processing of Si and SiGe for Photovoltaic Applications

Andriy Nadtochiy, Artem Podolian, Oleg Korotchenkov and Viktor Schlosser

Abstract

The usage of power ultrasound for sonochemical processing of Si wafers and thin layers of amorphous Si and SiGe alloys is described. Over the last decade different industries have become increasingly drawn to sonochemistry because it provides a green and clean alternative to conventional technologies, particular in the areas of processing of silicon-based materials for photovoltaic applications. Two techniques related to ultrasonic cleaning of Si wafers and sonochemical modification of Si, SiGe and a-Si/SiGe surfaces in hydrocarbon solutions of chloroform ($CHCl_3$) and dichloromethane (CH_2Cl_2) are discussed. The occurrence of cavitation and bubble implosion is an indispensable prerequisite for ultrasonic cleaning and surface processing as it is known today. The use of higher ultrasonic frequencies to expand the range of ultrasonic cleaning and processing capabilities is emphasized. Although exact mechanisms of an improved photoelectric behavior of Si-based structures subjected to power ultrasound are not yet clarified in many cases, the likely scenarios behind the observed photovoltaic performances of Si, SiGe and a-Si/SiGe surfaces are proposed to involve the surface chemistry of oxygen and hydrogen molecules as well hydrocarbon chains.

Keywords: silicon, germanium, surface passivation, dichloromethane, sonochemical, surface photovoltage, free carrier lifetimes

1. Introduction

The chemical effect of ultrasonic waves derives primarily from hot spots formed during acoustic cavitation in a chemical mixture. Due to locally achieved extreme conditions, an unusual chemical environment is attained in such experiments [1]. It is therefore not surprising that a growing interest in simple and cheap sonochemical syntheses of materials is observed, particularly in nanophases [2–4]. Given the fact that ultrasonic irradiation, or sonication, of reaction mixtures is easily controllable, sonochemical fabrication of high-quality materials becomes a particularly interesting subject. One of the goals of this article is to discuss the recent excitement about sonochemically modified semiconductor materials.

Quite recently, sonochemical techniques have been used in processing of semiconductor surfaces [5–14]. In this method, the main phenomenon is the acoustic cavitation, which enhances chemical reactions in a solution. The growth of cavitation bubbles occurs due to the diffusion of solute vapor in the volume of the bubbles. After the growth process, the bubbles collapse, breaking the chemical bonds on the material surface.

Over the last years different industries have therefore become increasingly drawn to sonochemistry because it provides a green and clean alternative to conventional technologies, particular in the areas of processing of silicon-based materials. The aim of this work is to provide a cohesive presentation of the related efforts. Two techniques related to ultrasonic cleaning of Si wafers and sonochemical modification of Si and SiGe surfaces in hydrocarbon solutions will be discussed. In both cases, the occurrence of cavitation and bubble implosion is necessary for ultrasonic cleaning and surface processing as it is known today. The use of higher ultrasonic frequencies to expand the range of ultrasonic cleaning and processing capabilities will be emphasized. Although exact mechanisms of an improved photoelectric behavior of Si-based micro- and nanostructures subjected to power ultrasound are not yet clarified in many cases, the likely scenarios behind the observed photovoltaic performances will be proposed to involve the surface chemistry of oxygen and hydrogen molecules as well hydrocarbon chains.

2. Processing and passivation of Si wafers used for solar cells

It is known that crystalline silicon (c-Si) based solar cells dominate the solar energy industry. The modern silicon wafer production technology and processing sequence is the most mature and including hundreds of flawless process steps [15]. With increasing the electronic quality of c-Si, the efficiency of c-Si solar cells improved considerably [16]. Most generally, free carriers generated by the incident sunlight would be efficiently collected far away from the fast recombination centers while they move towards the device terminals. In particular, carrier recombination at the cell surfaces should be avoided, especially in thin wafers.

It is convenient to define an effective surface recombination velocity (S_{eff}), which for a symmetrically passivated sample at a low injection level takes the form [17]:

$$S_{eff} = \sqrt{D\left(\frac{1}{\tau_{eff}} - \frac{1}{\tau_b}\right)} \tan\left(\frac{w}{2}\sqrt{\frac{1}{D}\left(\frac{1}{\tau_{eff}} - \frac{1}{\tau_b}\right)}\right), \tag{1}$$

where D is the minority carrier diffusivity, τ_{eff} is the effective lifetime, τ_b is the bulk lifetime of the wafer and w is the wafer thickness. If S_{eff} is rather large, one equates [17]:

$$\frac{1}{\tau_{eff}} = \frac{1}{\tau_b} + D\left(\frac{\pi}{w}\right)^2. \tag{2}$$

If the surface passivation is good and hence the surface recombination velocity is sufficiently small, the tangent term in Eq. (1) becomes linear and [17]:

$$\frac{1}{\tau_{eff}} = \frac{1}{\tau_b} + \frac{2S_{eff}}{w}. \tag{3}$$

Therefore, carrier recombination at the wafer surfaces restricts its effective lifetime thus posing inherent limitation of using c-Si in solar cells.

The surface itself terminates an atomic order in c-Si such that Si atoms that reside on the surface are not fully bonded to four Si neighbors. This yields dangling bonds, which form surface defects and thus reduce the efficiency of solar cells. Therefore, reducing the number of these defects is clearly a necessary prerequisite for manufacturing higher efficiency silicon solar cells.

In this respect, various surface-passivation layers have been employed. In particular, SiO_2 surface passivation coating layer has been proven to offer outstanding passivation [18–20]. In general, two oxide growth methods are employed termed as the dry and wet oxidation. Dry oxygen and water vapor are used in the former and latter cases, respectively. Dry oxidation typically forms thin oxide layers in practical structures due to perfect Si–SiO_2 interface developed in this case. In turn, wet oxidation yields greater growth rates, which is necessary for depositing thicker SiO_2 layers.

Furthermore, a hydrogenated amorphous silicon a-Si:H layer can saturate the dangling bonds by hydrogen termination [21–23]. The atomic hydrogen is also able to effectively passivate dangling bonds at the c-Si/SiO_2 interface thus drastically suppressing the interface state density and surface recombination velocities [24]. An important aspect is that the treatment in HF can produce an inversion layer on p-type Si surfaces while an accumulation of majority carriers is observed on n-type Si after treatment in either NH_4F or HF [25]. These both are due to positive charges induced by electronegative surface groups such as –H, –O–H and –F bonded to the Si surface atoms $\equiv Si_s-$ [26].

Most of the oxidizing solutions, e.g. H_2SO_4/H_2O_2, HCl, HNO_3, RCA-2 (the mixture of HCl, H_2O_2 and H_2O), lead to surface depletion of holes in p-Si and a weakly depleted majority carriers in n-Si that appear due to the positive fixed oxide charge. Surface processing in the RCA-1 solution, which contains the mixture of NH_4OH, H_2O_2 and H_2O, can be considered, in turn, noting a strong depletion in n-Si and a weak one on p-type Si surface thus implying the negative surface charge arisen from the dissociation of $\equiv$Si–OH groups in the SiO_2 film during the oxidation in the solution ($\equiv Si{-}OH \leftrightarrows \equiv Si{-}O^- + H^+$) [25, 26].

Using an amorphous silicon layer should also prove useful in Ge_xSi_{1-x}/Si structures [27]. It is seen in **Figure 1** that the surface photovoltage (SPV) is enhanced as the structure is covered with a-Si (curves 3 and 4 compared with curves 1 and 2 at time $t = 0$). Roughly a 10 times larger value of the SPV magnitude is observed in the

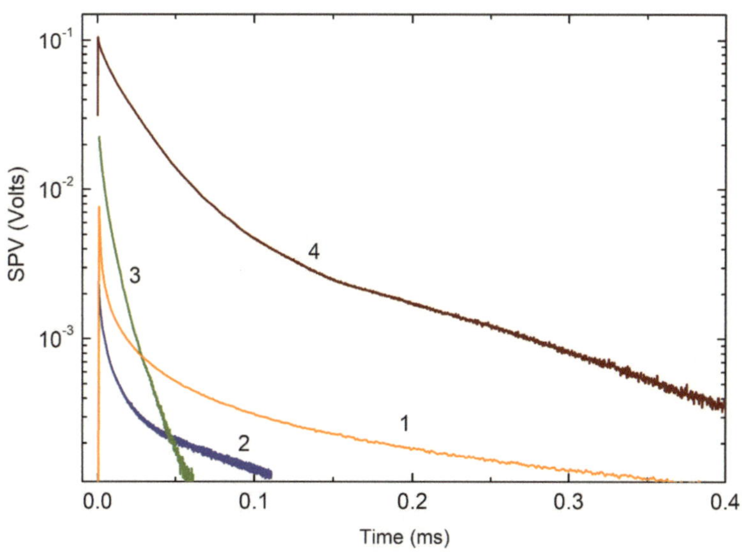

Figure 1.
Time-dependent SPV of Si wafer (1), structures of Ge_xSi_{1-x} islands on Si (2), 10 nm a-Si/Ge_xSi_{1-x}/Si (3), and 10 nm a-Si/Ge_xSi_{1-x}/Si annealed for 5 minutes at 400 °C in an O_2 ambient atmosphere (4). The concentration of Ge atoms in the islands is about 80%. Reprinted with permission from Podolian A, Nadtochiy A, Korotchenkov O, Romanyuk B, Melnik V, Popov V. Journal of Applied Physics. 2018;124:095703. Copyright 2018, AIP publishing.

capped Ge_xSi_{1-x}/Si structure (curve 3 at time $t = 0$) compared to that of bare Ge_xSi_{1-x}-on-Si islands (curve 2). This enlargement is even 5 to 10 times greater after subsequent sample annealing in O_2 (curve 4 at time $t = 0$).

Presuming the use of effective hydrogenation, a-Si:H/c-Si heterojunction solar cells (HET) represent one of the most promising solar cell structures that enable high efficiencies due to high open-circuit values coming from the excellent passivation properties of a-Si:H combined with the beneficial effect of the *a*-Si:H/c-Si heterojunction on the built-in voltage and reduced charge carrier loss at the interface [28]. HET cells also have reduced costs compared with systems installed today based on conventional silicon technologies.

Moreover, in contrast to dielectric passivation materials such as SiO_2 and amorphous Si nitride (a-SiN_x:H) [29], a-Si:H is simultaneously suitable for good passivation and electrical conduction. However, the surface passivation quality worsens both at low and high processing temperatures because of the porous medium grown in the amorphous silicon phase at excess amounts of hydrogen [30] and growing crystalline Si film instead of forming a-Si:H [31, 32], respectively. As a consequence, discrepancy exists in the literature as to the passivation ability of a-Si:H [33–37].

3. Sonochemistry: basic principles

The use of ultrasound for accelerating chemical reactions in liquid–solid heterogeneous systems is very attractive since ultrasound is capable of increasing the reactivity by more than a factor of 10^5 due to the fact that the reagent particles clash at such a high speed that they melt at the point of collision and generate microscopic flames in cold liquids [1, 38]. In ultrasonically irradiated slurries, turbulent flow and shock waves are produced by acoustic cavitation [39] resulting in many tiny gas bubbles. The bubbles expand and contract in accordance with the pressure oscillations of the ultrasonic wave. When the bubble radius is of a certain size and the acoustic amplitude is above a given threshold value, the bubbles collapse violently and the temperature inside a bubble increases dramatically due to the quasiadiabatic compression [40]. At the final stage of the collapse, the vapor, which often is water vapor, dissociates inside the collapsing bubble due to the high bubble temperature. This generates H and OH radicals as well as other kinds of oxidants, which are assumed to produce a variety of chemical reactions [3, 41–45]. The reactions involve the formation of primary radicals from the ultrasound-initiated dissociation of water within a collapsing cavity as

$$H_2O))) \rightarrow H + OH, \quad (4)$$

where the brackets stand for the sonolysis of water. The intermediate hydroxyl and hydrogen radicals can form H_2O_2 and O_2 products.

In aqueous media, these reactions occur in different regions surrounding the collapsed bubble. One of these regions is e.g. the interfacial liquid region between the cavitation bubbles and the bulk solution. The temperature in this region is lower than the one in the interior of the bubbles. The reaction is therefore a liquid phase reaction but the temperature is believed to be high enough to rupture chemical bonds. Apart from these oxidants, considerable concentrations of local hydroxyl radical have been reported [43, 46]. Another reactant region is the bulk solution. Here, the reaction between reactant molecules and OH or H takes place at ambient temperatures.

Since a quantitative analysis of the chemistry involved into the sonochemical reactions is yet difficult to perform [47], it is not certain whether or not the chemical effects indeed originate from acoustic cavitation. The implosive collapse of bubbles during cavitation produces local transient temperatures of about 5000 K and pressures of about 500 atm, with heating and cooling rates exceeding 10^{10} K/s [1, 38]. These conditions create high-velocity collisions between suspended particles and the estimated speed of colliding particles approaches almost the speed of sound in the liquid. Interestingly, sonochemical reactions in cavitational fields occur more slowly at elevated than at lower temperatures. Even so it is surprising, this counterintuitive property makes sense, because at higher temperatures more solvents vaporize into the bubble and, hence, cushion the collapse.

In cold liquids, ultrasound is able to drive reactions that normally occur only under extreme conditions. Examples [45] include intercalation, activation of liquid–solid reactions, and the synthesis of amorphous and nanophase materials. The sonochemical syntheses of nanophase metals, alloys, metal carbides, supported heterogeneous catalysts, and nano-colloids derives from the sonochemical decomposition of volatile organo-metallic precursors during cavitation, which produces clusters of a few nm in diameter.

Various types of sonochemical cells using ultrasound baths and ultrasound horn systems have been reported [48–50]. Most frequently, an ultrasonic transducer and the ultrasound horn are placed directly in the solution. One example is shown in **Figure 2**. An oscillating rf voltage from an oscillator (1 in **Figure 2**) with the amplitude of U_0 is amplified by an amplifier (2) and applied to a Langevin transducer, which consists of a piezoelectric transducer (3), back (4) and front (5) masses. The vibrating transducer is loaded to a glass flask (6) filled with a reactant solution (7) thus delivering an acoustic power at a resonance frequency of the transducer-solution system. The lowest-mode resonance frequency is defined by the solution thickness h such that $h = \lambda/2$ with λ the sound wavelength in the solution at the resonant frequency.

The samples can freely levitate near the pressure antinode zone at operating frequencies in the dozens kHz range. They tend to reside on the flask bottom when increasing the frequency up to several hundreds of kHz.

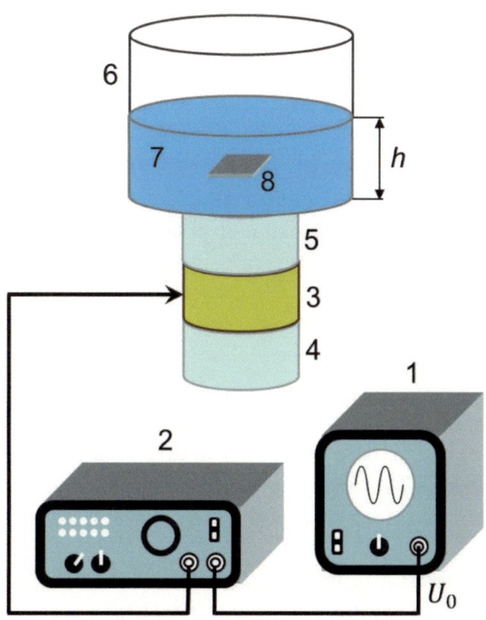

Figure 2.
Schematic of the sonochemical cell: 1 – Oscillator, 2 – Amplifier, 3 – Piezoelectric transducer, 4 – Back mass, 5 – Front mass, 6 – Boiling flask, 7 – Reactant solution, 8 – Sample.

4. Ultrasonic cleaning of Si wafers

Cleaning and conditioning of silicon wafer surfaces and Si/SiO_2 interfaces for the manufacturing of photovoltaic and microelectronic devices are in increasing demand with improved performance, reliability of these devices, scaling down to below 10 nm, incorporating extended metallization layers, employing epitaxial layers of compound SiGe and III-V semiconductors [51]. Clean wafer surfaces are crucial in high efficiency solar cell as well as in Ultra-Large Scale Integration (ULSI) fabrication processes, fin-shaped field effect transistors (FinFET), 3D NAND-stacked memory devices etc.

Deposition of monolayers and self-assembly of nanoparticles in multilevel structures requires the wafer surface to be completely free of any particulate contamination down to a nanometer scale. Moreover, installing a 1 MWp solar module generator implies that more than $2{\cdot}10^5$ Si wafers must be processed.

Furthermore, metallic contaminants on Si wafers cause substantial increase in leakage current in silicon p-n junctions and decrease the oxide breakdown voltage thus deteriorating the minority carrier lifetime [52, 53]. In particular, Cu and Al contaminants worsen the gate oxide integrity [54, 55]. To achieve ultraclean substrate surfaces with high reproducibility, it is also important to note that not only removal of contaminants is effective in improving the cell performance, but also prevention of their redeposition on the wafer surface. In this respect, dilute HF can be effectively used. However, some Cu and Al residues can be found on the wafer surface due to its hydrophobic nature [56].

The contamination of wafer surfaces by particle contaminants is one of the major problems in the industries. One way to increase the yield on fully processed silicon wafers is to use cleaning techniques specifically efficient to remove particle contaminants. Small particles are especially difficult to remove because they are strongly bounded to the substrate by electrostatic forces. It is therefore very important to find an effective way to remove particles from wafers without causing damage to the wafers.

A wide variety of cleaning methods are being used in wafer manufacturing such as brush or water-jet scrubbing of wafer surfaces employed prior to further immersion-type cleaning, scrubbing of rotating wafer surfaces between each processing step, adding cheating agents to the solution aiming to avoid metal adsorption onto Si wafers, cleaning in wet chemical baths, post treatment rinsing and many others [57]. They, however, are known to damage the wafer surface. Moreover, the chemical-type cleaning has inherent danger caused by residues from sulfuric acid, ammonium hydroxide, isopropyl alcohol and other chemical pollutants. For example, an immersion-type cleaning step widely used industrially utilizes RCA Standard Clean 1 (RCA-1) [58].

Wet-chemical processes are still the most widely used method for Si wafer cleaning in the semiconductor industry today [51]. The critical demands of surface purity raised by the International Technology Roadmap for Semiconductors (ITRS) [59] can generally be reduced by utilizing ultrasonic cleaning processes.

In the above context, a substrate independent cleaning process is highly desirable because, opposite to a chemical based cleaning process, it is equally well suited for different substrates and does not modify the surface through the etching, roughening, etc.

While keeping the compatibility with Si wafer standard processing steps, ultrasonic treatment of surfaces can be effective in passing several obstacles to achieving wafer cleaning mentioned above. Ultrasonic cleaning employs an ultrasonically activated liquid with a submerged wafer used to achieve or enhance the removal of surface contaminants [60]. Ultrasonic irradiation involves a variety of complex mechanisms, including mechanical vibration and appropriate pressure gradients, microcavitation bubbles that oscillate and dance around due to Bjerknes force [39, 61], acoustic streaming flows, etc. One of the most important aspects of using acoustic streaming is the effect of the frequency on the boundary layer [62]. Its thickness decreases and the streaming velocity increases with increasing the sonication frequency. These both remarkably increase the drag force and hence the particle removal efficiency. It is demonstrated that the acoustic streaming induced removal of foreign contaminants with sizes in the dozen nm range is accomplished at frequencies greater than 1 MHz, i.e. in a high-frequency sonication region [63]. The removal of contaminants having sizes down to ≈100 nm is possible at frequencies smaller than 1 MHz [64].

It has furthermore been demonstrated that the caustic etching process can be ultrasonically enhanced producing finer, more homogeneous textured surfaces [65]. This technique inhibits the sunlight reflection from the Si surface thus enhancing the performance of solar cells.

To demonstrate the capabilities of high-power ultrasound in powerfully manipulating surface species, consider the data reported in **Figure 3**. Here, following the method of cleaning from laser-induced cavitation bubbles [66], the wafer surface is covered with a thin layer of grease. This yields the optical transmission spectra indicative of organic contaminants, which are marked by several peaks in spectra 2 of **Figure 3**, such as C–CH_3 at 2960 cm^{-1}, –CH_2– at 2920 cm^{-1}, –CH at 2890 cm^{-1} and C–H stretching vibrations [–CH_3 and –$(CH_2)_n$–] at 2850 cm^{-1} [67]. It is seen in spectra 3 that organic contaminants are effectively removed from the wafer surface upon its exposure to ultrasonic cavitation with the peak acoustic intensity of about

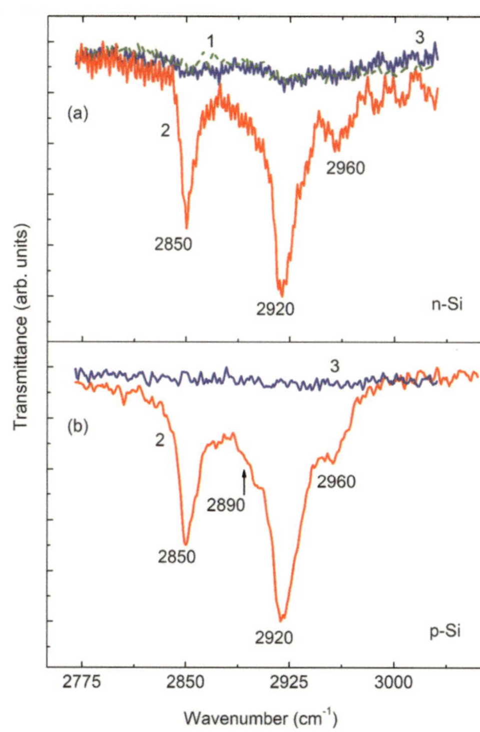

Figure 3.
FTIR spectra of n-Si (a) and p-Si (b) wafers, prior to surface greasing and applying the ultrasound (spectrum 1), covered with a thin layer of vaseline (2) and subsequently cleaned in an ultrasonic bath (3) during 15 min. Reprinted with permission from Podolian A, Nadtochiy A, Kuryliuk V, Korotchenkov O, Schmid J, Drapalik M, Schlosser V. Solar Energy Materials and Solar Cells. 2011;95:765–772. Copyright 2010, Elsevier.

400 W/cm^2 [68], so that the resulting absorption resembles the one taken before the wafer has been exposed to sonication (spectrum 1 in **Figure 3**).

Although cavitation bubble dynamics close to solid surfaces has been given particular attention [68], quite little is known about the streaming along the surface. Therefore, in attempting to explain the removal mechanism behind the cleaning effect observed in **Figure 3** there may be several potential ways. One is that the pressure gradients due to bubble implosion and acoustic streaming would bombard and remove organic contaminants on the silicon surface. The other is that some excited oxygen atoms produced by the sonochemical decomposition of the water adhere to the organic compounds, oxidize them finally decomposing into H_2O, O_2, H_2, CO and CO_2, having high vapor pressures allowing the lift-off from the wafer surface [68].

Typical forward and reverse bias $I - V$ characteristics of the wafers with and without organic contaminants are plotted in **Figure 4**. Here, distilled water and piranha (3:1 volume solution of H_2SO_4 and 30%-H_2O_2) are used as a cleaning liquid (curve 2 compared with curves 3 and 4 in **Figure 4**), and both chemical and sonochemical cleaning processes are contrasted (curve 3 compared with curve 4). It is interesting that the cleanings cause an overall decrease in the current through the wafer. This can, in part, be described by the removal of the organic contaminants from the wafer and appropriate quenching of the leaky currents between the basal wafer surfaces. The ultrasonic effect in piranha (curve 4 in **Figure 4**) is obviously greater than that in water (curve 2), as would be expected in reactive chemical agents (cf. curve 2 in **Figure 4**).

Perhaps it is best noted here that the cavitation processing affects a sub layer region beneath the wafer surface. Therefore, the air/oxide and oxide/wafer surface state or interface trap densities could be reduced significantly by this processing step.

This would have a similar effect on the photovoltaic response of the wafers since the photo-induced charge carriers are separated in the electric field of the surface space charge region. These carriers would partially screen the fixed surface or interface state charge thus reducing the surface band bending.

As one of the earliest attempt to manipulate the surface photovoltage, **Figure 5** illustrates the sonication effect on the SPV decays. The decay is seen to be faster for

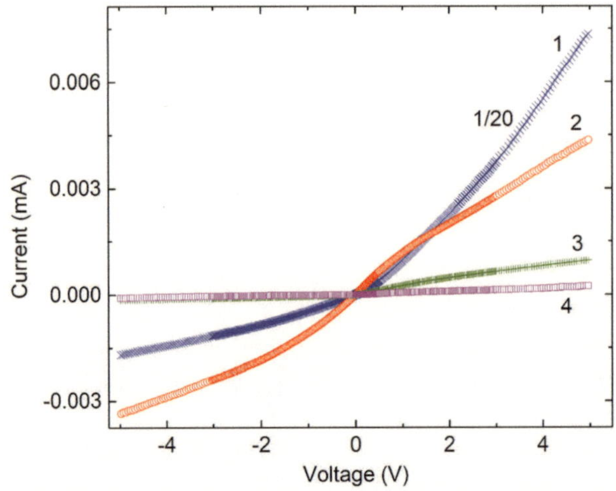

Figure 4.
$I - V$ curves of the Au Schottky contact to n-Si wafer, as-purchased (1), ultrasonically cleaned in distilled water (2), chemically (3) and ultrasonically (4) processed in a piranha bath. In each case, the cleaning time is 60 min. Reprinted with permission from Podolian A, Nadtochiy A, Kuryliuk V, Korotchenkov O, Schmid J, Drapalik M, Schlosser V. Solar Energy Materials and Solar Cells. 2011;95:765–772. Copyright 2010, Elsevier.

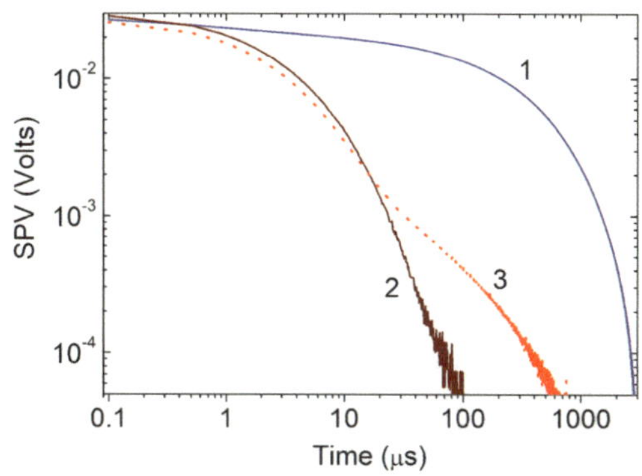

Figure 5.
SPV decays of n-Si, as-purchased (curve 1), ultrasonically cleaned in distilled water for 60 (2) and 120 (3) min. Reprinted with permission from Podolian A, Nadtochiy A, Kuryliuk V, Korotchenkov O, Schmid J, Drapalik M, Schlosser V. Solar Energy Materials and Solar Cells. 2011;95:765–772. Copyright 2010, Elsevier.

ultrasonically cleaned wafer (curve 2 compared with curve 1 in **Figure 5**), which can simply be attributed to an increased density of unsaturated dangling bonds on the wafer surface due to the cavitation-induced local removal of oxide from the silicon surface.

For sonication times increasing up to 90–150 min, the SPV decay is slightly worsened. For example, curve 3 in **Figure 5** exhibits a long-tail decay constituent at time instants greater than ≈20 μs. The involvement of the above surface states and interface traps is therefore reasonable to assume. In this context, for the initial decays at $t < 10$ μs, when the injected carrier concentration is large compared with the density of the trapping centers N_t, the recombination centers mainly determine the excess carrier lifetime. Once the concentration of these carriers becomes exceedingly small, particularly compared with N_t (at $t > 20$ μs, curve 3 in **Figure 5**), the SPV decay is determined by N_t. The initial decays at $t \leq 10$ μs in curves 2 and 3 are nearly identical, giving some indication of the importance of this cavitation processing step in obtaining clean wafer surfaces. Based upon these results, surface and interface trap generation is likely to be significant at prolonged sonication times, greater than 60–90 min.

5. Sonochemical passivation of Si and SiGe

A cogent resource for the burgeoning field of the surface passivation coating utilizing hydrocarbon chains [69], which can reduce the density of surface states and increase the recombination lifetime of the majority carriers. Different organic solvents can be used in practical ways for these purposes such as chloroform ($CHCl_3$) and dichloromethane (CH_2Cl_2). For example, silicone polymers were grown on the Si surfaces with gaseous $CHCl_3$ and Cu catalyst [70], brominated aromatic moieties were successfully prepared from KBr/H_2O_2 in sonochemically treated chloroform [71]. Recent work in applying ultrasound to chemical reactions demonstrates the promise of the sonochemical approach, yet the bromination of aromatic compounds is not achieved with simple mechanical stirring replacing sonication.

This topic is covered in more detail below for two types of samples. Type A sample is a Ge_xSi_{1-x} alloy layer, 100 nm thick, grown on a p-doped Cz-Si wafer. Type B sample is obtained by coating sample A with a 10-nm thick a-Si layer (see **Figure 6**). The Ge content x in the Ge_xSi_{1-x} layers is about 30 at.%.

In order to give an illustrating example for the differences in the effective lifetime τ_{eff}, quasi-steady-state measurements (QSSPC) [72] done in a-Si/Ge_xSi_{1-x}/Si sample are plotted in **Figure 7**. Here, an effective minority carrier lifetime τ_{eff} is obtained from the given by

$$J_{ph} = \frac{\Delta neW}{\tau_{eff}}, \tag{5}$$

where J_{ph} is the photogenerated current density and Δn is the excess carrier density. At low values of Δn ranging from about 10^{12} to 10^{13} cm^{-3}, **Figure 7(b)** shows≈ 4 times increase in the lifetime due to ultrasonic processing in chloroform (closed circles compared with open circles).

Regarding the role of amorphous Si coating layer (in sample B) and ultrasonic processing on the photovoltaic behavior, one can distinguish several maps of the surface-distributed SPV amplitudes U_0 and decay times τ shown in **Figures 8** and **9**.

It is seen in **Figure 8** that both the U(0) and τ distributions narrow but shift to smaller values once the Ge_xSi_{1-x}/Si structure (sample A) has been coated with a-Si layer (sample B). Faster decays in a-Si/Ge_xSi_{1-x}/Si can be accounted for by an increased number of fast recombination canters in sample B due to the deposited amorphous Si layer, which typically reduces the amplitude of the SPV response.

It is also seen in **Figure 8** that the sonication in chloroform allows for improved SPV performance. Indeed, the SPV decays are spread over much longer time scales to enlarge the SPV amplitude U_0 up to about 50% can be realized in samples, as observed in appropriate distributions, which are marked by B-sono and A-sono in **Figure 8**.

This effect is not pronounced in the samples sonochemically processed in distilled water; see **Figure 9**. Although the distributions of U_0 narrow, the SPV amplitude is itself quenched in the samples treated in distilled water (upper distributions in **Figure 9**). However, similar sonication-affected SPV decay dynamics is observed both in chloroform and distilled water (lower distributions in **Figures 8** and **9**).

The examples in water demonstrate that sonication provides a convenient tool to achieve surface cleaning, as reported previously [7, 73]. In this case, the assumption is based on the fact that (i) the cavitating bubbles are capable of locally removing the surface oxide layer affecting the dangling bonds on the bare Si surface, and (ii) the oxygen and hydrogen, decomposed in water by the presence of local strain fields and elevated temperatures inside a cavitating bubble, can micro-precipitate the Si wafer thus changing the recombination rate. These insights, combined with

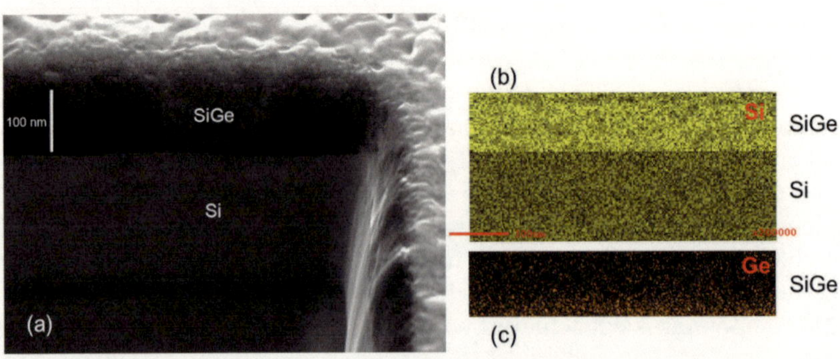

Figure 6.
(a) Cross-sectional scanning electron microscope (SEM) image of a Ge_xSi_{1-x} on Si layer covered with a 10 nm thick a-Si (sample B). (b) and (c) distributions of Si and Ge atoms near the interface mapped using a scanning auger microscopy technique. Reproduced with permission from Nadtochiy A, Korotchenkov O, Schlosser V. Physica Status Solidi (a). 2019;216:1900154. Copyright 2019, Wiley-VCH Verlag GmbH & Co. KGaA.

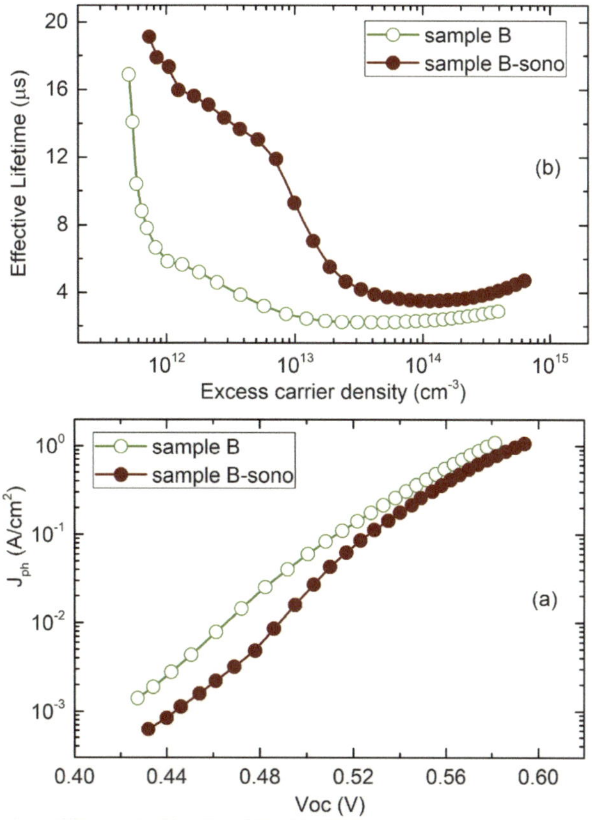

Figure 7.
Current–voltage curve (a) and variation of the effective lifetime τ_{eff} with excess carrier density Δn (b) for sample B before (open circles) and after sonication in chloroform (closed circles). Reproduced with permission from Nadtochiy A, Korotchenkov O, Schlosser V. Physica Status Solidi (a). 2019;216:1900154. Copyright 2019, Wiley-VCH Verlag GmbH & Co. KGaA.

complementary ultrasonic techniques that employ reactant solutions, make a significant contribution to developing a detailed picture of ultrasonic processing.

With respect to the reactant solutions that can be used, recent investigations report that SPV signal in Si can be significantly enhanced, by almost an order of magnitude, due to ultrasonic treatments in dichloromethane. Similar effect in CH_2Cl_2 can be observed for Ge_xSi_{1-x} surfaces exhibiting a 50% increase in the SPV amplitude [74].

The operating frequency range of sonochemical apparatus is typically up to dozens of kHz. A general working principle, which follows from the above guidelines, relies upon a specific assumption that the size of the cavitation bubble is inversely related to the frequency of ultrasound. Therefore, because the bubble size drops with increasing the ultrasonic frequency and the bubble implosions become less violent, the energy released by each imploding cavitation bubble decreases with the ultrasonic frequency. However, the number of the imploding events increases due to increased number of sound waves passing through the liquid at a higher frequency [60].

One may compare the data obtained in Ge_xSi_{1-x} with lower- and higher-frequency sonochemical processing in dichloromethane at about 25 kHz and 400 kHz, respectively. Etching in HF makes initial single-exponential decay nearly double-exponential. Sonication at 25 kHz slightly slows down the tail component of the decay while the higher-frequency processing at 400 kHz turns the SPV decay back into nearly single-exponential form.

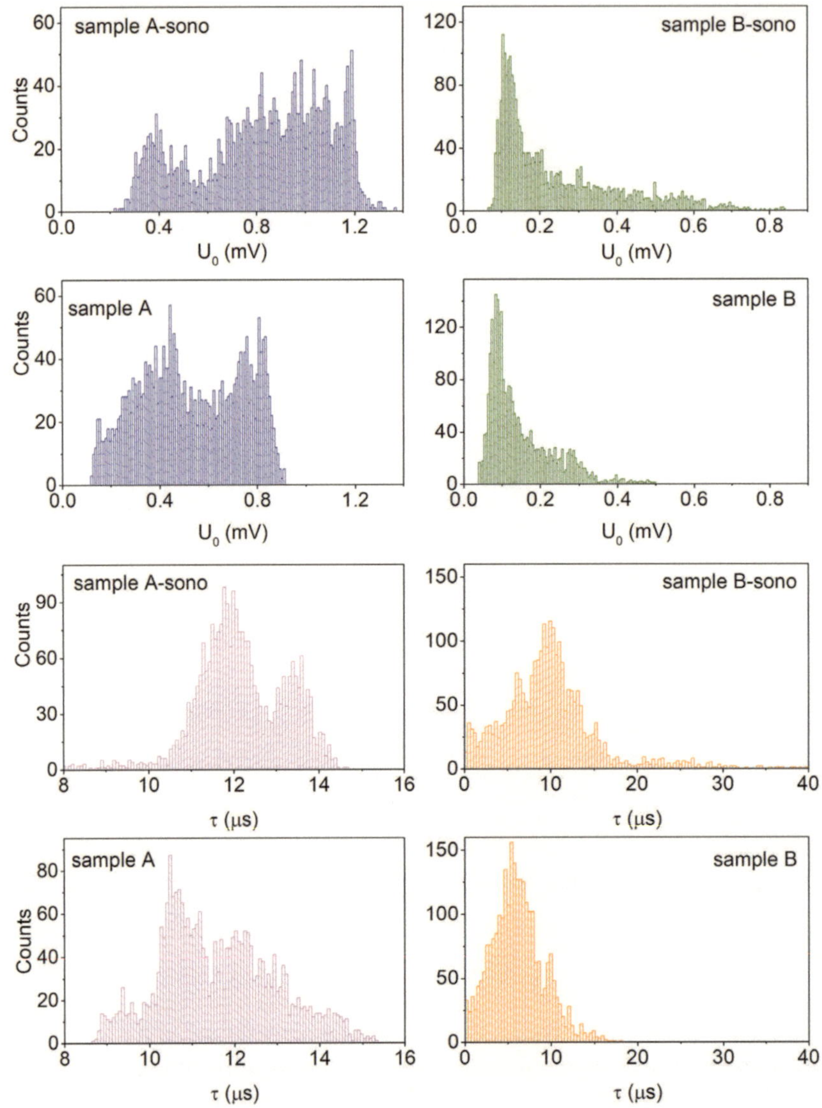

Figure 8.
Probability of occurrences of particular values of the SPV amplitude U_0 and decay time τ in samples a (Ge_xSi_{1-x}/Si) and B (a-Si/Ge_xSi_{1-x}/Si), which are measured by surface mappings of the SPV decays. The distributions marked by "sono" are taken after sonochemical treatment in chloroform during 1 min. Reproduced with permission from Nadtochiy A, Korotchenkov O, Schlosser V. Physica Status Solidi (a). 2019;216:1900154. Copyright 2019, Wiley-VCH Verlag GmbH & Co. KGaA.

The likely mechanism that has come to describe the observations is based on that both chloroform and dichloromethane can act as carbon sources. Being decomposed into hydrocarbon species due to extreme conditions in the solvents and at the etchant/solid interfaces, the sonicated reactants seem to saturate the dangling bonds revealed on the surface of Si and Ge_xSi_{1-x} alloys and hence to passivate the surface [14].

This is in accord with previous reports on the thermal decomposition of chloroform, which results in by-products of CCl_2, C_2Cl_4, Cl, H and HCl. When they react with metal (M) atoms, the reactions pathways are [75, 76].

$$CHCl_3(+M) \rightarrow CCl_2 + HCl\ (+M), \tag{6}$$

$$CCl_2 + CCl_2 \rightarrow C_2Cl_2 + Cl + Cl, \tag{7}$$

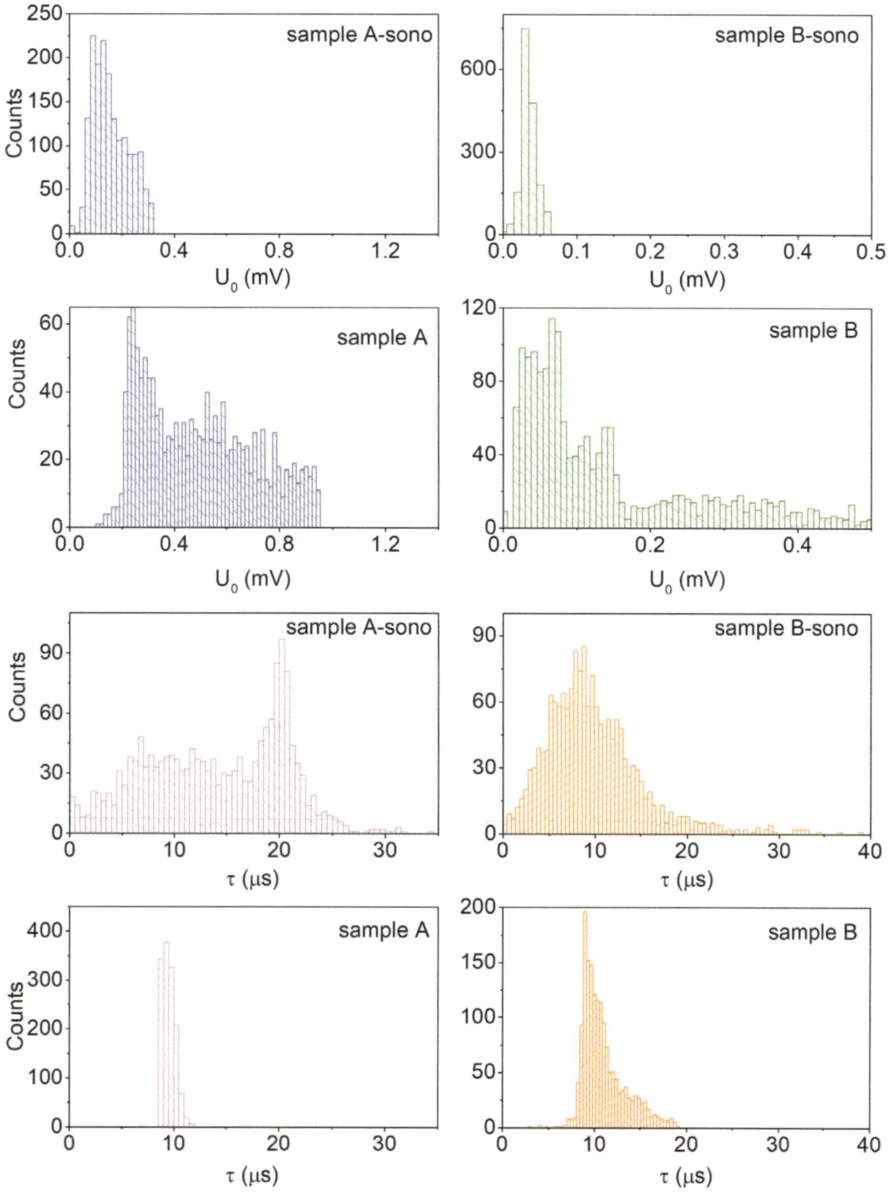

Figure 9.
Probability of occurrences of particular values of the SPV amplitude U_0 and decay time τ in samples a (Ge_xSi_{1-x}/Si) and B (a-Si/Ge_xSi_{1-x}/Si), which are measured by surface mappings of the SPV decays. The distributions marked by "sono" are taken after sonochemical treatment in distilled water during 1 min. Reproduced with permission from Nadtochiy A, Korotchenkov O, Schlosser V. Physica Status Solidi (a). 2019;216:1900154.

$$CCl_2 + M \rightarrow CCl + Cl + M, \tag{8}$$

$$CCl_2 + CCl_2 + M \rightarrow C_2Cl_4 + M, \tag{9}$$

$$HCl + M \rightarrow H + Cl + M, \tag{10}$$

The first reaction step given by Eq. (6) is the decomposition of $CHCl_3$, which is followed by secondary decomposition reactions in Eqs. (7)–(10).

The usual analysis approach for high temperatures achieved during the sonication process involves steps of radical formation, e.g., C_2 radicals:

$$CCl + CCl \rightarrow C_2 + Cl + Cl, \tag{11}$$

$$C_2Cl_2 \rightarrow C_2 + Cl + Cl, \tag{12}$$

Therefore, employing chloroform $CHCl_3$ or dichloromethane CH_2Cl_2, one gets the Si-H and C-Cl bonds that react yielding C-H species. This, in turn, resembles the chlorination/alkylation process that forms Si–alkyl converting Si–H into Si–C_nH_{2n+1} ($n \geq 1$). The alkyl chains on Si surfaces are known to provide low surface recombination velocities [77] thus featuring effective Si surface passivation [78].

In the model presented in **Figure 10**, presumed chemical reactions for the sonochemical surface passivation are made available. The above analysis assumes that the Si–H bond on the surface breaks up at high local temperatures and pressures inside the cavitation bubble. This produces highly reactive Si and Ge dangling bonds, as shown in **Figure 10(a)**. Being short-lived, they quickly react with the sonicated chloroform molecules. Next, molecular hydrocarbon and chlorine atoms cover the a-Si or Ge_xSi_{1-x} surface, as shown in **Figure 10(b)**. The wavy arrow illustrates that Si atoms can be released from the surface due to carbon atoms decomposed from chloroform (or dichloromethane). Finally, these carbon atoms at the surface create Si–C bonds and dangling carbon bonds being then saturated by the atoms of H and Cl. Some of them can meet activated carbon-containing molecules to form Si–C bonds.

In order to obtain the signatures of the chemical constituents, Fourier transformed-infrared (FTIR) spectroscopy is usually applied. FTIR transmittance spectra are shown in **Figure 11**. Among prominent infrared absorption peaks related to the Si–Si, Si–O, Ge–Ge, Ge–O and Ge–O–Si vibration modes, there are resolved bulk-like Si–H and Si–H_2 stretching modes at about 2000 and 2090 cm^{-1}, respectively, as well as a weak shoulder near 1880 cm^{-1} related to Ge–H vibrations [79, 80]. These results indicate that Si-H hydrides are present in the deposited a-Si and Ge_xSi_{1-x} films.

One also finds a spectral feature at about 670 cm^{-1} (arrow in the left-hand panel of **Figure 11**). This obviously strengthens in the hydrogenated a-Si film (spectrum 3). To account for this enlargement, one has to assume that this feature is related to the hydrogen complexes. In clear accord, the wagging modes near 640 cm^{-1} can be due to three bonding units of Si–H_n (n = 1, 2, 3) [81]. It is seen in spectrum 4 of **Figure 11** that the sonication quenches the 670 cm^{-1}, which is indicative of the

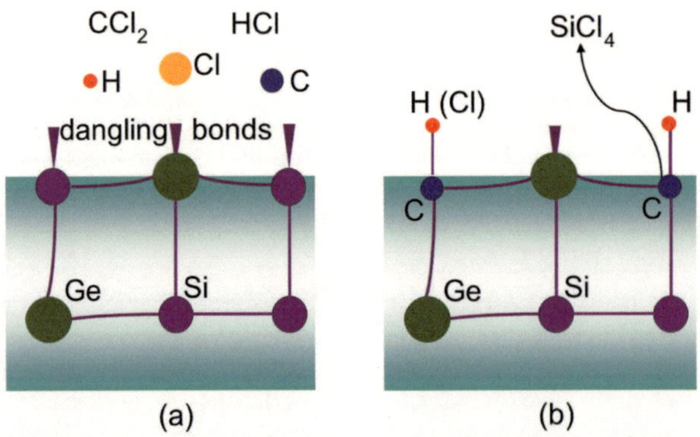

Figure 10.
How to passivate SiGe surface using chloroform reactants H, Cl, C, CCl2, HCl released in Eqs. (6)–(12). These remove Si atoms on the surface and saturate the dangling bonds. Reproduced with permission from Nadtochiy A, Korotchenkov O, Schlosser V. Physica Status Solidi (a). 2019;216:1900154. Copyright 2019, Wiley-VCH Verlag GmbH & Co. KGaA.

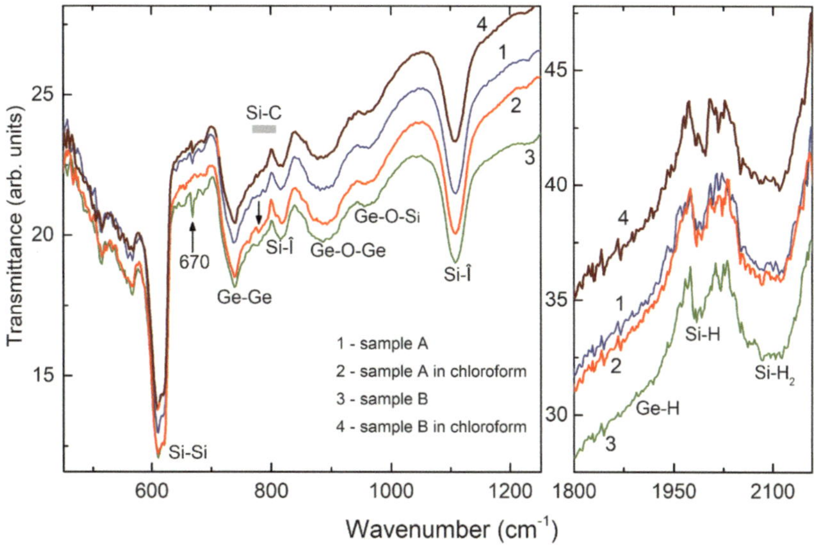

Figure 11.
FTIR spectra of samples Ge_xSi_{1-x}/Si (curve 1) and a-Si/Ge_xSi_{1-x}/Si (3), taken before ultrasonic processing and the ones obtained after the treatment in chloroform – Spectra 2 and 4, respectively. Reproduced with permission from Nadtochiy A, Korotchenkov O, Schlosser V. Physica Status Solidi (a). 2019;216:1900154. Copyright 2019, Wiley-VCH Verlag GmbH & Co. KGaA.

removal of H from the coating layers due to ultrasonic processing, supporting the pictorial view given in **Figure 10**.

6. Conclusions

Two techniques related to ultrasonic cleaning of Si wafers and sonochemical modification of Si, SiGe and a-Si/SiGe surfaces in hydrocarbon solutions of chloroform ($CHCl_3$) and dichloromethane (CH_2Cl_2) are outlined.

In spite of our lack of knowledge of the exact sonication mechanisms even in distilled water, this research field can be considered to be among potential candidates to develop a new class of environmental friendly cleaning steps in silicon-based technologies. Some progress has recently been made in understanding a unique potential capability of sonicated water in Si wafer cleaning processes. The underlying mechanisms related to the fundamental properties of cavitation and bubble implosion events, the role of a thin interphase layer between the bubble and the surface placed in the sonicated liquid can offer new far-reaching implications and importance for heterogeneous liquid–solid systems.

It is demonstrated that organic particle contaminants are effectively removed during the kHz-frequency sonication of crystalline Si wafers in distilled water over the first 40–60 min. When ultrasonically processing the wafers for treatment times less than ≈60 min at the peak acoustic intensity of about 400 W/cm^2, the dangling bonds at the air/oxide and oxide/wafer interface can be activated. That affects barriers of the free carrier migration at the interfaces, as revealed by the current–voltage curves, and acts as recombination centers, accelerating the surface photovoltage decays. A healing of the bonds may occur at longer cleaning times (from 60 to 120 min) with a partial recovery of the interfaces and a consequent reversing of the observed changes. The potential of using distilled water in environmental friendly and non-toxic ultrasonic cleaning step in crystalline Si wafer preparation is addressed.

In fact, the studies described above do not reveal information about the full complexity of subsurface defect distribution effects. Therefore, there remains a wide number of uncertainties, e.g., the fundamental problem of whether or not the ultrasonic processing exploiting high acoustic powers is capable of promoting effective cleaning without surface deterioration effects.

To improve the photovoltaic response of Si wafers, SiGe and amorphous silicon (a-Si)/SiGe surfaces, a sonochemical treatment in hydrocarbon solutions of chloroform ($CHCl_3$) and dichloromethane (CH_2Cl_2) can be employed. The use of the sonochemical reaction slows down the observed surface photovoltage decay and enhances its magnitude in SiGe and a-Si/SiGe thin layers grown on Si. The average surface-integrated photovoltage and decay time can increase up to 50%. This effect is not observed in distilled water, indicative of the fact that CH-containing radicals can lead to the observed improvements. It is suggested that the effect can be explained as follows. The hydrocarbon solution is decomposed and produces hydrocarbon chains, which are then decomposed further away into hydrogen and carbon. The reactive Si dangling bonds revealed on the surface of Si, a-Si or SiGe alloy layers are saturated by the hydrocarbon species to passivate the surface.

More work needs still to be done beyond the description of a very few links that have been highlighted above. In particular, the following experiments could pave the way for new mechanisms of surface passivation, activation of the interphase regions dangling bonds as well as cleaning of surfaces due to the ultrasonic processing.

Acknowledgements

The work at Kyiv was funded by the Ministry of Education and Science of Ukraine (0119U100303). Financial support from the University of Vienna is also acknowledged.

Conflict of interest

The authors declare no conflict of interest.

Author details

Andriy Nadtochiy[1], Artem Podolian[1], Oleg Korotchenkov[1] and Viktor Schlosser[2]*

1 Faculty of Physics, Taras Shevchenko National University of Kyiv, Kyiv, Ukraine

2 Department of Electronic Properties of Materials, Faculty of Physics, University of Vienna, Austria

*Address all correspondence to: viktor.schlosser@univie.ac.at

References

[1] Suslick KS. Sonochemistry. Science. 1990;247:1439–1445. DOI: 10.1126/science.247.4949.1439

[2] Suslick KS, Nyborg WL. Ultrasound: Its chemical, physical and biological effects. The Journal of the Acoustical Society of America. 1990;87:919–920. DOI: 10.1121/1.398864

[3] Gedanken A, Perelshtein I. Power ultrasound for the production of nanomaterials. In: Gallego-Juárez JA, Graff KF, editors. Power ultrasonics. Applications of high-intensity ultrasound. Amsterdam: Woodhead Publishing; 2015. p. 543–576. DOI: 10.1016/B978-1-78242-028-6.00018-1

[4] Sáez V, Mason TJ. Sonoelectrochemical synthesis of nanoparticles. Molecules. 2009;14: 4284–4299. DOI: 10.3390/molecules14104284

[5] Cobley A, Mason T. The evaluation of sonochemical techniques for sustainable surface modification in electronic manufacturing. Circuit World. 2007;33:29–34. DOI: 10.1108/03056120710776997

[6] Paniwnyk L, Cobley A. Ultrasonic surface modification of electronics materials. Physics Procedia. 2009;3: 1103–1108. DOI: 10.1016/j.phpro.2010.01.230

[7] Nadtochiy A, Podolian A, Korotchenkov O, Schmid J, Kancsar E, Schlosser V. Water-based sonochemical cleaning in the manufacturing of high-efficiency photovoltaic silicon wafers. Physica Status Solidi C. 2011;8:2927–2930. DOI: 10.1002/pssc.201084062

[8] Arruda LB, Orlandi MO, Lisboa-Filho PN. Morphological modifications and surface amorphization in ZnO sonochemically treated nanoparticles. Ultrasonics Sonochemistry. 2013;20: 799–804. DOI: 10.1016/j.ultsonch.2012.11.013

[9] Kang B-K, Kim M-S, Park J-G. Effect of dissolved gases in water on acoustic cavitation and bubble growth rate in 0.83 MHz megasonic of interest to wafer cleaning. Ultrasonics Sonochemistry. 2014;21:1496–1503. DOI: 10.1016/j.ultsonch.2014.01.012

[10] Savkina RK, Smirnov AB. Structured silicon surface via cavitation processing for the photovoltaic and biomedical application. In: Fesenko O, Yatsenko L, editors. Nanophysics, Nanophotonics, Surface Studies, and Applications. Springer Proceedings in Physics, v. 183; 2015. p. 291–303. DOI: 10.1007/978-3-319-30737-4_241

[11] Savkina RK, Gudymenko AI, Kladko VP, Korchovyi AA, Nikolenko AS, Smirnov AB, Stara TR, Strelchuk VV. Silicon substrate strained and structured via cavitation effect for photovoltaic and biomedical application. Nanoscale Research Letters. 2016;11:183. DOI: 10.1186/s11671-016-1400-2

[12] Skorb EV, Möhwald H. Ultrasonic approach for surface nanostructuring. Ultrasonics Sonochemistry. 2016;29: 589–603. DOI: 10.1016/j.ultsonch.2015.09.003

[13] Bai F, Wang L, Saalbach K-A, Twiefel J. A novel ultrasonic cavitation peening approach assisted by water jet. Applied Sciences. 2018;8:2218. DOI: 10.3390/app8112218

[14] Nadtochiy A, Korotchenkov O, Schlosser V. Sonochemical modification of SiGe layers for photovoltaic applications. Physica Status Solidi A. 2019;216:1900154. DOI: 10.1002/pssa.201900154

[15] May GS, Spanos CJ. Fundamentals of semiconductor manufacturing and

process control. New Jersey: Wiley; 2006. 463 p. ISBN: 978–0–471-78406-7

[16] De Wolf S, Descoeudres A, Holman ZC, Ballif C. High-efficiency silicon heterojunction solar cells: A review. Green. 2012;2:7–24. DOI: 10.1515/green-2011-0018

[17] Schroder DK. Semiconductor Material and Device Characterization. 3rd ed. Hoboken, New Jersey: Wiley; 2006. 790 p. ISBN: 978–0–471-73906-7

[18] Deal BE, Grove AS. General relationship for the thermal oxidation of silicon. Journal of Applied Physics. 1965; 36:3770–3778. DOI: 10.1063/1.1713945

[19] Aberle AG, Hezel R. Progress in low-temperature surface passivation of silicon solar cells using remote-plasma silicon nitride. Progress in Photovoltaics. 1997;5:29–50. DOI: 10.1002/(SICI)1099-159X(199701/02)5:1<29::AID-PIP149>3.0.CO;2-M

[20] Benick J, Zimmermann K, Spiegelmann J, Hermle M, Glunz SW. Fundamental losses in solar cells. Progress in Photovoltaics. 2011;19:361–365. DOI: 10.1002/pip.1024

[21] Pankove JI, Tarng ML. Amorphous silicon as a passivant for crystalline silicon. Applied Physics Letters. 1979;34:156–157. DOI: 10.1063/1.90711

[22] Lin Y, Feng M, Wang Z, Zeng Y, Liao M, Gong L, Yan B, Yuan Z, Ye J. Excellent passivation with implied open-circuit voltage of 710 mV for p-type multi-crystalline black silicon using PECVD grown a-Si:H passivation layer. Solar Energy. 2020; 211:753–758. DOI: 10.1016/j.solener.2020.10.023

[23] Balaji P, Dauksher WJ, Bowden SG, Augusto A. Improving surface passivation on very thin substrates for high efficiency silicon heterojunction solar cells. Solar Energy Materials and Solar Cells. 2020;216:110715. DOI: 10.1016/j.solmat.2020.110715

[24] Schmidt J, Peibst R, Brendel R. Surface passivation of crystalline silicon solar cells: Present and future. Solar Energy Materials and Solar Cells. 2018;187: 39–54. DOI: 10.1016/j.solmat.2018.06.047

[25] Angermann H. Characterization of wet-chemically treated silicon interfaces by surface photovoltage measurements. Analytical and Bioanalytical Chemistry. 2002;374:676–680. DOI: 10.1007/s00216-002-1450-4

[26] Angermann H, Henrion W, Röseler A. Wet-chemical conditioning of silicon: Electronic properties correlated with the surface morphology. In: Nalwa HS, editor. Silicon-Based Materials and Devices. San Diego: Academic Press; 2001; Volume 1, p. 268–299. DOI: 10.1016/B978-012513909-0/50009-X

[27] Podolian A, Nadtochiy A, Korotchenkov O, Romanyuk B, Melnik V, Popov V. Enhanced photoresponse of Ge/Si nanostructures by combining amorphous silicon deposition and annealing. Journal of Applied Physics. 2018;124:095703. DOI: 10.1063/1.5029948

[28] Coignus J, Baudrit M, Singer J, Lachaume R, Muñoz D, Thony P. Key issues for accurate simulation of a-Si:H / c-Si heterojunction solar cells. Energy Procedia. 2011;8:174–179. DOI: 10.1016/j.egypro.2011.06.120

[29] Duttagupta S, Lin F, Wilson M, Boreland MB, Hoex B, Aberle AG. Extremely low surface recombination velocities on low-resistivity n-type and p-type crystalline silicon using dynamically deposited remote plasma silicon nitride films. Progress in Photovoltaics. 2014;22: 641–647. DOI: 10.1002/pip.2320

[30] Koch C, Ito M, Schubert M. Solar Energy Materials and Solar Cells. 2001;

68:227–236. DOI: 10.1016/S0927-0248(00)00249-X

[31] Fujiwara H, Kondo M. Impact of epitaxial growth at the heterointerface of a-Si:H/c-Si solar cells. Applied Physics Letters. 2007; 90:013503. DOI: 10.1063/1.2426900

[32] De Wolf S, Kondo M. Abruptness of a-Si:H/c-Si interface revealed by carrier lifetime measurements. Applied Physics Letters. 2007;90:042111. DOI: 10.1063/1.2432297

[33] Fujiwara H, Kaneko T, Kondo M. Application of hydrogenated amorphous silicon oxide layers to c-Si heterojunction solar cells. Applied Physics Letters. 2007;91:133508. DOI: 10.1063/1.2790815

[34] Mueller T, Schwertheim S, Scherff M, Fahrner WR. High quality passivation for heterojunction solar cells by hydrogenated amorphous silicon suboxide films. Applied Physics Letters. 2008;92:033504. DOI: 10.1063/1.2837192

[35] Mueller T, Schwertheim S, Fahrner WR. Crystalline silicon surface passivation by high-frequency plasma-enhanced chemical-vapor-deposited nanocomposite silicon suboxides for solar cell applications. Journal of Applied Physics. 2010;107:014504. DOI: 10.1063/1.3264626

[36] Rattanapan S, Watahiki T, Miyajima S, Konagai M. Improvement of rear surface passivation quality in p-type silicon heterojunction solar cells using boron-doped microcrystalline silicon oxide. Japanese Journal of Applied Physics. 2011;50:082301. DOI: 10.1143/JJAP.50.082301

[37] Xiao SQ, Xu S, Zhou HP, Wei DY, Huang SY, Xu LX, Sern CC, Guo YN, Khan S. Amorphous/crystalline silicon heterojunction solar cells via remote inductively coupled plasma processing. Applied Physics Letters. 2012;100:233902. DOI: 10.1063/1.4721642

[38] Suslick KS, Didenko Y, Fang MM, Hyeon T, Kolbeck KJ, McNamara III WB, Mdleleni MM, Wong M. Acoustic cavitation and its chemical consequences. Philosophical Transactions of the Royal Society A: Mathematical, Physical and Engineering Sciences. 1999;357:335–353. DOI: 10.1098/rsta.1999.0330

[39] Leighton TG. The Acoustic Bubble. London: Academic; 1994. 613 p. ISBN: 0–12–441920-8

[40] Yasui K. Influence of ultrasonic frequency on multibubble sonoluminescence. The Journal of the Acoustical Society of America. 2002;112:1405–1413. DOI: 10.1121/1.1502898

[41] Gutierrez M, Henglein A, Ibanez F. Radical scavenging in the sonolysis of aqueous solutions of iodide, bromide, and azide. The Journal of Physical Chemistry. 1991;95:6044–6047. DOI: 10.1021/j100168a061

[42] Kumar RV, Koltypin Yu, Xu XN, Yeshurun Y, Gedanken A, Felner I. Fabrication of magnetite nanorods by ultrasound irradiation. Journal of Applied Physics. 2001;89:6324–6328. DOI: 10.1063/1.1369408

[43] Prasanthkumar KP, Rayaroth MP, Alvarez-Idaboy JR. Insights into the mechanism of hydroxyl radical mediated oxidations of 2-aminopurine: A computational and sonochemical product analysis study. The Journal of Physical Chemistry B. 2020;124:6245–6256. DOI: 10.1021/acs.jpcb.0c03974

[44] González-García J, Sáez V, Tudela I, Díez-Garcia MI, Esclapez MD, Louisnard O. Sonochemical treatment of water polluted by chlorinated organocompounds. A review. Water. 2010;2:28–74. DOI: 10.3390/w2010028

[45] Xu H, Zeiger BW, Suslick KS. Sonochemical synthesis of nanomaterials. Chemical Society Reviews. 2013;42:2555–2567. DOI: 10.1039/c2cs35282f

[46] Birkin PR, Power JF, Leighton TG, Vinçotte AML. Cathodic electrochemical detection of sonochemical radical products. Analytical Chemistry. 2002;74:2584–2590. DOI: 10.1021/ac010964o

[47] Suslick KS. Sonochemistry. In: Kirk-Othmer Encyclopedia of Chemical Technology, v. 26. New York: Wiley; 1998. p. 517–541.

[48] Compton RG, Eklund JC, Marken F. Sonoelectrochemical processes. A review. Electroanalysis. 1997;9:509–522. DOI: 10.1002/elan.1140090702

[49] Brett C. Sonoelectrochemistry. In: Vives AA, editor. Piezoelectric Transducers and Applications. Heidelberg, Berlin: Springer-Verlag; 2008; Chapter 15, p. 399–411. DOI: 10.1007/978-3-540-77508-9

[50] Mason TJ, Lorimer JP, Bates DM. Quantifying sonochemistry: casting some light on a 'black art'. Ultrasonics. 1992;30:40–42. DOI: 10.1016/0041-624X(92)90030-P

[51] Kern W. Overview and evolution of silicon wafer cleaning technology. In: Reinhardt KA, Kern W, editors. Handbook of Silicon Wafer Cleaning Technology. 3rd ed. Elsevier: Oxford; 2018. p. 3–85. DOI: 10.1016/B978-0-323-51084-4.00001-0

[52] Chung HY, Kim YH, Cho HY, Lee BY, Yoo HD, Lee SH. Collection efficiency of metallic contaminants on Si wafer by vapor-phase decomposition-droplet collection. Analytical Sciences. 2001;17:653–658. DOI: 10.2116/analsci.17.653

[53] Lee D-H, Kim H-T, Jang S-H, Yi J-H, Choi E-S, Park J-G. Effect of organic acids in dilute HF solutions on removal of metal contaminants on silicon wafer. Microelectronic Engineering. 2018;198:98–102. DOI: 10.1016/j.mee.2018.06.012

[54] Ramappa DA, Henley WB. Effects of copper contamination in silicon on thin oxide breakdown. Journal of the Electrochemical Society. 1999;146:2258–2260. DOI: 10.1149/1.1391924

[55] Lim SW, Machuca F, Liao H, Chiarello RP, Helms RC. Effect of initial Al contamination on ultrathin gate oxides. Journal of the Electrochemical Society. 2000;147:1136–1140. DOI: 10.1149/1.1393325

[56] Ryuta J, Yoshimi T, Kondo H, Okuda H, Shimanuki Y. Adsorption and desorption of metallic impurities on Si wafer surface in SC1 solution. Japanese Journal of Applied Physics. 1992;31:2338–2342. DOI: 10.1143/JJAP.31.2338

[57] Osaka T, Hattori T. Influence of initial wafer cleanliness on metal removal efficiency in immersion SC-1 cleaning: Limitation of immersion-type wet cleaning. IEEE Transactions on Semiconductor Manufacturing. 1998;11:20–24. DOI: 10.1109/66.661280

[58] Kern W, Puotinen DA. Clean solutions based on hydrogen peroxide for use in silicon semiconductor technology. RCA Review. 1970;31:187–206.

[59] Hoefflinger B. ITRS: The international technology roadmap for semiconductors. In: Hoefflinger B, editor. Chips 2020. The Frontiers Collection. Springer: Berlin; 2011. p. 161–174. DOI: 10.1007/978-3-642-23096-7_7

[60] Fuchs FJ. Ultrasonic cleaning and washing of surfaces. In: Gallego-Juárez JA, Graff KF, editors. Power ultrasonics. Applications of high-intensity ultrasound. Amsterdam: Woodhead Publishing; 2015. p. 577–609.

DOI: 10.1016/B978-1-78242-028-6.00019-3

[61] Ellenberger J, van Baten JM, Krishna R. Exploiting the Bjerknes force in bubble column reactors. Chemical Engineering Science. 2005;60: 5962–5970. DOI: 10.1016/j.ces.2005.03.036

[62] Sapozhnikov OA. High-intensity ultrasonic waves in fluids: nonlinear propagation and effects. In: Gallego-Juárez JA, Graff KF, editors. Power ultrasonics. Applications of high-intensity ultrasound. Amsterdam: Woodhead Publishing; 2015. p. 9–35. DOI: 10.1016/B978-1-78242-028-6.00002-8

[63] Gale GW, Busnaina AA. Removal of particulate contaminants using ultrasonics and megasonics: A review. Particulate Science and Technology. 1995;13:197–211. DOI: 10.1080/02726359508906678

[64] Busnaina AA, Gale GW. Roles of cavitation and acoustic streaming in megasonic cleaning, J. Particulate Science and Technology. 1999;17:229–238. DOI: 10.1080/02726359908906815

[65] Kim JM, Kim YK. The enhancement of homogeneity in the textured structure of silicon crystal by using ultrasonic wave in the caustic etching process. Solar Energy Materials and Solar Cells. 2004;81:239–247. DOI: 10.1016/j.solmat.2003.11.019

[66] Ohl C-D, Arora M, Dijkink R, Janve V, Lohse D. Surface cleaning from laser-induced cavitation bubbles. Applied Physics Letters. 2006;89: 074102. DOI: 10.1063/1.2337506

[67] Endo M, Yoshida H, Maeda Y, Miyamoto N, Niwano M. Infrared monitoring system for the detection of organic contamination on a 300 mm Si wafer. Applied Physics Letters. 1999;75: 519–521. DOI: 10.1063/1.124434

[68] Podolian A, Nadtochiy A, Kuryliuk V, Korotchenkov O, Schmid J, Drapalik M, Schlosser V. The potential of sonicated water in the cleaning processes of silicon wafers. Solar Energy Materials and Solar Cells. 2011;95:765–772. DOI: 10.1016/j.solmat.2010.10.019

[69] Heinz H, Pramanik C, Heinz O, Y Ding, Mishra RK, Marchon D, Flatt RJ, Estrela-Lopis I, Llop J, Moya S, Ziolo RF. Nanoparticle decoration with surfactants: Molecular interactions, assembly, and applications. Surface Science Reports. 2017;72:1–58. DOI: 10.1016/j.surfrep.2017.02.001

[70] Swalen JD, Allara DL, Andrade JD, Chandross EA, Garoff S, Israelachvili J, McCarthy TJ, Murray R, Pease RF, JF Rabolt, Wynne KJ, Yu H. Molecular monolayers and films. Langmuir. 1987;3: 932–950. DOI: 10.1021/la00078a011

[71] Lévêque J-M, Fujita M, Bosson A, Sohmiya H, Pétrier C, Komatsu N, Kimura T. Secondary sonochemical effect on Mo-catalyzed bromination of aromatic compounds. Ultrasonics Sonochemistry. 2011;18:753–756. DOI: 10.1016/j.ultsonch.2010.11.006

[72] Sinton RA, Cuevas A. Contactless determination of current–voltage characteristics and minority-carrier lifetimes in semiconductors from quasi-steady-state photoconductance data. Applied Physics Letters. 1996;69:2510–2512. DOI: 10.1063/1.117723

[73] A Nadtochiy, Korotchenkov O, Drapalik M, Schlosser V. Effects of ultrasonic cleaning on carrier lifetimes and photovoltage in monocrystalline silicon. Solid State Phenomena. 2011; 178–179:221–225. DOI: 10.4028/www.scientific.net/SSP.178-179.221

[74] Shmid V, Podolian A, Nadtochiy A, Yazykov D, Semen'ko M, Korotchenkov O. Photovoltaic characterization of Si and SiGe surfaces sonochemically treated in

dichloromethane. Journal of Nano- and Electronic Physics. 2020;12:01023. DOI: 10.21272/jnep.12(1).01023

[75] Kumaran SS, Su M-C, Lim KP, Michael JV, Klippenstein SJ, DiFelice J, Mudipalli PS, Kiefer JH, Dixon DA, Peterson KA. Experiments and theory on the thermal decomposition of $CHCl_3$ and the reactions of CCl_2. Journal of Physical Chemistry A. 1997;101:8653–8661. DOI: 10.1021/jp971723g

[76] Chun YS, Lee ES, Jeong MG, Lim DS. Synthesis and characteristics of chloroform-treated silicon carbide-derived carbon layers. RSC Advances. 2016;6:96669–96675. DOI: 10.1039/C6RA22768F

[77] Royea WJ, Juang A, Lewis NS. Preparation of air-stable, low recombination velocity Si(111) surfaces through alkyl termination. Applied Physics Letters. 2000;77:1988–1990. DOI: 10.1063/1.1312203

[78] Hunger R, Fritsche R, Jaeckel B, Jaegermann W, Webb LJ, Lewis NS. Chemical and electronic characterization of methyl-terminated Si(111) surfaces by high-resolution synchrotron photoelectron spectroscopy. Physical Review B. 2005; 72:045317. DOI: 10.1103/PhysRevB.72.045317

[79] Mui K, Smith FW. Optical dielectric function of hydrogenated amorphous silicon: Tetrahedron model and experimental results. Physical Review B. 1988;38:10623–10632. DOI: 10.1103/PhysRevB.38.10623

[80] Kessels WMM, Marra DC, van de Sanden MCM, Aydil ES. In situ probing of surface hydrides on hydrogenated amorphous silicon using attenuated total reflection infrared spectroscopy. Journal of Vacuum Science & Technology A. 2002;20:781–789. DOI: 10.1116/1.1469012

[81] Brodsky MH, Cardona M, Cuomo JJ. Infrared and Raman spectra of the silicon-hydrogen bonds in amorphous silicon prepared by glow discharge and sputtering. Physical Review B. 1977;16: 3556–3571. DOI: 10.1103/PhysRevB.16.3556

Section 3

Emerging Photovoltaics

Chapter 10

Mixed 2D-3D Halide Perovskite Solar Cells

Alaa E. Abd El-Samad, Radwa S. Mostafa,
Hager H. Zeenelabden, Menahtullah M. Mabrouk,
Ahmed Mourtada Elseman, Nasr Gad, Mostafa El-Aasser
and Mohamed M. Rashad

Abstract

The 3D-perovskite halides have gained a considerable reputation versus their counterpart semiconductor materials since they achieved a remarkable high-power conversion efficiency of 25.2% within a decade. Perovskite solar cells also have some problems as lattice degradation and sensitivity against moisture, oxygen, and strong irradiation. The perovskite instability is the drawback in front of this emerging technology towards mass production and commercialization. 2D-perovskites, with the general formula $A_2B_{n-1}M_nX_{3n+1}$, have been recently introduced to overcome some of the drawbacks of the stability of 3D-perovskites; however, this is at the expense of sacrificing a part of the power conversion efficiency. Mixed 2D/3D perovskites could solve this dilemma towards the way to high stability-efficiency perovskites. The research is expected to obtain highly stable and efficient mixed 2D/3D perovskite solar cells in the few coming years. This chapter reviews 2D-perovskites' achieved progress, highlighting their properties, current trends, challenges, and future prospects.

Keywords: 2D, 3D, perovskite solar cells, stability, efficiency

1. Introduction

Perovskites are among the essential material science topics in the last decades due to their low-cost, solution-processed devices and exceptional optoelectronic properties [1–28]. The most studied compositions are represented by the formula ABX_3 (organic cation A is larger than the metal cation B, and X is a halide anion). For example, methylammonium lead tri-iodide ($CH_3NH_3PbI_3$ or $MAPbI_3$), the other halide variants such as $CH_3NH_3PbBr_3$ and mixed halides, $CH_3NH_3PbI_{3-x}Cl_x$ [29–31]. The most advantages of 3D-perovskite ($CH_3NH_3PbI_3$) are combining direct bandgap with high molar extinction coefficient ($\approx 10^4$–10^5 M^{-1} cm^{-1}), low trap densities, low exciton binding energies ($\approx$10–50 meV), which cause long-range free-carrier diffusion lengths ($\approx$100 nm). The perovskite is unique in such a way that its efficiency boosted up from 3.8% to 23.7% in just few years as compared to all other types of traditional solar cells. However, the lack of durability of these materials (hydrophilic cations) due to thermal instability and degradation upon exposure to humidity, U.V. radiation, and the electric field is still a significant barrier to commercialization [32–34]. 3D-perovskite solar cells' performance decreases due to ion migration and segregation, the ionic nature of the materials, and their low formation energies, making them vulnerable to water-induced hydrolysis [35]. The instability issues associated with perovskite materials have been overcome by using additives, introducing intermediate phases, encapsulating the layers (to avoid spreading Pb toxicity into the external environment), etc. [36–38]. Several studies have also shown that mixed cations and halides tend to enhance perovskites' stability and efficiency.

A 2D Ruddlesden–Popper perovskite has a general formula of $A_2B_{n-1}M_nX_{3n+1}$, where "A" represents a large-sized organic cation [39]. Incorporating a bulky organic cation (2D) into the 3D-perovskite layer's crystal lattice passivates the vulnerable 3D-perovskite against oxygen and moisture intrusion, resulting in enhanced stability while maintaining the efficiency of 3D-Perovskite solar cell [40]. 2D-perovskites are very stable but have larger bandgaps and higher exciton binding energies than 3D-perovskites. On the other hand, the exceptional stability under heat and light soaking conditions of low-dimensional perovskites makes them essential to protect the highly efficient 3D [41]. Phenyl ethyl ammonium iodide (PEAI) as a bulky cation in 2D-perovskite was investigated and fabricated as a layered 2D/3D structure demonstrating an impressive PCE 20.1% with 85% PCE retention after 800 h in ambient conditions by Cho et al. [40]. The perovskite absorber composition has been optimized using a 2D-perovskite, and stable performance over 12000 h without the HTM was observed [42].

The 2D-perovskite solar cells (PSCs) have shown superiority over 3D ones, such as improved stability towards humidity and light, improved processability, long-term durability, and higher chemical versatility. All this makes 2D-perovskites a promising alternative and one that has attracted substantial attention over the past few years. The ability of 2D-perovskites to incorporate bigger, less volatile, and generally hydrophobic organic cations; which makes the materials with improved thermal and chemical stability. Furthermore, the ability to use an enormous variety of organic cations and various metals, halides, and combinations of all of the above make this family of materials can be employed in different applications such as solar cells and others [43, 44]. Though there are several approaches to stabilize the 3D-perovskite, the most common one is cation tuning. The bigger cations beyond the Goldsmith tolerance limit produce low dimensional perovskites at least 2.5 eV, restricting photons conversion to less than 500 nm. 2D-perovskites are very stable, but unfortunately they have larger bandgaps and higher exciton binding energies ($\approx$300 meV), penalizing output photovoltage and, therefore, the power conversion efficiencies [45, 46]. Improved instability in metal halide PSCs is one of the most interesting issues to open the door for them towards commercialization. The reduced-dimensional perovskites (RDPs) have shown increase in ambient and operating lifetimes of PSCs. In other words 2D/3D heterostructure PSCs consisting of a thin layer of RDPs a top and a 3D active layer, improved both stability and efficiency compared to pure 3D counterparts [31, 47, 48], as shown in **Figure 1(a)**. Indeed, 2D/3D engineering aimed to combine both advantages, namely the outstanding optoelectronic properties of the 3D-perovskite and the 2D Ruddlesden–Popper phase's high robustness. Recently, based on the 2D/3D heterostructured lead halide perovskites, high-efficiency stable PSCs with an average PCE of 17.5 ± 1.3% was demonstrated. The "post-burn-in" efficiency could be retained over 80% after the air's operation for 1000 h and encapsulated process for 4000 h. Therefore, the performances of PSCs are brought closer to meeting the commercial requirements [49].

In this chapter, the 2D, 3D, and 2D/3D hybrid systems for perovskites will be discussed, focusing on the crystal structure, optoelectronic properties, synthesis methods, and layer orientation. Finally, application in regular or inverted structure

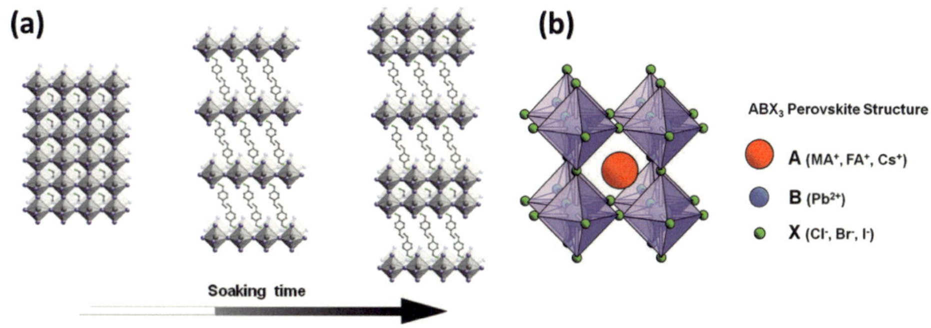

Figure 1.
(a) Dimensional mixed 2D/3D increases the soaking time and the stability of perovskite solar cells. Reprinted with permission from Ref. [31]. (b) The perovskite unit cell consists of an a cation (red) at the center, B cations in the corners (blue), and X anions (green) on the edges. The B cation forms an octahedral with the surrounding X anions (all eight octahedral are shown). Reprinted with permission from Ref. [50].

PSCs which remain to be addressed are herein highlighted while giving the outline on the perspectives of 3D and 2D/3D perovskites for high efficiency stable solar cells.

2. 3D-perovskites

2.1 Structures

Perovskite materials generally contain a cubic unit cell with the general formula ABX_3. Cation A, which is larger than cation B, is in the center of the unit cell. The B cations are in every corner of the unit cell; Cation B also serves as the center of an octahedron with an X anion surrounding cation B, corner-sharing between each cation B. As shown in **Figure 1(b)**, the full picture of cation A is surrounded by eight octahedra, each of which contains a cationic center B and anions X. In this orientation, the cubic structure of the perovskite has 6-fold the coordination number for cation A and 12-fold the coordination number for cation B. It should be noted that the ionic radii are quite crucial in maintaining a stable cubic unit cell.

2.2 Applications in regular or inverted structured perovskite solar cells

As mentioned, before we have n-i-p typical structure and p-i-n inverted structure. In 2018, Hua Dong et al. [51] applied $CH_3NH_3PbI_3$ film in highly efficient inverted planar heterojunction perovskite solar cells obtaining an efficiency of 17.04%. In 2020, Shuai Gu et al. [52] applied tin and mixed lead in tin halide perovskite tandem solar cells with a power conversion efficiency (PCE) over 25%.

2.3 Device fabrication methods

There are two ways to fabricate a PSC, solution and vacuum processing: (i) Spin coating is a solution deposition technique that uses high rotation speeds, as shown in **Figure 2(a** and **b)** [18, 26, 28]. A device rotates the substrate while a drop of the precursor solution is placed on the substrate. The high speed distributes the solution evenly on the substrate. After the material is deposited, the substrate is heated to evaporate and remove the solvent. This step is called the annealing step and the perovskite film is formed after removing the solvent. There are two methods of spin coating: a one-step and a two-step spin coating. In single-stage spin-coating, the

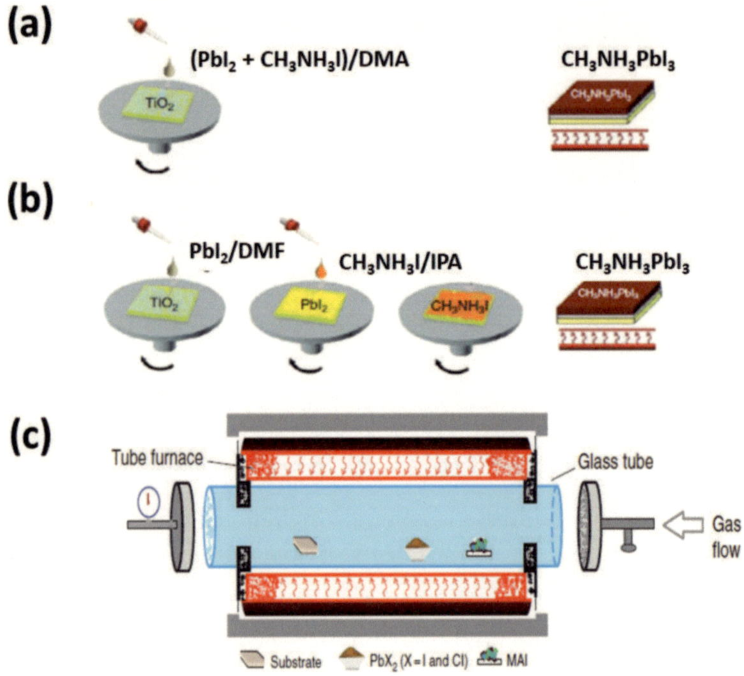

Figure 2.
(a) Method of one-step spin coating. (b) Method of two-step spin coating (c) a diagram of the CVD technique. Reprinted with permission from Ref. [53, 55]

solution contains all the chemicals deposited on the substrate. With a two-stage spin coating, organic and inorganic chemicals are deposited separately on the substrate. For example, $CH_3NH_3PbI_3$ perovskite material can be deposited onto a substrate by one-step or two-step spin-coating methods. In the one-step process, CH_3NH_3I and PbI_2 are mixed in solution with the solvent (dimethyl formamide (DMF)), and the solution is spin-coated onto the substrate. In the two-step method, PbI_2 is dissolved in the solvent (DMF) and spin-coated onto the substrate. Then, CH_3NH_3I is dissolved in the solvent (isopropanol (IPA)) and spin-coated onto the PbI_2-coated substrate [7, 53].

(ii) Vacuum treatment is a technique in which a CVD (Chemical Vapor Deposition) machine is used to achieve high temperatures in a glass housing [54]. Gases can flow through the pipe ends through the glass holder. This property is commonly used to achieve desired pressures or add reactive gas to the system. CVD has a temperature gradient along the tube so that the positions near the center are warmer than the positions near the ends of the tube. This temperature gradient is a critical aspect of the CVD manufacturing technique. The technology begins with the selection of the materials for the solar cell. The substrate is placed near the end of the tube and the materials for the solar cell are placed in a solid phase towards the center. When the CVD machine reaches the appropriate temperature, the solids in the center of the tube evaporate. An inert carrier gas such as argon flows through the tube and pushes the vaporized solids away from the substrate. The substrate has a lower temperature and causes condensation of the evaporated materials when they meet the substrate. This causes the materials to deposit on the substrate and form a thin layer. A diagram of the CVD technique is shown in **Figure 2(c)** [55].

The CVD technique can be used in several steps to deposit each material layer separately if desired. CVD technology has two distinct advantages over spin coating and other processes. First, the film layer produced is exceptionally clean since no solvent was used in the process, eliminating the possibility of impurities being added to the solvent. Second, the process is relatively easy to scale up for large-scale manufacture. In addition, perovskite films produced with this technique have good uniformity, non-porous films, large grain sizes, and a long carrier life. Two parameters are used to optimize the vacuum deposition technique, deposition time and temperature. Deposition time determines the amount of material deposited on the substrate, so film thickness is the main effect of deposition time. The climate is essential for vacuum deposition technology. The temperature should be high enough to vaporize the materials, but most importantly, the temperature should not be high enough to melt the substrate. Adjusting the position of the substrate and solid materials in the tube will help heat the materials sufficiently and prevent the substrate from melting. However, the temperature and location of placement must be used together to achieve the desired results [56].

3. 2D-perovskites

Despite the high performance of the 3D-perovskite [57], which qualified it to be a strong competitor to the various other types of solar cells, the stability or the ability of 3D to resist various factors of humidity, heat, and so on represent a critical issue in the direction of the possibility of becoming commercial [48]. Although the researchers' focus was first on 3D, they turned to 2D to solve the stability problem that plagues 3D [58]. In the next section, we are going to talk about the structure of the 2D-perovskites, their optoelectronic properties, preparation methods, layers orientation, and applications in regular or inverted structure PSCs.

3.1 Structures

The general chemical formula of the 2D-perovskite is $A_2B_{n-1}M_nX_{3n+1}$, where A can be a monovalent or divalent organic cation that intercalates between the inorganic $A_{n-1}B_nX_{3n+1}$ 2D sheets works as a spacer between the inorganic cation as shown in **Figure 3(a)**. n is the thickness or the number of the inorganic layers and (n = 1 at the divalent A, and n = 2 at the monovalent A) [43].

2D-halide perovskite layers are conceptually obtained by cutting along the crystallographic planes <100>, <110> or < 111> of the 3D-perovskite structure [59] as shown in **Figure 3(b)**, so we can classify the 2D perovskite depending on cutting the shape of the 3D-perovskite into <100>, <110>, and < 111> – oriented perovskites. Cutting layers along <110> direction (can be derived from the face diagonal) and along <111> direction (can be derived from the body diagonal) are less common in 2D-halide perovskites. Unlike these two types, <100> perovskites are the most common type of 2D-halide perovskites and are commonly used in solar cells. The general formula of <100> – oriented 2D-perovskites is $A_2B_{n-1}M_nX_{3n+1}$, and their inorganic sheets are obtained by taking n-layers along the 100 direction of the 3D-perovskites. The <100> – oriented 2D-perovskites can be divided into two commonly used types. The first is Ruddlesden-Popper (R.P.), and the second is Dion-Jacobson (D.J.) [60, 61].

In Ruddlesden-Popper (R.P.), the most used and studied type (owing to its superior ambient stability [62]) has the chemical formula $A_2B_{n-1}M_nX_{3n+1}$. Each inorganic layer is confined between bilayers of bulky ammonium cations. The relatively weak van der Waals forces between the alkyl chains separating the layers generate a 2D structure. In 2017, Xiaoyan Gan and co-workers fabricated a 2D-perovskite $(PEA)_2(MA)_{n-1}Pb_nI_{3n+1}$ (phenylethylammonium = PEA, n = 1, 2, 3) with incorporation of TiO_2 nanorod arrays into a solar cell harvesting efficiency of 3.72% [60] with a structure of glass/FTO/TiO_2 as compact layer/$(PEA)_2(MA)_{m-1}Pb_mI_{3m+1}$/spiroO-MeTAD/Au as shown in **Figure 4**.

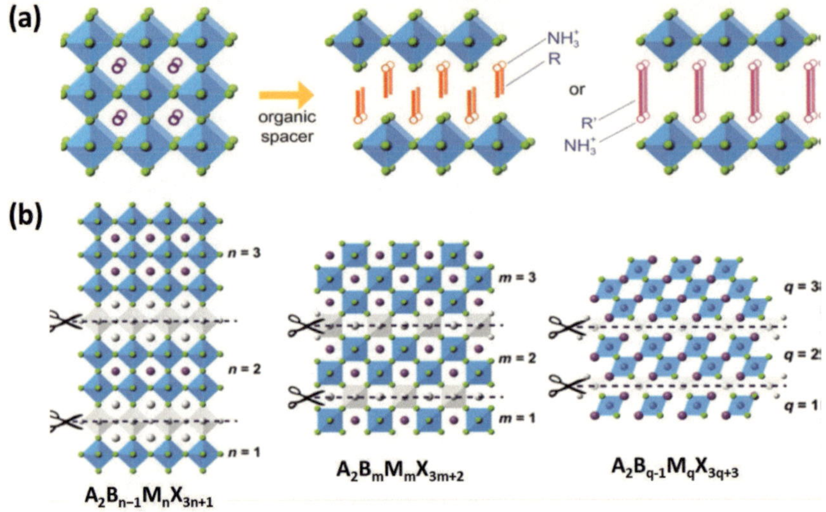

Figure 3.
(a) Crystal structures of a 3D-perovskite and the 2D-hybrid perovskite with monovalent and divalent ammonium cations. Reprinted with permission from Ref. [43]. (b) Cuts along <100>, <110> and < 111> directions and the 2D-perovskites that result from such cuts. Reprinted with permission from Ref. [43].

Dion–Jacobson (D.J.) perovskites adopt a general formula $AB_{n-1}M_nI_{3n+1}$. This type has layered structures where the stacking of inorganic layers is unique as they lay precisely on top of another, and this is quite the opposite of Ruddlesden-Popper (R.P.) [63]. The difference between R.P. and D.J. is shown in **Figure 5**.

In 2018, Sajjad Ahmad and co-workers developed a series of Dion-Jacobson phase 2D-perovskites that record a cell power conversion efficiency of 13.3% with high stability. Unencapsulated devices retain over 95% efficiency upon exposure to ambient air (40–70% relative humidity [R.H.]) for 4,000 hours, damp heat (85°C and 85% R.H.) for 168 hours, and continuous light illumination for 3,000 hours. This device is more stable than the R.P. counterpart, as shown in **Figure 6**. The

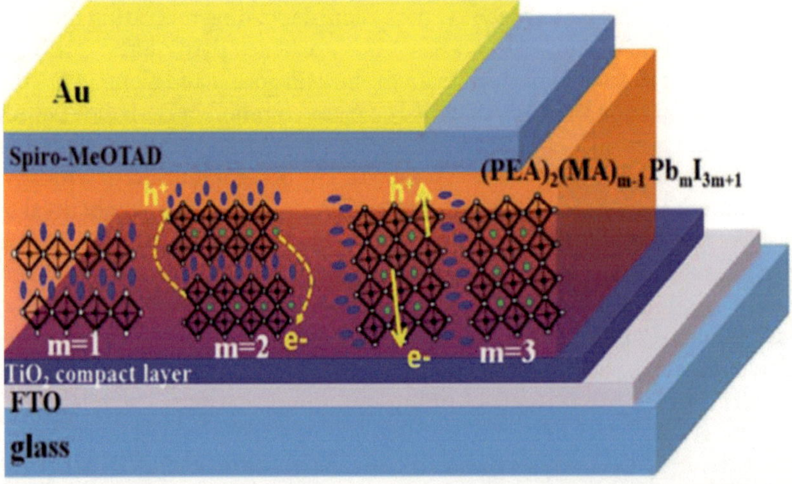

Figure 4.
A schematic device architecture of planar $(PEA)_2(MA)_{m-1}Pb_mI_{3m+1}$ based perovskite solar cell, consisting of glass/FTO/TiO_2CL/$(PEA)_2(MA)_{m-1}Pb_mI_{3m+1}$/spiroOMeTAD/Au. Reprinted with permission from Ref. [60].

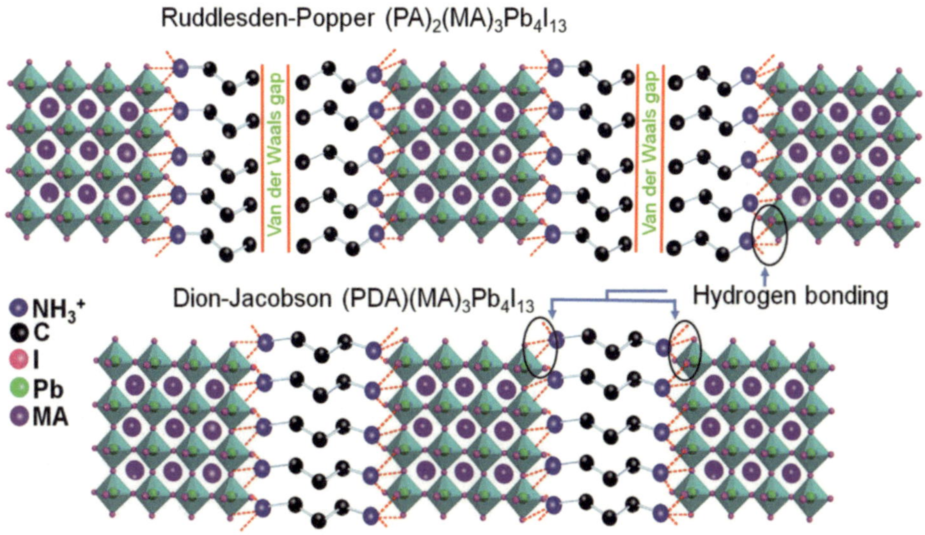

Figure 5.
Illustration of R.P. and D.J. phase 2D-layered perovskites. Reprinted with permission from Ref. [64].

improved device stability over the R.P. counterpart is attributed to alternating hydrogen bonding interactions between diammonium cations and inorganic slabs, strengthening the 2D-layered perovskite structure [64].

3.2 Applications in regular or inverted structure perovskite solar cells

The 2D-perovskites have many applications; it is used in solar cells, light emitting diodes, etc. Here we will concentrate on their applications in the n-i-p normal structure and p-i-n inverted structure solar cells as follows: In 2015, Cao and co-workers fabricated a 2D-perovskite thin-film and recorded an efficiency of 4.02% with a regular structure n-i-p with a device structure FTO/TiO_2/2D perovskite/spiro-OMeTAD/Au [65]. In 2016, Hsinhan Tsai and co-workers reported a photovoltaic efficiency of 12.52% with no hysteresis for an inverted structure solar cell FTO/PEDOT:PSS/$(BA)_2(MA)_3Pb_4I_{13}$/PCBM/Al [66]. In 2018, Xinqian Zhang and co-workers fabricated a vertically orientated highly crystalline 2D $(PEA)_2(MA)_{n-1}Pb_nI_{3n+1}$, n = 3, 4, 5) films with the assistance of an ammonium thiocyanate (NH_4SCN) additive. Planar-structured PSC with the device structure of ITO/PEDOT:PSS/$(PEA)_2(MA)_4Pb_5I_{16}$ (n = 5)/$PC_{61}BM$/BCP/Ag was fabricated. They got an efficiency up to 11.01% with the optimized NH_4SCN addition at n = 5 [67]. In the same year, Chunqing Ma and co-workers fabricated a 2D-PSC with the device configuration ITO/PEDOT:PSS/$PDAMA_3Pb_4I_{13}$/C_{60}/BCP/Ag and recorded an efficiency of 13% with $PDAMA_3Pb_4I_{13}$ (PDM: propane-1,3-diammonium) as the 2D-perovskite layer using p-i-n inverted structure [67].

3.3 Optoelectronic properties

In this subsection, we will discuss some of the 2D-perovskite's optoelectronic properties, like how reducing dimensionality can affect the bandgap, exciton's influence on charge transport, and so on. In the 2D-perovskites, the large-sized organic cation interlayers can restrict or limit the charge carriers. These organic interlayers act as dielectric regulators, determining the electrostatic force on the electron–hole pairs. The alternating arrangement of inorganic sheets and

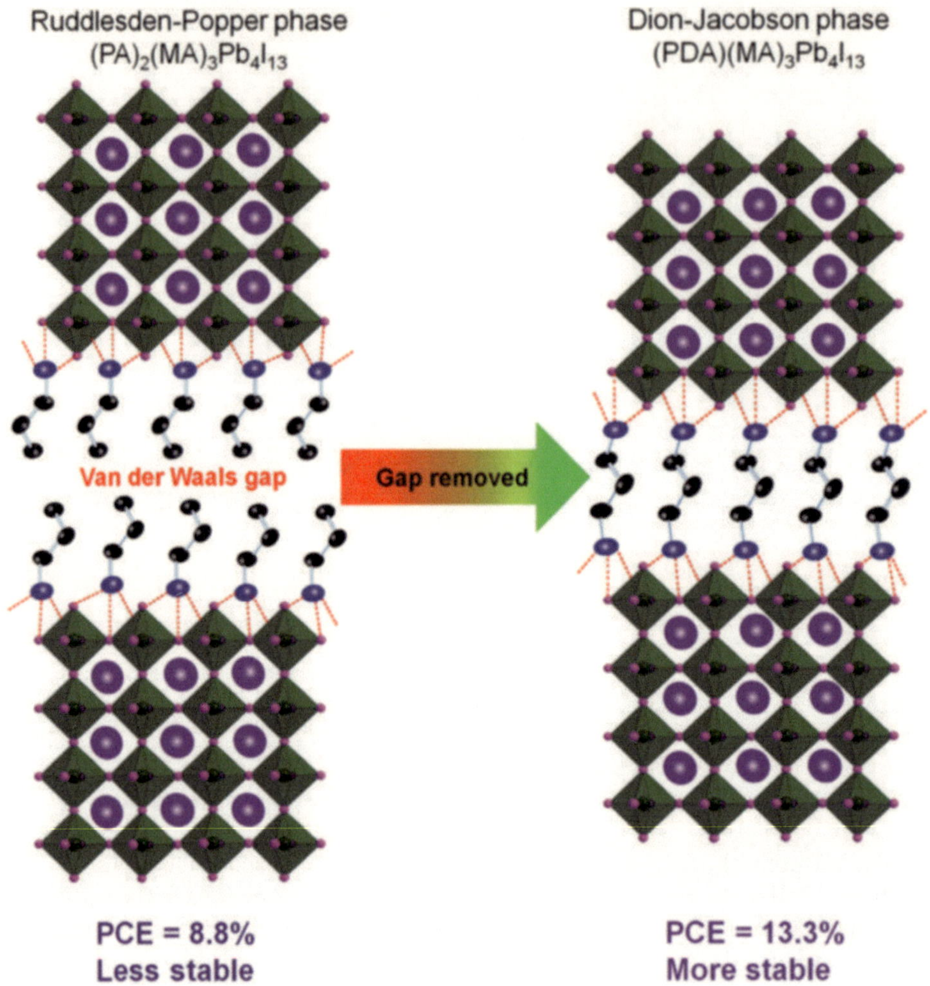

Figure 6.
The D.J. device is more stable and more efficient than the R.P. counterpart. Reprinted with permission from Ref. [64].

bulky organic interlayers results in a multiple-quantum-well (MQW) electronic structure [68]. In other words, the inorganic slabs serve as the potential "wells" while the organic layers function as the potential "barriers" [10] as shown in **Figure 7**. The high organic and inorganic dielectric contrast leads to a huge electron–hole binding energy (Eb) in 2D perovskites [69, 70]. The confinement effect of 2D-perovskites directly affects the bandgap. For an R.P. hybrid perovskite, the bandgap depends on the good width of layer thickness [71]. The total bandgap energy is determined by the base 3D-structure and extra quantization energies of the electron–hole [72]. The optical bandgap of 2D-perovskite generally decreases as the "n" value increases. For example, the bandgap value for the 2D $(CH_3(CH_2)_3NH_3)_2(CH_3NH_3)_{n-1}Pb_nI_{3n+1}$ perovskites decreases with increased layer thickness from 2.24 eV (at n = 1) to 1.52 eV (at n = ∞) due to the quantum-confinement effects associated with the dimensional increase [14]. This flexibility of bandgap tuning can facilitate various optoelectronic applications with targeted optical bandgap materials like tandem solar cells. Those in tandem solar cells, the upper absorber layer needs to have a higher bandgap than the bottom one [73].

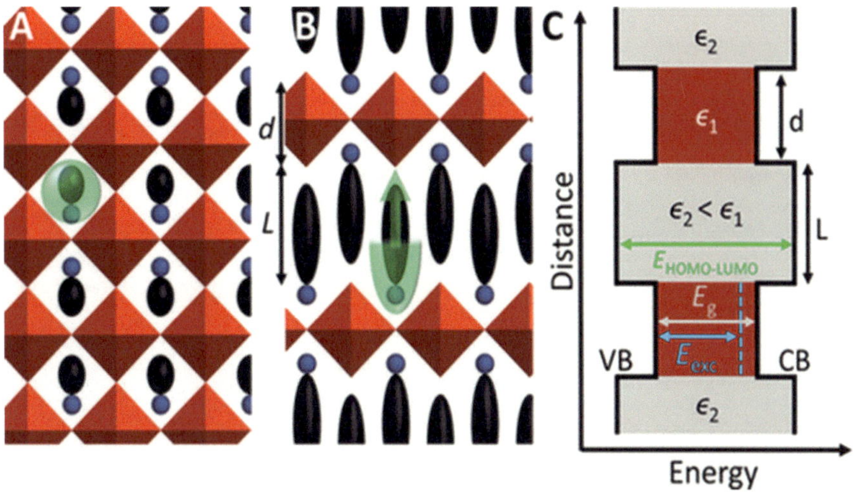

Figure 7.
(A) A schematic of a projection of the 3D-hybrid perovskite, showing an inorganic network of corner-sharing metal halide octahedra (red) with interstitial organic cations (blue, black). (green) highlighting the restriction on cation size. (B) Schematic of a projection down the c-axis of the 2D-hybrid perovskite, showing the alternation of organic (blue, black) and corner-sharing inorganic (red) layers for n = 1 with inorganic layer thickness d and organic layer thickness L. (green), highlighting the restriction solely on the cross-sectional area, but not the length of the organic cation. (C) Energy diagram corresponding to the 2D-structure in (B). Labeling of the valence band (V. B), conduction band (C. B), electronic band gap E_g (gray) and the optical band gap E_{exc} (blue) of the inorganic framework, and the larger HOMO-LUMO gap of the organic cations (green). The organic framework (gray regions) has a dielectric constant ϵ_2, which is smaller than the dielectric constant ϵ_1 of the inorganic framework (red areas). Reprinted with permission from Ref. [70].

Quasi-two-dimensional perovskites have been shown to have strong excitonic effects, and their structure generally shows a large exciton binding energy (E_b) of several hundreds of meV. This improves the interaction between electrons and holes compared to the exchange in 3D-perovskites [74]. The sizeable binding energy E_b in low "n" 2D-perovskites may be detrimental for charge separation in solar cells. So, the considerable binding energy of excitons is one of the main reasons for declining performance standards.

3.4 Device fabrication of 2D-perovskite films

The finite preparation methods of 2D-perovskite films are different from the multiple preparation methods of 3D-perovskite films. *One-step spin-coating methods* are the most used to prepare 2D-perovskite films [75]. In this method, organics and metal halides are dissolved in solvents, e.g., DMF or (DMSO/GBL(1/1)) on substrates. By adjusting the ratio of the precursors, the dimension of perovskite is changed. 2D-perovskite films in both n-i-p and p-i-n structures are fabricated using one-step spin-coating methods as $(PEA)_2(MA)_{n-1}Pb_nI_{3n+1}$, $(BA)_2(MA)_{n-1}Pb_nI_{3n+1}$ and $(PEI)_2(MA)_{n-1}Pb_nI_{3n+1}$ [65, 76, 77]. The *fast deposition-crystallization procedure* was also introduced to fabricate 2D-perovskite by dropping antisolvent, e.g., chlorobenzene, during the spin-coating process. It is shown, homogeneous nuclei are formed immediately and grow up slowly. Finally, dense and uniform films are obtained without oversize grains that may destroy the morphology [78]. The preferred *hot-cast method* was introduced for $(BA)_2(MA)_3Pb_4I_{13}$ films on PEDOT:PSS substrate [66]. The FTO/PEDOT:PSS substrate was heated before the precursor solution was spin-coated on it. **Figure 8** shows the photograph of $(BA)_2(MA)_3Pb_4I_{13}$ films that were prepared on substrates with different hot-cast temperatures from room temperature (R.T.) to 150°C. The films became dark and shiny with lower

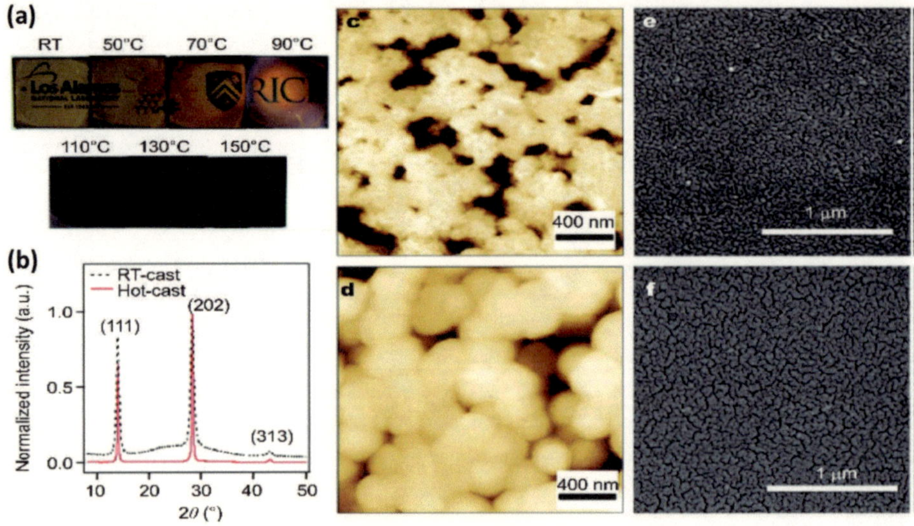

Figure 8.
(a) The $(BA)_2(MA)_3Pb_4I_{13}$ films cast from room temperature (R.T.) to 150°C, (b) GIXRD spectra, (c, d) AFM images, and (e, f) SEM images of films prepared by traditional room-temperature-cast method. Reprinted with permission from Ref. [66].

pinhole density with rising temperature, resulting in high-efficiency devices. The inorganic layers in 2D-perovskite films prepared by the hot-cast method have a preferential orientation vertical to the substrate with excellent crystallinity and few carrier traps, which favors the charge transport.

Another solution vapor annealing method was used to prepare high-quality $(PEA)_2PbBr_4$ nanosheets at room temperature as an emitting layer for LEDs [79]. Using a precursor solution of DMF comprising PEABr and $PbBr_2$ (2/1) was spin-coated on the top of the ITO/PEDOT:PSS substrate and the productive sample was placed face down on the edge of a glass dish without contacting it. In the following step, the dish was transferred into a lidded beaker filled with DMF to form a closed space with DMF vapor at 30°C for several minutes before the DMF vapor diffused under the $(PEA)_2PbI_4$ film and contacted with it. The film was removed out into open air rapidly and heated at 100°C for 10 min while turning into purple. By using DMF vapor annealing, the small and compact $(PEA)_2PbI_4$ perovskite grains was recrystallized into sized nanosheets equally distributed on the substrate, which has larger grain size, higher crystallinity, and higher P.L. intensity (higher photoluminescence quantum yield (PLQY) due to the quantum confinement), comparing with unprocessed films. The nanosheet-LEDs showed a longer P.L. lifetime (much longer than the 3D-perovskite $MAPbI_3$). They exhibited a small leak of current and low turn-on voltage, the external quantum efficiency (EQE) was 20 times higher than poly-LEDs.

Sequential deposition method The sequential deposition is used to prepare a sequential dipping process and sequential spin-coating of 2D-perovskite films [80]. The dipping process was used to fabricate quasi-2D and quasi 3D-perovskite films with spin-coating $(IC_2H_4NH_3)_2PbI_4$ layers on mp-TiO_2 substrates [81]. It is observed, immersing the films into MAI solution with a specific concentration for different times (1–5 min), transferring the films into cleaning fluid (2 mL isopropanol mixed with 10 mL methylbenzene) to remove the MAI residual. With increasing dipping time, MA^+ cations in solution passed through $(IC_2H_4NH_3)_2PbI_4$ films and increased inorganic PbI_4 layers' thickness. Then, the nanoparticles regularly grew up and interconnected with each other in film morphology (uniform films without residuals), leading to sharper diffraction peaks. In the previous studies, ITO/NiO_x/mixed-dimensional perovskite/PCBM/bis-C_{60}/Ag devices were prepared using the sequential spin-coating method with mixed-cations perovskite $FAxPEA_{1-x}PbI_3$ as the active layer [81]. With observing, a PbI_2 layer was spin-coated on NiO_x substrate, followed by the $PEA_xFA_{1-x}I$ solution loaded on it for 15 s before spin-coating. After thermal treatment, the mixed-cations $FA_xPEA_{1-x}PbI_3$ perovskite films were formed, with migrating PEA^+ cations to lattice surfaces and grain boundaries of 3D-perovskite $FAPbI_3$ to form quasi 3D- rather than quasi 2D-perovskite at room temperature. This self-assembly organic shell could prevent perovskite crystals from ambient moisture and passivated the surface defects to enhance the device performance and quasi 3D-perovskite's stability (the transition energy transformed from black phase to yellow phase). In summary, the 2D-perovskite films were commonly fabricated using one-step spin-coating method (simple process and low cost), with small-n members contrary to their 3D-counterparts but increasing "n" showed pre- or post-treatment is necessary such as hot-cast, antisolvent or solution vapor annealing processes for better crystallization. On the other hand, sequential deposition was used for obtaining efficient charge collection and extraction, dense and uniform films (due to the superior PbI_2 framework for crystal growth). In sequential spin-coating, the large cations were not likely to enter into the lattice but pack on the surface of 3D-perovskite grains, forming a quasi 3D-structure. For future large-scale fabrication, the solution process was not enough for high-quality. Still, doctor blading, pressure-processing method, and so on were more appropriate as long as high-quality crystal is needed [82]. Melt processing is another innovative technique with an excellent quality. It was implemented in 2017 to fabricate lead iodide based 2D-perovskites using PEA derivatives [83]. Although it had not been used for device fabrication yet, however it is a promising one that it exhibits high phase purity, crystallinity, and potential crystal orientation control under ambient air, but its disadvantage is the used toxic solvents in processing.

3.5 Layer orientation

The device performance has been enhanced using the vertical alignment of the inorganic sheets of the 2D-perovskites concerning the substrate. The vertically oriented inorganic slabs provided a direct pathway for charge transport between layers, whereas the bulky organic separators act as electrical insulators hindering out-of-plane conduction of charges [63]. 2D single-layered (n = 1) halide perovskites have shown to align horizontally to the substrate on which they were grown. When "n" was greater than 1, competition arose between horizontal and vertical to the perovskite layers alignment (caused by BA and MA cations). When n = 1–4, it was found that the devices with $n \geq 3$ had a better performance than lower "n", as shown in previous studies [84, 85]. It showed that for $BA_2MA_3Pb_4I_{13}$, the nucleation process and film growth orientation occurred at the liquid-air interface rather than at the liquid–liquid or liquid–substrate interfaces (since surface tension made nucleation and growth at the liquid-air interface more favorable) and that the initial nuclei are oriented in vertical configuration [86]. With increasing "n", the probability of obtaining an n-homogeneous-film decreased, giving rise to a mixture of a 3D-like dominant phase with some 2D-perovskite phases. Finally, films with n = 5–10 tend to align perpendicular to the substrate (**Figure 9**) [87]. This crystal orientation was examined using a scanning electron microscope (SEM), X-ray diffraction (XRD), and grazing incidence wide-angle X-ray scattering (GIWAXS) tests. XRD spectra of $(BA)_2(MA)_{n-1}Pb_nI_{3n+1}$ (n = 1, 2, 3, 4, ∞), with (n = 1) showed (00 k) peaks implying the preferential growth along (110) direction, in (n = 2), (0 k0) reflections occurred in addition to (111) and (222) in

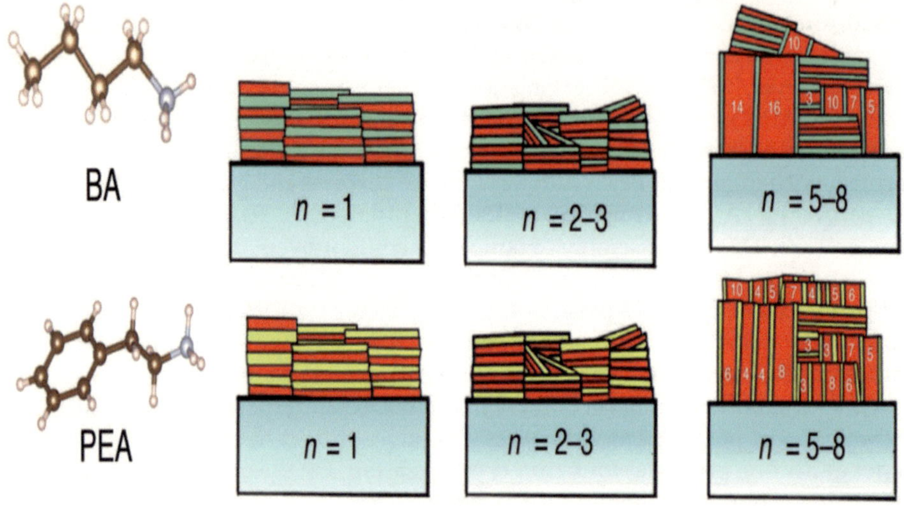

Figure 9.
The generalized concept of the Q.W. morphology in both PEA-based and BA-based spin-cast films. Reprinted with permission from Ref. [87].

the obtained XRD spectra corresponding to in-plane and out-of-plane orientations. In contrast, (111) and (222) reflections were noticed for n = 3 and n = 4 quasi 2D-perovskite films (110) and (220) analogously for n = ∞, which implies the vertical growth orientation (n = 3, 4), increasing towards 3D-perovskite; The preferential layer alignment vanished because few BA cations doped in 3D-perovskite and no influence on orientations **(Figure 9)** [65]. From (SEM, GIWAXS) it was concluded that the hot-cast films grew along certain orientations, confirmed by the most remarkable reflections of (111) and (202) planes. The inorganic crystal plates $<(MA)_{n-1}Pb_nI3_{n+1}>^{2-}$ are perpendicular to the substrate, forming continuous charge transfer channels favorable to charge transport for optoelectronic applications [66].

4. Mixed 2D/3D perovskites

Downsides of 2D-perovskites can be reduced by mixing 2D with the ordinary 3D-perovskites to form 2D/3D perovskite. 2D/3D designing intends to combine the advantages of 3D-perovskite (high optoelectronic properties) and 2D-perovskite (high stability) [88] to produce an efficient and stable perovskite material that could contribute to the advancement of PSCs towards commercial industrialization. So, owing to these reasons, the 2D/3D perovskite attracted the researcher's attention during the last few years [89].

4.1 Optoelectronic properties

The 2D/3D, in comparison to 2D-perovskite, has higher charge mobility, less non-radiative charge recombination, smaller bandgap as illustrated in **Figure 10**, higher power conversion efficiency, long-term stability, and in some cases, do not even need encapsulation.

The number of inorganic layers (n) affects the performance parameters of 2D/3D perovskite. The higher the (n) number, the better the performance

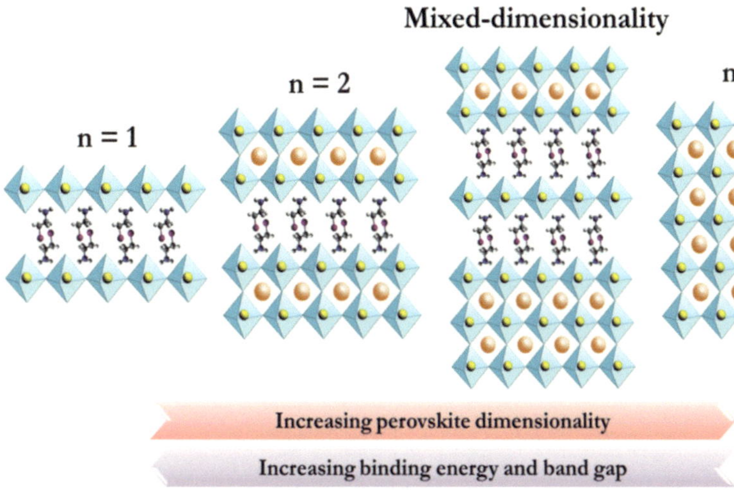

Figure 10.
A schematic illustrates the effect of increasing the dimensionality on the bandgap and the binding energy. Reprinted with permission from Ref. [89].

parameters (as the 2D/3D perovskite gets closer to 3D). Moreover, 2D in 2D/3D can work as a capping layer to protect the 3D layer from air, moisture, heat; It serves as a hydrophobic encapsulation layer. In 2014, Karunadasa et al. introduced the mixed dimensional perovskite in the PSCs for the first time [77]. In this work [77], a mixture of phenylethylammonium (PEA) and MA cations was prepared to obtain a Ruddlesden–Popper structure of $(PEA)_2(MA)_2[Pb_3I_{10}]$ at (n = 3) with an achieved efficiency of 4.73% [77]. The bandgap of the corresponding 2D and 3D-perovskite was 2.10, and 1.63 eV, respectively. After that, Sargent et al. investigated the efficiency and stability of $(PEA)_2(MA)_{n-1}[Pb_nI_{3n+1}]$ perovskites with higher n (n = 6, 10, 40, 60, and ∞) [78], the stability of this 2D/3D perovskites was improved compared to the 3D equivalents. The encapsulated device with lower value of n (i.e., closer to 2D) has the highest stability, but the devices with n < 40 have poor performances due to the lower carrier mobility and high radiative recombination losses. For perovskites with (n < 10), a wider bandgap and lower carrier transport lead to inferior performances. On the other hand, the perovskite with (n = 60) recorded the best efficiency (η =15.3%).

4.2 Device fabrication methods

The 2D/3D multidimensional perovskite can be synthesized via many ways: one-step deposition, two-step deposition, anti-solvent method, the self-assembly method, vapor-assisted solution deposition approach [90], etc. In the one-step deposition process, the 2D and 3D precursors are mixed, and the layers are grown simultaneously. In a two-step process, the 3D layer is first deposited then the 2D-perovskite is grown on top of it in a consecutive step. In 2019, Zhang and his co-workers reported a 2D/3D perovskite by post-treated n-butylamine iodide (BAI) and the residual PbI_2 on a one-step deposited $MAPbI_3$ film. They added a thin 2D-perovskite layer on the top of the 3D-perovskite and grain boundaries. The formed 2D/3D perovskite has the stability of 2D (after three months the perovskite still support 80% of its initial efficiency) and the high performance of 3D (V_{OC} = 1.09 V, J_{SC} = 22.55 mA/cm^2 a, FF =0.74, PCE = 18.3%) [91]. The anti-solvent method has been used as an ordinary recipe for getting a 3D-perovskite film [92]. The anti-solvent (e.g., C.B., toluene, etc.)

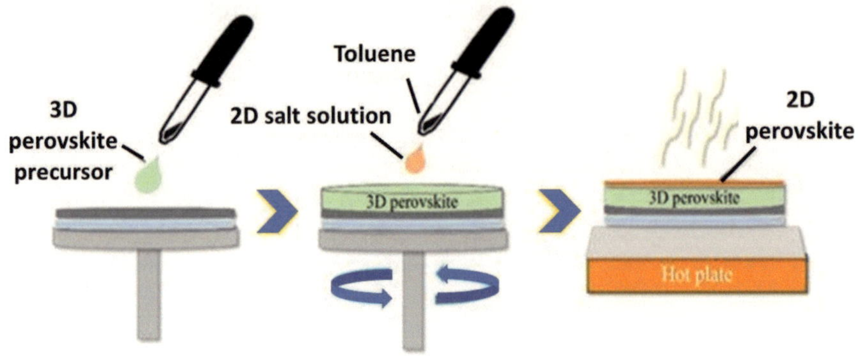

Figure 11.
The process of the anti-solvent method. Reprinted with permission from Ref. [40].

is employed as a second solvent to cause instantaneous nucleation for the 3D film's fast crystallization. When the anti-solvent is used as a solvent for the 2D organic salt, the in-situ 2D growth method can be practically combined with the usual anti-solvent method, as illustrated in **Figure 11**. The anti-solvent containing the organic salt for 2D is directly dripped on the substrate, which is wet with a precursor solution of the 3D-perovskite. The dripping process of the anti-solvent is followed by a one-step process. Toluene containing phenylethylammonium iodide (PEAI) was used as the anti-solvent for a 3D/2D perovskite structure [93].

In the low-pressure vapor assisted solution process, the larger organohalide (PEA) doped with a metal halide (PbI_2) is first spin-coated on the substrate and then reacts with the smaller organohalide (MAI) vapor in a low-pressure oven [94].

4.3 Applications in regular or inverted structure perovskite solar cells

Planer structures are always employed, n-i-p standard structure and p-i-n inverted structure. The researchers used 2D/3D perovskite in these two structures as follows: In 2019, Abbas and his co-workers [95] developed a way to add a 2D-layered perovskite with $MAPbI_3$ bulk 3D-perovskite by a vertically oriented 2D-layered mixed with a bulk 3D-perovskite they select the anilinium ion as the sizeable organic cation. In this research, they used a normal n-i-p structure and recorded an efficiency up to 16%. The planar solar cell heterojunction with ITO/PEDOT-PSS/perovskite/PCBM/BCP/Ag was fabricated [95]. In 2020, Jia Zhang and Bin Hu recorded a 15.93% efficiency with 2D/3D Pb-Sn alloyed perovskite. They applied the inverted structure p-i-n on their work with the solar cell structure (ITO/PEDOT:PSS/$PEA_xMA_{1x}Pb_{0.5}Sn_{0.5}I_3$/$PC_{61}BM$/PEI/Ag) [96].

5. Conclusion

Multi-dimensional 2D/3D hybrid perovskites have proven to be one of the most promising approaches to improving PCE and stability than pure 3D PSCs. Many novel 2D-perovskites can be synthesized in terms of materials engineering and consistent unknown properties studied. Besides, the recently proposed computational theoretical study would speed up selecting and applying 2D materials. In particular, long-chain alkyl cations seem to be suitable for stabilizing perovskite materials based on 2D-Pb-Sn or Sn. The performance can be improved by technical composition and structure. Besides, it remains essential to study the mechanism of formation of 2D-perovskite layers, which allows us to produce 2D-perovskites with high phase purity and out-of-plane orientation and to cultivate 2D/3D perovskite materials in a more controllable approach. Also, a thorough understanding of the chemical or physical interaction between 2D and 3D perovskite materials, 2D/3D perovskite materials, and neighboring charge transport layers will help achieve efficient and stable PSCs. As the bandgap increases, 2D-perovskite could be ideal for tandem solar cells to drive energy generation. In summary, the development strategy for multi-dimensional 2D/3D perovskites offers an opportunity for efficient and stable PSCs that enable ingenious modification of PSCs and inform their future marketing.

Acknowledgements

The authors acknowledge Science and Technology Development Fund (STDF) and (CMRDI) for their support to this study through the project grant no. 25250 in Egypt.

Conflict of interest

The authors declare no conflict of interest.

Author details

Alaa E. Abd El-Samad[1,2], Radwa S. Mostafa[1], Hager H. Zeenelabden[1,2], Menahtullah M. Mabrouk[3], Ahmed Mourtada Elseman[1*], Nasr Gad[2], Mostafa El-Aasser[2] and Mohamed M. Rashad[1]

1 Electronic and Magnetic Materials Department, Advanced Materials Division, Central Metallurgical Research and Development Institute (CMRDI), Cairo, Egypt

2 Faculty of Science, Physics Department, Ain Shams University, Cairo, Egypt

3 Faculty of Computer Science, Physics Department, Ain Shams University, Cairo, Egypt

*Address all correspondence to: amourtada@cmrdi.sci.eg

References

[1] Stranks SD, Eperon GE, Grancini G, Menelaou C, Alcocer MJ, Leijtens T, et al. Electron-hole diffusion lengths exceeding 1 micrometer in an organometal trihalide perovskite absorber. Science. 2013;342(6156): 341-344.

[2] Sum TC, Mathews N, Xing G, Lim SS, Chong WK, Giovanni D, et al. Spectral features and charge dynamics of lead halide perovskites: origins and interpretations. Acc Chem Res. 2016;49(2):294-302.

[3] De Wolf S, Holovsky J, Moon S-J, Löper P, Niesen B, Ledinsky M, et al. Organometallic halide perovskites: sharp optical absorption edge and its relation to photovoltaic performance. The journal of physical chemistry letters. 2014;5(6):1035-1039.

[4] Xu CY, Hu W, Wang G, Niu L, Elseman AM, Liao L, et al. Coordinated Optical Matching of a Texture Interface Made from Demixing Blended Polymers for High-Performance Inverted Perovskite Solar Cells. ACS Nano. 2020;14(1):196-203.

[5] Wang G, Liao LP, Elseman AM, Yao YQ, Lin CY, Hu W, et al. An internally photoemitted hot carrier solar cell based on organic-inorganic perovskite. Nano Energy. 2020;68:104383.

[6] Selim MS, Elseman AM, Hao Z. ZnO Nanorods: An Advanced Cathode Buffer Layer for Inverted Perovskite Solar Cells. ACS Applied Energy Materials. 2020;3(12):11781-11791.

[7] Elseman AM, Zaki AH, Shalan AE, Rashad MM, Song QL. TiO2 Nanotubes: An Advanced Electron Transport Material for Enhancing the Efficiency and Stability of Perovskite Solar Cells. Industrial & Engineering Chemistry Research. 2020;59(41):18549-18557.

[8] Elseman AM, Xu C, Yao Y, Elisabeth M, Niu L, Malavasi L, et al. Electron Transport Materials: Evolution and Case Study for High-Efficiency Perovskite Solar Cells. Solar RRL. 2020;4(7):2000136.

[9] Elseman AM, Luo L, Song QL. Self-doping synthesis of trivalent Ni2O3 as a hole transport layer for high fill factor and efficient inverted perovskite solar cells. Dalton Transactions. 2020;49(40):14243-14250.

[10] Yang XD, Han JJ, Wang G, Liao LP, Xu CY, Hu W, et al. Robust perovskite-based triboelectric nanogenerator enhanced by broadband light and interface engineering. Journal of Materials Science. 2019;54(12): 9004-9016.

[11] Sanad MMS, Elseman AM, Elsenety MM, Rashad MM, Elsayed BA. Facile synthesis of sulfide-based chalcogenide as hole-transporting materials for cost-effective efficient perovskite solar cells. Journal of Materials Science: Materials in Electronics. 2019;30(7):6868-6875.

[12] Sajid S, Elseman AM, Wei D, Ji J, Dou S, Huang H, et al. NiO@carbon spheres: A promising composite electrode for scalable fabrication of planar perovskite solar cells at low cost. Nano Energy. 2019;55:470-476.

[13] Sajid S, Elseman AM, Wei D, Ji J, Dou S, Huang H, et al. Corrigendum to "NiO@Carbon spheres: A promising composite electrode for scalable fabrication of planar perovskite solar cells at low cost" [Nano Energy, 55 (2019) 470-476]. Nano Energy. 2019;58:94-95.

[14] Liu DB, Wang G, Niu LB, Chen LJ, Liu DY, Rao X, et al. Energy Level Bending of Organic-Inorganic Halide Perovskite by Interfacial Dipole. physica

status solidi (RRL) – Rapid Research Letters. 2019;13(7):1900103.

[15] Hu W, Xu CY, Niu LB, Elseman AM, Wang G, Liu B, et al. High Open-Circuit Voltage of 1.134 V for Inverted Planar Perovskite Solar Cells with Sodium Citrate-Doped PEDOT:PSS as a Hole Transport Layer. ACS Appl Mater Interfaces. 2019;11(24):22021-22027.

[16] Elseman AM, Selim MS, Luo L, Xu CY, Wang G, Jiang Y, et al. Efficient and Stable Planar n-i-p Perovskite Solar Cells with Negligible Hysteresis through Solution-Processed Cu2 O Nanocubes as a Low-Cost Hole-Transport Material. ChemSusChem. 2019;12(16):3808-16.

[17] Elseman AM, Sajid S, Shalan AE, Mohamed SA, Rashad MM. Recent progress concerning inorganic hole transport layers for efficient perovskite solar cells. Appl Phys A. 2019;125(7).

[18] Wei D, Ma F, Wang R, Dou S, Cui P, Huang H, et al. Ion-Migration Inhibition by the Cation-pi Interaction in Perovskite Materials for Efficient and Stable Perovskite Solar Cells. Adv Mater. 2018;30(31):e1707583.

[19] Shalan AE, Barhoum A, Elseman AM, Rashad MM, Lira-Cantú M. Nanofibers as Promising Materials for New Generations of Solar Cells. In: Barhoum A, Bechelany M, Makhlouf A, editors. Handbook of Nanofibers. Cham: Springer International Publishing; 2018. p. 1-33.

[20] Sajid S, Elseman AM, Ji J, Dou S, Wei D, Huang H, et al. Computational Study of Ternary Devices: Stable, Low-Cost, and Efficient Planar Perovskite Solar Cells. Nano-Micro Letters. 2018;10(3).

[21] Sajid S, Elseman AM, Huang H, Ji J, Dou S, Jiang H, et al. Breakthroughs in NiOx-HTMs towards stable, low-cost and efficient perovskite solar cells. Nano Energy. 2018;51:408-424.

[22] Sajid AME, Jun Ji, Shangyi Dou, Hao Huang, Peng Cui, Dong Wei, Meicheng Li. Novel hole transport layer of nickel oxide composite with carbon for high-performance perovskite solar cells. Chin Phys B. 2018;27(1): 17305-017305.

[23] Elseman AM, Sharmoukh W, Sajid S, Cui P, Ji J, Dou S, et al. Superior Stability and Efficiency Over 20% Perovskite Solar Cells Achieved by a Novel Molecularly Engineered Rutin-AgNPs/Thiophene Copolymer. Adv Sci (Weinh). 2018;5(11):1800568.

[24] Elseman AM, Shalan AE, Sajid S, Rashad MM, Hassan AM, Li M. Copper-Substituted Lead Perovskite Materials Constructed with Different Halides for Working (CH3NH3)2CuX4-Based Perovskite Solar Cells from Experimental and Theoretical View. ACS Appl Mater Interfaces. 2018;10(14):11699-11707.

[25] Elseman AM. Organometal halide perovskites thin film and their impact on the efficiencyof perovskite solar cells. Coatings and Thin-Film Technologies: IntechOpen; 2018.

[26] Elseman AM, Shalan AE, Rashad MM, Hassan AM. Experimental and simulation study for impact of different halides on the performance of planar perovskite solar cells. Mater Sci Semicond Process. 2017;66:176-185.

[27] Rashad MM, Elseman AM, Hassan AM. Facile synthesis, characterization and structural evolution of nanorods single-crystalline (C4H9NH3)2PbI2X2 mixed halide organometal perovskite for solar cell application. Optik. 2016;127(20): 9775-9787.

[28] Elseman AM, Rashad MM, Hassan AM. Easily Attainable, Efficient Solar Cell with Mass Yield of Nanorod Single-Crystalline Organo-Metal Halide Perovskite Based on a Ball Milling

Technique. ACS Sustainable Chemistry & Engineering. 2016;4(9):4875-4886.

[29] Saparov B, Mitzi DB. Organic–inorganic perovskites: structural versatility for functional materials design. Chem Rev. 2016;116(7): 4558-4596.

[30] Wells A. Structural Inorganic Chemistry, 739 Oxford. UK: Clarendon Press; 1984.

[31] Teale S, Proppe A, Jung EH, Johnston A, Parmar DH, Chen B, et al. Dimensional Mixing Increases the Efficiency of 2D/3D Perovskite Solar Cells. The Journal of Physical Chemistry Letters. 2020.

[32] Jiang P, Xiong Y, Xu M, Mei A, Sheng Y, Hong L, et al. The influence of the work function of hybrid carbon electrodes on printable mesoscopic perovskite solar cells. The Journal of Physical Chemistry C. 2018;122(29):16481-16487.

[33] Leguy AM, Hu Y, Campoy-Quiles M, Alonso MI, Weber OJ, Azarhoosh P, et al. Reversible hydration of CH3NH3PbI3 in films, single crystals, and solar cells. Chem Mater. 2015;27(9):3397-3407.

[34] Christians JA, Miranda Herrera PA, Kamat PV. Transformation of the excited state and photovoltaic efficiency of CH3NH3PbI3 perovskite upon controlled exposure to humidified air. J Am Chem Soc. 2015;137(4):1530-1538.

[35] Urbani M, de la Torre G, Nazeeruddin MK, Torres T. Phthalocyanines and porphyrinoid analogues as hole-and electron-transporting materials for perovskite solar cells. Chem Soc Rev. 2019;48(10):2738-2766.

[36] Wang H, Zhao Y, Wang Z, Liu Y, Zhao Z, Xu G, et al. Hermetic seal for perovskite solar cells: An improved plasma enhanced atomic layer deposition encapsulation. Nano Energy. 2020;69:104375.

[37] Meng H, Shao Z, Wang L, Li Z, Liu R, Fan Y, et al. Chemical composition and phase evolution in DMAI-derived inorganic perovskite solar cells. ACS Energy Letters. 2019;5(1):263-270.

[38] Kang Y-J, Kwon S-N, Cho S-P, Seo Y-H, Choi M-J, Kim S-S, et al. Antisolvent additive engineering containing dual-function additive for triple-cation p–i–n perovskite solar cells with over 20% PCE. ACS Energy Letters. 2020;5(8):2535-2545.

[39] Yao Q, Xue Q, Li Z, Zhang K, Zhang T, Li N, et al. Graded 2D/3D Perovskite Heterostructure for Efficient and Operationally Stable MA-Free Perovskite Solar Cells. Adv Mater. 2020:2000571.

[40] Choi H-S, Kim H-S. 3D/2D bilayerd perovskite solar cells with an enhanced stability and performance. Materials. 2020;13(17):3868.

[41] Jodlowski AD, Roldán-Carmona C, Grancini G, Salado M, Ralaiarisoa M, Ahmad S, et al. Large guanidinium cation mixed with methylammonium in lead iodide perovskites for 19% efficient solar cells. Nature Energy. 2017;2(12):972-979.

[42] Grancini G, Roldán-Carmona C, Zimmermann I, Mosconi E, Lee X, Martineau D, et al. One-Year stable perovskite solar cells by 2D/3D interface engineering. Nature communications. 2017;8(1):1-8.

[43] Ortiz-Cervantes C, Carmona-Monroy P, Solis-Ibarra D. Two-Dimensional Halide Perovskites in Solar Cells: 2D or not 2D? ChemSusChem. 2019;12(8):1560-1575.

[44] Chen P, Bai Y, Lyu M, Yun JH, Hao M, Wang L. Progress and

Perspective in Low-Dimensional Metal Halide Perovskites for Optoelectronic Applications. Solar Rrl. 2018;2(3):1700186.

[45] Misra RK, Cohen BE, Iagher L, Etgar L. Low-dimensional organic–inorganic halide perovskite: structure, properties, and applications. Chem SusChem. 2017;10(19):3712-3721.

[46] Slavney AH, Smaha RW, Smith IC, Jaffe A, Umeyama D, Karunadasa HI. Chemical approaches to addressing the instability and toxicity of lead–halide perovskite absorbers. Inorg Chem. 2017;56(1):46-55.

[47] Rodríguez-Romero J, Sanchez-Diaz J, Echeverría-Arrondo C, Masi S, Esparza D, Barea EM, et al. Widening the 2D/3D Perovskite Family for Efficient and Thermal-Resistant Solar Cells by the Use of Secondary Ammonium Cations. ACS Energy Letters. 2020;5(4):1013-1021.

[48] Wang Z, Lin Q, Chmiel FP, Sakai N, Herz LM, Snaith HJ. Efficient ambient-air-stable solar cells with 2D–3D heterostructured butylammonium-caesium-formamidinium lead halide perovskites. Nature Energy. 2017;2(9):17135.

[49] Di Giacomo F, Fakharuddin A, Jose R, Brown TM. Progress, challenges and perspectives in flexible perovskite solar cells. Energy & Environmental Science. 2016;9(10):3007-3035.

[50] Lu C-H, Biesold-McGee GV, Liu Y, Kang Z, Lin Z. Doping and ion substitution in colloidal metal halide perovskite nanocrystals. Chem Soc Rev. 2020;49(14):4953-5007.

[51] Dong H, Wu Z, Xi J, Xu X, Zuo L, Lei T, et al. Pseudohalide-Induced Recrystallization Engineering for CH3NH3PbI3 Film and Its Application in Highly Efficient Inverted Planar Heterojunction Perovskite Solar Cells. Adv Funct Mater. 2018;28(2):1704836.

[52] Gu S, Lin R, Han Q, Gao Y, Tan H, Zhu J. Tin and mixed lead–tin halide perovskite solar cells: progress and their application in tandem solar cells. Adv Mater. 2020:1907392.

[53] Selim MS, Elseman AM, Hao Z. ZnO Nanorods: An Advanced Cathode Buffer Layer for Inverted Perovskite Solar Cells. ACS Applied Energy Materials. 2020.

[54] Wang DN, White JM, Law KS, Leung C, Umotoy SP, Collins KS, et al. Thermal CVD/PECVD reactor and use for thermal chemical vapor deposition of silicon dioxide and in-situ multi-step planarized process. Google Patents; 1991.

[55] Meléndrez M, Solis-Pomar F, Gutierrez-Lazos C, Flores P, Jaramillo A, Fundora A, et al. A new synthesis route of ZnO nanonails via microwave plasma-assisted chemical vapor deposition. Ceram Int. 2016;42(1):1160-1168.

[56] Shahidi S, Moazzenchi B, Ghoranneviss M. A review-application of physical vapor deposition (PVD) and related methods in the textile industry. The European Physical Journal Applied Physics. 2015;71(3):31302.

[57] Grancini G, Nazeeruddin MK. Dimensional tailoring of hybrid perovskites for photovoltaics. Nature Reviews Materials. 2019;4(1):4-22.

[58] Lin H, Zhou C, Tian Y, Siegrist T, Ma B. Low-dimensional organometal halide perovskites. ACS Energy Letters. 2017;3(1):54-62.

[59] Shi E, Gao Y, Finkenauer BP, Coffey AH, Dou L. Two-dimensional halide perovskite nanomaterials and heterostructures. Chem Soc Rev. 2018;47(16):6046-6072.

[60] Gan X, Wang O, Liu K, Du X, Guo L, Liu H. 2D homologous organic-inorganic hybrids as light-absorbers for planer and nanorod-based perovskite solar cells. Sol Energy Mater Sol Cells. 2017;162:93-102.

[61] Zhang F, Lu H, Tong J, Berry JJ, Beard MC, Zhu K. Advances in two-dimensional organic–inorganic hybrid perovskites. Energy & Environmental Science. 2020;13(4):1154-1186.

[62] Chen Y, Yu S, Sun Y, Liang Z. Phase engineering in quasi-2D Ruddlesden–Popper perovskites. The journal of physical chemistry letters. 2018;9(10):2627-2631.

[63] Mao L, Ke W, Pedesseau L, Wu Y, Katan C, Even J, et al. Hybrid Dion–Jacobson 2D lead iodide perovskites. J Am Chem Soc. 2018;140(10): 3775-3783.

[64] Ahmad S, Fu P, Yu S, Yang Q, Liu X, Wang X, et al. Dion-Jacobson phase 2D layered perovskites for solar cells with ultrahigh stability. Joule. 2019;3(3):794-806.

[65] Cao DH, Stoumpos CC, Farha OK, Hupp JT, Kanatzidis MG. 2D homologous perovskites as light-absorbing materials for solar cell applications. J Am Chem Soc. 2015;137(24):7843-7850.

[66] Tsai H, Nie W, Blancon J-C, Stoumpos CC, Asadpour R, Harutyunyan B, et al. High-efficiency two-dimensional Ruddlesden–Popper perovskite solar cells. Nature. 2016;536(7616):312-316.

[67] Zhang X, Wu G, Fu W, Qin M, Yang W, Yan J, et al. Orientation regulation of phenylethylammonium cation based 2D perovskite solar cell with efficiency higher than 11%. Advanced Energy Materials. 2018;8(14):1702498.

[68] Davy MM, Jadel TM, Qin C, Luyun B, Mina G. Recent progress in low dimensional (quasi-2D) and mixed dimensional (2D/3D) tin-based perovskite solar cells. Sustainable Energy & Fuels. 2020.

[69] Guo Z, Wu X, Zhu T, Zhu X, Huang L. Electron–phonon scattering in atomically thin 2D perovskites. ACS nano. 2016;10(11):9992-9998.

[70] Straus DB, Kagan CR. Electrons, excitons, and phonons in two-dimensional hybrid perovskites: connecting structural, optical, and electronic properties. The journal of physical chemistry letters. 2018;9(6):1434-1447.

[71] Gao P, Nazeeruddin MK. Dimensionality engineering of hybrid halide perovskite light absorbers. Nature communications. 2018;9(1):1-14.

[72] Sichert JA, Tong Y, Mutz N, Vollmer M, Fischer S, Milowska KZ, et al. Quantum size effect in organometal halide perovskite nanoplatelets. Nano Lett. 2015;15(10):6521-6527.

[73] Pohle L. US photovoltaic patents: 1991--1993. National Renewable Energy Lab., Golden, CO (United States); 1995.

[74] Manser JS, Christians JA, Kamat PV. Intriguing optoelectronic properties of metal halide perovskites. Chem Rev. 2016;116(21):12956-13008.

[75] Cortecchia D, Dewi HA, Yin J, Bruno A, Chen S, Baikie T, et al. Lead-free MA2CuCl x Br4–x hybrid perovskites. Inorg Chem. 2016;55(3):1044-1052.

[76] Yao K, Wang X, Xu Y-x, Li F, Zhou L. Multilayered perovskite materials based on polymeric-ammonium cations for stable large-area solar cell. Chem Mater. 2016;28(9): 3131-3138.

[77] Smith IC, Hoke ET, Solis-Ibarra D, McGehee MD, Karunadasa HI. A layered hybrid perovskite solar-cell absorber with enhanced moisture stability. Angew Chem. 2014;126(42): 11414-11417.

[78] Quan LN, Yuan M, Comin R, Voznyy O, Beauregard EM, Hoogland S, et al. Ligand-stabilized reduced-dimensionality perovskites. J Am Chem Soc. 2016;138(8):2649-2655.

[79] Liang D, Peng Y, Fu Y, Shearer MJ, Zhang J, Zhai J, et al. Color-pure violet-light-emitting diodes based on layered lead halide perovskite nanoplates. ACS nano. 2016;10(7): 6897-6904.

[80] Li N, Zhu Z, Chueh CC, Liu H, Peng B, Petrone A, et al. Mixed cation FAxPEA1–xPbI3 with enhanced phase and ambient stability toward high-performance perovskite solar cells. Advanced Energy Materials. 2017;7(1):1601307.

[81] Koh TM, Shanmugam V, Schlipf J, Oesinghaus L, Müller-Buschbaum P, Ramakrishnan N, et al. Nanostructuring mixed-dimensional perovskites: a route toward tunable, efficient photovoltaics. Adv Mater. 2016;28(19):3653-3661.

[82] Ma S, Cai M, Cheng T, Ding X, Shi X, Alsaedi A, et al. Two-dimensional organic-inorganic hybrid perovskite: from material properties to device applications. Science China Materials. 2018;61(10):1257-1277.

[83] Li T, Dunlap-Shohl WA, Han Q, Mitzi DB. Melt processing of hybrid organic–inorganic lead iodide layered perovskites. Chem Mater. 2017;29(15):6200-6204.

[84] Tsai H, Asadpour R, Blancon J-C, Stoumpos CC, Even J, Ajayan PM, et al. Design principles for electronic charge transport in solution-processed vertically stacked 2D perovskite quantum wells. Nature communications. 2018;9(1):1-9.

[85] Liao Y, Liu H, Zhou W, Yang D, Shang Y, Shi Z, et al. Highly oriented low-dimensional tin halide perovskites with enhanced stability and photovoltaic performance. J Am Chem Soc. 2017;139(19):6693-6699.

[86] Quintero-Bermudez R, Gold-Parker A, Proppe AH, Munir R, Yang Z, Kelley SO, et al. Compositional and orientational control in metal halide perovskites of reduced dimensionality. Nature materials. 2018;17(10):900-907.

[87] Wu C-G, Chiang C-H, Chang SH. A perovskite cell with a record-high-V oc of 1.61 V based on solvent annealed CH 3 NH 3 PbBr 3/ICBA active layer. Nanoscale. 2016;8(7):4077-4085.

[88] Gharibzadeh S, Abdollahi Nejand B, Jakoby M, Abzieher T, Hauschild D, Moghadamzadeh S, et al. Record Open-Circuit Voltage Wide-Bandgap Perovskite Solar Cells Utilizing 2D/3D Perovskite Heterostructure. Advanced Energy Materials. 2019;9(21):1803699.

[89] Krishna A, Gottis S, Nazeeruddin MK, Sauvage F. Mixed dimensional 2D/3D hybrid perovskite absorbers: the future of perovskite solar cells? Adv Funct Mater. 2019;29(8):1806482.

[90] Yusoff ARbM, Nazeeruddin MK. Low-Dimensional Perovskites: From Synthesis to Stability in Perovskite Solar Cells. Advanced Energy Materials. 2018;8(26):1702073.

[91] Zou Y, Cui Y, Wang H-Y, Cai Q, Mu C, Zhang J-P. Highly efficient and stable 2D–3D perovskite solar cells fabricated by interfacial modification. Nanotechnology. 2019;30(27):275202.

[92] Xiao M, Huang F, Huang W, Dkhissi Y, Zhu Y, Etheridge J, et al. A fast deposition-crystallization

procedure for highly efficient lead iodide perovskite thin-film solar cells. Angew Chem Int Ed. 2014;53(37):9898-9903.

[93] Bai Y, Xiao S, Hu C, Zhang T, Meng X, Lin H, et al. Dimensional Engineering of a Graded 3D–2D Halide Perovskite Interface Enables Ultrahigh Voc Enhanced Stability in the p-i-n Photovoltaics. Advanced Energy Materials. 2017;7(20):1701038.

[94] Li MH, Yeh HH, Chiang YH, Jeng US, Su CJ, Shiu HW, et al. Highly Efficient 2D/3D Hybrid Perovskite Solar Cells via Low-Pressure Vapor-Assisted Solution Process. Adv Mater. 2018;30(30):1801401.

[95] Abbas MS, Hussain S, Zhang J, Wang B, Yang C, Wang Z, et al. Orientationally engineered 2D/3D perovskite for high efficiency solar cells. Sustainable Energy & Fuels. 2020;4(1):324-330.

[96] Zhang J, Hu B. Revealing photoinduced bulk polarization and spin-orbit coupling effects in high-efficiency 2D/3D Pb–Sn alloyed perovskite solar cells. Nano Energy. 2020;76:104999.

Chapter 11

A New Generation of Energy Harvesting Devices

Byunghong Lee and Robert Bob Chang

Abstract

This chapter has been mainly focused on the development and fabrication of various nanostructured materials for electrochemical energy conversion, specially, third generation (3rd) thin film photovoltaic system such as organic dye or perovskite -sensitized Solar Cells. Enormous efforts have been dedicated to the development of a variety of clean energy, capable of harvesting energy of various forms. Among the various energy forms, electrochemical devices that produce electric energy from chemical energy have received the most attention as the most promising power sources. In the majority of cases, researchers who come from the different background could engage on certain aspects of the components to improve the photovoltaic performances from different disciplines: (i) chemists to design and synthesize suitable donor–acceptor dyes and study structure–property relationships; (ii) physicists to build solar cell devices with the novel materials, to characterize and optimize their performances, and to understand the fundamental photophysical processes; and (iii) engineers to develop new device architectures. The synergy between all the disciplines will play a major role for future advancements in this area. However, the simultaneous development of all components such as photosensitizers, hole transport layer, photoanodes and cost effective cathode, combined with further investigation of transport dynamics, will lead to Photovoltaic cells, 30%. Herein, in this book, with taking optimized processing recipe as the standard cell fabrication procedure, imporant breakthough for each components is achieved by developing or designing new materials, concepts, and fabrication technique. This book report the following studies: (i) a brief introduction of the working principle, (ii) the detailed study of the each component materials, mainly including TiO_2 photoanode under the category of 0D and 3D structures, strategies for co-sensitization with porphyrin and organic photosensitizers, and carbon catalytic material via controlled fabrication protocols and fundamental understanding of the working principles of electrochemical photovoltaic cell has been gained by means of electrical and optical modelling and advanced characterization techniques and (iii) new desgined stratages such as the optimization of photon confinement (iv) future prospects and survival stratagies for sensitizer assisted solar cell (especially, DSSC).

Keywords: photovoltaic cell, DSSC, perovskite, TiO_2, photosensitizer, carbon, photonic crystal

1. Introduction

The search for grean sources of enegy is considered one of the priorities in today's societies and occupies many policy makers' agendas. Excitonic Solar cell such as organic dye or perovskite -sensitized solar cells (DSSC or PSC) is showing up as a promising solar harvesting technology that has bright future. Unlike conventional silicon p-n junction solar cell, an excitonic solar cell (ESC) can be modeled as a unipolar-junction cell and made from low-cost materials that do not need to be highly purified but still work well using simple manufacturing processes [1]. For example, similar to the photosynthetic process in plants where chlorophyll absorbs photons but does not participate in charge transfer, the photoreceptor and charge carrier are implemented by different components in solar cell. This separation of functions leads to lower purity demands on raw materials and consequently makes exitonic solar cell a low-cost alternative. Nevertheless, to compete with the future PV market, ESC should foucus on the "Golden Triangle" issues, i.e., increasing light-to-electric energy conversion efficiency, enhancing long-term stability, and decreasing device cost [2].

Since the last third decades, scientists have devoted a great deal of effort on DSSCs' four important components: photosensitizer, photoanodes, electrolytes, and counter electrodes; some significant processes have been achieved. The sensitization of semiconductors using dyes dates back to 19th century. It showed that the photosensitivity can be extended to longer wavelengths by adding a dye to silver halide emulsions [3, 4]. Grätzel has then extended the concept to the DSSC by adsorption of dye molecules on the nanocrystalline TiO_2 electrodes [5]. This breakthrough was due to large surface area of the mesoporous TiO_2 that allowed anchoring significantly high amount of dye molecules ($\sim$0.13 mmol/cm^2) onto it; thereby increasing the absorption cross-section. Since the first successful demonstration of DSSC over three decades ago, a wealth of DSSC components have been investigated for enhancing energy conversion efficiency. State of the art DSSCs achieve more than 11% energy efficiency allied to good performance under any atmospheric condition and low irradiance. Moreover, the possibilities of transparency and multicolored cells are truly a competitive advantages of DSSCs. However, for stimulating industrial market, performance of DSSCs must be improved close to the efficiency values of the conventioanl silicon solar cells.

Since 2012, organic–inorganic perovskites solar have attracted considerable research attention for photovoltaic device due to its superb light-harvesting characteristics and electrical properties. Within relatively a short time, this organic–inorganic lead halide perovskite has yielded photovoltaic efficiencies of 25.5%, being the highest-performing solution-processed solar cell on record and displacing technologies such as DSSCs and organic photovoltaics (OPVs) [6]. In addition, the use of perovskite materials as light absorbers and hole transporting mateiral (HTM) in solid-state sensitized solar cells has opened up a new direction for the high-efficiency thin-film solar cells. Despite the best candidate at satisfying the need for high efficiencies, concerns surrounding the long-term stability in an ambient atmosphere as well as the water-soluble toxic lead components in the archetypal perovskite, $APbX_3$ (A = methylammonium, formamidium, cesium, X = I^-, Br^{--}, Cl^-), have the potential risk for the environment issue. After around 10 years of intense PSC researches, several physical chemistry still have questions for the basic mechanism. In this aspect, the physical and chemical operation of DSSC has becom more clear and a good example of a ESC system where the function of the overall device is better than prodicted from the sum of the properties of its components. Therefore, this book address imporant breakthough's our laboratory research for DSSC with taking optimized processing recipe as the standard cell fabrication procedure.

In order to meet this requiremnt, it will be necessary to alter at least two of the three major components simultaneously. In the majority of cases, however, researchers who come from the different background could engage on certain aspects of the components to improve the photovoltaic performances from different disciplines: (1) chemists or material scientists worked for developing photosensitizer and charge transport materials; (2) physicists to eluciate photovoltaic properties and working mechanism; and (3) engineers to develop process and device structure for module level production. However, the simultaneous development of all components such as new photosensitier or new classed strucutral perovskite materials with intrinsic stability and beneficial optoelectronic properties, solid state HTM and photoanodes, combined with further investigation of transport dynamics, will lead to ESCs with efficiencies exceeding 30%.

This report the following studies: (i) a brief introduction of the working principle, (ii) the detailed study of the each component materials, mainly including TiO_2 photoanode under the category of 0D and 3D structures, strategies for co-sensitization with porphyrin and organic photosensitizers, and carbon catalytic material via controlled fabrication protocols and fundamental understanding of the working principles of electrochemical photovoltaic cell has been gained by means of electrical and optical modeling and advanced characterization techniques and (iii) new desgined stratages such as the optimization of photon confinement (iv) future prospects and survival stratagies for sensitizer assisted solar cell (especially, DSSC).

2. Fundamentals of dye-sensitized solar cells

2.1 Operation principle

The principle of operation of a DSSC is well documented in the literature [7]. The simplified principle of the thin layer DSSC is shown in **Figure 1**. A light-harvesting Ru complex potosensitizers adsorbed on the surface of a porous nano-crystalline film composed of a wide bandgap metal oxide such as TiO_2, ZnO or SnO_2 absorb incident photon flux. The photosensitizers are exited from the ground state (D) to the excited state (D*) owing to the metal to ligand charge transfer (MLCT) transition (Eq. (1)). The exited electrons are injected into the conduction band of the TiO_2 electrode, resulting in the oxidation of the photosensitizer (Eq. (2)).

$$\mathrm{D\ (adsorbed\ on\ TiO_2)} + h\nu \rightarrow \mathrm{D*\ (adsorbed\ on\ TiO_2)} \tag{1}$$

$$\mathrm{D*\ (adsorbed\ on\ TiO_2)} \rightarrow \mathrm{D^+\ (adsorbed\ on\ TiO_2)} + \mathrm{e^-\ (injected)} \tag{2}$$

Injected electrons in the conduction band of TiO_2 are transported between TiO_2 nanoparticles with diffusion toward the back contact (TCO) and consequently reach the counter electrode through the external load and wiring. The oxidized photosensitizer (D^+) accepts electrons from the I^- ion redox mediator, regenerationg the ground state (D), and I^- is oxidized to the oxidized state, I_3^- (Eq. (3)). The injected eletrons may recombine either with oxidized sensitier at the TiO_2 (Eq. (4)).

$$\mathrm{D^+\ (adsorbed\ on\ TiO_2)} + 3/2\,\mathrm{I^-} \rightarrow \mathrm{D\ (adsorbed\ on\ TiO_2)} + \tfrac{1}{2}\,\mathrm{I^{-3}} \tag{3}$$

$$\mathrm{D^+\ (adsorbed\ on\ TiO_2)} + \mathrm{e^-\ (TiO_2)} \rightarrow \mathrm{D\ (adsorbed\ on\ TiO_2)} \tag{4}$$

The oxidized redox mediator, I_3^-, diffuses toward the counter electrode and is rereduced to I^- ions.

$$\mathrm{I^{-3}} + 2\mathrm{e^-\ (TiO_2)} \rightarrow 3\mathrm{I^-\ (anode)} \tag{5}$$

The primary energy conversion process in DSSCs is a photoinduced charge separation at the metal oxide/dye/electrolyte interface.

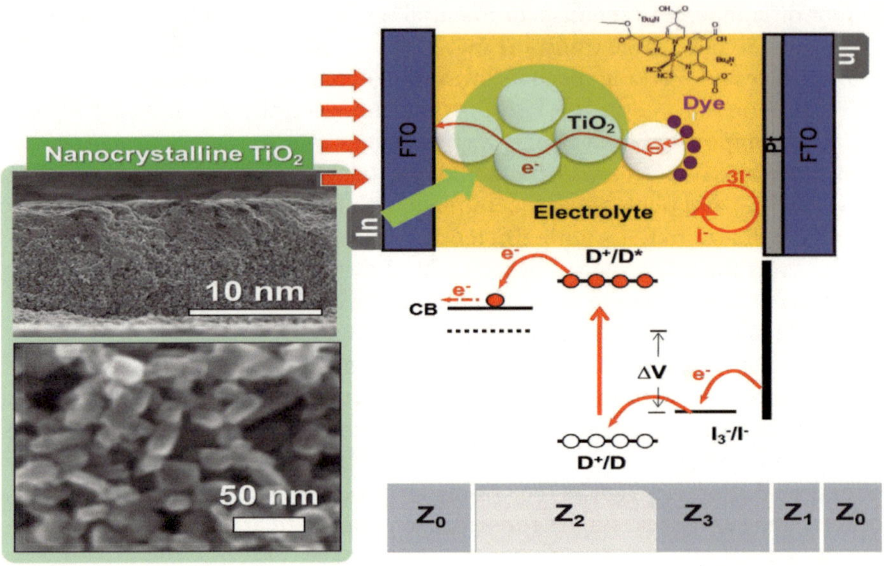

Figure 1.
Schematic diagram and principle of operation of proto-typed dye sensitized solar cells (inserted in the nanocrystalline TiO_2 film).

In order to elucidate DSSC working phenomenon, the physical kinetics and dynamics of charge transfer motion have been investigated in detail by many researchers, both experimentally and via computational modeling. Here, the electron transfer dynamics taking place at the oxide/dye/electrolyte interface for DSSCs are presented in **Figure 2** [7–9]. This figure also contains the information for several competing loss pathways, shown as red arrows. These loss pathways include decay of the dye excited state to ground, and charge recombination of injected electrons whith dye cations and with the redox couple.

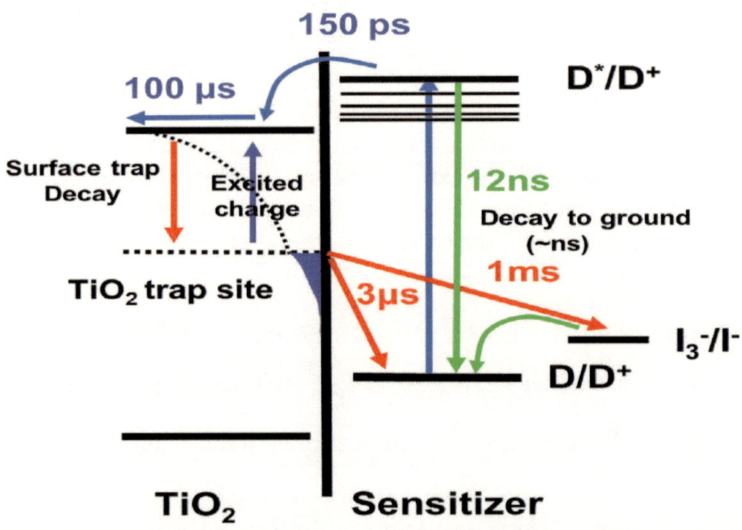

Figure 2.
State diagram of kinetics and energetics of a traditional DSSC. The positive charge transfer processes are indicated by blue color. The competing loss pathways of recombination with TiO2 trap sites and dye surface are shown in red. Reprinted from [8, 10].

2.2 Dye/metal oxide interface

A monolayer dye molecules attached to the surace of the nanocrystalline film. The high surface area of nanocrystalline film ensured s a large concentration of the light absorber leading to good light haversting are critical to be efficient DSSCs. In addition, how and where the photosensitizer bound to the surface of TiO_2 is another important issues since unexpected bonding between photosensitizer and the TiO_2 surface lead to degrade electron injection performance. In the case of Ru complex, one carboxylic ligand provides good anchoring to metal oxide surface (Long-term stability against moisture condition can be improved by using phosphonic acid) as well as good electronic coupling of the excited states of photosensitizer with TiO_2 conduction band states [11].

To be the effective electron transfer, the lowest unoccupied molecular orbital (LUMO) level of excited dye molecules is appropriately lower than the highest unoccupied molecular orbital (HUMO) level of the metal oxide. Under illumination, the photoxcited sensitizers are injected into the metal oxide and diffused from the dye/metal oxide interface to substrate. In an efficient DSSC, a time scale of femtoseconds to picoseconds nanoseconds of injection process is witnessed on the strong electronic coupling between photoexicted sensitizer and metal oxide. The injected electrons experience fast recombination process and thermalisation down to at the electron Fermi level of the electrode [12].

2.3 Dye/electrolyte interface

After ultrafast electron injection from the photoexcited dye into the conduction band of TiO_2, the dye is in its oxidized state and must be reduced by an electron donor in the electrolyte for regeneration. The standard electron donor is iodide. For many typed dye, high value of regeneration efficiency, which give the probability that an oxidized dye is regenerated by an electron donor in the electrolyte rather than by recombination with an electron in the TiO_2, have been estimated in iodide. The reaction mechanism from reduction of oxidized sensitizer (D^+) by iodide follows:

$$D^+ + I^- \rightarrow (D\cdots I) \tag{6}$$

$$(D\cdots I) + I^- \rightarrow (D\cdots I_2 - \bullet) \tag{7}$$

$$(D\cdots I_2{}^{-\bullet}) \rightarrow D + I_2 - \bullet \tag{8}$$

$$I_2 - \bullet \rightarrow I_3{}^- + I^- \tag{9}$$

Eq. (6) is most likely a one-electron transfer reaction between D^+ and I^-. (ε ($I^\bullet/I^-$) = +1.33 V vs. NHE in aqueous solution.) The dye regeneration is firstly occured from the formation of a (D···I) complex since the redox potential ($\boldsymbol{\varepsilon}$) of the iodine radical reacted to the sensitizer (ε (D···I) = +1.23 V vs. NHE in acetonitrile) shows a relatively lower potential of 0.1 V. The dye regeneration is firstly occured from the formation of a (D···I) complex since the redox potential ($\boldsymbol{\varepsilon}$) of the iodine radical reacted to the sensitizer (ε (D···I) = +1.23 V vs. NHE in acetonitrile) shows a lower potential of 0.1 V. Dissociation of ground-state dye D and $I_2{}^{-\bullet}$ is secondly occurred from the formation of a ($D\cdots I_2{}^{-\bullet}$) complex.Finally, $I_2{}^{-\bullet}$ is by the composition of triiodide and iodide. The second-order rate constant for this reaction is about $2.3 \times 10^{10}\ M^{-1}\ s^{-1}$ in acetonitrile.

The oxidized sensitizer (specially, the most commonly reported Ru complex dye, named as N719) is regeneated by the iodide/iodine redox couple in the liquid electrolyte in the range of 100 ns to 10 μs. The different regenation time can be explained by iodide concentration, the presence of additivies such as identity of caion salt (lithium ions) and tertbutylpyridine (tBP). To be the effective DSSC, the recombination process of electrons with either oxidized dye molecules or acceptors in the electrolyte must be minimized compared with regeneration process, which usually happens on a time scale of about 1 μs [7, 13–15].

2.4 TiO_2 interface for electron transport

Unlike the typical the electron transport mechanism in bulk semiconductors, the electron transport mechanism of TiO_2 in DSSC need to be extend from the properties of the individual nanoparticles to the particle connectivity or electronic coupling between the particles, and the geometrical configuration of the assembly. Interestingly, the mesoporous nanocrystalline TiO_2 layer exhibit the highly efficient charge tranport through the nanocrystalline TiO_2 layer, while the low inherent conductivity of the film ca. bulk mobility of TiO_2 ($1cm^2/Vs$), ZnO ($200cm^2/Vs$), and SnO_2 ($250cm^2/Vs$) as well as the presence of disorder from randomly arranged metal [16]. The main consequence of disorder in the electronic structure of the material is the appearance of localized states. This puzzling phenomenon has been explained by multiple-trapping model, in which the diffusion of conduction band electrons is affected by the trapping-detrapping events [17–20]. Electron transport in nanostructured oxide films impregnated with a highly concentrated electrolyte is believed to occur mainly by diffusion. It is generally accepted that this diffusional transport is influenced by the existence of electron localized states or traps in the semiconductor.

In the TiO_2 nanoparticles, electrons undergo a number of processes. A main process is the transport of electrons in the extended states of the conduction band (CB), where the concentration of free electrons n_c is defined as

$$n_c = n_0 e^{qV/k_B T} \tag{10}$$

here n_0 is an equilibrium concentration in the effective density of states of the CB and k_B is the Boltzmann constant, q denotes elementary charge and T is the temperature unut. In a steady state, transport process is under the control of displacing electrons in the TiO_2 CB with the trap in equilibrium occupation.

TiO_2 nanoparticles exhibit numerous localized states in bandgap, which can capture and release electrons to the transport level. The position of Fermi level (E_{Fn}) plays an important role in calculating the probability of electron capture since trapping events below E_{Fn} are quiescent from almost fully occupied traps, while empty traps above the E_{Fn} happens in reverse.

A localized state at energy E_t sets free electrons at a rate;

$$\tau_t^{-1} = \beta_n N_c e^{\frac{E_t - E_c}{k_B T}} \tag{11}$$

where β_n is the time constant for electron capture (which is independent of trap depth). Since τ_t increases exponentially with the depth of the state in the bandgap, the slowest trap is the deepest unoccupied trap. The effect of traps is dominant whenever a transient effect is induced, in which the position of the Fermi level is modified, with the correspondent need for traps release. Therefore, traps also become a dominant aspect of the recombination of electrons in a DSSC, that is a charge transfer to the electrolyte or hole conductor. The specific density of localized states (DOS) in the bandgap of TiO_2, named as $g(E)$, can be also established by capacitance techniques that provide the chemical capacitance, that is defined as follows:

$$g(E) = \frac{N_L}{k_B T_0} \exp\left(\frac{E - E_c}{k_B T_0}\right) \tag{12}$$

Here N_L is the total DOS, and T_0 is a parameter with temperature units that determines the depth of the distribution, that is alternatively expressed as a coefficient $\alpha = T/T_0$. If we consider that electrons can only be transported via the CB, then we need to consider how the density of electrons in the CB.

The carrier mobility is associated with random motion of free carriers, which is directed linked to the conductivity and resistivity. The expression of the electron conductivity σ_n associated to free carrier transport is [21].

$$\sigma_n = \frac{q^2 D_0}{k_B T} n_c \tag{13}$$

Here D_0 is the free electron diffusion coefficient. The conductivity can be expressed by the temperature-dependent features using the framework of multiple trapping where electron displacement occure via extended states at the CB edge with the free electron diffusion coefficient. This result has been measured by the common techniques such as intensity modulated photocurrent spectroscopy (IMPS), transient photocurrent under a small perturbation of the illumination, or transport resistance of impedance spectroscopy [22–24].

2.5 Electrolyte/counter electrode interface

In theory, the maximum photovoltage of the DSSC is determined by the energy difference between the redox potential and the Fermi level of the metal oxide semiconductor. However, the output voltage under load is usually less than the open circuit voltage the theorical expected. This voltage loss is mainly attributed the the overall overpotential of electrolyt and counter electrode (CE) interface. The mass-transfer overpotential is largely affected by the ionic conductivity of electrolytes and the transportation of mediator species from the CE to the photoanode. The kinetic overpotential or charge-transfer overpotential can be determined by electrocatalytic properties of the CE surface toward mediator reduction [25]. To be effective CE, it should exhibit excellent conductivity and inhibit high electrocatalytic activity for reduction of the redox couple. The effect of electric field and transport by ionic migration can be negligible because of high ionic conductivity and ionic strength of liquid electrolyte. With increasing viscosity such as gel, quasi- solid and solid state electrolyte, an inadequate flux of redox components lead to sacrifice the photocurrent of the DSSC. In the case of the iodide/triiodide electrolyte, charge transport corresponds to formation and cleavage of chemical bonds:

$$I_3^- + I^- \rightarrow I^- \cdot \cdot I_2 \cdot \cdot I^- \rightarrow I^- + I_3^- \tag{14}$$

The CE must exhibit catalytically fast reaction and low overpotential because the counter electrode reduces the redox species. (triiodide →iodide) Platinum (Pt) materials is widely used as a suitable catalyst for reaction 1.14 to replenish the reduced species in the redox electrolyte. However, Pt is very expensive and rare so they have limited potential for commercial use. Therefore substantial researches are under way to develop inexpensive alternatives materials for larger commercialisation prospects. The charge transfer reaction between the counter electrode surface and the electrolyte occurs the charge-transfer resistance (R_{CT}) in the DSSCs. The small R_{CT} will give a higher fill factor (FF), which lead to a high conversion efficiency.

3. Optimizing material and configuration for typical DSSCs

Numerous publications and review articles on DSSC have appeared in the literature for certain aspects of the DSSC fabrication or performance during the past three decades [26–30]. Proto-type module cells for manufacturing are actively under development [31–35]. This book reviews systematical parameters via controlled experiments (such as materials preparation, processing, device fabrication and assembly, measurements and device modeling) and the interfacial properties introduced by electrochemical impedance spectroscopy (EIS) for making better solar performance [36–38]. The inherent material properties such as impurity, surface properties in the DSSC are studied from this powerful tool for studying the kinetics of charge transport and electron–hole recombination. Furthermore, this book suggests other ways to improve cell efficiency externally from the photon confinement.

3.1 Understanding DSSC operation model by EIS

As mentioned in Section 1. diverse electrochemical processes take place in a DSSC during the cell operation. The I-V curves provide the information on the basic parameters such as short-circuit current (I_{SC}), open-circuit potential (V_{OC}), Fill factor (*FF*), and cell efficiency (η). The impedance spectroscopy measurements are widely used for investigating the properties of a broad class of material system and device. It provides essential information on carrier transport and recombination. In EIS measurement, a direct current (DC) signal is applied to the cell by a small sinusoidal alternating current (AC) perturbation under steady light illumination. From the measured current response, the magnitude of the impedance for amplitude and phase shift is determined as a function of modulation frequency. From the relevant equivalent circuit, the measured data are fitted by some software (such as Zview or Gamry Echem Analyst etc) and the charge transfer kinetics in a DSSC can be estimated.

In this book, for a deep comprehensive understanding of the device operation, Adachi model are employed to figure out the key parameters that control the efficiency of a DSSC [39–42]. **Figure 1** shows the schematic diagram of DSSC structure. In this system, the light comes from the indium or fluorine doped tin oxide (ITO or FTO) coated transparent conductive glass, the sheet resistance as small as 10 Ω/sq. is required, as an anode electrode (left) [43]. An insulating blocking layer and mesoporous metal oxide film (TiO_2 ZnO, SnO_2 etc.) is deposited on top of bottom electrode. Next, dye molecules are covered by the surface of mesoporous metal oxide and iodine electrolyte is interpenetrated into metal oxide/dye layer. The platinum coated FTO substrate is used as a catalyst for iodide reduction as the counter electrode (right).

The kinetic behavior of charge trap and de-trap mechanism from electrochemical and photoelectrochemical processes affects the different capacitive and resistive element of faradaic impedance. EIS data generally shows Nyquist and Bode plots. An general plot shows x-axis the real impedance (Z') versus the y-axis the imaginary impedance (Z") in the complex plane. From EIS analysis, the complex impedance *(Z)* of each component express Z_0, Z_1, Z_2, and Z_3, respectively. (i.e. $\mathbf{Z_0}$ is the contact impedance of conductive glass; $\mathbf{Z_1}$ is the Pt-catalyzed counter electrode impedance; $\mathbf{Z_2}$ is the complex impedance for the interfacial resistivity among metal oxide, dye molecule, and the iodine electrolyte; $\mathbf{Z_3}$ is the Warburg impedance for diffusion of tri-iodide ions). The resulting data from impedance is directly linked to the information on materials quality, internal, interfacial properties of the DSSC.

Suppose the complex impedance is the same as its resistance. ($Z = R$) R_o represents the ohmic contact resistance from the FTO and the metal. The impedance of the electron transfer at the Pt counter electrode (Z_1) can be modeled as an RC parallel circuit, and simply expressed as:

$$Z_1 = \frac{1}{\frac{1}{r_{p1}} + i\omega C_{p1}} \tag{15}$$

where, r_{p1}, C_{p1} describes the resistance and capacitance at the Pt coated CE, respectively.

$\boldsymbol{Z_3}$ represents the finite Warburg impedance contributes to the diffusion impedance for the diffusion of tri-iodide ions in the electrolyte at the low frequency region with a frequency maxima of ω_{z3} [40, 44].

$$Z_3 = R_D \frac{1}{\sqrt{\frac{i\omega}{(D_1/\delta^2)}}} \tanh\sqrt{\frac{i\omega}{(D_1/\delta^2)}} \; where \; R_D = \frac{k_B T}{m^2 q^2 A v C^* D_1 \delta} \tag{16}$$

$\boldsymbol{D_1}$ and $\boldsymbol{\delta}$ represent the diffusion coefficient of I_3^- and the thickness of the liquid film respectively. $\boldsymbol{R_D}$ is the DC resistance of impedance of diffusion of tri-iodide. The number of electrons transferred in each reaction, m, is 2 in this case. A_v and C^* are the Avogadro number and the concentration of I_3^- in the bulk, respectively.

The impedance $\boldsymbol{Z_2}$ of the charge transport through diffusion into the mesoporous TiO_2 and recombination at the TiO_2/dye/electrolyte interface shows in the middle semicircle of the EIS graph. As seen in **Figure 2**, photoexcited electrons in the CB will decay and excite at rate $\boldsymbol{k_1}$ and $\boldsymbol{k_2}$ into surface trap states of the TiO_2 as well as the back reaction in the iodine electrolyte at a rate of ($\boldsymbol{k_r}$). The charge transfer kinetics for injection, diffusion, collection, trapping, de-trapping and recombination of electrons in the TiO_2 of the DSSC can be calculated by [39, 40]:

$$\frac{\partial n(x,t)}{\partial t} = D_{cb} \frac{\partial^2 n(x,t)}{\partial x^2} - k_1 n(x,t) + k_2(x,t) + G\,(x,t) \tag{17}$$

$$\frac{\partial N(x,t)}{\partial t} = -k_2 N(x,t) - k_r N^2 + k_1 n \tag{18}$$

Here $\boldsymbol{n}$ is the excess electron density in the CB of the TiO_2 under illumination, N is the excess electron density of the trap sites, $\boldsymbol{D_{cb}}$ describes the diffusion coefficient of an electron in the CB and the function **G** is the generation rate of electrons injected into the TiO_2 [40]. By varying the potentials, the excess CB electron density $n(x,t)$ will be expressed as followed [39].

$$n(x,t) = n_S(x,t) + \Delta n(x) e^{i\omega t} \tag{19}$$

$$N(x,t) = N_S(x,t) + \Delta N(x) e^{i(\omega t + \varphi)} \tag{20}$$

Here, n_s and N_s describe the steady state electron densities (Δn and ΔN are the amplitudes of the modulated component) in the CB and the trap state, respectively.

Defining,

$$D_{\text{eff}} = D_{\text{cb}} \frac{k_2}{k_2} \tag{21}$$

$$k_{\text{eff}} = 2N_s k_r \tag{22}$$

$$\gamma^2 = \frac{k_{\text{eff}}}{D_{\text{eff}}} + \frac{i\omega}{D_{\text{eff}}} \tag{23}$$

and using the following boundary conditions at

$$x = 0, qD_{\text{eff}}\left(\frac{\partial \Delta n}{\partial x}\right) = \frac{\Delta I}{A} = \Delta J \tag{24}$$

$$x = L, \frac{\partial \Delta n}{\partial x} = 0 \tag{25}$$

the impedance Z_2 is obtained by Kern et al. as [39].

$$Z = -S\frac{1}{qA}\frac{1}{D_{\text{eff}}\gamma\sqrt{\frac{1}{k_{\text{eff}}}}}\frac{1+e^{2\gamma L}}{1-e^{2\gamma L}} \tag{26}$$

$$S = \frac{K_B T}{qn_s}\sqrt{\frac{1}{k_{\text{eff}}}} \tag{27}$$

By defining

$$\omega_{\text{d}} = \frac{D_{\text{eff}}}{L^2}, \omega_{\text{k}} = k_{\text{eff}} \text{ and } \gamma L = \sqrt{\frac{\omega_{\text{k}}}{\omega_{\text{d}}} + \frac{i\omega}{\omega_{\text{d}}}} \tag{28}$$

The equivalent impedance Z_2 of Bisquert [39] is obtained as follows:

$$Z_2 = R_{\omega}\left(\frac{1}{(\omega_{\text{k}}/\omega_{\text{d}})(1+i\omega/\omega_{\text{k}})}\right)^{1/2} \coth[(\omega_{\text{k}}/\omega_{\text{d}})(1+\text{i}\omega/\omega_{\text{k}})]^{1/2} \tag{29}$$

$$R_w = \frac{k_B T}{q^2 A n_s}\frac{L}{D_{\text{eff}}} = Con\frac{L}{D_{\text{eff}}}, R_k = \frac{\omega_{\text{d}}}{\omega_{\text{k}}} \times R_{\omega} = Con\frac{1}{Lk_{\text{eff}}} \tag{30}$$

Here, $\boldsymbol{\omega}$ is modulation frequency (s^{-1}) (with $\omega_{\text{k}} = k_{\text{eff}}$), $\mathbf{R_w}$ is the electron transport resistance, and $\boldsymbol{R_k}$ is the charge transfer resistance related to recombination of electrons at the TiO_2/electrolyte interface. The relation can be expressed by $R_{\text{k}} = (\omega_{\text{d}}/\omega_{\text{k}})\cdot R_{\text{w}}$.

The total impedance (**Zs**) of the DSSC can be calculated by the summation of Z_1, Z_2, Z_3, and the external resistance, Z_0,

$$Zs = Z_0 + Z_1 + Z_2 + Z_3. \tag{31}$$

From experimental (L, A, δ) and EIS data (the maxima values of ω_{Z1}, ω_{Z2}, ω_{Z3} of the semi-circle diameters along the Z' axis), necessary information on charge transport kinetics can be determined (**Figure 3**).

3.2 Materials preparation

3.2.1 TiO_2 nanoparticle

The highly crystalline anatase TiO_2 nanoparticles (NPs) was synthesized by 2-step autoclaving technique (**Figure 4(a)**). A pH controlled TiO_2 suspension was prepared by commercial TiO_2 power. (P25, Degussa) The TiO_2 suspension is placed in a total volume of 60 mL in a Teflon-lined stainless steel autoclave (125 mL

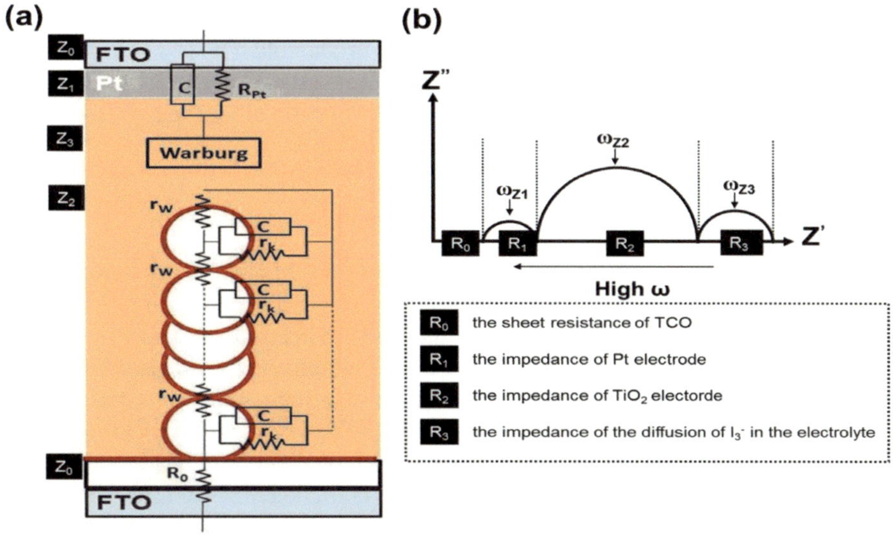

Figure 3.
(a) A electrical equivalent and (b) an illustrative Nyquist plot of a DSSC with R_o, R_1, R_2, and R_3, and each peak frequency maxima of ω_{Z1}, ω_{Z2}, and ω_{Z3}, respectively. Reprinted from [26, 51].

volume, Parr Instrument Co.) and heated at 240°C for 12 h. The resulting powders are dried at ~80°C in a conventional drying oven for 24 hour (**Figure 4(b)**). The pure Anatase colloidal TiO_2 nanoparticle was obtained by autoclaving the low-pH titanate suspension at 240°C for 12h (**Figure (4c)**) [45].

3.2.2 Photosensitizer (purifed N719 dye)

For photosensitization studies, the calcined TiO_2 nanoparticle electrode were immersed in the ethanol solution containing purified $3x10^{-4}$ M cis-di(thiocynato)-N,N′-bis(2,2′-bipyridyl-4-caboxylic acid-4′-tetrabutylammonium carboxy late) ruthenium (II) (N719, Solaronix) for 18 h at room temperature [46]. Commercial N719 dye may not produce high efficiency because of impurities. Therefore, purification is required. The N719 complex is firstly dissolved in water with

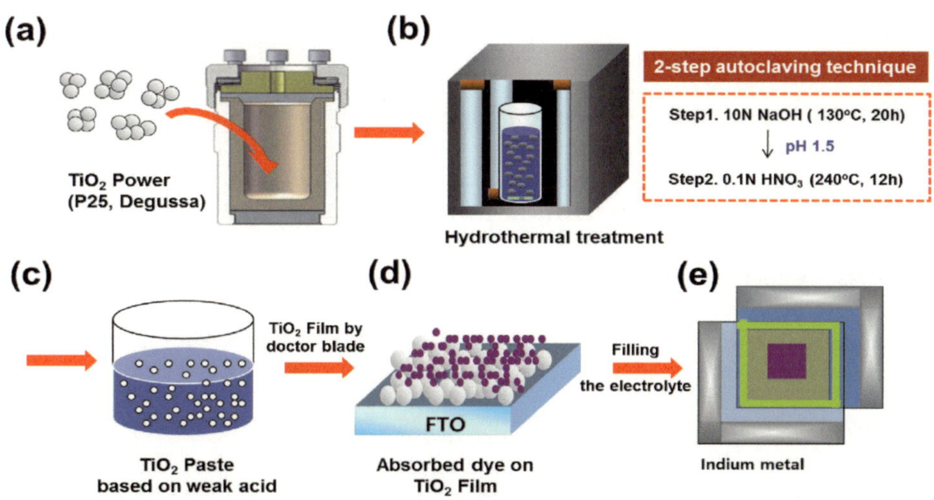

Figure 4.
Scheme showing a typical procedure for highly crystalline TiO_2 nanoparticle based DSSC. Reprinted from [26].

tertrabutylammonium hydroxide and then purified three times with chromatography. (Sephadex LH-20) The pH values below 4.5 is controlled by 0.02 M HNO_3. The titration is carried out slowly over a period of three hours. Then, the solution is kept at -20°C for 15 hours. After allowing the flask to warm to 25°C, the precipitated complex is collected on a glass frit and air dried.

3.2.3 Liquid electrolyte

he liquid electrolyte was prepared by dissolving 0.6 M of 1- butyl-3-methylimidazolium iodide (**BMII**), 0.03 M of iodine, 0.1 M of guanidinium thiocyanate (**GuSCN**) and 0.5 M of 4-tert-butylpyridine (**tBP**) in acetonitrile and valeronitrile (85:15 v/v). Most additives are understood at a fairly phenomenological level, and their effects are often attributed to modification of redox couple potential, band shifts of the semiconducting electrode material, effects of surface blocking, or surface dye organization. The study for additives have been reviewed extensively elsewhere [47, 48]. For example, it was found that smaller size Li^+ ions cause a shift in the TiO_2 conduction band edges toward more positive potentials than larger size 1,2-dimethyl-3-hexylimidazolium ions. As a result, an electrolyte containing Li^+ ions produced a lower photovoltage and at the same time higher photocurrent than that containing imidazolium cations because of an alteration of both the energy and number of excited state levels of the dye that participate in electron injection [46]. The effects of tBP were studied in more detail in DSSCs showing that both band edge shift and increased electron lifetime play a role [49].

Even if tBP and Li^+ ions play a different role for solar cell performance, both additives are widely used in liquid electrolyte. As another effective additive, adding GuSCN can be also good choice. It shifts the conduction band edge of TiO_2 toward lower energies, suggesting adsorption of the cation onto the TiO_2 surface. Coadsorbants, adsorbed onto the TiO_2 during the dye adsorption procedure, can have similar effects as additives in the electrolytes [50].

3.2.4 Catalytic layer

The cathode-electrode was produced by coating F:SnO_2 glass with a thin layer of a 5 mM solution of H_2PtCl_6 in isopropanol and was heated at 400°C for 20 min. Best performance and long-tern stability has been achieved with very low Pt-loadings (5 μg cm^{-2}), so that the counter electrode remains transparent. Charge transfer resistances of less than 1 Ω cm^2 can be achieved.

3.3 DSSC fabrication for conventional typed cell

A paste of anatase hydrothermal TiO_2 powder was made by stirring with the mixture 0.5 g of anatase-TiO_2 NPs, 100 μl of Triton X-100, 0.2 g of polyethylene glycol (PEG, Fluka, Mw = 20,000) into 3 ml acetic acid (0.1 M). The TiO_2 paste is coated on a FTO glass by doctor blade technique. The thickness of the TiO_2 film was measured by a surface profiler (TENCOR. P-10) [26, 51]. **Figure 4(d)** and **(e)** gives an illustrative sequence of steps that were used in the fabrication of our solar cells. The top and bottom electrodes were sandwiched together with thermal melt polymer film (**Figure 4e**).

3.3.1 TiO_2 Nanoparticles prepared by 2-step hydrothermal treatment

Anatase TiO_2 NPs were investigated by using a field-emission scanning electron microscope (SEM, S4800, Hitachi) and JEOL-2010 TEM (JEOL, Japan) equipped with an energy dispersive spectrometer (EDS) to investigate the TiO_2 NPs and determine

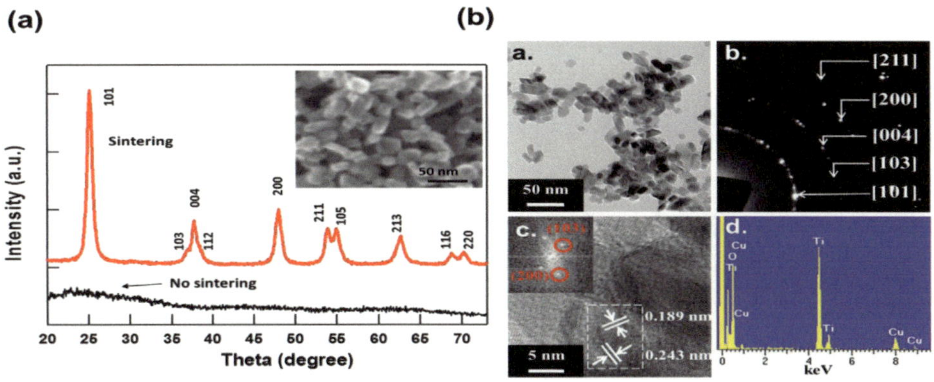

Figure 5.
(a) X-ray diffraction patterns of hydrothermal TiO_2 nanoparticle after calcination at 5000C. Insert shows the SEM images of nanopaticle after calcination. (b) a.TEM, b. SEAD pattern, c. HR-TEM bright field images and d. EDX analysis of TiO_2 nanocrystals prepared by 2-step autoclaving technique scheme showing a typical procedure for highly crystalline TiO_2 nanoparticle based DSSC.

the compositions of the samples at 200 kV. The NPs were also characterized by x-ray diffraction (D/Max-A, Rigaku) measurements. **Figure 5(a)** shows a X-ray scan of the Anatase TiO_2 NPs before and after sintering as described below. The inset shows a typical SEM photo of the NPs used in the experiment. These results reveal broad diffraction peaks at 25.3°, 37.8°, 48.2°, 53.9° and 55.2° were observed, which can be indexed to the (101), (004), (200), (105) and (211) reflections of anatase TiO_2. **Figure 5(b)** provides the high resolution transmission microscopic (TEM) images and selected area electron diffraction pattern of the TiO_2 NPs. It provides the corresponding electron diffraction pattern taken from the TiO_2 NPs showing that the NPs are crystalline Anatase TiO_2. **Figure 5(b)** gives the high-resolution TEM (HRTEM) image of NPs indicating that the TiO_2 NPs are single-crystalline. Lattice of 0.189 nm and 0.243 nm corresponding to (200) and (103) planes of the tetragonal TiO_2, respectively, have been resolved. The inset is the corresponding fast Fourier transform (IFFT) image of selected area in **Figure 5(b)c**. In **Figure 5(b)d** the EDS taken from TiO_2 NPs shows that the NPs are composed of Ti and O. The Cu peak comes from the grid. From these measurements, we concluded that our starting materials were pure Anatase.

3.3.2 Study for the optimal TiO_2 film thickness

To find the optimized condition for hydrothermal treated TiO_2 film, the cell efficiency as a function of the TiO_2 film thicknesses is studied. In **Figure 6** (top), I give plots of V_{oc}, J_{sc}, FF, and η as functions of film thicknesses. The efficiency per NP increases almost linear with decreasing film thickness until about 2 μm. From these measurements, the optimal TiO_2 film thickness is around 11.5 μm with our cell architect and fabrication procedures. It should also be noted that the highest open circuit voltage and fill factor are achieved at the thinnest active layer thickness, while the short circuit current density increases with increasing film thickness (for thickness below 15 μm). From this thickness study, we then varied other cell parameters to optimize the overall cell efficiency. In **Figure 6** (bottom) we plot cell efficiency times the number TiO_2 NP as a function of NP film thickness. it is shown that the efficiency per TiO_2 NP increases dramatically as the active NP layer decreases. This shows the large inherent loss of charges in these DSSCs.

In **Figure 7**, AC impedance measurements with best-fit model curves of the cell is used for analyzing function on the different TiO_2 NP film thicknesses. Form the Bode phase plots, the negative shift of the frequencies of the main peaks with an

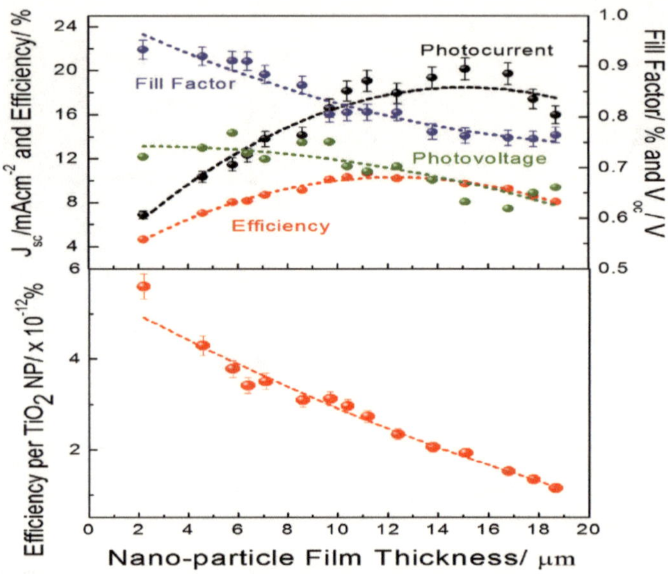

Figure 6.
Relationship of DSSC device performance (top) and efficiency per TiO_2 NP (bottom) depending on film thickness (top). Reprinted from [26, 52].

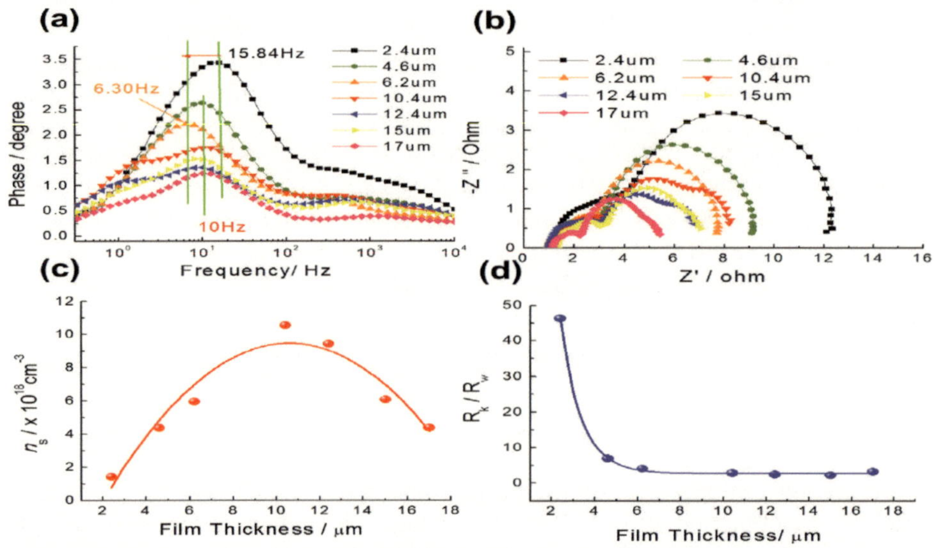

Figure 7.
*AC impedance measurement **(a)** bode, **(b)** Nyquist plots, **(c)** electron density (n_s) and **(d)** the relation between the electron recombination resistance (R_k) and the electron diffusion resistance (R_w) of cells with different TiO_2 NP film thicknesses. The solid curves are from a best fit model.*

increase in the film thickness. In a simple equation, since ω_{max} is inversely associated with the life time of electron $\tau = 1/(2\pi f)$, the decrease in ω_{max} indicated a reduced rate for the charge-recombination process in DSSC. Hence, electrons with longer τ values were prevented from recombining, characterized by a larger charge transfer resistance. In the aspect of ω_{max}, about 6 μm thick film show the longest electron lifetime and relatively small total series resistance, leading to high *Voc* and *FF*. However, in the case of N719 dye, the dye absorption at that thickness is not enough to reach the maximum performance. The more detailed phenomenon can be understood by the further consolidated impedance model we suggested.

All cases of samples were equal in the total cell thicknesses (***L***). Therefore, there is a tradeoff relation between cell gaps of dye coated TiO_2 film and electrolyte, for example, for the thicker thin films of TiO_2 film, the electrolyte spacing is smaller. The resulting data from our model shows specific characteristics worth noting; (1) In the case of ultra-thin TiO_2 film, two distinctive circles can be found and the impedance **Z3** is dominated rather than that of Z1 and Z2 in the low-frequency region. (2) For thicker TiO_2 layers, the contribution from Z_2 is more pronounced (3) The estimated the diffusion coefficient in the electrolyte ($\boldsymbol{D_1}$) is the highest for the thinnest TiO_2 layer [26]. From this model, low k_{eff}, high R_k/R_w, high D_{eff} and high n_s are necessary condition to attain highly efficient DSSC. As seen in **Figure** 7 **(c,d)**, over 10 μm thick film exhibit the improved electron density ($\boldsymbol{n_s}$) and high electron diffusion coefficient (D_{eff}), but the recombination time is shorter than the time for diffusion across the TiO_2 layer ($R_k < < R_w$, $w_d < < w_k$). As a result, we conclude about 11.5 μm is the optimized film thickness in my system. The detailed physical values are summarized in **Table 1**.

3.4 Interfacial modification

There are numerous efforts to improve cell performance from different modifications techniques such as different semiconductors, dyes, or ionic conductors or on changing its nanostructures [53, 54]. Most of modification works show a tradeoff relation between the short-circuit current and open-circuit voltage. This can be explained the the modification of components affect sensibility of charge-transport and recombination dynamics [55]. Therefore, this has been a big challenge in the field of DSSC design. In this book, effective surface passivation and treatment method provide on how it possible to contribute on the cell performance.

3.4.1 $TiCl_4$ treatment

The best-known techiqnue to improve the performance of the solar cells is a post-treatment of the TiO_2 film with a solution in $TiCl_4$ is grown onto an extra layer of TiO_2 nanoparticles constituting the film. The $TiCl_4$ treatment results in an improvement in photocurrent, normally between 10% and 30%. Depending on the quality of the TiO_2 used to make the initial film, the extrema of the improvement can be from <5% to >200% [35, 56]. The largest improvements come when using the poorest quality TiO_2 films. **Figure 8(a)** show the SEM images and XRD patterns for $TiCl_4$ treated TiO_2 film. When the $TiCl_4$ exposure condition is increased, the

Film Thickness (μm)	D_{eff} (10^{-5} cm^2s^{-1})	k_{eff} (Hz)	R_k/R_w	*Con* (Ωcms^{-1})	R_d (Ω)	n_s (10^{18} cm^{-3})	D_1 (10^{-6} cm^2s^{-1})	V_{OC} (V)	J_{sc} (mA/cm^2)	*FF* (%)	EFF (%)
2.4	2.10	31.6	46.2	0.368	19	1.41	42.0	0.932	4.851	72.1	3.26
4.6	0.56	14.2	6.79	0.122	11.7	4.36	7.2	0.919	7.252	73.9	4.93
6.2	0.64	10	4.04	0.084	8.1	5.94	4.5	0.910	8.029	76.8	5.61
10.4	2.79	20	2.76	0.048	7.1	10.5	3.2	0.802	11.61	73.0	6.80
12.4	2.58	10	2.38	0.055	8.3	9.39	0.09	0.808	12.54	72.2	7.32
15.0	3.29	10	2.14	0.093	6.9	6.06	0.56	0.762	12.46	68.6	6.51
17.1	9.05	10	3.13	0.133	3.6	4.36	0.21	0.764	11.19	65.7	5.62

Table 1.
Parameters for the best fit of the impedance data for the different thickness. Measured in ***Figure 6****.*

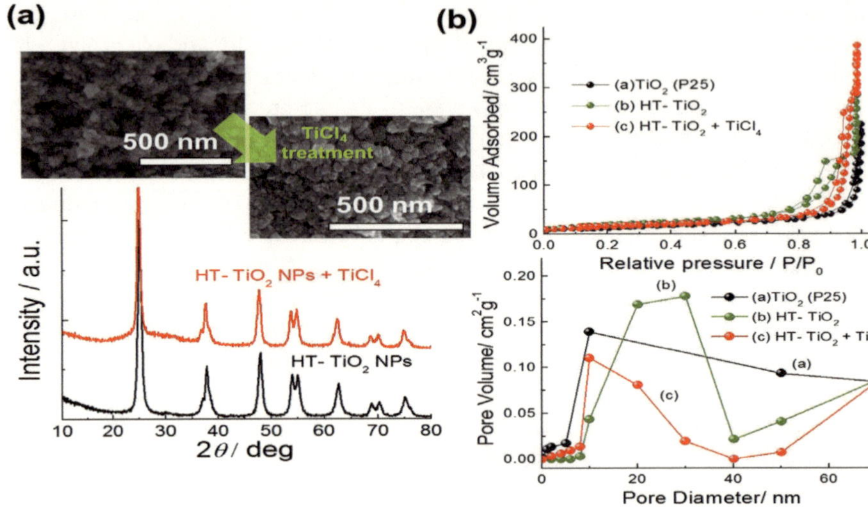

Figure 8.
(a) XRD pattern and SEM images of TiO_2 film with and without post-treatment with $TiCl_4$ (b) BET and BJH analysis of the pristine TiO_2, and hydrothermal treated TiO_2 film with and without post-treatment with $TiCl_4$ (note: HT-TiO_2 is a hydrothermal treated TiO_2 nanoparticles).

XRD intensity of the rutile peak increased [57]. However, under well-controlled condition, no obvious different in the rutile content (20–25%) can be seen between treated and untreated TiO_2 from XRD analysis., while an increased size of TiO_2 NPs and densely packed TiO_2 NPs film is observed. **Figure 8(b)** shows Brunauer–Emmett–Teller (**BET**) and Barrett–Joyner–Halenda (**BJH**) pore-size distribution plots of the pristine TiO_2 NP (P25), 2-step hydrothermalized (HT)-TiO_2 NP and $TiCl_4$ treated HT-TiO_2 NP. Hydrothermal treated samples show similar type-IV isotherms, which are representative of mesoporous solids.

The 2-step HT-TiO_2 NP show about 18.5% increased surface areas and 43.6% widened cumulative pore volume compared with the pristine TiO_2 NP (P25). However, $TiCl_4$ treated samples show ~16.7% decreased surface area compared with pristine TiO_2 film. However, the loss in actual electrode surface area after $TiCl_4$ treatment is small because of the increase in mass of approximately 10.3% TiO_2 volume on the electrode. From these observations it follows that, despite the substantial decrease in BET surface area, the film thickness is not affected. Therefore, the porosity must have decreased, as is shown in **Table 2**. In spite of the decreased surface area, the $TiCl_4$ treated TiO_2 film morphology is observed by about a 40% higher dye absorption at the 480 nm. (see **Figure 9(a)**) For more accurate experiment, the quantity of TiO_2 NPs surface-bound sensitizers was measured by desorption process [35]. UV–vis absorption spectra is used for calculating the number of desorbed sensitizer molecules (set as the extinction coefficient (ε) of the N719 sensitizer is about 3.748 x 10^{-3} cm^{-1} M^{-1} at 535 nm). It is estimated that roughly 8.8% and 34.6% more dye molecules are attached to the surface of $TiCl_4$ treated TiO_2 ($\approx 3.97 \times 10^{-8}$ $molmg^{-1}$) compared to HT-TiO_2 NP ($\approx 3.65 \times 10^{-8}$ $molmg^{-1}$) and commercial TiO_2 NP ($\approx 2.95 \times 10^{-8}$ $molmg^{-1}$), respectively. The TiO_2 surface after the $TiCl_4$ treatment provides more specific binding sites, leading to reduce the fraction of the TiO_2 surface area that is inaccessible for the dye due to sterical constraints [56]. The enhanced dye loading results in an improvement in photocurrent and indeed, the $TiCl_4$ treatment is a 28.5% increase in the photocurrent along with a decreased in the fill factor (7.4%) and open circuit voltage (2.3%). (see in **Figure 9(b)** and more discussed in cell properties part (iii)).

DOI: http://dx.doi.org/10.5772/intechopen.94291

		BET & BJH		Absorption Properties			Electrical Properties				Solar Properties			
		Surface area (m^2g^{-1})	Pore Volume (cm^3g^{-1})	TiO_2 mass (mg/cm^2)	λ_{max} (at 498 nm)	# of mole. ($\times10^{-8}$ mol /mg)	D_{eff} ($\times 10^{-5}$ cm^2s^{-1})	R_k/R_w	n_s ($\times10^{18}$ cm^{-3})	R_{total} (Ω)	V_{OC} (V)	J_{sc} (mA/cm^2)	*FF* (%)	EFF (%)
(a)	TiO_2	53.29	0.348	2.3	0.267	2.95	0.84	1.72	4.81	41.6	0.796	9.725	70.6	5.46
(b)	HT- TiO_2	65.33	0.607	2.6	0.410	3.65	2.58	2.38	9.39	31.5	0.815	12.38	72.2	7.28
(c)	w $TiCl_4$	55.98	0.445	2.9	0.504	3.97	3.24	2.14	11.2	23.7	0.796	15.92	66.8	8.47
(d)	w CF_4 (10 min)	—	—	2.3	0.390	3.48	3.45	3.13	5.15	29.3	0.852	10.29	76.0	6.67
*	w CF_4 (30 min)	—	—	1.4	0.211	2.45	2.11	2.84	3.24	58.9	0.838	6.52	75.1	4.10
(e)	w $TiCl_4/CF_4$	—	—	2.6	0.432	3.72	2.59	2.52	8.14	19.8	0.835	13.68	74.5	8.50
(f)	w $CF_4/TiCl_4$	—	—	2.6	0.503	3.94	4.12	2.48	10.4	13.1	0.813	15.75	71.8	9.20

Table 2.
*Parameters for the best fit of the impedance data for the different interfacial modificated DSSC film. Measured in **Figures 9** and **12**.*

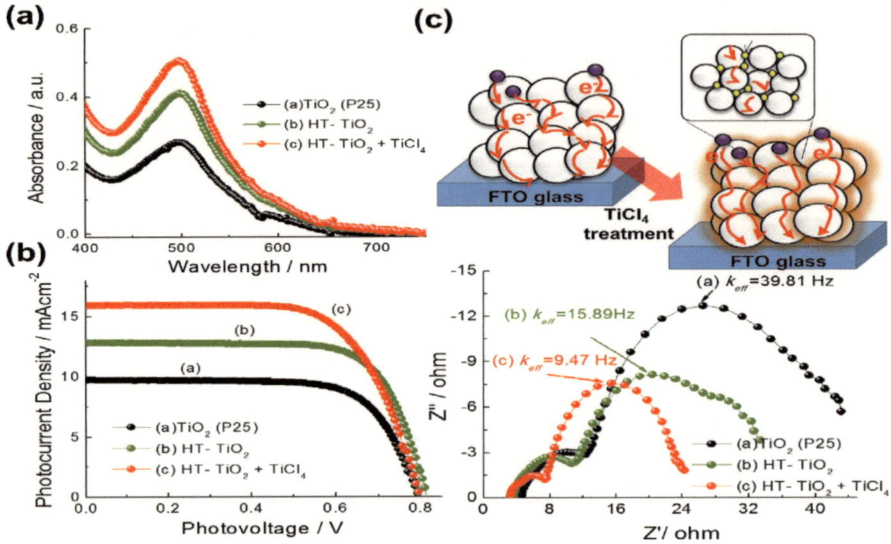

Figure 9.
(a) Absorbance curve of aqueous dye solutions after desorption (b) JV characteristics and (c) EIS analysis from the pristine TiO_2, and hydrothermal treated TiO_2 film with and without post-treatment with $TiCl_4$ (inserted in simple illustration of $TiCl_4$ treatments).

3.4.2 Fuorine plasma etching

As mentioned above, the best efficiency of cell can be found at about 11.5 μm thicked TiO_2 film. This resulting data shows mediately the importance of electrolyte's infiltration all of the way into the TiO_2 film. To minimize the effect on the panetration issues, plama etching techinque is introduced for widening the channels in TiO_2 films [51]. **Figure 10(a)** displays a schematic diagram of the plasma etching system. The detailed etching condition can be found in our eariler paper [51]. Aa ehching gas, CF_4/O_2 of gas mixture (CF_4 and O_2) is used by generating the fluorine atoms in the plasma through complicated chemical reaction paths [58–62]. The plasma etching reaction is followed by:

$$CF_4 + O \rightarrow COF_2 + 2F \tag{32}$$

$$TiO_2(s) + F(g) \rightarrow TiO_xF_y(s) + TiF_4(s, g) + O_2(g) \tag{33}$$

TiO_2 surface is subsequently ethced by free fluorine gas and several titanium fluoroxy compounds (ultimately TiF_4) are developed [61, 63]. Along with the etching process, it was expected that the surface of the NP would be populated with fluorine atoms and bonded to Ti sites as discussed below.

The evidence of etched TiO_2 surface can be found by X-ray photon spectroscopy (XPS) measurement. The peak is located at binding energies 458.5 eV (Ti $2p^{3/2}$) and 464.2 eV (Ti $2p^{1/2}$), respectively, which correspond to the signals characteristic for the Ti^{4+} state of titanium [64, 65]. This indicates that TiO_2 are formed. (see in **Figure 10(b)**). After plasma treatment the peak locations move to the somewhat higher binding-energies of 458.7 and 464.4 eV. The direct bonding with titanium and fluorine makes it possible to move higher binding-energy due to fluorine exhibit more electronegative than oxygen. From XPS analysis, the post-plasma F 1 s peak shows is also indicative of the fluorine being bonded directly to the titanium. The majority of the F 1 s peak (99%) is found at 684.8 eV which has been responsible for $TiOF_2$. The minor (1%) energy shoulder at 686.8 eV is ascrbied to the replacement of oxygen lattice site in TiO_2 by fluorine [66]. The fluorine shows at

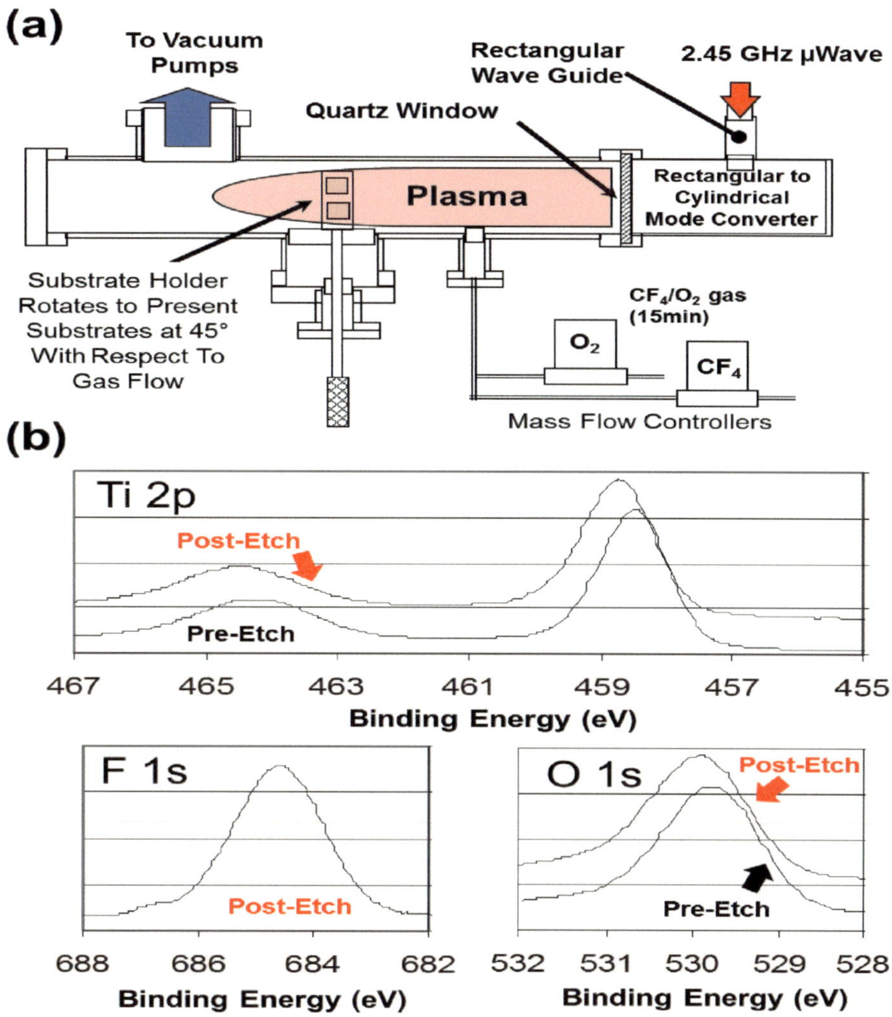

Figure 10.
(a) Schematic diagram of the plasma etching system (b) pre- and post-etch Ti 2p, post-etch F 1 s XPS spectrum, and pre- and post-etch XPS O 1 s spectra. Reprinted from [51].

~14% the oxygen level. Two O 1 s peaks are observed prior to plasma treatment. The major peak with ~90% of the total oxygen is measured at a binding energy of 529.8 eV and 531.4 eV for Ti-O and Ti-OH bonding, respectively [67]. The plasma treatment make the TiO_2 peak move a little higher binding-energy of 529.9 eV.

Figure 11(a) shows SEM images of pre and post fluorine-treated films. It is clear from these images that fluorine effectively opens the channels among the NPs of the TiO_2 film. The variance of film thickness and weight as a function of plasma etching time can be seen in **Figure 11(b)**. With increased etching time, quantitative analysis (EDS) and morphology (SEM) of TiO_2 surface reveal the plasma expand the pores of the TiO_2 film surface. As the pore start to open up (~10 min), the surface etching process of removed TiO_2 and incorporated fluorine into the film is accelerated by active fluorine species. The detailed information can be found in our eailr paper [51].

As seen in **Figure 12(a)**, the plasma etched TiO_2 surface can rectify the interfacial property by decreased the electron-triiodine recombination rates taking place on the parts of the dye uncovered TiO_2 surface. According to CF_4 etching process, it function as diminishing the surface defects such as oxygen vacancies since fluorine atom comprised of stronger affinity for electrons compared with oxygen atom can

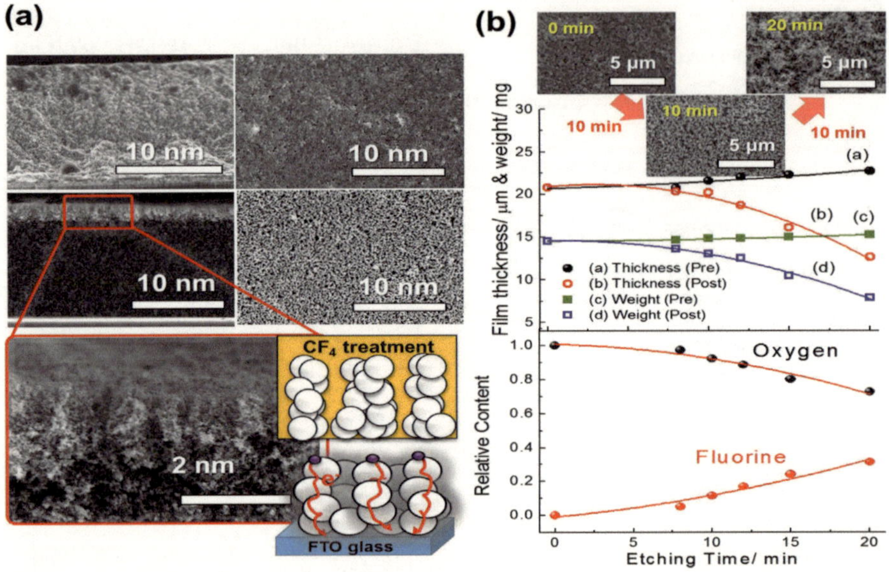

Figure 11.
(a) SEM images of top and side views of pre-etch film and CF_4-etched TiO_2 film. (inserted in simple illustrations of etched film) (b) film thickness, weight and film morphologies images (top) and relative content of oxygen and fluorine in TiO_2 NPs (bottom) as a function of etching time. Reprinted from [51].

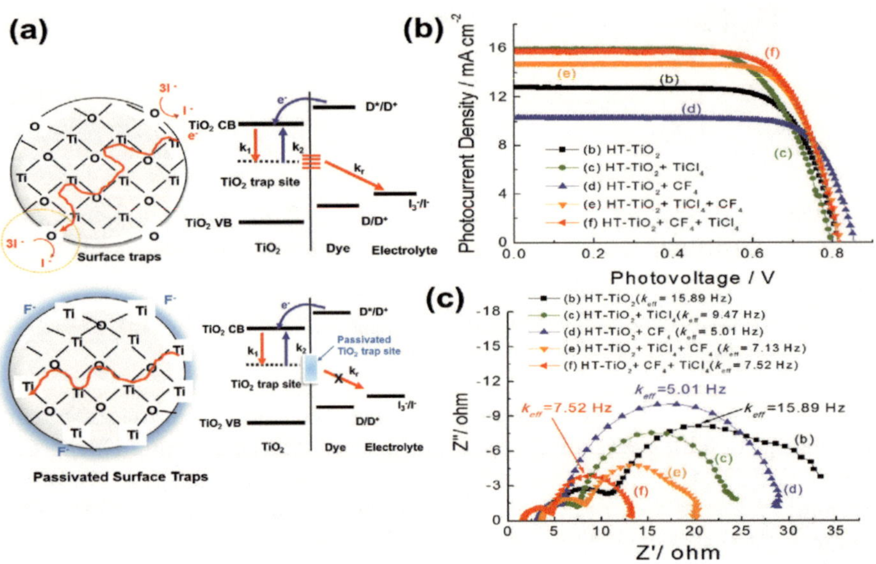

Figure 12.
(a) Schematic illustration for TiO_2 surface interfacial mechanism (b) JV curve and (c) AC impedance measurements of cells with different surface treatments.

enfeeble the bond in a titanium and oxygen [68]. Therefore, the TiO_2 surface comprised of less surface defects may play important role in a improvement of the open-circuit potential, V_{oc}, of the solar cell.

3.4.3 Cell properties for interfacial modification

The detailed solar cell research studied a series of five cells, each with a layer of 11.6 μm thick TiO_2 NP film. Each of the four samples were treated differently prior to dye and electrolyte infiltration. As a first series of cells, TiO_2 film prepared for cell (a) commercial and cell (b) hydrothermal treatment is studied. Next, for understanding the influence on interfacial modified TiO_2 surface, post-treated cells with cell (c) with $TiCl_4$ treatment and Tare cell (c) with fluorine etching are investigated. **Figure 12(b)** and **(c)** give, respectively, the $J - V$ and the impedance measurements of each cells. Using the kinetic model discussed in section and extrapolating the parameters for the best fit to each of the measured impedance curves, summarized the results in **Table 2** for the electrical data for all cells. The model calculation and data fitting provide some physical insight into the differences in the transport properties and effects due to plasma etching of the TiO_2 NP films of the cells. Cell (b) can be considered as the control for the other three cells. There is no treatment to the TiO_2 NP film in this case. In general, J_{sc} can be approximated by the expression;

$$J_{sc} = q\eta_{lh}\eta_{inj}\eta_{cc}I_0 \tag{34}$$

where q is the elementary charge, $\boldsymbol{\eta_{lh}}$ is the light harvesting efficiency of a cell, η_{inj} is the charge-injection efficiency of the excited dye into the TiO_2, $\boldsymbol{\eta_{cc}}$ is charge collection efficiency, and $\boldsymbol{I_0}$ is the incident photon flux [69]. From this equation, it is clearly that the short-circuit current density (***Jsc***) is directly proportional to the value of $\boldsymbol{\eta_{lh}}$ related to numerous specific anchoring sites on TiO_2 surface for dye absorption and light scattering events for optical absorption as an external property. As we mentioned earlier, the η_{cc} is related to the charge transfer kinetics and this value can be estimated by comparing the charge transport and recombination time constants when both values are measured by EIS measurements.

Cell (c) has a 25.5% increase in the D_{eff}, leading to about 19.3% increased electron density (n_s) compared with the cell (b). With increasing electron density, deeper traps become filled, and trapping/detrapping events occur more frequently in shallower traps, leading to faster transport. The increase in the photocurrent density is well explained by the higher electron density. However, the value of R_k/R_w related to the recombination related value is decreased by about 10%, leading to a decrease Voc. Cell (d) with the plasma etched device shows the resulting values for about 45.2% lower the charge density value (n_s,), 31.5% higher the interfacial recombination rate (R_k/R_w) and 33.4% increased D_{eff} rather than that of the unetched sample (cell (b)). Although the fitted charge transfer properties on etched sample are substantial, the overall performance of cell is infinitesimal because the efficiency just increased by about 8% with etching.

These changes show that the etching has significant effect on the electron recombination and transport properties of the cell, but the overall effect is not pronounced because the cell efficiency only increased by 8% with etching. As suggested in **Figure 12(a)**, the fluorine etching can help to minimize the electron loss between TiO_2 surface and reduced iodide, leading to the highest V_{oc} and FF value (~0.852 V and 75.1%). The decreased electron transport properties are attributed to the less interconnection between NPs compared to the case of cell (b). However, when a film is treated in the much more time (~30 min), the morphology changed dramatically, and the pin-hole of TiO_2 film formed. This result can be confirmed by the dye desorption experiments and the weight loss of TiO_2 from etching process. While this situation should improve the dye molecule attachments to the NPs and allow further penetration of the molecules into the TiO_2 film, as indicated by a large increased in the values of D_{eff} in **Table 2**.

Finally, we believe that combination of $TiCl_4$ and fluorine etching post-treatment with the opposite physical properties make it possible to increase both the V_{oc} and J_{sc}, giving an efficiency higher than that of cell (b). Therefore, cell (e) is treated with $TiCl_4$ coating with same treatment as cell (c), followed by fluorine etching with same treatment as cell (c), and cell (f) with the TiO_2 NP film etched by fluorine first and followed by $TiCl_4$ treatment and then high-temperature. As we expected, cell (f) has the lowest series resistance (i.e., $R_{total} = R_0 + R_1 + R_2 + R_3$) and therefore the highest efficiency. The contribution to this increase of cell efficiency comes primarily from the 27.7% increase in $J_{sc,}$ no loss in the V_{oc} and FF value. It also notice from our model that the effective electron diffusion coefficient is a significant improved by as much as 59%, the interfacial recombination (k_{eff}) is decreased by 52.7%, which leads to large R_k compared with the cell (b). These changes are quite significant, and as a result, the cell efficiency was increased nearly 26%.

4. Photon management using three-dimensional photonic crystals

Besides the interfacial modification, cell performance can be improved by external source (such as anti-reflection film and photonic crystal etc.) as reflecting passed photons back into the absorbing film [26, 70]. For the photon trapping, two approaches for geometrical or wave optics are employed [71]. In this book, photon management concept using 3-dimensional photonic crystals (**PhC**) is introduced. Photonic crystals are highly ordered materials with a periodically modulated dielectric constant. The presence of photonic band gap can be confining and controlling the propagation of light propagation of light, a band of frequencies in which light propagation in the photonic crystal is forbidden [72]. Therefore, we can manage to reflect, transmit, and diffract light for specific wavelengths by appropriate designing the crystal structure. In fact, a diffuse scattering layer of large TiO_2 colloids [73] is typically introduced for this specific purpose [74]. There are successful demonstrations for enhancement in performance of silicon photovoltaic cells based on the realization of coherent scattering processes [75–81]. However, these concepts cannot be easily realized in DSSC. Herein, this book takes the most efficient cells and measure how much of the income light is actually being absorbed by the N719 dye as photons transit across the solar cell.

4.1 Photon management effect on DSSCs

For 3D PhC layer, the vertical deposition technique for polystyrene (PS) opal templates is employed [55] (see in **Figure 13(a)**). The experimental procedures are reported in earlier work [82, 83]. **Figure 13(b)** is a SEM micrograph of a ZnO PhC (or inverse opal) layer showing both the side view (cross sectional, top) and top view (bottom).

The optical property can be estimated by using UV- vis- NIR spectrophotometer in a wavelength range of 185 ~ 3300 nm. The quantum efficiency and transmission of the N719 cell with a thickness of about 11.5 μm as a function of the wavelength is shown in **Figure 13(c)** [76]. N719 dye displays a broad absorption spectral coverage in the region 400 ~ 700 nm, but a large proportion of infrared (IR) spectral range, which constitutes almost half the energy of the sun's radiation, cannot be utilized.

To make up for insufficient photon absorption, a reflector by either using a Ag mirror or a stack of the PhC was simply attached to the bottom cell or cathode electrode (**Figure 14(a)** and **(b)**). As seen in this illustration, there are two physical mechanisms by which wave optics approaches (based on reflector and photonic crystals) can improve light-trapping: reflection and diffraction. Firstly, Ag film can provide high reflection property like a mirror. Distributed Bragg reflectors exhibit high index contrast and they can reflect light over a extensive wavelengths region and incident light angles [84]. Likewise, a PhC can reflect incident light from broad

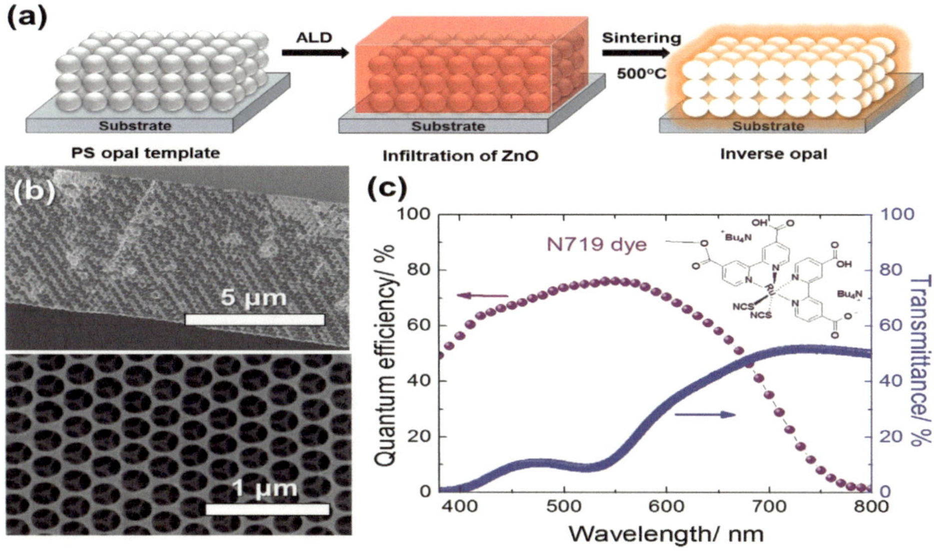

Figure 13.
(a) Fabrication procedure of the 3D PhC, (b) SEM micrograph of a typical ZnO PhC (or inverse opal) layer, showing both the side (cross sectional) view (top) and the top view (bottom) (c) quantum efficiency measurement of the N719 dye and the transmittance measurement of DSSCs using the same dye.

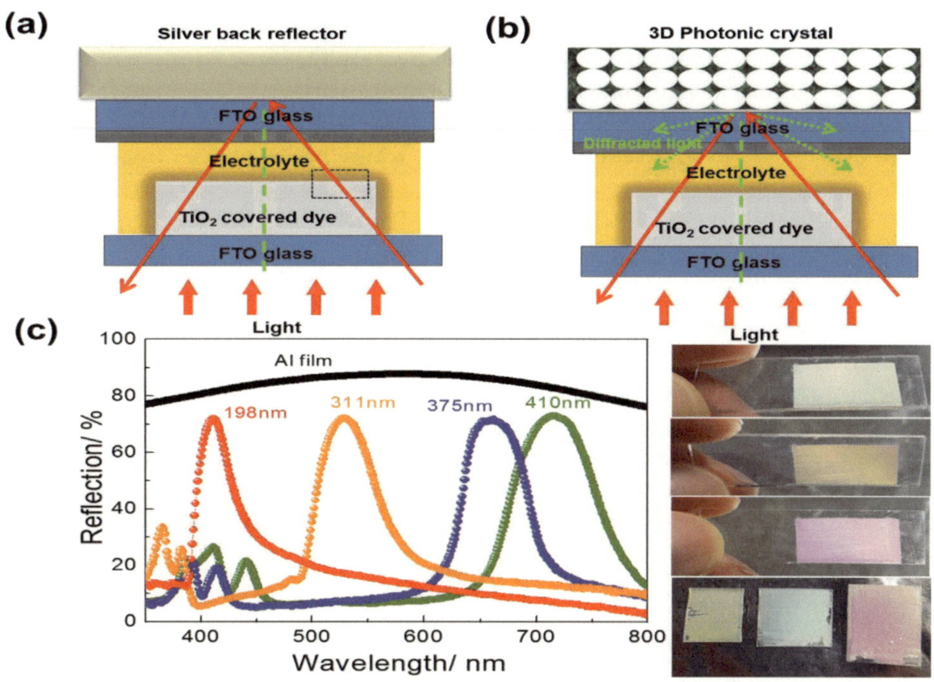

Figure 14.
Schematic of a DSSC (a) with a typical geometric optic concept of reflection to trap light. And (b) with wave optics (3D PhC) to trap light with reflection and diffraction (c) reflection spectra of Ag film and 3D PC with different sphere diameters (198, 311, 375, and 410 nm).

range of angles for frequencies and polarizations within the photonic crystal in that it can reflect light within the bandgap incident from any angle or medium.

Secondly, wave optics-based devices was developed for diffracting incoming beams into indirect angles according to Bragg's law [85]. The diffraction originated from an interface improve light trapping with the distance increment that light must go forward to the front surface of the cell as well as diffracted beam are internal reflected back into the solar cell when the angle of critical angle overcome the diffracted beam [86].

Figure 14(c) shows the reflection spectra of Ag film and 3D PC with 198, 311, 375, and 410 nm diameters. The reflectivity of the Ag film is more than 80% in wavelengths ranging from 400 to 800 nm; clearly, this Ag film has better reflectivity than the 3D PC. The red line represents the spectrum from the 198 nm inverse opal sphere; a reflection peak can be observed at 410 nm corresponding to the lowest photonic band gap (**PBG**). The orange line represents the spectrum from the 311 nm inverse opal sphere that shows the main reflection peak at 526 nm corresponding to the lowest PBG and additional reflection spectra peaks at 382 and 365 nm. The blue line shows the reflection spectrum of the 375 nm inverse opal sphere: it comprises a reflection peak at 661 nm corresponding to the fundamental PBG and additional reflection spectra peaks at 415 and 390 nm. The green line shows the reflection spectrum of the 410 nm inverse opal sphere that has a main reflection peak at 715 nm corresponding to the lowest PBG and additional reflection spectra peaks at 442 and 411 nm. The 3D PC show reflectivity peaks amplitudes of around 73% at the lowest PBG and around 25% at the high order PBG. This implies that we can recycle the photons back into the DSSC for further absorption and processing. The peak positions can be related to the sphere diameter and the effective refractive index of the medium using Bragg's law, $\lambda_{max} = 2n_{eff} d_{111}$, where d_{111} is the 111 lattice spacing and $\boldsymbol{n_{eff}}$ is the effective refractive index of the medium. Furthermore, the differences in the frequency of the reflection peaks are a result of the different sizes of the inverse opal spheres because the same effective refractive indices (ZnO) of the medium were used in all the experiments. Thus, the 3D PC can be devised to present a Bragg peak that matches the absorption band of the ruthenium dye but has a significant effect on longer wavelengths, thereby increasing the absorption efficiency.

4.2 Measurements and modeling of DSSCs with 3D PhC

The effects of selective light trapping, which was caused by the reflection and diffraction by 3D PhCs, on the solar-to-electric conversion efficiencies and AC impedance measurements of the cell, are analyzed by measuring the photocurrent-density-voltage (*J-V*) curves under simulated sunlight radiations of 1000 Wm^{-2} intensity. From a series of $J - V$ curves, samples coupling with Ag and PhC reflector display the increased photocurrent (seen in **Figure 15(a)**) The corresponding impedance measurements are plotted in **Figure 15(b)**, and the solid curves

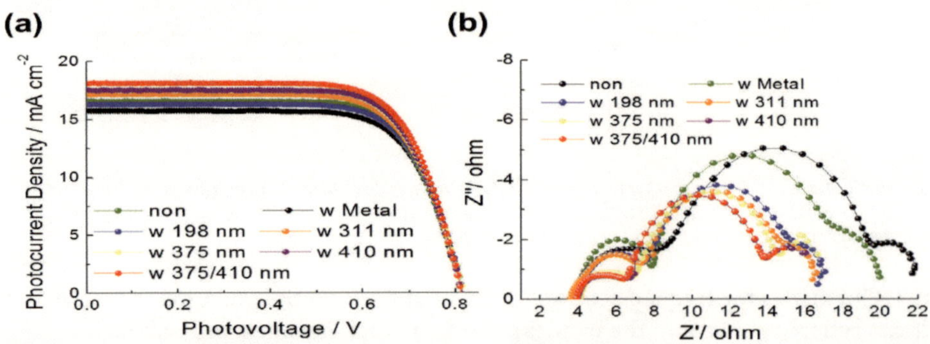

Figure 15.
(a) $J - V$ characteristics and (b) Nyquist plots of DSSC attached Ag metal and different sized PhC reflection (film thickness ~ 11 μm).

		D_{eff} (10^{-5} cm^2s^{-1})	k_{eff} (Hz)	τ_{eff} (ms)	R_k/R_w	*Con* (Ωcms^{-1})	R_d (Ω)	n_s (10^{18} cm^{-3})	D_1 (10^{-6} cm^2s^{-1})	V_{OC} (V)	J_{sc} (mA/cm^2)	*FF* (%)	EFF (%)
non		2.153	7.52	132	2.48	0.067	3.3	10.4	0.48	0.815	15.74	71.5	9.18
w Metal		2.273	7.52	132	2.54	0.069	3.0	10.6	0.06	0.818	16.58	71.2	9.64
w 3D PhC (nm)	198	2.175	7.52	132	2.60	0.064	3.0	10.5	0.06	0.815	16.00	71.7	9.36
	311	2.506	7.52	132	2.97	0.065	3.5	10.8	0.01	0.813	17.16	70.9	9.89
	375	2.725	7.94	125	3.06	0.068	3.0	10.9	0.12	0.818	17.51	70.9	10.1
	410	2.807	7.94	125	3.15	0.068	4.3	10.8	0.01	0.813	17.52	71.8	10.2
	Double (375/410)	2.647	7.94	125	2.97	0.068	3.7	11.1	0.12	0.812	18.10	71.4	10.5

Table 3.
*Parameters for the best fit of the impedance data and Photovotaic for the different sized 3D PhC attached DSSC film. Measured in **Figure 15**.*

represent the best fits from our modeling calculations. The fitting parameters are tabulated in **Table 3**. (the photoanode comprised 11 μm thick TiO_2 film and approximately 0.332 cm^2 active area with mask). The DSSC with the Ag film and 3D PCs showed higher short-circuit current densities and solar-to-electric conversion efficiencies than the traditional DSSC. For the Ag film, the short-circuit current densities were increased by approximately 0.84 mA/cm^2; the open-circuit voltage and fill factor were almost identical; and the solar-to-electric conversion efficiency was enhanced by 5% .

In the case of the DSSCs with 3D PhCs, the solar-to-electric conversion efficiencies were enhanced by 1.94, 7.78, 10.7, 11.3 and 14.4% for the 198 nm, 311 nm, 375 nm, 410 nm, and double layer (375/410 nm), respectively, as compared to the traditional typed DSSC. The 3D PC with 198 nm sized PhC showed the lowest efficiency enhancements because the traditional DSSC absorbed most of the light, this corresponds to the reflection peak of this PhC, as shown in **Figures 13(c)** and **14(c)**. The 375 nm and 410 nm sized PhC showed high enhancements in the solar-to-electric conversion efficiencies because the traditional DSSC transmitted more than 50% of the light; this corresponds to the reflection peaks of these PhC. This enhancement in the solar-to-electric conversion efficiencies is higher than that in the Ag film even though the Ag film has a considerably higher reflection intensity; this can be explained by the diffraction effect of the PhC as shown in **Figure 14(b)**. The PhC with 375/410 nm double layers showed highest enhancements in the solar-to-electric conversion efficiencies because the double layer PhC has a significant overlap with the quantum efficiency spectra of the ruthenium dye; moreover, the diffraction effect of the PhC is also present. From **Table 3**, the multilayer whose PBG has a large overlap with the quantum efficiency spectra of the ruthenium dye leads to larger enhancements in the photocurrent. Furthermore, the PhC -based DSSC exhibits considerably higher short circuit photocurrents than the traditional DSSC and is much more effective than the conventional Ag film.

The generation of photocurrent is primarily influenced by the light absorption of the dye. Therefore, when coupling PhC with the back surface of the DSSC, light absorption is increased attributed to the reflection and diffraction of light. This directly reflect the enhancement of the short-circuit current densities and the overall efficiency, while the open-circuit voltage, fill factor is unaffected. The DSSC with the PhC of 375, 410 nm and double layer diameters showed higher conversion efficiency than that with the Ag film; this can be explained not only by the reflection but also the diffraction effect in the DSSCs. As a result, we demonstrated here that recycling of photons by PhC is an effective way to increase the cell efficiency.

5. Nanostructures TiO_2 electrode in DSSC from 0 to 3 dimensions

As the main core of DSSCs, the nanocrystalline morphology of the photoelectrode film is critical for the efficient operation. Large efforts have been paid to optimize the morphology of the nanostructured photoanodes with improved architectures for high dye loading and fast electron transport. Although 2-step HT-NPs films treated with the interfacial modification have been regarded as a paradigm of porous photoelectrode for use in DSSCs, there are still challenges to boost the photovoltaic performance or competitiveness. Therefore, the metal oxides in their nanoform can be synthesized under various morphologies with different shapes and sizes, thus offering the possibility for modulating their properties. In literature, a variety of preparation techniques, such as sol–gel, [87] hydrothermal/solvothermal, [88] electrochemical anodization, [89] electrospinning, [35] spray pyrolysis, [90] and atomic layer deposition [91], have been developed and applied

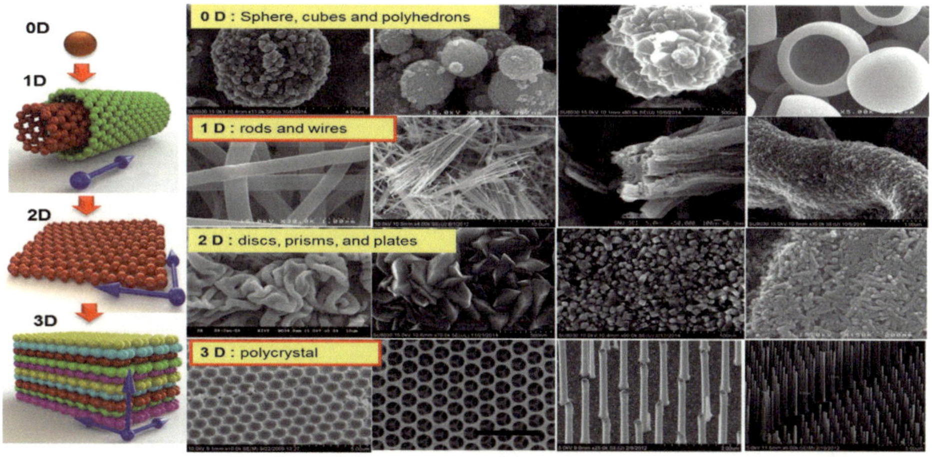

Figure 16.
SEM images of different types of 0D 1D 2D 3D- TiO_2. Produced in our lab.

to obtain different morphologies in photoanode materials. In this study, a variety of nanostructures from zero dimensional (0D) to three dimensional (3D) has been tested. Electro-spinning and spraying technique has been adopted to prepare the the different dimensional metal oxide semiconductor film by controlling the polymer host and solvents. The detailed results can be seen in followed section (**Figure 16**).

5.1 Titanium dioxide (TiO_2) nanomaterials as a photoelectrode

Various typed metal oxide semiconductors such as TiO_2, ZrO_2, SnO_2, ZnO, Nb_2O_5, Fe_2O_3, Al_2O_3, (binary compounds) and and ternary compounds such as $SrTiO_3$ and Zn_2SnO_4 can all act as photoelectrodes in DSSC due to their electronic structures which are characterized by a filled valence band and an empty conduction band [7, 92]. Among these heterogenous semiconductors, TiO_2 is the most widely used photoelectrode material because of the large amount of grain boundary and the large interface between the TiO_2 surface, dye and the electrolyte in the solar cell, further defects in the crystal structure are expected. In TiO_2, the Ti ions are in a distorted octahedral environment and formally have a $Ti^{4+}(3d^0)$ electronic configuration. The different obital hybridized structure of valence band (VB) and conduction bands (CB) at Ti 3d states occur the decreased the transition probability of electrons to the VB when the electron–hole recombination probability is reduced [92]. Therefore, in view of the electronic configuration and recombination probability TiO_2 are the best choice as photoelectrodes among 3d transition metal oxides. TiO_2 exsist naturally four commonly known crystalline polymorphs, i.e., anatase (tetragonal, E_g = 3.23 eV), rutile (tetragonal, E_g = 3.05 eV), brookite (orthorhombic, E_g = 3.26 eV) and metastable forms (monoclinic, orthrhombic, cotunnute) [93]. These structures can be described in terms of chains of TiO_6 octahedra. The crystal structures differ in the distortion of each octahedron and by the assembly pattern of the octahedra chains [94]. The crystalline strucuture of TiO_2 can be seen in **Figure 17** [95]. Both rutile and anatase have tetragonal structure with a = 0.46 nm and c = 0.29 nm (rutile); a = 0.3782 nm and c = 0.9502 nm (anatase). Brookite has orthorhombic structure with a = 0.5456 nm, b = 0.9182 nm, and c = 0.5143 nm and it is very hard to synthesize in the laboratory, while the rutile and anatase can be easily prepared. For solar cell application, anatase structure is more prferred because of ~0.1 eV higher the Fermi level, lower recombination rate of electron–hole pairs and lower formation temperature [96].

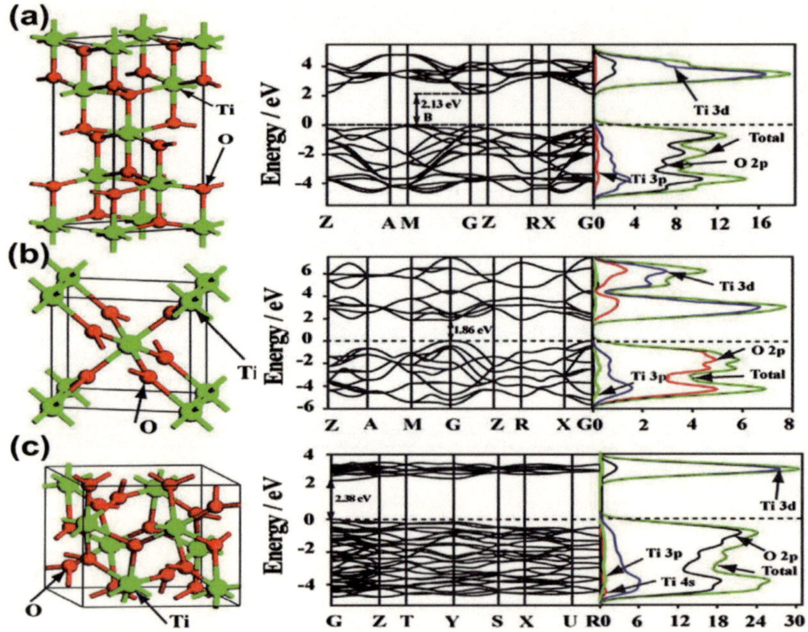

Figure 17.
The schematic TiO_2 structure and the DOS calculated band structure for (a) anatase, (b) rutile and (c) brookite TiO_2. The big green spheres represent Ti atoms and the small red spheres represent O atoms. Reprinted from [95].

Thermodynamic calculations based on calorimetric data predict that rutile is the stablest phase at all temperatures, exhibiting lower total lower total free energy than metastable phases of anatase and brookite. The small differences in the Gibbs free energy (4 ~ 20 kJ/mole) between the three phases suggest that the metastable polymorphs are almost as stable as rutile at normal pressures and temperatures [97]. The physical and chemical properties of TiO_2 nanocrystals are affected not only by the intrinsic electronic structure, but also by their size, shape, organization, and surface properties. For example, if the particle sizes of the three crystalline phases are equal, anatase is most thermodynamically stable phase below 10 ~ 15 nm, brookite is most stable between 11 ~ 35 nm, and rutile is most stable at sizes greater than 35 nm [98]. Herein, interesting morphologies and properties have recently attracted considerable attention and many nanostructured TiO_2 materials, such as nanotubes, nanorods, nanofibers, nanosheets, and interconnected architectures, mesoporus material such as inverse opal and photonic crystal have been fabricated and applied in PV devices [35, 83, 99, 100]. In order to be effective photoelectode, several parameters such as morphologies, pore volume, and the crysltalinity of TiO_2 influence the charge transport and recombination processes.

5.2 0-dimensional (0D) titanium dioxide (TiO_2) Nanosphere

Particles which are more or less spherical in shape like fullerenes, quantum dots, nano-onions, nanoparticles, etc. are considered as 0D. These 0D nanomaterials are sized at nanoscale level in all three dimensions and must be amorphours or crystalline; single or polycrystalline; and composed of single- or multichemical element. Within 0D materials, all electrons are fully confined and their length equals the width. Due to confinement of both electrons and holes, the lowest energy optical transition from the valence to conduction band will increase in energy, effectively increasing the band gap. A 0D TiO_2 nanosphere (**NS**) film is noticed as a effective photoelectrode because of its structural advantages such as high surface area, submicro- or meso-porous structure for light scattering function and better infiltration of electrolyte. There are several different efforts to produce spheres [101, 102]. Nevertheless, there are limitations for the quality control, large-scale production and flexibility. In this book, a simple electro-spraying technique is introduced [103, 104]. As a first step, well-dispersed TiO2 suspensions is very important to make a continuous fluid jet. **Figure 18** show the three phase statues during E-sparying process: (i) TiO_2 suspension zone, which strong electric field extracts droplets from Taylor cone (the non-sphere formation), (ii) mixed zone of TiO_2 suspension and solid, which non-spheres and spheres are formed by solvent evaporates, leading to the droplet shrink, and (iii) solidified TiO_2 zone, depositing the TiO_2 NSs onto the conductive glass.

In this systmem, the formation of tight cluster sphere filled sphere is caused by the ultrafast evaporation of alcholic solvent under an electric field. For the desireable sphere typed TiO_2 film, it is necessary to controll several parameters such as electric field, feed rate, a size of tip and a distance between the nozzle and FTO substrate. It is noted that the the sphere size is controlled by changing the TiO_2 concentration and the mixture of solvent in the dispersion solution. In order to find the optimazed condition, TiO_2 SPs with the different concentration are tested at 6 μm thick film. **Figure 19(a)** show the SEM results for E-sprayed film prepared from the different concentration at 1 wt%, 3 wt%, 5 wt% and 10 wt%. With increasing the concentration, the diameters of sphere are almost linearly increased, while the number of molecules calculated from UV–vis absorption spectra of desorbed sensitizers reveal that a TiO_2 NSs with 5 wt% have about 11.6% and 13.9% higher value than that of 1 wt% and 10 wt% TiO_2 NSs film, respectively (see **Figure 19(b)**).

Experiemntally, E-spraying at the low weight percent of TiO_2 make it hard to achieve the competely formed sphere over 8 μm thick film. Therefore, in this reaserch, about 5 wt% TiO_2 suspension in EtOH is used as the best condition. The detailed phase diagram accroding to the electrospraying parameters can be seen in the literature [105]. **Figure 19(c)** displays the thickness profiler on the various

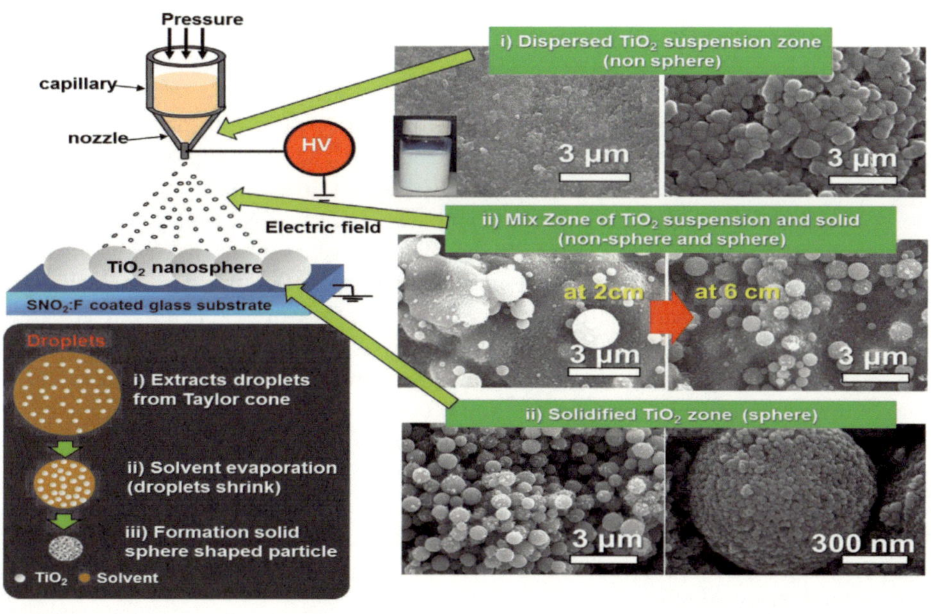

Figure 18.
The schematic diagram of the formation of hierarchically structured 0D TiO_2 Nanosphere (NS).

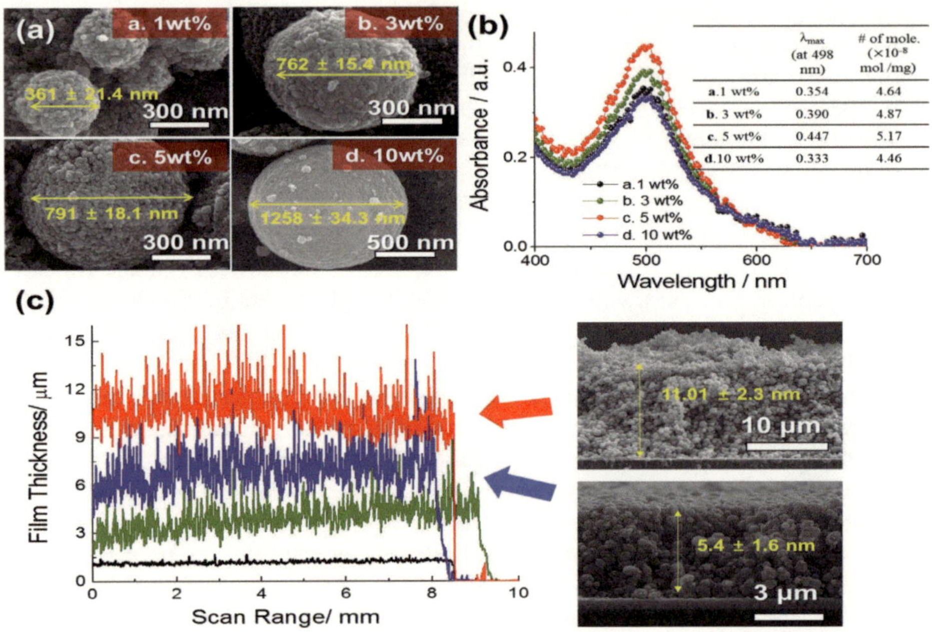

Figure 19.
(a) SEM images and (b) absorbance curve of aqueous dye solutions after desorption and (c) the different thickness of E-sprayed TiO_2 spheres at the different concentration a. 1 wt%, b. 3 wt%, c. 5 wt% and d. 10 wt%.

thickness of TiO_2 NSs photoelectrode film. In an E-spraying process, the film can control the feeding volume. **Figure 19(c)** shows the cross-sectional image of an E-sprayed TiO_2 SPs film that demonstrates the uniform shape of the spheres from bottom to top. However, with increasing film thickness, TiO_2 NSs photoelectrode tends to increase the film roughness since intrinsic insulator properties of TiO_2 can influence on an electric field film during depositing process.

Figure 20(a) and **(b)** show SEM images of a distribution of mesoporous TiO_2 NSs in the anode electrode of E-sprayed TiO_2 NSs film. This film exhibits a range of diameters between 100 $\sim$ 800 nm, formed from nucleation and crystallization of 10 nm TiO_2 particles. Also, the existence of such mesopores improved the electrolyte penetration. With the $TiCl_4$ treatement, a filled sphere with a relatively rough surface is generated. (seen in **Figure 20(b)**) This is benefical for improving the interconnection between primary particles inside the TiO_2 NSs. **Figure 20(c)** shows a TEM image of the spheres with the crystalline nano particles. The surface area and porosity of both TiO_2 SPs and NPs film can be estimated by N_2 adsorption–desorption isotherm at 77 K (see in **Figure 20(d)**). The SPs films show a type-IV isotherm as well as an increase in the adsorbed amount at high relative pressure, indicating the existence of mesopores in the sample [106]. The measured specific surface area of TiO_2 SPs is ~188.47 m^2g^{-1}, ~2.9 higher than that of similar amount of TiO_2 NPs electrode (65.33 m^2g^{-1}). Furthermore, the cumulative BJH mesopore volume and maximum pore radius is 1.194 cm^3g^{-1} and 10.05 nm, respectively, while hydrothermal nanoparticle film shows the pore volume of 0.607 cm^3g^{-1} and maximum pore radius of 7.50 nm. In the case of $TiCl_4$ treated SP, the surface area and the cumulative pore volume is decreased by as much as 65.3% and 65.2%, respectively compared with the non-treated SPs film. The amounts of dye adsorbed onto each of the pristine- and $TiCl_4$ treated TiO_2 SP layers at 11 μm were 5.80×10^{-6} and 8.19×10^{-5}, respectively.

It shows about 2.1 times increased dye absorption properties compared with the NP. This high density of dye molecules onto TiO_2 NS can be explained by the high

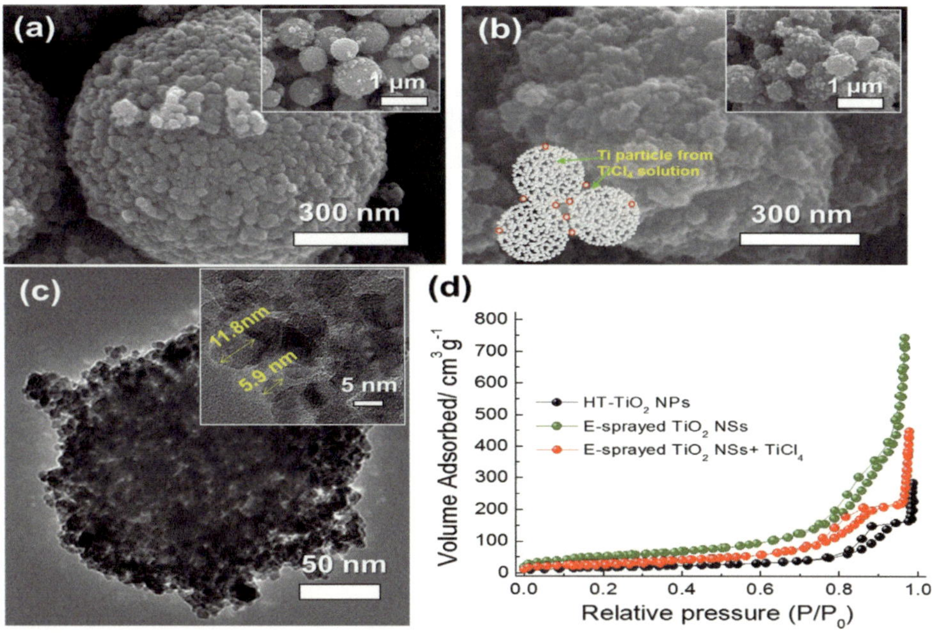

Figure 20.
SEM images of E-sprayed (a) TiO_2 NSs and (b) $TiCl_4$ treated TiO_2 NSs (c) TEM images of the TiO_2 NSs, which themselves are formed from crystallized ~10 nm particle. (d) N_2 adsorption–desorption isotherms of TiO_2 NPs and NSs.

intrinsic surface area of each TiO_2 NSs surface as well as the crack free surface morphology. In general, casting TiO_2 NPs from a polymer supporter for the thick TiO_2 layer onto substrates is by far the preferred method of depositing electrodes for DSSCs. Therefore, heating process at temperature ranging from 450–500°C is necessary to burn out the residual organic binder and to enhance the inter-particle connection between NPs. In this process, cracks are formed at the interface with TiO_2 NPs, leading to deteriorate dye absorption. (see **Figure 21(a)**). In the case of E-sprayed TiO_2 SPs from binder-free, the surface of the e-sprayed TiO_2 layer is uniform and without cracks, unlike the surfaces produced using paste methods as shown in **Figure 21(b)**.

The *J–V* curves, EIS analysis and IPCE data and detailed information of dye-sensitized NPs and SPs film of the same thickness with mask (0.220 cm^2) at 1 sun light intensity are shown in **Figure 21(c)** and **(d)** and **Table 4**, respectively. Their photocurrent density is almost the same (10.6 mA/cm^2), and V_{oc} of the SP-based solar cell (0.831 V) is only about 2% higher than that of the anatase-based cell (0.815 V). The respective overall energy conversion efficiencies of the NPs and SPs -based cells are a 7.28% and 7.46%. The predominant increase of cell efficiency at the TiO_2 SPs film can be obtained by $TiCl_4$ treatment because of the enhancement of the interconnection between TiO_2 SPs, which illustrated in **Figure 20(b)**. The $TiCl_4$ treated sample represents a 21.0% increase in the energy conversion over the non-treated SPs. This increment of SPs films is about twice higher than that of NPs because the presence of intrinsic microporous SPs structure, which make its more open structure increase amount of I_3^- a in the pores. From the suggested impedance models, I confirm the electron diffusion coefficient rate of triiodide D_1 for $TiCl_4$ treated SP film is 4.3×10^{-7} cm^2S^{-1}, 53% higher than that obtained from $TiCl_4$ treated SP film, at 2.1×10^{-7} cm^2S^{-1}. This increase is in good agreement with the increase of V_{oc} and *FF* observed in **Figure 21(c)**. Hence the SPs based cell plays a key role in attaining the higher cell efficiency with solid state electrolyte.

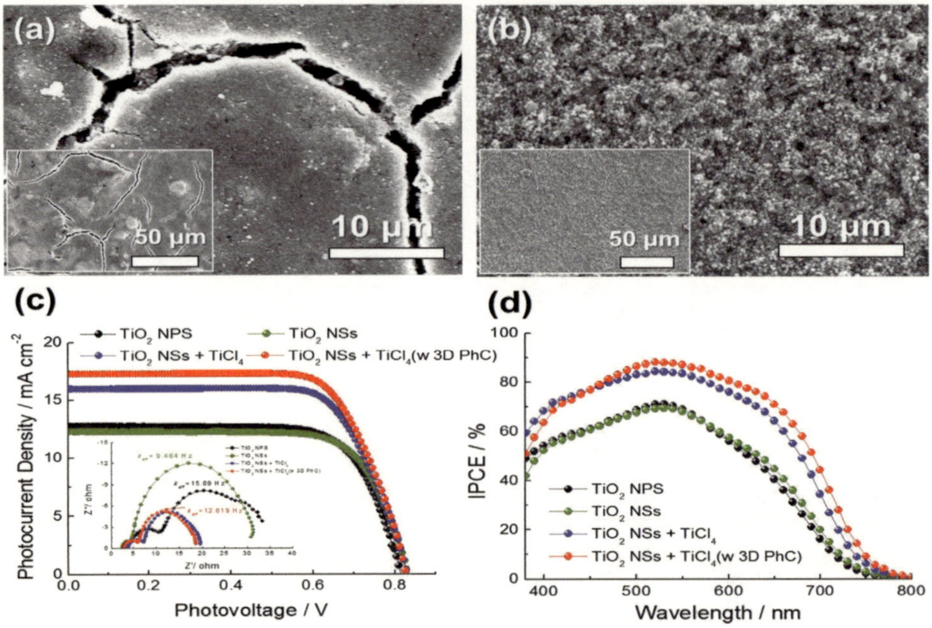

Figure 21.
SEM images of TiO2 film (a) doctor bladed and (b) E-sprayed TiO_2 film. (c) and (d) JV and IPCE data for TiO_2 NSs films.

Interestingly, despite of the decreased charge density of SPs film, the photocurrent density is nearly the same. This reason can be explained by scattering effect on the SPs film. **Figure 20(d)** shows the IPCE curves for each sample range from 350 to 800 nm. The films with SPs layer the entire IPCE curve is slightly shifted upward in the region between 550 and 800 nm. With $TiCl_4$ treatment, the increase is more pronounced, which is good agreement with those reported by Sommeling [56]. The effect of scattering clearly results in a much larger improvement of the red response than originates from the higher dye loading.

5.3 1-dimensional (1D) titanium dioxide (TiO_2) nanorod

One-dimensional (1D) nanostructures (e.g., nanowires, nanorods, nanotubes, nanobelts, cauliflower-shaped structures) have been widely considered to have superior electron transport characteristics compared to conventional nanorod (NP) -based systems. Eariler reported research have demonstrated that highly efficient TiO_2 nanorod-based NR DSSCs and compare charge transport properties to those of typical TiO_2 nanoparticle-based (NP, commercially available P-25) NP-DSSCs [35]. The electrospun TiO_2 NR based photoelectrodes exhibited about 2 times larger the pore volume and ~2.5 times more dye coverages at the same weight than those of TiO_2 NP. Therefore, NR based cell indicated about 40% higher efficiencies than NP-DSSCs attribied to ~8 times slower electron–hole recombination in NR-DSSCs. With the $TiCl_4$ post-treatment, further enhancement of electron diffusion coefficient, a charge collection efficiency and the efficiency can be demonstrated [35].

5.4 2-dimensional (2D) titanium dioxide (TiO_2) nanosheets

Two-dimensional (2D) anisotropic nanostructures of metal oxides and semiconductor possess pronounced quantum surface effects and dramatic changes in electronic structures and thus in the physical and chemical properties, which are

		BET & BJH		Absorption Properties		Electrical Properties					Solar Properties			
		Surface area (m^2g^{-1})	Pore Volume (cm^3g^{-1})	λ_{max} (at 498 nm)	# of mole. ($\times 10^{-8}$ mol /mg)	D_{eff} ($\times 10^{-5}$ cm^2s^{-1})	R_k/R_w	n_s ($\times 10^{18}$ cm^{-3})	D_1 ($\times 10^{-6}$ cm^2s^{-1})	R_{total} (Ω)	V_{OC} (V)	J_{sc} (mA/cm^2)	*FF* (%)	EFF (%)
NP	—	65.33	0.607	0.410	3.65	2.58	2.38	9.39	0.32	31.5	0.815	12.38	72.2	7.28
	w $TiCl_4$	55.98	0.445	0.504	3.97	3.24	2.14	11.2	0.21	23.7	0.796	15.92	66.8	8.47
SP	—	188.47	1.365	0.514	5.80	1.83	6.79	6.07	0.78	33.0	0.831	12.28	73.1	7.46
	w $TiCl_4$	114.0	0.915	0.772	8.06	3.29	3.39	9.40	0.45	19.2	0.829	15.96	71.4	9.45
	w $TiCl_4$/ 3D	—	—	—	—	3.49	3.13	10.0	0.43	18.2	0.827	17.24	71.3	10.1

Table 4.
*Parameters for the best fit of the impedance and Photovotaic data for doctor bladed and E-sprayed TiO_2 film. Measured in **Figure 21**.*

anticipated to inspire new fundamental and technological research [107]. In my earlier research, TiO_2 or ZnO nanofiber mat is used for employing DSSC. [108, 109] For empolying 2D TiO_2 film for DSSCs, two different technique are used for TiO_2 photoelectrode: (1) Nanofiber (NF) mats produced by electrospinning: (2) Nanotubes (NT) film from 2-step hydrothermal method. As see in **Figure 22(a)**, TiO_2/poly(vinly acetate, PVAc) composite fiber mats are directly electrospun onto a FTO glass and then polymer binders are removed by two step process: (i) a THF solvent tratement for melting PVAc polmyer, (ii) calcination this mat at 500°C for 30 min. With the post treatment, the adhesion issues between TiO_2 NFs or NFs and FTO glass can be overcome. The TiO_2 NTs film is prepared from adjusting pH values [110]. Following this method, the well-growed NT poweders are deposited by mixing a mixture of PEO/PEG polymer binder as described in a section of "***DSSC Fabrication for Conventional typed Cell.***"

The detailed surface and cross sectional images can be confirmed by the SEM images (1.22(b)). as can bmodified 2-step hydrothermal process. **Figure 22(c)** show the nitrogen adsorption–desorption isotherm and Barret–Joyner–Halenda (BJH) pore size distribution plot of the NF mats and NT film, respectively. The specific surface area and pore volume of both samples is very small and it is difficult to obtain a sufficient photocurrent density. In agreement with the BET analysis, the JV curve of all cells display an obviously low photocurrent density, leading to the poor cell efficiency.

5.5 3-dimensional (3D) titanium dioxide (TiO_2)

In the section of "***Photon Management using Three-Dimensional Photonic Crystals***", the effective use of three-dimensional photonic crystals (3D PhC) for photon management in our cells was demonstrated. Here, 3D PhC is used as a photoelectrode for DSSCs. An important approach to enhance DSSC efficiency is to increase the path length of light by enhancing light scattering in TiO_2 films by

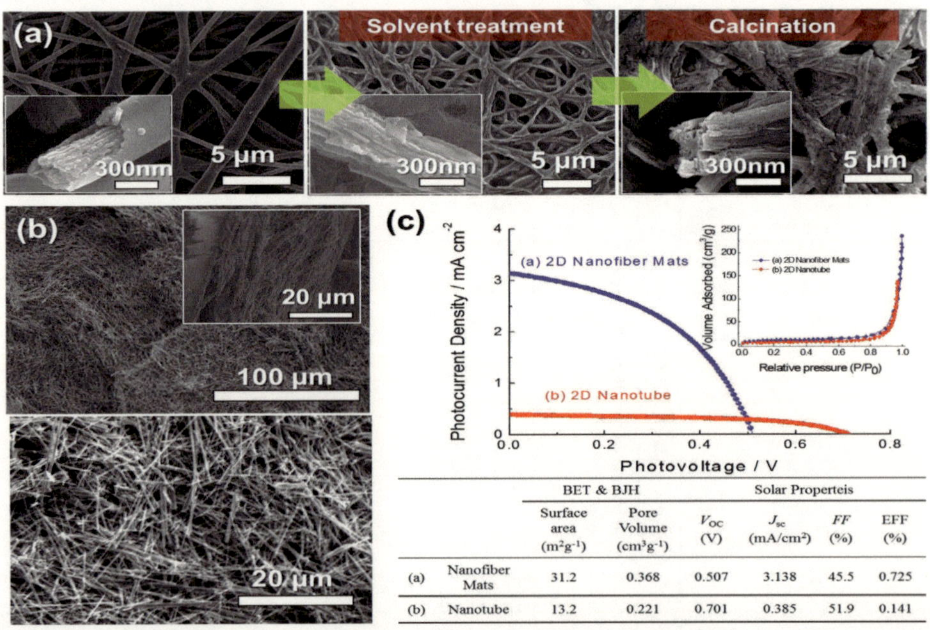

		BET & BJH		Solar Properteis			
		Surface area (m^2g^{-1})	Pore Volume (cm^3g^{-1})	V_{OC} (V)	J_{sc} (mA/cm^2)	FF (%)	EFF (%)
(a)	Nanofiber Mats	31.2	0.368	0.507	3.138	45.5	0.725
(b)	Nanotube	13.2	0.221	0.701	0.385	51.9	0.141

Figure 22.
2D TiO_2 film (a) TiO_2 nanofiber (NF) film prepared by E-spraying technique (b) TiO_2 nanotube (NT) film by hydrothermal treatment. (c) JV and BET analysis of each cells.

coating the large particles onto the small sized of TiO_2 NPs [111]. To understand the effect on the PhC coupled DSSCs that feature enhanced photocurrent over a large spectral region, a double layered photoanode integrates a high surface mesoporous underlayer TiO_2 with an optically active TiO_2 PhC overlayer is fabricated as illustrated in **Figure 23(a)**. For the FTO/3D TiO_2 PhCs structure, the polystyrene (PS) opal template of sphere size 370 nm was prepared using a vertical deposition technique onto the nc-TiO_2 film and then TiO_2 is coated by ALD. The inverse opal structure can be made by ion etching (CF_4 5 min).

Figure 23(b) shows the J-V curves for the different sized 3D PhC at liquid electrolyte. (as seen above) When the size of hole is increased from 198 nm to 410 nm, the V_{oc} values increased from 0.787 V to 0.849 V. The 311 nm 3D PhC bilayer exhibits the best PCE, which is a 46.1% increase in J_{sc} (8.657 mA/cm^2) and 36.2% increase in PCE (5.102%) compared to the NPs layers (7.064 mA/cm^2 of J_{sc}, 3.258% of PCE) The significant improved *PCE* of 3D PhC indicates that the 3D PhC top layer is electrically connected and contributes to light harvesting over the entire spectrum.

Moreover, large porosity structures of the inverted 3D TiO_2 PhCs have enabled good infiltration for high-viscosity electrolytes. For solid state cell fabrication, we have chosen ($P_{1,4}I$)-doped succinonitrile plastic crystal electrolyte [112, 113]. This electrolyte has the best performance for solid state DSSC as reported in the literature [114]. At a room temperature, this compound electrolyte has a high conductivity of 3.3 mS/cm and fast ion transport of iodine and triiodine in its plastic phase of 3.7×10^{-6} and 2.2×10^{-6} cm^2/Vs, respectively. The observed fast ion transport in this solid material can be seen as a decoupling of diffusion and shear relaxation times, which probably originates from local defect rotations in the succinonitrile plastic crystal [115–117]. Thus, this plastic electrolyte showed best high cell efficiency among other competing electrolyte materials. **Figure 23(c)** top shows the molecular structure of the compound. For better cell performance, we firstly injected the liquid state electrolyte into TiO_2 NP film and evaporated this electrolyte by putting it in the dry oven at 80°C for 12 hour. Next, the plastic electrolyte was heated until it became a liquid (>70°C) and injected into the warmed sandwiched cells. After the cells included the plastic electrolyte was cooled down to room

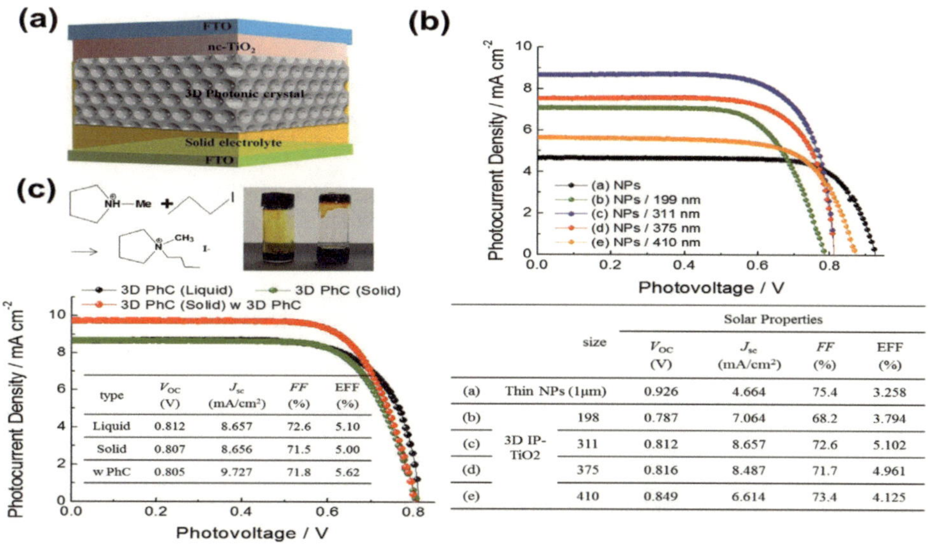

type	V_{OC} (V)	J_{sc} (mA/cm^2)	FF (%)	EFF (%)
Liquid	0.812	8.657	72.6	5.10
Solid	0.807	8.656	71.5	5.00
w PhC	0.805	9.727	71.8	5.62

	size		Solar Properties			
			V_{OC} (V)	J_{sc} (mA/cm^2)	FF (%)	EFF (%)
(a)	Thin NPs (1μm)		0.926	4.664	75.4	3.258
(b)	3D IP-TiO2	198	0.787	7.064	68.2	3.794
(c)		311	0.812	8.657	72.6	5.102
(d)		375	0.816	8.487	71.7	4.961
(e)		410	0.849	6.614	73.4	4.125

Figure 23.
3D TiO_2 film (a) schematic design (b) JV characteristics of 3D PhC TiO_2 at the different size (c) solid state results.

temperature, a waxy solid was obtained. **Figure 23(c)** shows the JV curve for solid state of 311 nm 3D PhC bilayer. Interestingly, no obvious efficiency change could be seen in all typed solid state electrolyte. The PCE of the solid state system is found to be 2% of that of the liquid electrolyte system. In conclusion, the design of 3D PhC bilayer film enables effective dye sensitization, electrolyte infiltration and charge collection because the layers are in direct physical and electronic contact, light harvesting in specific spectral regions was significantly increased by the 3D PhC effect and PC-induced resonances. This approach should be useful in solid-state devices where pore infiltration is a limiting factor as well as in weakly absorbing photovoltaic devices.

6. Photosensitizer for DSSC: state of the art.

6.1 Ruthenium (Ru) complex based photosensitizer

Since the advent of DSSCs, ruthenium-series dyes are considered as the most powerful photosensitizer because of the following reason: their strong metal to ligand charge transfer transition (MLCT) process, a broad absorption spectrum, suitable excited and ground state energy levels, relatively long excited-state life-time, and good (electro) chemical stability and power conversion efficiency (PCE) >11% [7]. In order to improve the efficiency of DSSCs, the sensitizer should be panchromatic, that is, absorb photons from the visible to near-infrared (NIR) region of the solar spectrum while maintaining sufficient thermodynamic driving force for the electron injection and dye regeneration process. Many efforts have been made to change the change the ligands of Ru complexes: [118, 119] As seen in **Figure 24**, ruthenium(II) – polypyridyl complexes such as the thiocyanato deriva-tive, *cis*-$(SCN)_2$ bis (2,2′-bipyridyl-4,4′-dicarboxylate) ruthenium(II), (coded as **N3**), the doubly protonated form, $(Bu_4N)_2(Ru(dcbpyH)_2(NCS)_2)$, (named **N719**)

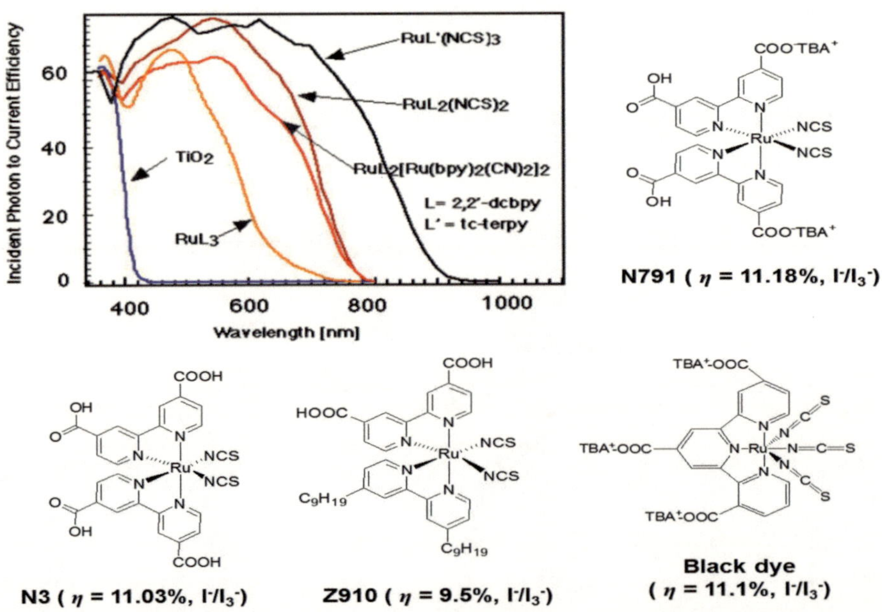

Figure 24.
Incident photon to current conversion efficiency as a function of the wavelength for the standard ruthenium sensitizers N3 (red line), N719 (brown) the black dye N749 (black curve), and the blank nanocrystalline TiO2 film (blue curve). Reprinted with permission from [7, 120, 121].

and the Ru center has three thiocyanato ligands and one terpyridine ligand substituted with three carboxyl groups, (also called the "**black dye**") are widely used as reference and high efficiency sensitizers for DSSCs. The amphiphilic heteroleptic ruthenium sensitizer, known as **Z907**, reported noticeable thermal stability with stable 7% energy conversion efficiency [114]. For optimized DSSC with N3 or N719, the certified IPCE values are ~80% for wavelengths ~650 nm.

However, the IPCE increases only gradually from the absorption onset to shorter wavelengths due to relatively low extinction coefficients ($1.40 \times 10^4\ M^{-1}\ cm^{-1}$) [7, 47]. In addition, this class of compounds contains expensive ruthenium metal and requires careful synthesis and tricky purification steps. Therefore, efforts in the synthesis of sensitizers for DSSCs step forward to the metal-free organic donor–acceptor (D–A) dyes system.

6.2 Porphyrins based sensitizer

The idea for mimicking the light harvesting processes based on chlorophyll occurring photosynthetic reaction centres inspire the research for porphyrins based sensitizer. The porphyrin-based dyes exhibit large absorption coefficients in the visible and infrared region as well as their rigid molecular structures consisted of four meso and eight β reaction sites, which can control their properties [122–124]. The designed porphyrin dyes with a π-conjugated link at the β-position of the porphyrin ring enhance the electronic coupling of the dye with the surface of TiO_2, reaching η =7.1% of a PCE [125]. Numerous series of porphyrins are reported for the DSSC application as an effective sensitizer. of DSSC synthetized. Among them, a class of sensitizer consisting of a push–pull porphyrins with an electron-donating diarylamino group and an electron-withdrawing carboxyphenylethynyl anchoring group shows the outstanding solar properties. For example, advances in optimization of the device performance for a zinc porphyrin sensitizer (YD2-oC8) co-sensitized with an organic dye (Y123) using a cobalt-based electrolyte to enhance photovoltage of the device attained an unprecedented power conversion efficiency of η = 12.3% [126]. (see **Figure 25(a)**) Further improvement can be found at

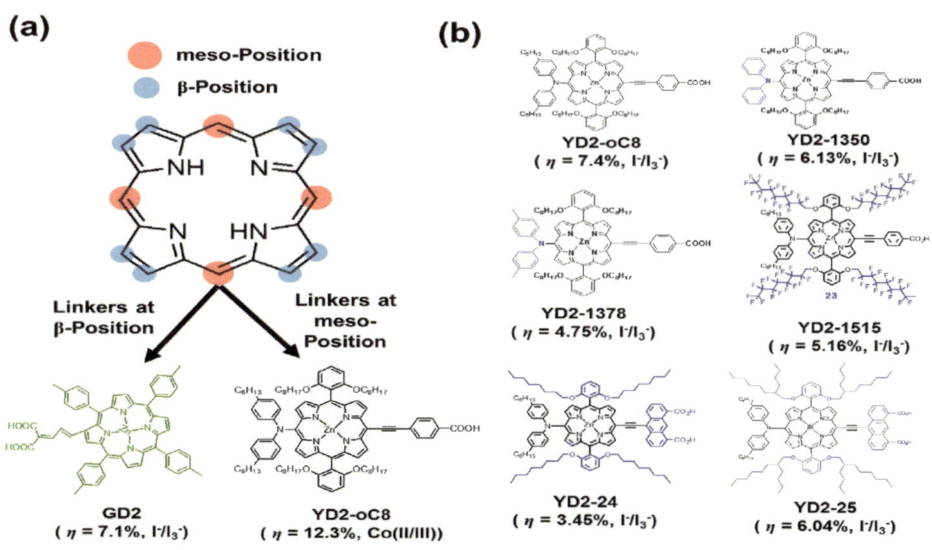

Figure 25.
(a) Typical structure of a porphyrin showing the four meso- and the eight β-positions to be functionalized for porphyrin-sensitized solar cells. Reprinted from [122, 123] (b) solar performance for the different series of porphyrin based DSSC at liquid electrolyte and TiO_2 NSs.

incorporation of the proquinoidal benzothiadiazole (BTD) unit into the functionalization of the porphyrin core with the bulky bis(2′,4′-bis(hexyloxy)-(1,1′-biphenyl)-4-yl)amine donor and a 4-ethynylbenzoic acid yielded the green dye, which exhibited a slightly improved PCE of 13% [127]. The detailed physical and structural studies responsible for recent advances of the porphyrin-based DSSCs have been reviewed in several reports [122–124]. Thanks to Yeh group's support, various porphyrin with the different structure based sensitizers can be tested in my optimized condition and the detailed performance listed in **Figure 25(b)**. Interestingly, the cell performance is gradually improved upon light exposure and heat treatment. Specially, after 90 min light exposure, cell performance increased from 6.12% to 7.71% attributed to the increase of J_{sc} value (see **Figure 26(a)-(c)**). The improvement can be explained by the different charge recombination process. According to Mori et al., Li^+ ions are removed from the TiO_2 surface and replaced with $DMPIm^+$ ions under light exposure [128, 129]. This process is found to enhance the electron lifetime by decreasing charge recombination with the redox mediator (**Figure 26(d)**). This can be explained by initial limited injection and fast charge recombination processes. As a result, this process enhances the cell performance by decreasing recombination with the redox mediator. However, about 20.1% improved cell efficiency by light exposure indicate YD2-oC8 sensitizer exhibit the extra open space at the TiO_2 SP surface. Therefore, the best device performance in our system show about 7.7% of energy efficiency at the I^-/I_3^- liquid electrolyte and TiO_2 NSs samples.

6.3 Tetrathienoacenes (TTAs) based sensitizers

Metal-free organic dyes have attracted much attention to researchers due to chemical or optical versatility, environmental compatibility and potential cost reductions [7]. In spite of these attractions and efforts, a little lower cell performance and difficulty in synthesis cannot meet the requirement for commercialization.

A donor-π-acceptor (D-π-A) structure is a promising strategy for metal-free organic sensitizers because of the effective photoinduced intramolecular charge

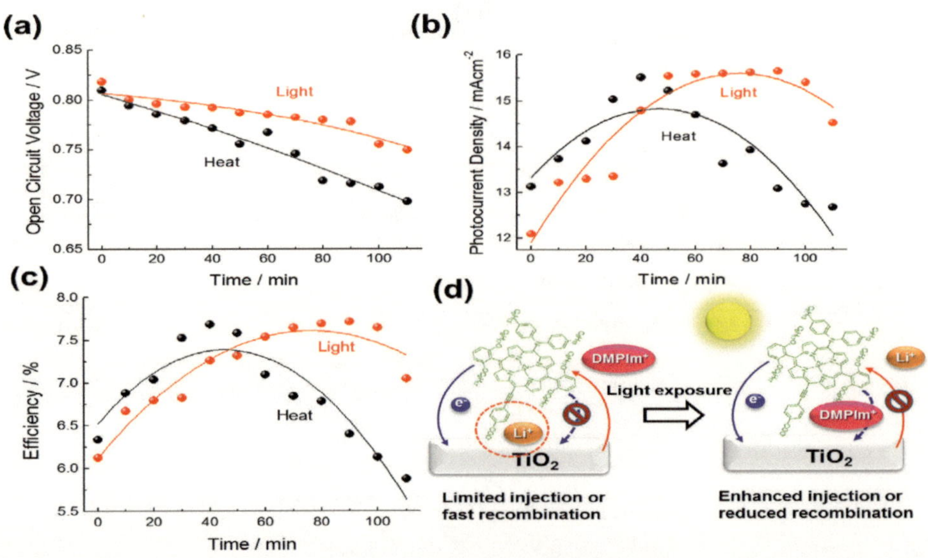

Figure 26.
Solar cell performance (a) Voc, (b) Jsc and (c) eff versus light exposure time (d) schematic representation of light-induced cation exchange. Reprinted with permission from [129].

transfer property [7, 130]. Thanks to support from Chen's group, a new series of organic dye based on tetrathienoacene are applied for DSSC [105, 131]. In this design, triphenylamine electron-donor (D) unit and cyanoacrylic acid electron-acceptor (A) unit is connected to an electron rich a lipophilic dihexyloxy-substituted thiophene-based fused tetrathienoacene. (**TPA-TTAR-TA**); (1) the triphenylamine unit composed of a large conjugated tertiary amine system is known as a strong electron donor in its initial state as well as act as stabilize the charge-transferred state [132]. (2) The cyanoacrylic acid, which is an anchoring group to bind to a TiO_2 surface functions as an effective electron acceptor. (3) As the π-system, we used fused tetrathienoacene cores produced by the newly designed one-pot synthetic routes [133]. Fused-thiophenes offer the attraction of good charge transport properties with extensive molecular conjugation and strong intermolecular S···S interactions, [134, 135] which might enhance the efficiency of DSSC. (see in **Figure 27(a)**).

Several fused thiophene derivatives have already been demonstrated to have excellent charge transport performance. For example, dithienothiophene (DTT) and tetrathienoacene (TTA) based OTFTs exhibited mobilities up 0.42 (p-channel) and 0.30 (n-channel) $cm^2V^{-1}\,s^{-1}$, respectively [136]. Relative to organic semiconductors, the potential of fused thiophene-based DSSCs has not been well explored until recently, and only for a limited range of TT and DTT materials. To the best of our knowledge, the first example of a TT-based small molecule DSSC with a PCE of 7.8% was reported by *P. Wang* and M. Gratzel et al. in 2008 [137]. For DTT, was reported by the same team in the same year with a PCE of 8.0% [138]. Presumably, due to high coplanarity of poly fused thiophenes may lead to aggregation of dye molecules on nanocrystals, caused a dissipative intermolecular charge transfer, and then rendered an unfavorable effect on the cell efficiency. As a result, the more conjugated TTA-based small molecules have never been explored for DSSCs yet. Nevertheless, in terms of tuning the energy-level of chromophores to attain a better capability of panchromatic light-harvesting, conjugated TTAs elevate the HOMO and lower a suitable LUMO compared to TT and DTT based DSSCs. Red shifted absorption with high molar coefficient and better charge transport (*vide infra*)

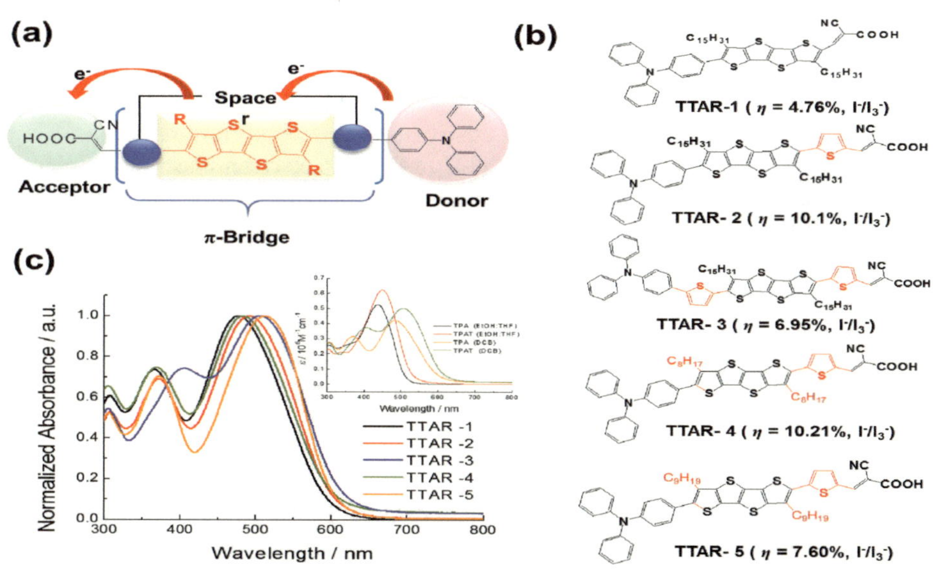

Figure 27.
(a) Schematic representation of the donor-π-bridge-acceptor molecular dye design concept and (b) the chemical structures with solar performance and (c) UV–vis absorption spectra of dyes 1–5 and their corresponding molar absorption coefficients measured in o-DCB in concentration of 10^{-5} M of TTAR series dyes.

might enhance the efficiency of DSSC thus offering TTA a potential good conjugation unit for a new organic dye. The detailed synthetic route and procedures aren't described in my thesis.

The effects of thiophene introduction between the bridge and the donor and/or acceptor moieties can be systematically understood. Starting from the simplest **TTAR** structure, **TTAR-1**, a thiophene spacer is either inserted on the acceptor side between TTAR and cyanoacrylic acid to yield **TTAR-2**, and finally on both acceptor and donor sides to yield **TTAR-3**. These π-extended systems should enhance panchromatic light-harvesting, as the result of the uplifted TTA HOMO and lower LUMO compared to TT and DTT based molecules. In addition, to maintain adequate dye processability, a long alkyl substituent ($R = n\text{-}C_{15}H_{31}$) on the TTA core in all sensitizers are introduced, which also suppresses dye aggregation and charge recombination on the TiO_2 surface. (**Figure 27(b)**). The linear C15-alkyl chain substituent not only prevents dye aggregation but also curtails charge recombination. However, it does not efficiently suppress intermolecular π-π interactions when the molecules assemble on the TiO_2 surface. Therefore, the TTA unit effectiveness can be compared with two different branched alkyl group (b-C_8H_{17}), **TTAR-4** and (n-C_9H_{19}) **TTAR-5**.

The optical absorption spectra of the five organic dyes are displayed in **Figure 27 (c)**, indicating the presence of two strong absorptions from a charge-transfer band (~482 nm (**TTAR-1**), ~498 nm (**TTAR-2**), ~513 nm (**TTAR-3**), ~485 nm (**TTAR-4**), ~519 nm (**TTAR-5**)) and a higher energy π–π* transition (~340 nm (**TTAR-1**), ~369 nm (**TTAR-2**), ~402 nm (**TTAR-3**), ~373 nm (**TTAR-4**), ~375 nm (**TTAR-5**)). Significantly, **TTAR-2 ~ 5** with one or two thiophenes show red-shifted the absorption onset compared to **TTAR-1** due to their extended conjugation lengths. Presumably the reduced molecular co-planarity of **TTAR-4** due to the branched alkyl substituents may perturb the conjugation efficiency of the chromophore, and hence hypsochromically shift the absorption maximum wavelength. Electronic structure and solar performance of each dye from cyclic voltagramms (CV) and the B3LYP/6-31G** level of density functional theory (DFT) is shown in **Table 5**. Among 5 dyes, **TTAR-4** show the best cell efficiency (η) as high as 10.21%, attributable to the relatively lower aggregation of the dye due to the bulky 3-methyl-5,5-dimethylhexyl substituents. Generally, dye aggregation on the surface of TiO_2 reduces the electron lifetime and facilitates charge recombination in DSSCs fabricated using metal–organic or organic dyes [139].

To understand the exciton dynamics in dyes, femtosecond time-resolved photoluminescence (FTR-PL) was used [140, 141]. The electron injection efficiency can be calculated by the followed equation, $\eta = (1/\tau_1)/(1/\tau_2 + 1/\tau_3)$, where τ_1 is the electron injection time from the dye to TiO_2, τ_2 is the exaction-exciton annihilation time, and τ_3 is the lifetime of an exciton in the dye [141]. The calculated time constants and electron injection efficiency for the dye coated TiO_2 film are summarized in **Table 5**. The electron injection efficiency of 5 dyes shows almost 97% at $\lambda = 420$ nm, which is well matched results for IPCEs data (>93%) at the same wavelength.

6.4 Multi-sensitized DSSC

Designed sensitizers with higher molar extinction coefficient have allowed significant improvement of cell performance. Nevertheless, possible efficiency boosts can be obtained by extending the dye absorption range while simultaneously avoiding a negative impact on other parameters, such as ground- and excited-state redox potential, intensity of absorption, or stability. Many series of dyes have been developed to find an efficient absorber with high extinction coefficients particularly

DOI: http://dx.doi.org/10.5772/intechopen.94291

Dye	λ_{max} (nm)	E_g (eV)	Energy level (eV)		Energy level (eV)		Photoexcited Electron Dynamics			Electron injection efficiency (%)	Solar Performance			
			E_{HOMO}[a]	E_{LUMO}[a]	E_{HOMO}[b]	E_{LUMO}[b]	τ_1 (ps)	τ_2 (ps)	τ_3 (ps)		Voc (V)	Jsc (mA/cm^2)	FF (%)	EFF(%)
TTAR-1	482	2..08	−5.16	−3.08	−5.01	−2.69	0.68	4.35	23.56	97.2	0.893	10.1	68.1	6.15
TTAR-2	490	2.08	−5.22	−3.14	−5.06	−2.78	0.74	3.08	19.30	96.3	0.833	16.5	73.7	10.1
TTAR-3	513	2.04	−5.09	−3.05	−4.95	−2.80	0.84	2.77	15.07	94.7	0.832	11.8	70.3	6.91
TTAR-4	485	1.92	−5.19	−3.27	−5.02	−2.71	0.71	3.37	24.74	97.2	0.811	17.5	72.0	10.2
TTAR-5	519	1.91	−5.20	−3.29	−4.99	−2.71	0.77	4.59	29.86	97.5	0.781	15.5	62.0	7.67

[a] *Measured in o-DCB at a concentration of* $10^{-5}\ M^{-1}$.
[b] $E_{HOMO-CV} = -(4.8 + E_{ox})$ *where* E_{ox} *= Onset potential of the first oxidation peak when the Fc/Fc + internal standard is referenced to 0.0 V.*

Table 5.
Summary of optical, electrical properties, solar properties of dyes. Reprinted from [141].

in the solar spectrum from 350 ~ 1,000 nm that is also environmentally benign and inexpensive to use. However, it's a big challenge. Therefore, scientists are tried to mix two or more dyes with different absorption range [142–144]. Although this concept is very sensible, the multi-sensitizers system should meet some essential conditions: (i) strong molar extinction coefficients to minimize the thickness of the mesoscopic TiO_2 film; (ii) a suitable structure to avoid unfavorable dye aggregation; (iii) to reduce the recombination of electrons in the TiO_2 film with I_3^- and other acceptors materials through the formation of a compact molecule monolayer covering the bare TiO_2 surface. With the continuous efforts, the co-sensitized DSSC has achieved an 11.5% efficiency record with the liquid state electrolyte, [145] but the energy conversion efficiency of mixed sensitizer still lags behind a champion cell made from the single dye. The reason is that the formation of molecular aggregates onto the TiO_2 surface having a finite number of anchoring sites as well as the deactivation of dye excited states due to energy or electron transfer processes between the different sensitizers [146].

In this book, three sensitizers with the different main absorption position and D-π-A framework (a Zn based porphyrin dye (YD2-oC8) [126] and the more conjugated TTA-based small molecules (TTAR) [141], which hold records of solar performance as a single dye) are applied by modified stepwise approach. For the efficient cosensitizaion, a good understadning of the intermolecular interactions such as matchable size, shape, and orientations as well as compensating light-harvest between or among the coadsorbed dyes must take precedence. In addition dye aggregation on the surface of TiO_2 significantly must be avoided because the formation of dye aggregates significantly decrease the efficiency of electron injection. Attached long alkoxyl chains at phenyl group is plausible strategies to suppress effectively the dye aggregation. The key structural feature of YD2-oC8 involves long alkoxyl chains in the *ortho*-positions of the *meso*-phenyls so as to envelope effectively the porphyrin ring to decrease the degree of dye aggregation and also protect the porphyrin core for retarded charge recombination [126].

For supporting the lack of light-harvesting ability beyond 700 nm, near-IR dyes, named as YDD6, is employed [147]. The ortho-substituted alkoxyl chains in YDD6 can be contributed as not only a light-harvesting ability extending beyond 800 nm but also failed to prevent dye aggregation for this porphyrin. A weakly absorption around $\lambda = 520$ nm at the YD2-oC8 can be improved by TTAR dye molecules. In addition, several papers reported the co-sensitization of different molecular sizes allowed a better surface coverage, yielding a high short circuit current density (J_{sc}) and open circuit voltage (V_{oc}) and resulting in a high PCE [147–149]. Similarly, the covering empty space on TiO_2 NS surface demonstrated from the light soaking test lead to enhance the efficiency. Therefore, insering small sized TTAR dye molecules (estimated molecular length = 24.9 Å) into the gaps within the YD2-oC8 (about 167 Å) saturated TiO_2 surface can help to boost cell performance.

The possibility of dye aggregation of TTAR are test with various amounts of chenodeoxycholic acid (CDCA) co-adsorbed on the photoanodes. The effect of CDCA concentration on the photovoltaic properties of the DSSCs fabricated with **TTAR-4** was first investigated, and the results are shown in **Figure 28(a, left)**. Although dye aggregation should be suppressed by CDCA incorporation among the dye molecules, this would also decrease the dye coverage on the TiO_2, leading to decrease J_{sc} and cell performance. The **TTAR-4** may be sufficient to suppress aggregation, and adsorption of the added CDCA may compete with dye adsorption on the TiO_2. The electron-transport resistance (R_{ct}) and electron lifetimes (τ_e) as a function of the concentration of CDCA can be obtained by EIS analysis (see **Figure 28(a, right)**).

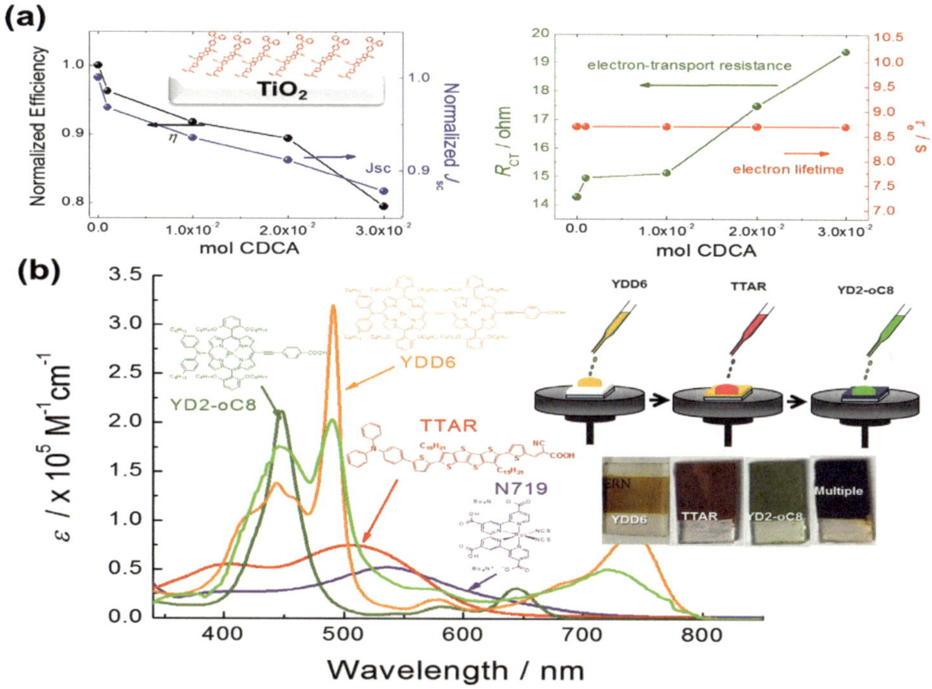

Figure 28.
*(a) The normalized efficiency (η) and normalized photocurrent density (Jsc) from the photocurrent density-voltage (J-V) characteristics (left) and the electron-transport resistant (R_{CT}) and electron lifetime (τ_e) fitted from EIS analysis of the DSSCs using **TTAR-4** dye with various CDCA concentrations, measured under 1 sun illumination (b) molar absorption coefficients measured in o-$C_6H_4Cl_2$ in a concentration of 10^{-5} M (insert: Schematic process for cosensitized system).*

For the effective co-sensitization deposition, there are two well known approaches: (i) the cocktail approach uses a mixture of dye solutions with certain molar ratios of the two dyes' [147, 150] and (ii) the stepwise approach accomplishes sequential adsorptions of the two sensitizers [151, 152]. In our experiement, the fabrication of a co-sensitized onto the TiO_2 sphere film is performed by the modified stepwise deposition method. (see **Figure 28(b)**) Considering the enegetic band position, the TiO_2 electrode is firstly immersed into a 1 M **TTAR** solution in a mixture of 1,2-dichlorobenzene (DCB) and ethanol (volume ratio: 1:10) and kept at room temperature for 1 h. After spin-coating at 3000 rpm for 45 sec, a small quantity of the dye solutions (YD2-oC8) prepared from the same ratio of the mixture solvent (DCB: EtOH) is then dropped onto TTAT/TiO_2 sphere at room temperature and left for 5 min before spin coating at 3000 rpm for 60 sec. The well covered sensitizer layer could be obtained by repeating the coating procedure until the optimized condition. After YD2-oC8 sensitizer coating, as-synthesized YDD6 solution dissovled in the mixture solvent (DCB:EtOH) is deposited onto YD2-oC8/TTAR/ TiO_2 sphere in the same procesure. **Figure 28(b)** shows the molar extinction coefficient of individual and multiple sensitized system. The absorption spectra of YD2-oC8 is well matched in the literature [122, 152]. The YD2-oC8 observed along with the Soret band (400–520 nm, log ε/M^{-1} cm^{-1} = 5.33 at 448 nm) and Q-band absorption (550 ~ 600 nm, log ε/M^{-1} cm^{-1} = 4.49 at 644), while the absorption spectrum of TTAR is found in the range of 350 ~ 560 nm with the molar extinction coefficients at 449.5 nm (log ε/M^{-1} cm^{-1} = 5.01). As a near IR dye, the absorption spectrum of YDD6 shows a broad split feature for the Soret band (380–550 nm, log ε/M^{-1} cm^{-1} = 5.51 at 491 nm) and Q-band absorption (550–800 nm, log $\varepsilon/$ M^{-1} cm^{-1} = 4.94 at 741 nm) Therefore, this dye can help to cover the lack of light-harvesting ability beyond 700 nm. As a result, the combination of complementary porphyrin and organic dyes produces a panchromatic spectral feature to promote the performance of DSSC.

Figure 29(a) and **(b)** show the current–voltage characteristics and the corresponding IPCE action for the TTAR, YD2-oC8, YDD6 and Multiple (TTAR/YD2-oC8/YDD6) sensitized DSSC, respectively. The photovoltaic parameters are listed in **Table 6**. An impressive cell performance of ca 11.2% (J_{sc} = 18.6 mAcm^{-2}, V_{oc} = 0.818 V, FF = 72.8%, 998 mW cm^{-3}) is attainable by increasing J_{sc} value of the TTAR and YD2-oC8 (each 14.3 mAcm^{-2}) to ca. 18.6 mAcm^{-2} when the multiple sensitization is used. For the multisensitized DSSCs, the IPCE spectra showed two major differences compared to the individual dye DSSCs: no dip is observed in the visible region as the efficiency remained above 80% from 400 to 650 nm compared with YD2-oC8 dye, As compared with the corresponding IPCE maxima of the single dye systems, the IPCE value of the multisensitized cells remain between 75–85% from 410 nm to 670 nm and current respond is respond above 710 nm. This effect is consistent with the absorption spectral feature shown in **Figure 34(b)**. To understand the charge transport kinetics of cell, electrochemical impedance spectroscopy (EIS) is next performed from my fitting model (see in **Figure 29(c)**). The estimated total resistance (R_{IR}) of multi-sensitized DSSC at 1 ~ 0.3 MHz frequency region was about 18.51 Ω, which is about 1.58, 2.23 and 3.41 times lower values than that of TTAR, YD2-0C8 and YDD6, respectively. From simulated data, the multi-sensitized DSSC displays about 40.2%, 74.9% and 84.9% lower interfacial recombination rate in TiO_2, k_{eff} and 0.8%, 129% and 294% higher R_k/R_w value rather than those of TTAR, YD2-oC8 and YDD6 based DSSC. The higher interfacial recombination rate of porphyrin series sensitizer compared with TTAR sensitizer reveals recombination between iodide and oxidized electron at the TiO_2 open space surface between dyes. For similar reasons, a multisensitized DSSC give rise to a denser packing and coverage and lead to the highest k_{eff}. The schematic illusting melucular

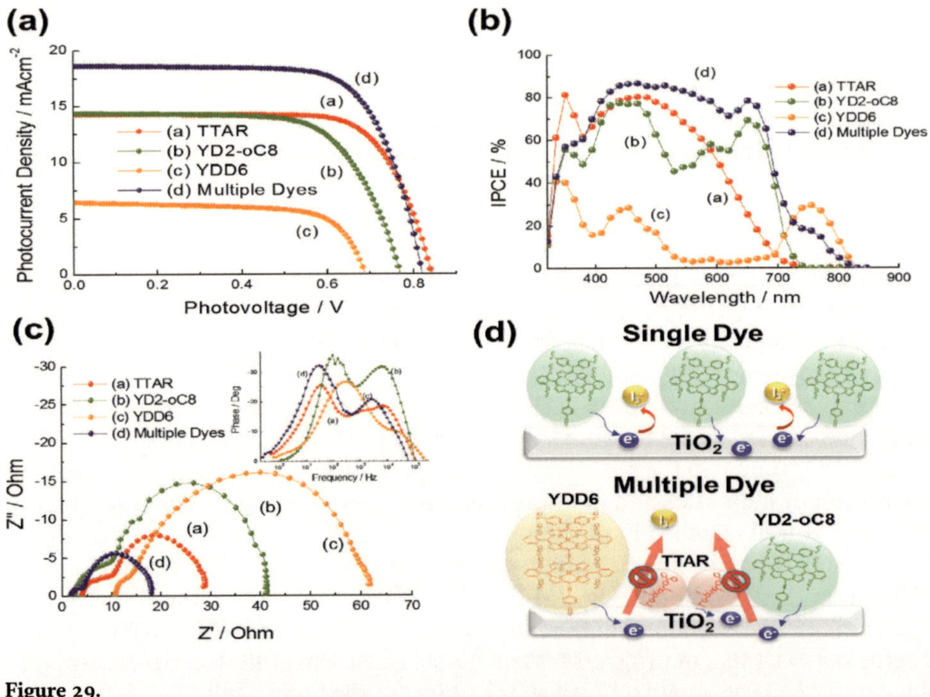

Figure 29.
(a) photocurrent density–voltage (J-V) characteristics (b) IPCE analysis and (c) EIS analysis for the individual and multiple sensitized DSSC (d) Schematic illustrating molecular structure and adsorption sites of YDD6, YD2-oC8 and TTAR dyes on TiO_2.

DOI: http://dx.doi.org/10.5772/intechopen.94291

	D_{eff} (10^{-5} cm^2s^{-1})	k_{eff} (Hz)	τ_{eff} (ms)	R_k	R_w	R_k/R_w	*Con* (Ωcms^{-1})	R_d (Ω)	n_s (10^{18} cm^{-3})	R_{IR} (Ω)	V_{OC} (V)	J_{sc} (mA/cm^2)	*FF* (%)	EFF (%)
TTAR	1.918	19.9	314.7	8.3	3.3	2.5	0.102	3.5	5.378	29.40	0.842	14.3	72.7	8.73
YD2-oC8	2.014	47.5	132.1	10.3	9.3	1.10	0.302	4.3	1.822	41.21	0.769	14.3	67.5	7.45
YDD6	1.142	79.1	79.42	7.9	12.3	0.64	0.387	5.7	1.121	63.21	0.687	6.42	68.7	3.03
Triple	1.952	11.9	527.1	6.1	2.4	2.52	0.044	2.9	12.07	18.52	0.818	18.6	72.8	11.2

Table 6.
Parameters for the best fit of the impedance and Photovotaic data for a single dye (TTAR, YD2-oC8 and YDD6) and multi sensitized DSSC. Measured in ***Figure 29****.*

strucutre for single and multi senstized film can be see in **Figure 29(d)**. Therefore, the charge density in TiO_2 conduction band (n_s) with multisensitized system was increased by about 124%, 562% for TTAR and YD2-oC8 single dye. The relative competition between electron–hole recombination and electron diffusion is conveniently described by the electron diffusion length, L_n, the relation $L_n = (D_{eff} \times \tau_{eff})^{1/2}$. The effective diffusion length L_n of the conduction band electrons for single-(TTAR, YD2-oC8, YDD6) and multi-senstitized DSSC can be calculated from this equation. The $\boldsymbol{L_n}$ for all samples is estimated to be ~24.6 μm of TTAR, 16.3 μm of YD2-oC8, 9.52 of YDD6 and ~ 32.1 μm of TTAR/ YD2-oC8/YDD6, respectively. This calculated data suggests that multisenstitized sensitizer leads to enhance the collection of being photogenerated electrons in comparison to those in single sensitizer. The detailed parameters can be seen in **Table 6**.

7. Electrolyte and solid state hole transport material

The liquid electrolytes possess some important features such as easy preparation, high conductivity, low viscosity, and good interfacial wetting between electrolytes and electrodes and thus high conversion efficiency for the DSSCs [153, 154]. Today, the best working redox-couple known so far is the iodide/triiodide system. The unique performance of I^-/I_3^- based liquid electrolytes is mainly attributed to the favorable penetration into the nanoporous semiconductor film, very fast dye regeneration, and relatively slow recombination losses through reaction with injected photoelectrons. Cobalt based redox mediators bring the concomitant improvements in the V_{oc}, which produced the highest efficiency of 13% for traditional DSSCs [127]. However, there are several negative features limiting industrial application following reason: (1) iodine is extremely corrosive toward metals such as copper or silver, which are used as current collectors in some DSSCs; (2) acetonitrile as a main solvent has a relatively high vapor pressure, which makes proper encapsulation of the cells challenging; (3) the I_3^- ion absorbs a significant part of visible light, stealing photons from the sensitizing dye. These drawbacks can be potentially remedied via the use of solid-state hole transport materials (HTM). The premise of the effective solid state DSSC is that viscous HTM materials would penetrate into the all the deep lying empty spaces in the porous TiO_2 network, and form a continuous film that connects these filled pores all the way up to the back electrode. Hence, our sphere typed TiO_2 electrode give many advantages for solid state electrolyte system and used by default.

In our research, three different classes of solid state electrolyte are investigated (i) iodide based plastic crystal electrolyte prepared by mixing synthesized N-methyl-N-butylpyrrolidinium iodide ($P_{1,4}I$), I_2, and succinonitrile [51]; (ii) a novel cross-linkable organiosiloxane cross-linkable molecule, 4,4′-bis((*p*-trichlorosilylpropylphenyl)phenylamino) biphenyl ($TPDSi_2$) [155]; (iii) perovskite typed inorganic materials, $CsSnI_3$ and Cs_2SnI_6. (this part did not deal with this book) [104, 156]. $P_{1,4}I$ based electrolyte exhibit the highest value, reaching about 9% using a masked frame measurement technique [51]. The detailed information is explained in earlier section.

As an organic molecule system, several different systems have been proposed including melt processing, [157] in-situ polymerizastion, [158] use of low T_g materials, and even use of small molecule/polymeric blends [159]. Among them, silane chemistry based cross linkable hole transport material 4,4′-bis((*p*-trichlorosilylpropylphenyl)phenylamino biphenyl) ($TPDSi_2$) is applied for organic HTM. Silane chemistry has not been previously employed in DSSC, but our group and others have exploited their property of forming a robust cross-linked network of charge carriers for use its various optoelectronic devices. Both the performance and the stability of the devices improved dramatically when $TPDSi_2$ was used the interfacial layer in OLED (organic light emitting diode) and OPV (organic photovoltaic) devices. The $TPDSi_2$ molecule has two key components-widely used hole transport moteity triphenyldiamine (TPD) and a trichlorosilyl ($SiCl_3$) pendant. The $SiCl_3$ subgroups can be easily introduced into a charge conducting moiety like TPD via the highly efficient catalytic hydrosilylation reaction. The Si-Cl bonds are very stable during storage under inert atmosphere, but they are readily hydrolyzed by hydroxyl (OH-) groups. Moisture content of ambient atmosphere and surface OH groups that are present on oxides surfaces of ITO (indium tin oxide), SiO_2 (silicon dioxide) and TiO_2 [160, 161] substrates, along with water molecules that are physiadsorped on the surface, are the relevant sources of such OH groups. Once the $SiCl_3$ groups come in contact with the OH groups, they get hydrozyzed at a rapid pace and inadvertently, Si-Cl groups from two different molecules get hydrozyzed by the two OH groups of the same water molecule. The very strong Si-O-Si bond hence formed, covalently linking two $TPDSi_2$ molecules. Alternatively, Si-O-Si bonds are formed when two neighboring silanol groups undergo condensation reaction. The Si-O-Si bonds will extend over the length and breadth of the film in all directions and hence form a tightly held network of TPD units. While $TPDSi_2$ is a very soluble organic solid, when it is cast onto a film allowed to hydrolyze, it will form a heavily cross-linked network that is rigid, insoluble and rugged. Our strategy in this study was to fill up the TiO_2 pores with the soluble small molecule $TPDSi_2$ and then to allow the molecules to cross link and form an extensive network that is in contact with both the deep lying dye molecules and the back electrode and hence shuttle holes from the former to the latter with great efficiency. **Figure 30(a)** shows the schematic image and the chemical structure of DSSCs utilizing P3HT and $TPDSi_2$ organic seminconductor as HTM. A mesoporous TiO_2 nanocrystlline film is

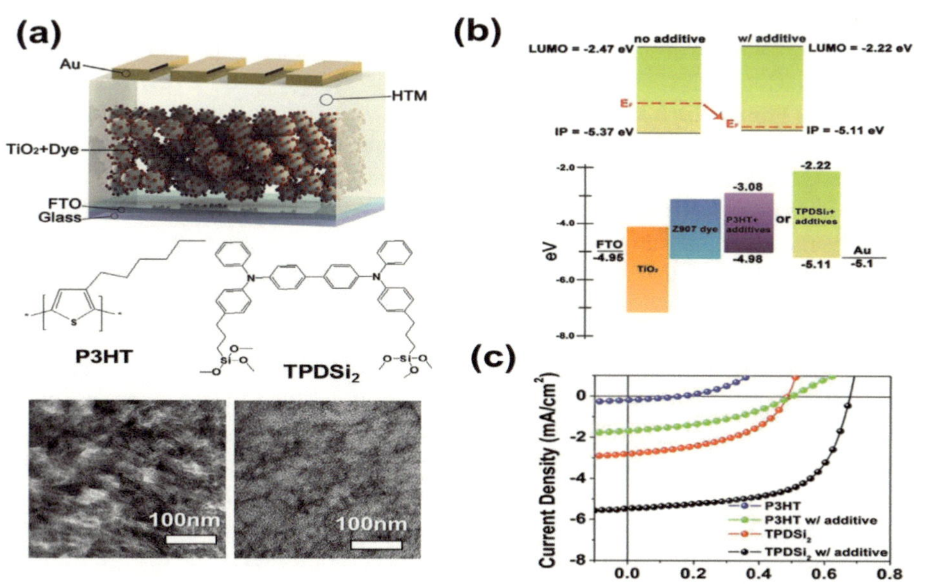

Figure 30.
(a) Device architecture (top)of ss-DSSC and chemical structures with transmission electron microscope image (bottom) of P3HT and $TPDSi_2$ (b) illustration of p-doping effect on TPDSi2 by additive treatment (top) and complete energy diagram of FTO/TiO_2/Z907/HTM/Au ssDSSC (bottom) (c) photovoltaic performance of Z907 dye based ssDSSCs employing P3HT and $TPDSi_2$ as HTMs with and without additive treatment.

deposited on top of FTO-coated glass substrate. Z907 dye molecules are attached to the surface of TiO_2 to form a bicontinuous nanocomposite active layer, which is then subsequently infiltrated by the organic HTM. Finally, a Au counter electrode is deposited to form contact with HTM and encapsulates the device. In this system, the HTM extracts hole from photo-oxidized sensitizer molecules and transfers the holes to the Au counter-electrode, allowing complete dye regeneration. TEM specimens are prepared by transferring spin-coated thin films from poly(3,4-ethylenedioxythiophene):poly(styrenesulfonate) (PEDOT:PSS)-coated Si substrates onto lacey carbon TEM grids. P3HT exhibits typical semicrystalline polymer features. Such relatively large, continuous domains as a result of polymer aggregation are indicative of poor HTM infiltration into pores of TiO_2/sensitizer. On the contrary, $TPDSi_2$ molecular HTM shows significantly smaller features, while forming homogenous and organized networks. Such features are could be advantageous for large-scale processability of $TPDSi_2$ as an efficient HTM.

Figure 30(b) shows a scheme on the possible change of electronic structure by additive treatment by the UPS spectra of P3HT before and after additive treatment. The detailed information can be seen in **Table** 7. In HOMO emission regions, the HOMO band onsets also show no significant difference, corresponding to IPs of 4.94 and 4.98 eV for P3HT before and after additive treatment, respectively. This result is in perfect agreement with CV data (**Table** 7). However, for $TPDSi_2$ films, the additive treatment result in a clear shift both in high binding energy cutoff (HBEC) and HOMO emission regions. In particular, the HBEC position shifts to lower binding energy after additive treatment, whereas the HOMO cutoff shifts close to 0 eV. This effect could indicate a strong p-doping effect of additives on $TPDSi_2$. Scholin et al. [162] observed a similar effect of Fermi level shift to move closer to the HOMO level on Spiro-OMeTAD by Li-TFSI doping. Here, by using the same substrates and a common vacuum level, the shift of Fermi energy to the move closer to the HOMO position is confirmed. In addition, IPs also shift from 5.36 to 5.12 eV after additive treatment, which is in excellent agreement with CV-derived HOMO energies.

The overall energy alignment including all DSSC device components, specifically, FTO/TiO2/Z907/HTMs (P3HT + additives or $TPDSi_2$ + additives)/Au is presented as **Figure 30(b)**.

The exact energetic alignment of HTMs are carefully derived from optical absorption, cyclic voltammetry and ultraviolet photoemission spectroscopy and is discussed below. To carefully investigate the charge transport properties of HTMs, both TFT and space charge limited current (SCLC) measurements are performed. For TFT mobility measurements, bottom-gate/top-contact configurations were employed for all devices fabricated by spin-coating of a 5 mg/mL chlorobenzene solution onto Si/SiO_2 substrates using Au source and drain contacts. All TFT and SCLC mobilies are summarized in **Table** 7. For P3HT, typical p-type TFT behavior was observed both in pristine films and films treated with additives. A noticeable increase in p-type mobility from 4.3×10^{-3} to 1.1×10^{-2} is observed. However, $TPDSi_2$ exhibit relatively low TFT performance, as a result of disconnected domains in the lateral direction. In real DSSC condition, hole transport through the HTMs to the counter electrode can significantly influence charge recombination and device performance. Therefore, SCLC method which directly measures the charge transport in the direction perpendicular to substrate can be employed as a good indicator mimicking real device conditions [163]. For the cross-linked $TPDSi_2$ HTM, it show about 2.8 time higher p-type hole mobility than that of pristine, which is indicative of favorable charge transport properties in DSSCs. Therefore, power conversion efficiency (PCE) of DSSCs of >2% is achieved when using standard amphiphilic dye and TPDSi2 as HTM (see **Figure 30(c)**).

DOI: http://dx.doi.org/10.5772/intechopen.94291

Sample	E_{HOMO}[a] (eV)	IP[b] (eV)	Eopt,gap[c] (eV)	E_{LUMO}[d] (eV)	μ_{TFT} (cm^2/Vs)	μ_{SCLC} (cm^2/Vs)	V_{oc} (V)	J_{sc} (mA/cm^2)	FF (%)	PCE (%)
P3HT	?	−4.94	1.91	−3.03	4.3×10^{-3}	1.0×10^{-3}	0.166	0.38	32.1	0.02
P3HT + additives	?	−4.98	1.90	−3.08	1.1×10^{-2}	2.3×10^{-3}	0.497	1.69	39.4	0.33
$TPDSi_2$	−5.31	−5.37	2.90	−2.47	NA	4.6×10^{-4}	0.485	2.81	46.7	0.64
$TPDSi_2$ + additives	−5.09	−5.11	2.90	−2.22	NA	2.8×10^{-3}	0.683	4.97	60.3	2.05

[a]*Determined by CV.*
[b]*Determined by UPS.*
[c]*Determined at the UV absorption onset.*
[d]*ELUMO = IP + $E_{opt,gap}$.*

Table 7.
Summary of electrochemical, optical, charge transport and photovoltaic data for the HTMs used in this study.

8. Carbonaceous cathode for DSSC

In this part, we show our effort on developing the counter electrode (CE). Generally, CE mainly functions as reducing the redox species from I_3^- to I_2. A suitable redox charge mediator should effectively perform the function of shuttling the generated positive charge away from the light absorbing sensitizer residing on the semiconductor surface to the CE, thus completing the electrical circuit. For being an effective CE, it exhibit high electrical conductivity and electrocatalytic activity toward I_3^-reduction and corrosion resistant to iodide/triiodide electrolyte [164]. To date, a platinum (Pt) is the most common catalyst material for DSSC [165]. However, its cost remains a concern for the large-scale commercialization of the DSSC. Therefore, numerous researchers are trying to find new CE materials. Carbonaceous materials such as graphite [166, 167] carbon black [168, 169], activated carbon [170], hard carbon sphere [171], carbon nanotube [172], fullerene and graphene [173], conductive polymers [174], metal compounds [175] and composites have been developed and tested as promising counter electrodes. So far, carbonaceous materials are regarded as the most attractive option.

In this book, we introduce large-effective-surface-area polyaromatic hydrocarbon (LPAH) for DSSC. A detailed description for generating LPAH species has been reported in an earlier publication [176]. The catalytic properties of LPAH can be calculated by EIS method. Device symmetric structure and the Randles-typed equivalent circuit model consisting of charge transfer resistance (R_{ct}), a constant phase element (CPE) and a series resistance (R_s) can be seen in **Figure 31**. Here, the R_{ct} is a barrier for the charge transfer process at the LPAH/electrolyte interface [177]. The CPE is the interfacial capacitance, considering the roughness of the electrodes. In more detail, the impedance of CPE is described as $Z_{CPE} = B(i\omega)^{-\beta}$ $(0 \leq \beta \leq 1)$ where, ω is the angular frequency, B and β are frequency-independent parameters of the

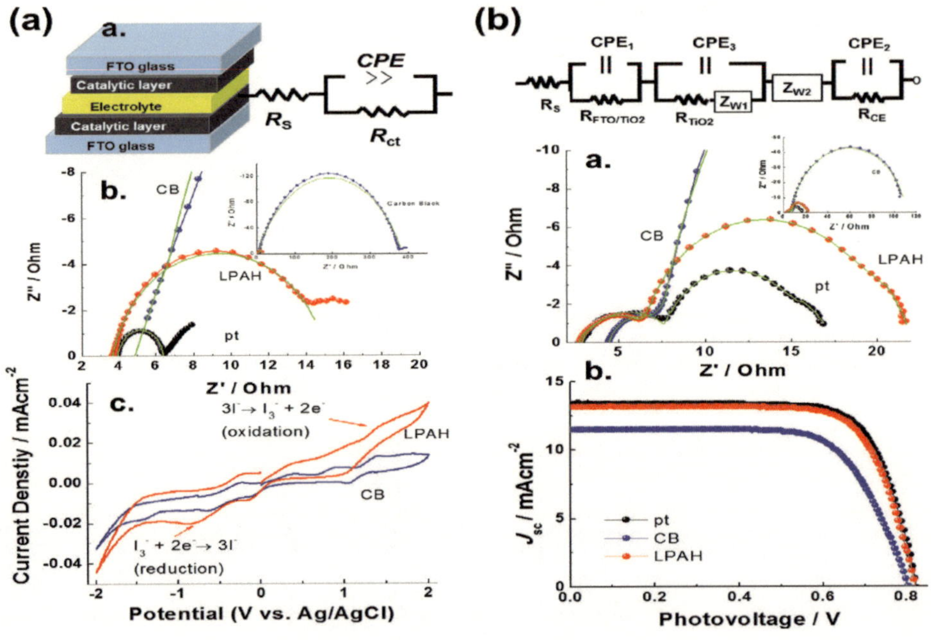

Figure 31.
(a) a. Equivalent circuit of a symmetrical cell used to fit the impedance spectra; b. Nyquist plots of different counter electrode catalytic materials (Platinum, carbon black and LPAH) prepared with identical electrodes; c. CVs of carbon black and LPAH films. (b) a. Nyquist plot of different counter electrodes. Equivalent circuit of the complete DSSC is given in the inset and b. J–V Characteristics of different counter electrode catalytic materials.

CPE. β indicated the capacitance of CPE and the deviation from the semicircle probably due to the porosity of electrode surface, respectively [41, 44, 178]. R_s indicate the ohmic resistance of the electrolyte, the conductive glass and the carbon layer. In order to confirm the catalytic mechanism operating in an electrochemical system with I^-/I_3^- redox, cyclic voltammetry (CV) measurement was carried out. A carbon electrode was used as the working electrode, a Pt coil as the counter electrode, and an Ag/Ag^+ electrode as the reference electrode. **Figure 31(a).c** shows the CV curves obtained using CB and LPAH (versus Ag/AgCl). The negative current was assigned to the reduction of I_3^- to I^- and positive current was the oxidation of I^- at the carbon surface, respectively [179]. The LAPH sample showed a negative peak potential (−0.82 V) and a much higher current density (−0.02 mA/cm^2), indicating a better reduction rate comparable to that CB (−1.4 V, −0.017 mA/cm^2). In cyclic voltammetry, the anode peak and the cathodic peak are related to the redox couple (I^-/I_3^-) reactions and they can be used to estimate the diffusion coefficient (D) using the Randles-Sevcik equation. The Randles-Sevcik equation: $i_p = (2.687 \times 10^5)n^{3/2}\upsilon^{1/2} D^{1/2}AC$ where, i_p is the peak current (A), n is number of electrons transferred in the redox event (usually 1), υ is scan rate in V/s, A is the area of working electrode, C is the bulk concentration of analyte in mol/cm^3. In the anodic and cathodic reaction, the calculated D values of 1.05×10^{-4} cm^2/s and 1.76×10^{-4} cm^2/s for LPAH were larger than 6.70×10^{-5} cm^2/s and 6.3×10^{-5} cm^2/s for CB. This result indicates that the LPAH would speed up the I^-/I_3^- redox reaction ascribed to a higher catalytic activity, leading to a more efficient DSSC.

In general, the equivalent circuit of the complete solar cell may be represented as indicated in **Figure 31(b)** [37, 176]. From right to left, it demonstrates impedance for charge transfer at electrolyte/catalytic materials-FTO interface, diffusion of I_3^- species in the electrolyte, electron transport and electron capture by the I_3^- at the TiO_2/electrolyte interface and the electron transport at the FTO/TiO_2 interface, respectively. These components can be simplified from proposed in DSSC model: $R_{FTO/TiO2}$ is the resistance of the FTO/TiO_2 contact and CPE_1 is the capacitance of this interface. TiO_2 network consists of a diffusion element Z_{W1} that is in series connected with the charge-transfer element R_{TiO2}, the two being in parallel with a capacitive (constant phase angle) element CPE_3. Z_{W2} is the Warburg impedance describing the diffusion of I_3^- in the electrolyte. R_{CE} is the charge-transfer impedance at the counter electrode, and CPE_2 is the double layer capacitance at the electrolyte/catalytic materials-FTO interface [180].

The catalytic activity described the exchange current density (J_0) using followed equation,

$$J_0 = RT/\eta FR_{ct}.$$

Here, J_0 is a kinetic parameter that depends on the reaction and on the electrode surface upon which the electrochemical reaction occurs, R is the gas constant, T is temperature (here T = 300 K), n is the gas constant and F is Faraday constant [44]. From the value of J_0, we can know how easily the electrochemical reaction can occur on the electrode surface. The R_{ct} of LPAH based CE shows approximately 35 times lower value at ~3 μm thick film than that of a symmetric CB electrode at ~8 μm thick film, leading a better energy efficiency at the DSSC (see **Figure 31(b)**). Furthermore, the relatively thin film of LPAH can be expected by reducing the internal series resistance of devices. The shifting the peaks of Bode phase at the high frequency region of the cells supported this expectation [181]. Consequently, LPAH would speed up the I^-/I_3^- redox reaction, leading to improve the fill factor and cell efficiency.

Figure 31(c).b described fitted Nyquist plot and *J-V* curves of the different CE (Pt, CB and LPAH) on DSSCs. The detailed parameters for internal resistance (R_{IR})

sample	d^* (μm)	BET & BJH analysis			Symmetric Cell				Compete DSSC (w/o mask)				
		Surface area (m^2/g)	Pore Volume (cm^3/g)	Pore diameter (nm)	CPE:B (S.s $^{\beta}$)	CPE: β	R_{CT} (Ωcm^2)	J_o (mA/cm^2)	V_{OC} (V)	J_{sc} (mA/cm^2)	FF (%)	EFF (%)	R_{IR}^* (Ohm)
(a) pt		—	—	—	2.9×10^{-5}	0.87	0.61	22.0	0.824	13.4	73.8	8.12	17.1
(b) CB	7.5 ± 0.8	69.6	0.094	5.93	2.3×10^{-5}	0.76	70.3	0.190	0.804	11.5	68.9	6.35	111
(c) LPAH	3.2 ± 0.6	216.8	0.474	13.3	2.3×10^{-3}	0.85	2.12	6.31	0.820	13.1	73.2	7.89	21.9

Table 8.
Characteristic of different counter electrode materials (platinum, carbon black and LPAH). Reprinted from [176].

and photovoltaic properties are summarized in **Table 8**. The internal resistance (R_{IR},) can be calculated by the sum obtained from each resistance, $R_{IR} = R_0 + R_1 + R_2 + R_3$. Although LPAH counter electrode has somewhat lower catalytic property compared with the Pt counter electrode, V_{oc}, fill factor is close to Pt based DSSC electrode due to relatively similar R_{IR} values. Finally, LPAH based DSSC show an overall energy conversion efficiency of 9.3% without mask, which is higher than the 7.5% achieved for CB counter electrode devices reported recently [169]. Therefore, we believe LPAH is a good candidate as a next catalytic material.

9. Optimization of the DSSC performance for having maximum performance

Based on these fundamental achievements, our efforts are headed for achieving an energy efficiency of over 12.3% by combining new materials and concept. For a photoanode, the ITO NR array with over 3 μm spacing and 10 μm thickness is used as the 3D cell [52]. In this design, it is very important how well the TiO_2 nanoparticles have infiltrated among the ITO nanowire. In our earlier research, TiO_2 solution including polymer binder is air-sprayed into ITO NWs. Although this method is easy and efficient enough to fill TiO_2 NPs into ITO NW, it formed surface defect (such as crack) on the surface of TiO_2 film during sintering process (see **Figure 32**). Through Electro-spraying process, crack free TiO_2 film is deposited into

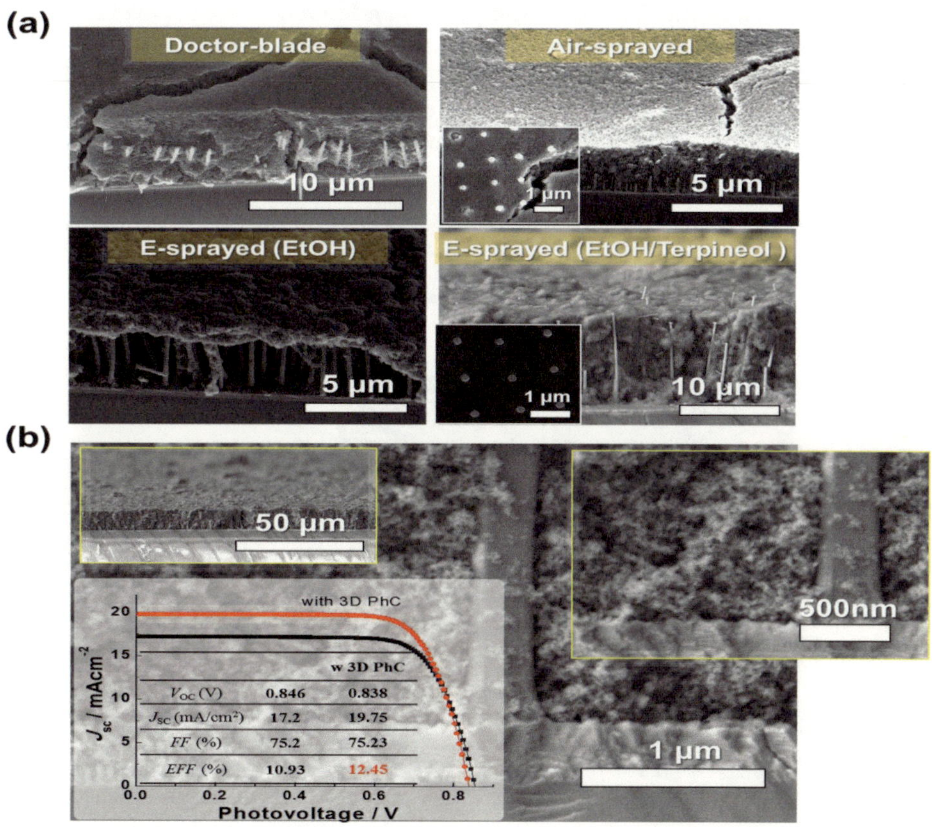

Figure 32.
(a) SEM images of TiO_2/ITO NRs on the different deposition technique; doctor-blade, air-sprayed from TiO_2 solution including polymer binder and E-sprayed technique from EtOH and EtOH/terphineol (b) E-sprayed TiO_2/ITO NRs film inserted in the JV characteristic.

ITO NW. For an electro-spraying, aqueous solvent in the hydrothermal treated TiO_2 nanoparticle solution is replaced by ethanol and then alpha-terpineol is added into the ethanol solution of TiO_2 particles. The replaced solution was loaded into a syringe equipped with a 27-gauge stainless steel needle. The spaying rate (25 μL/min) was controlled using a syringe pump. The electric field (12–15 kV) was applied between a metal orifice and the aluminum foil at a distance of 10 cm using a power supply and was electrosprayed onto ITO NWs substrate. The TiO_2 coated electrode was gradually calcined under an air flow at 150°C for 15 min, at 320°C for 10 min, at 500°C for 30 min. As seen in **Figure 32(a)**, TiO_2 sphere solution prepared from pure EtOH formed the film on the top of ITO NWs, which make it difficult to penetrate into the ITO NWs. Therefore, adding alpha-terpineol solvent with viscous and low evaporation rate into TiO_2 solution, TiO_2 solution is directly dropped into highly charged ITO NWs in between and spreading through x-y-z moving robot system. As a result, completely filled and crack free TiO_2 film in ITO NWs can be obtainable (see **Figure 32(b)**). A Jsc of 17.2 $mAcm^{-2}$, a Voc of 0.846 V, and a FF of 0.75.2% are derived from the JV curve with purified N719 dye and liquid electrolyte, thus giving an overall power conversion efficiency (η) of 10.9% under illumination with standard AM 1.5G simulated sunlight (1000 $mWcm^{-2}$). While attached with a 3D PhC, the efficiency is further improved to 12.45% (Jsc = 19.75 $mAcm^{-2}$, Voc = 0.838 V, FF = 75.2%). The dramatically improved PCE can be achieved by using cosensitization system with a TTAR/ YD2-oC8/YDD6 and 3D PhC designed structure.

Figure 33(a) presents the solar performance for the best cell. Form an effective approach to enhance the light-harvesting ability and to retard the charge recombination, the maximum PCE is about 13.26%, with short-circuit current density (Jsc) =22.3 $mAcm^{-2}$, open circuit voltage (Voc) = 0.811 V, fill factor (FF) = 73.3% and maximum power (P_{max}) = 1.99 mW. In conclusion, our concept and designed material and concept lead to significant promotion of the overall performance of a DSSC.

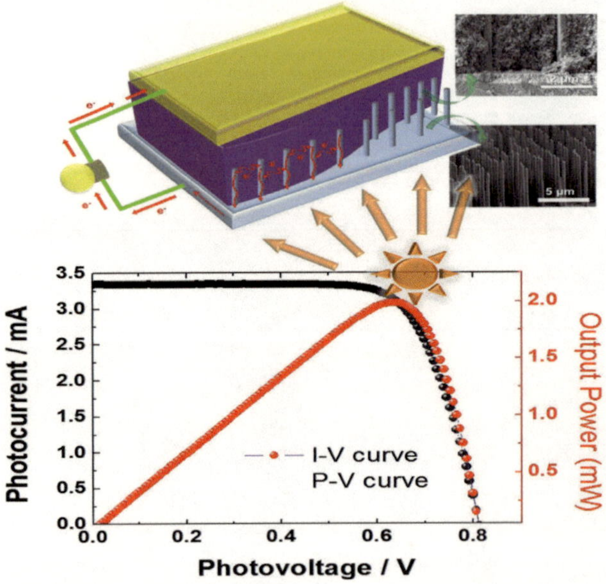

Irradiance (W/sqm)	997.9358
Temperature (°C)	28.976
V_{oc} (V)	0.81139
I_{sc} (A)	0.003346
J_{sc} (mA/cm^2)	22.305
P_{max} (mW)	1.99
V_{mp} (V)	0.63550
I_{mp} (A)	0.00313
Eff (%)	13.264
FF (%)	73.306
R_s (Ohm)	31.1916
R_{sh} (Ohm)	68542.36

Figure 33.
I-V and P-V curve for the best performance DSSC with the cosensitization and 3D PhC.

10. Future prospects of ESC

Compared with traditional silicon solar cells, the DSSC device promises to be less expensive (the ease of fabrication and cost effectiveness of materials, which do not need to be highly purified process), thinner, more flexible, and amenable to a wide range of lighting conditions, all of which makes it viable and efficient solar cells for the future. Therefore, they said that ESC along with its good performance will definitely replace c-Si dominated photovoltaic markets very sooner. However, although each material is extremely inexpensive, the cost of silicon continues to fall as well. Silicon photovoltaics module are a mature technology and their costs have continued to reduce from US \$1.52 W^{-1} in 2010 to just US\$ 0.39 W^{-1} in 2018, with module efficiencies ranging from 15–20% and lifetime guaranteed to 25 years [182]. Unfortunately, the ESC technology is not so trivial that the highest module efficiency, the fastest production methods, or the lowest materials cost necessarily provides the best module solution. In addition, long term stability is the big challenge. Therefore, for surviving the future photovoltaic market, we believe ESCs have to develop the system of transparency with esthetics ability.

Building integrated photovoltaic (BIPV) technology has become an emerging research hotspot of solar PV technology because it intends to achieve "zero energy building (ZEB)" consumption through transformations of buildings from energy consumers to sustainable energy producers [183]. A sustainable building can minimize energy consumption, while at the same time supplying its own energy demand through self-generation. PV was generally installed on building's roofs, but it's very harsh to meet strict requirements of ZEB regulation. The potential power of rooftop in the United States is approximately estimated as 1400 TWh/yr. (0.16TW) at about 16% of module efficiency cell, nearly 40% of the total electricity generation of the US [184]. In buildings rooftop system with conventional PV application, the available area for PV installation is limited and it cannot fulfill the building's energy needs. On the other hands, transparent or semi-transparent PV windows have a tremendous potential to increase harvesting area as well as reduce the annual electricity consumption for cooling and heating. Therefore, we believe transparent PV is the most promising future energy system and research efforts for minimizing the tradeoff relation between the average visible transparency (AVT) and power conversion efficiency (PCE) can decide ESC's fate [185].

11. Summary

The main aim of this book is to put a comprehensive review how to improve and what kind of factors are influence on cell performance. Fundamental and professional understanding of the DSSC has been gained by means of experimental, electrical and optical modeling and advanced characterization techniques. In this book, TiO_2 photoanode under the category of 0D and 3D structures, organic photosensitizer, solid state transporting materials and catalytic carbon as a material field was studied by using various optical and electrical tools. The combined research efforts have led to the important technical achievements: by controlling the nanocrystal structure, size, shape, organization and interface of titanium dioxide, we have made great progress in controlling the light harvesting, charge transfer and transport properties in the devices and have greatly boosted the performance of the devices.. By using specially designed photonic crystals to confine the photons in the cells, the overall cell efficiency can reach to about 13.26%.

Acknowledgements

The Northwestern University authors acknowledge the support of the Argonne-Northwestern Solar Energy Research (ANSER) Center, an Energy Frontier Research Center funded by the U.S. Department of Energy (DOE), Office of Science, Basic Energy Sciences (BES) under Award Number DE-SC0001059. This work made use of the EPIC, Keck-II, and SPID facilities of Northwestern University's NUANCE Center, which has received support from the Soft and Hybrid Nanotechnology Experimental (SHyNE) Resource (NSF ECCS), the MRSEC program (NSF DMR-1121262 and 1542205) at the Materials Research Center, the International Institute for Nanotechnology (IIN); the Keck Foundation; and the State of Illinois,

Author details

Byunghong Lee[1,2]* and Robert Bob Chang[2]

1 Future Energy Research Team, Hyundai Motor Group, Uiwang-si, Gyenggi-do, South Korea

2 Department of Materials Science and Engineering, Northwestern University, Evanston, Illinois, USA

*Address all correspondence to: redboho@gmail.com

References

[1] Gregg, B.A., *Excitonic Solar Cells.* The Journal of Physical Chemistry B, 2003. **107**(20): p. 4688–4698.

[2] Snaith, H.J., *Perovskites: The Emergence of a New Era for Low-Cost, High-Efficiency Solar Cells.* The Journal of Physical Chemistry Letters, 2013. **4** (21): p. 3623–3630.

[3] Becquerel, A.E., *Mémoire sur les effets électriques produits sous l'influence des rayons solaires.* Comptes Rendus des Séances Hebdomadaires, 1839. **9**: p. 561–567.

[4] Moser, J., *Notiz über Verstärkung photoelektrischer Ströme durch optische Sensibilisirung.* Monatshefte für Chemie und verwandte Teile anderer Wissenschaften, 1887. **8**(1): p. 373–373.

[5] O'Regan, B. and M. Gratzel, *A low-cost, high-efficiency solar cell based on dye-sensitized colloidal TiO2 films.* Nature, 1991. **353**(6346): p. 737–740.

[6] Best Research-Cell Efficiency Chart https://www.nrel.gov/pv/cell-efficiency.html

[7] Hagfeldt, A., et al., *Dye-Sensitized Solar Cells.* Chemical Reviews, 2010. **110** (11): p. 6595–6663.

[8] Listorti, A., B. O'Regan, and J.R. Durrant, *Electron Transfer Dynamics in Dye-Sensitized Solar Cells.* Chemistry of Materials, 2011. **23**(15): p. 3381–3399.

[9] Martinson, A.B.F., et al., *New Architectures for Dye-Sensitized Solar Cells.* Chemistry – A European Journal, 2008. **14**(15): p. 4458–4467.

[10] Fabregat-Santiago, F., et al., *Characterization of nanostructured hybrid and organic solar cells by impedance spectroscopy.* Physical Chemistry Chemical Physics, 2011. **13**(20): p. 9083–9118.

[11] Bae, E., et al., *Effects of Surface Anchoring Groups (Carboxylate vs Phosphonate) in Ruthenium-Complex-Sensitized TiO2 on Visible Light Reactivity in Aqueous Suspensions.* The Journal of Physical Chemistry B, 2004. **108**(37): p. 14093–14101.

[12] Thavasi, V., et al., *Controlled electron injection and transport at materials interfaces in dye sensitized solar cells.* Materials Science and Engineering: R: Reports, 2009. **63**(3): p. 81–99.

[13] Wang, X. and D.M. Stanbury, *Oxidation of Iodide by a Series of Fe(III) Complexes in Acetonitrile.* Inorganic Chemistry, 2006. **45**(8): p. 3415–3423.

[14] Wang, P., et al., *Charge Separation and Efficient Light Energy Conversion in Sensitized Mesoscopic Solar Cells Based on Binary Ionic Liquids.* Journal of the American Chemical Society, 2005. **127** (18): p. 6850–6856.

[15] Pelet, S., J.-E. Moser, and M. Grätzel, *Cooperative Effect of Adsorbed Cations and Iodide on the Interception of Back Electron Transfer in the Dye Sensitization of Nanocrystalline TiO2.* The Journal of Physical Chemistry B, 2000. **104**(8): p. 1791–1795.

[16] Tiwana, P., et al., *Electron Mobility and Injection Dynamics in Mesoporous ZnO, SnO2, and TiO2 Films Used in Dye-Sensitized Solar Cells.* ACS Nano, 2011. **5** (6): p. 5158–5166.

[17] J. Frank, A., N. Kopidakis, and J.v.d. Lagemaat, *Electrons in nanostructured TiO2 solar cells: transport, recombination and photovoltaic properties.* Coordination Chemistry Reviews, 2004. **248**(13–14): p. 1165–1179.

[18] Bisquert, J., et al., *Determination of Rate Constants for Charge Transfer and the Distribution of Semiconductor and Electrolyte Electronic Energy Levels in Dye-Sensitized Solar Cells by Open-Circuit Photovoltage Decay Method.*

Journal of the American Chemical Society, 2004. **126**(41): p. 13550–13559.

[19] Nelson, J. and R.E. Chandler, *Random walk models of charge transfer and transport in dye sensitized systems.* Coordination Chemistry Reviews, 2004. **248**(13–14): p. 1181–1194.

[20] Ansari-Rad, M., Y. Abdi, and E. Arzi, *Monte Carlo Random Walk Simulation of Electron Transport in Dye-Sensitized Nanocrystalline Solar Cells: Influence of Morphology and Trap Distribution.* The Journal of Physical Chemistry C, 2012. **116**(5): p. 3212–3218.

[21] Bisquert, J., *Interpretation of electron diffusion coefficient in organic and inorganic semiconductors with broad distributions of states.* Physical Chemistry Chemical Physics, 2008. **10**(22): p. 3175–3194.

[22] Bisquert, J., E. Palomares, and C.A. Quiñones, *Effect of Energy Disorder in Interfacial Kinetics of Dye-Sensitized Solar Cells with Organic Hole Transport Material.* The Journal of Physical Chemistry B, 2006. **110**(39): p. 19406–19411.

[23] Jennings, J.R., et al., *Dependence of Dye-Sensitized Solar Cell Impedance on Photoelectrode Thickness.* The Journal of Physical Chemistry C, 2012. **116**(1): p. 1556–1562.

[24] Mao, X., et al., *High Efficiency Dye-sensitized Solar Cells Constructed with Composites of TiO2 and the Hot-bubbling Synthesized Ultra-Small SnO2 Nanocrystals.* Scientific Reports, 2016. **6** (1): p. 19390.

[25] Wu, J., et al., *Counter electrodes in dye-sensitized solar cells.* Chemical Society Reviews, 2017. **46**(19): p. 5975–6023.

[26] Lee, B., et al., *Materials, Interfaces, and Photon Confinement in Dye-Sensitized Solar Cells.* The Journal of Physical Chemistry B, 2010. **114**(45): p. 14582–14591.

[27] Grätzel, M., *Dye-sensitized solar cells.* Journal of Photochemistry and Photobiology C: Photochemistry Reviews, 2003. **4**(2): p. 145–153.

[28] Thavasi, V., et al., *Controlled electron injection and transport at materials interfaces in dye sensitized solar cells.* Materials Science & Engineering R-Reports, 2009. **63**(3): p. 81–99.

[29] Gratzel, M., *Photoelectrochemical cells.* Nature, 2001. **414**(6861): p. 338–344.

[30] Frank, A.J., N. Kopidakis, and J. van de Lagemaat, *Electrons in nanostructured TiO2 solar cells: transport, recombination and photovoltaic properties.* Coordination Chemistry Reviews, 2004. **248**(13–14): p. 1165–1179.

[31] Goncalves, L.M., et al., *Dye-sensitized solar cells: A safe bet for the future.* Energy & Environmental Science, 2008. **1**(6): p. 655–667.

[32] Chen, C.Y., et al., *Highly Efficient Light-Harvesting Ruthenium Sensitizer for Thin-Film Dye-Sensitized Solar Cells.* Acs Nano, 2009. **3**(10): p. 3103–3109.

[33] Toivola, M., et al., *Nanostructured dye solar cells on flexible substrates - Review.* International Journal of Energy Research, 2009. **33**(13): p. 1145–1160.

[34] Sastrawan, R., et al., *New interdigital design for large area dye solar modules using a lead-free glass frit sealing.* Progress in Photovoltaics: Research and Applications, 2006. **14**(8): p. 697–709.

[35] Lee, B.H., et al., *Charge Transport Characteristics of High Efficiency Dye-Sensitized Solar Cells Based on Electrospun TiO2 Nanorod Photoelectrodes.* The Journal of Physical Chemistry C, 2009. **113**(51): p. 21453–21457.

[36] Fabregat-Santiago, F., et al., *Decoupling of Transport, Charge Storage, and Interfacial Charge Transfer in the Nanocrystalline TiO2/Electrolyte System by Impedance Methods.* The Journal of Physical Chemistry B, 2001. **106**(2): p. 334–339.

[37] Wang, Q., J.-E. Moser, and M. Gratzel, *Electrochemical Impedance Spectroscopic Analysis of Dye-Sensitized Solar Cells.* The Journal of Physical Chemistry B, 2005. **109**(31): p. 14945–14953.

[38] Fabregat-Santiago, F., et al., *Correlation between Photovoltaic Performance and Impedance Spectroscopy of Dye-Sensitized Solar Cells Based on Ionic Liquids.* The Journal of Physical Chemistry C, 2007. **111**(17): p. 6550–6560.

[39] Adachi, M., et al., *Determination of Parameters of Electron Transport in Dye-Sensitized Solar Cells Using Electrochemical Impedance Spectroscopy.* The Journal of Physical Chemistry B, 2006. **110**(28): p. 13872–13880.

[40] Kern, R., et al., *Modeling and interpretation of electrical impedance spectra of dye solar cells operated under open-circuit conditions.* Electrochimica Acta, 2002. **47**(26): p. 4213–4225.

[41] Bisquert, J., et al., *Theoretical models for ac impedance of finite diffusion layers exhibiting low frequency dispersion.* Journal of Electroanalytical Chemistry, 1999. **475**(2): p. 152–163.

[42] Bisquert, J., *Theory of the Impedance of Electron Diffusion and Recombination in a Thin Layer.* J. Phys. Chem. B, 2002. **106**(2): p. 325–333.

[43] Han, L.Y., et al., *Improvement of efficiency of dye-sensitized solar cells by reduction of internal resistance.* Applied Physics Letters, 2005. **86**(21).

[44] Hauch, A. and A. Georg, *Diffusion in the electrolyte and charge-transfer reaction at the platinum electrode in dye-sensitized solar cells.* Electrochimica Acta, 2001. **46**(22): p. 3457–3466.

[45] Tsai, C.-C. and H. Teng, *Structural Features of Nanotubes Synthesized from NaOH Treatment on TiO2 with Different Post-Treatments.* Chemistry of Materials, 2005. **18**(2): p. 367–373.

[46] Park, N.-G. and K. Kim, *Transparent solar cells based on dye-sensitized nanocrystalline semiconductors.* physica status solidi (a), 2008. **205**(8): p. 1895-1904.

[47] Nazeeruddin, M.K., et al., *Conversion of light to electricity by cis-X2bis(2,2'-bipyridyl-4,4'-dicarboxylate) ruthenium(II) charge-transfer sensitizers (X = Cl-, Br-, I-, CN-, and SCN-) on nanocrystalline titanium dioxide electrodes.* Journal of the American Chemical Society, 1993. **115**(14): p. 6382–6390.

[48] Schlichthörl, G., et al., *Band Edge Movement and Recombination Kinetics in Dye-Sensitized Nanocrystalline TiO2 Solar Cells: A Study by Intensity Modulated Photovoltage Spectroscopy.* The Journal of Physical Chemistry B, 1997. **101**(41): p. 8141–8155.

[49] Boschloo, G., L. Häggman, and A. Hagfeldt, *Quantification of the Effect of 4-tert-Butylpyridine Addition to I-/I3- Redox Electrolytes in Dye-Sensitized Nanostructured TiO2 Solar Cells.* The Journal of Physical Chemistry B, 2006. **110**(26): p. 13144–13150.

[50] Kopidakis, N., N.R. Neale, and A.J. Frank, *Effect of an Adsorbent on Recombination and Band-Edge Movement in Dye-Sensitized TiO2 Solar Cells: Evidence for Surface Passivation.* The Journal of Physical Chemistry B, 2006. **110**(25): p. 12485–12489.

[51] Lee, B., et al., *Optimizing the Performance of a Plastic Dye-Sensitized Solar Cell.* The Journal of Physical

Chemistry C, 2011. **115**(19): p. 9787–9796.

[52] Lee, B., et al., *Three Dimensional Indium–Tin-Oxide Nanorod Array for Charge Collection in Dye-Sensitized Solar Cells.* ACS Applied Materials & Interfaces, 2014. **6**(20): p. 17713–17722.

[53] *Front Matter A2 - Nishinaga, Tatau*, in *Handbook of Crystal Growth (Second Edition)*. 2015, Elsevier: Boston. p. iii.

[54] *Tapping Solar Power With Perovskites.* Chemical & Engineering News Archive, 2014. **92**(8): p. 10–15.

[55] *Index A2 - Nishinaga, Tatau*, in *Handbook of Crystal Growth (Second Edition)*. 2015, Elsevier: Boston. p. 1157–1182.

[56] O'Regan, B.C., et al., *Influence of the TiCl4 Treatment on Nanocrystalline TiO2 Films in Dye-Sensitized Solar Cells. 2. Charge Density, Band Edge Shifts, and Quantification of Recombination Losses at Short Circuit.* The Journal of Physical Chemistry C, 2007. **111**(37): p. 14001–14010.

[57] Lee, S.-W., et al., *Effects of TiCl4 Treatment of Nanoporous TiO2 Films on Morphology, Light Harvesting, and Charge-Carrier Dynamics in Dye-Sensitized Solar Cells.* The Journal of Physical Chemistry C, 2012. **116**(40): p. 21285–21290.

[58] Pearton, S.J. and D.P. Norton, *Dry Etching of Electronic Oxides, Polymers, and Semiconductors.* Plasma Processes and Polymers, 2005. **2**(1): p. 16–37.

[59] Rauf, S., *Model for photoresist trim etch in inductively coupled CF[sub 4]/O [sub 2] plasma.* Journal of Vacuum Science & Technology B: Microelectronics and Nanometer Structures, 2004. **22**(1): p. 202–211.

[60] Mogab, C.J., A.C. Adams, and D.L. Flamm, *Plasma etching of Si and SiO[sub 2]—The effect of oxygen additions to CF [sub 4] plasmas.* Journal of Applied Physics, 1978. **49**(7): p. 3796–3803.

[61] Fracassi, F. and R. Dagostino, *CHEMISTRY OF TITANIUM DRY ETCHING IN FLUORINATED AND CHLORINATED GASES.* Pure and Applied Chemistry, 1992. **64**(5): p. 703–707.

[62] Norasetthekul, S., et al., *Dry etch chemistries for TiO2 thin films.* Applied Surface Science, 2001. **185**(1–2): p. 27–33.

[63] d'Agostino, R., et al., *Plasma etching of Ti in fluorine-containing feeds.* Journal of Applied Physics, 1992. **71**(1): p. 462.

[64] Biesinger, M.C., et al., *Resolving surface chemical states in XPS analysis of first row transition metals, oxides and hydroxides: Sc, Ti, V, Cu and Zn.* Applied Surface Science, 2010. **257**(3): p. 887–898.

[65] C.D. Wagner, W.M.R., L.E. Davis, J. F. Moulder and G.E. Muilenburg, ed. *Handbook of X-Ray Photoelectron Spectroscopy*. ed. 2nd. 1979, Perkin-Elmer Corp: Eden Prairie, MN.

[66] Li, D., et al., *Fluorine-doped TiO2 powders prepared by spray pyrolysis and their improved photocatalytic activity for decomposition of gas-phase acetaldehyde.* Journal of Fluorine Chemistry, 2005. **126**(1): p. 69–77.

[67] Yang, H. and X. Zhang, *Synthesis, characterization and computational simulation of visible-light irradiated fluorine-doped titanium oxide thin films.* Journal of Materials Chemistry, 2009. **19** (37): p. 6907–6914.

[68] Tang, J., H. Quan, and J. Ye, *Photocatalytic Properties and Photoinduced Hydrophilicity of Surface-Fluorinated TiO2.* Chemistry of Materials, 2006. **19**(1): p. 116–122.

[69] Zhu, K., et al., *Enhanced Charge-Collection Efficiencies and Light Scattering*

in Dye-Sensitized Solar Cells Using Oriented TiO2 Nanotubes Arrays. Nano Letters, 2006. **7**(1): p. 69–74.

[70] Kumar, K.R.T., M. Ramakrishna, and G.D. Sukumar, *A review on PV cells and nanocomposite-coated PV systems.* International Journal of Energy Research, 2018. **42**(7): p. 2305–2319.

[71] Bermel, P., et al., *Improving thin-film crystalline silicon solar cell efficiencies with photonic crystals.* Optics Express, 2007. **15**(25): p. 16986–17000.

[72] Shao, J., G. Liu, and L. Zhou, *12 - Biomimetic nanocoatings for structural coloration of textiles*, in *Active Coatings for Smart Textiles*, J. Hu, Editor. 2016, Woodhead Publishing. p. 269–299.

[73] Kay, A. and M. Gratzel, *Low cost photovoltaic modules based on dye sensitized nanocrystalline titanium dioxide and carbon powder.* Solar Energy Materials and Solar Cells, 1996. **44**(1): p. 99–117.

[74] Ito, S., et al., *High-efficiency organic-dye-sensitized solar cells controlled by nanocrystalline-TiO2 electrode thickness.* Advanced Materials, 2006. **18**(9): p. 1202–+.

[75] Johnson, D.C., et al. *Advances in Bragg stack quantum well solar cells.* in *International Conference on Physics, Chemistry and Engineering of Solar Cells*. 2004. Badajoz, SPAIN: Elsevier Science Bv.

[76] Nishimura, S., et al., *Standing Wave Enhancement of Red Absorbance and Photocurrent in Dye-Sensitized Titanium Dioxide Photoelectrodes Coupled to Photonic Crystals.* Journal of the American Chemical Society, 2003. **125**(20): p. 6306–6310.

[77] Rogers, J.A., M. Meier, and A. Dodabalapur, *Using printing and molding techniques to produce distributed feedback and Bragg reflector resonators for plastic lasers.* Applied Physics Letters, 1998. **73** (13): p. 1766–1768.

[78] Takahashi, S., et al., *Direct creation of three-dimensional photonic crystals by a top-down approach.* Nat Mater, 2009. **8** (9): p. 721–725.

[79] Noda, S., et al., *Full Three-Dimensional Photonic Bandgap Crystals at Near-Infrared Wavelengths.* Science, 2000. **289**(5479): p. 604–606.

[80] Bielawny, A., et al., *Intermediate reflectors for enhanced topcell performance in photovoltaicthin-film tandem cells.* Opt. Express, 2009. **17**(10): p. 8439–8446.

[81] Wijnhoven, J.E., et al., *Preparation of Photonic Crystals Made of Air Spheres in Titania.* Science, 1998. **281**(5378): p. 802–804.

[82] Dae-Kue, H., et al., *Efficiency Enhancement in Dye-Sensitized Solar Cells by Three-Dimensional Photonic Crystals.* Applied Physics Express, 2012. **5**(12): p. 122301.

[83] Hwang, D.-K., B. Lee, and D.-H. Kim, *Efficiency enhancement in solid dye-sensitized solar cell by three-dimensional photonic crystal.* RSC Advances, 2013. **3** (9): p. 3017–3023.

[84] Schubert, M., et al., *Distributed Bragg reflector consisting of high-and low-refractive-index thin film layers made of the same material.* Applied Physics Letters - APPL PHYS LETT, 2007. **90**.

[85] Vos, W.L., et al., *Strong effects of photonic band structures on the diffraction of colloidal crystals.* Physical Review B, 1996. **53**(Copyright (C) 2010 The American Physical Society): p. 16231.

[86] Bielawny, A., et al., *3D photonic crystal intermediate reflector for micromorph thin-film tandem solar cell.* physica status solidi (a), 2008. **205**(12): p. 2796–2810.

[87] Bahramian, A., *High Conversion Efficiency of Dye-Sensitized Solar Cells Based on Coral-like TiO2 Nanostructured*

Films: Synthesis and Physical Characterization. Industrial & Engineering Chemistry Research, 2013. **52**(42): p. 14837–14846.

[88] Wu, H.-P., et al., *Hybrid Titania Photoanodes with a Nanostructured Multi-Layer Configuration for Highly Efficient Dye-Sensitized Solar Cells.* The Journal of Physical Chemistry Letters, 2013. **4**(9): p. 1570–1577.

[89] Mir, N., et al., *Optimizing TiO2 Nanotube Top Geometry for Use in Dye-Sensitized Solar Cells.* Chemistry – A European Journal, 2012. **18**(38): p. 11862–11866.

[90] Huo, J., et al., *Mixed Solvents Assisted Flame Spray Pyrolysis Synthesis of TiO2 Hierarchically Porous Hollow Spheres for Dye-Sensitized Solar Cells.* Industrial & Engineering Chemistry Research, 2013. **52**(32): p. 11029–11035.

[91] Son, H.-J., et al., *Dye Stabilization and Enhanced Photoelectrode Wettability in Water-Based Dye-Sensitized Solar Cells through Post-assembly Atomic Layer Deposition of TiO2.* Journal of the American Chemical Society, 2013. **135** (31): p. 11529–11532.

[92] Jose, R., V. Thavasi, and S. Ramakrishna, *Metal Oxides for Dye-Sensitized Solar Cells.* Journal of the American Ceramic Society, 2009. **92**(2): p. 289–301.

[93] Carp, O., C.L. Huisman, and A. Reller, *Photoinduced reactivity of titanium dioxide.* Progress in Solid State Chemistry, 2004. **32**(1–2): p. 33–177.

[94] Linsebigler, A.L., G. Lu, and J.T. Yates, *Photocatalysis on TiO2 Surfaces: Principles, Mechanisms, and Selected Results.* Chemical Reviews, 1995. **95**(3): p. 735–758.

[95] Zhang, J., et al., *New understanding of the difference of photocatalytic activity among anatase, rutile and brookite TiO2.* Physical Chemistry Chemical Physics, 2014. **16**(38): p. 20382–20386.

[96] Kavan, L., et al., *Electrochemical and Photoelectrochemical Investigation of Single-Crystal Anatase.* Journal of the American Chemical Society, 1996. **118** (28): p. 6716–6723.

[97] Zhang, H. and J. F. Banfield, *Thermodynamic analysis of phase stability of nanocrystalline titania.* Journal of Materials Chemistry, 1998. **8**(9): p. 2073–2076.

[98] Barnard, A.S. and H. Xu, *An Environmentally Sensitive Phase Map of Titania Nanocrystals.* ACS Nano, 2008. **2** (11): p. 2237–2242.

[99] Grimes, C.A. and G.K. Mor, *TiO2 Nanotube Arrays: Synthesis, Properties, and Applications*. 2009: Springer US.

[100] Wen, P., et al., *Single Nanocrystals of Anatase-Type TiO2 Prepared from Layered Titanate Nanosheets: Formation Mechanism and Characterization of Surface Properties.* Langmuir, 2007. **23** (23): p. 11782–11790.

[101] Liao, J.-Y., et al., *Effect of TiO2 morphology on photovoltaic performance of dye-sensitized solar cells: nanoparticles, nanofibers, hierarchical spheres and ellipsoid spheres.* Journal of Materials Chemistry, 2012. **22**(16): p. 7910–7918.

[102] Wang, H.-E., et al., *Rapid Microwave Synthesis of Porous TiO2 Spheres and Their Applications in Dye-Sensitized Solar Cells.* The Journal of Physical Chemistry C, 2011. **115**(21): p. 10419–10425.

[103] Lee, B., et al., *Air-Stable Molecular Semiconducting Iodosalts for Solar Cell Applications: Cs2SnI6 as a Hole Conductor.* Journal of the American Chemical Society, 2014. **136**(43): p. 15379–15385.

[104] Lee, B., et al., *Solution processing of air-stable molecular semiconducting*

iodosalts, Cs2SnI6-xBrx, for potential solar cell applications. Sustainable Energy & Fuels, 2017.

[105] Hwang, D., et al., *Electrospray Preparation of Hierarchically-structured Mesoporous TiO2 Spheres for Use in Highly Efficient Dye-Sensitized Solar Cells.* ACS Applied Materials & Interfaces, 2011. **3**(7): p. 2719–2725.

[106] Kruk, M. and M. Jaroniec, *Gas Adsorption Characterization of Ordered Organic–Inorganic Nanocomposite Materials.* Chemistry of Materials, 2001. **13**(10): p. 3169–3183.

[107] Ma, R. and T. Sasaki, *Nanosheets of Oxides and Hydroxides: Ultimate 2D Charge-Bearing Functional Crystallites.* Advanced Materials, 2010. **22**(45): p. 5082–5104.

[108] Kim, I.-D., et al., *Dye-sensitized solar cells using network structure of electrospun ZnO nanofiber mats.* Applied Physics Letters, 2007. **91**(16): p. 163109.

[109] Kim, I.-D., et al., *Ultrasensitive Chemiresistors Based on Electrospun TiO2 Nanofibers.* Nano Letters, 2006. **6**(9): p. 2009–2013.

[110] Tsai, C.-C. and H. Teng, *Structural Features of Nanotubes Synthesized from NaOH Treatment on TiO2 with Different Post-Treatments.* Chemistry of Materials, 2006. **18**(2): p. 367–373.

[111] Kroon, J., et al., *Nanocrystalline Dye-Sensitized Solar Cells having Maximum Performance.* Progress in Photovoltaics, 2007. **15**: p. 1–18.

[112] MacFarlane, D.R., et al., *Pyrrolidinium Imides: A New Family of Molten Salts and Conductive Plastic Crystal Phases.* The Journal of Physical Chemistry B, 1999. **103**(20): p. 4164–4170.

[113] Dai, Q., D.R. MacFarlane, and M. Forsyth, *High mobility I- / I3- redox couple in a molecular plastic crystal: A potential new generation of electrolyte for solid-state photoelectrochemical cells.* Solid State Ionics, 2006. **177**(3–4): p. 395–401.

[114] Wang, P., et al., *Ambient Temperature Plastic Crystal Electrolyte for Efficient, All-Solid-State Dye-Sensitized Solar Cell.* Journal of the American Chemical Society, 2004. **126**(42): p. 13590–13591.

[115] Dai, Q., et al., *Rapid I−/I3− Diffusion in a Molecular-Plastic-Crystal Electrolyte for Potential Application in Solid-State Photoelectrochemical Cells.* Angewandte Chemie, 2005. **117**(2): p. 317–320.

[116] Cardini, G., R. Righini, and S. Califano, *Computer simulation of the dynamics of the plastic phase of succinonitrile.* The Journal of Chemical Physics, 1991. **95**(1): p. 679–685.

[117] Hawthorne, H.M. and J.N. Sherwood, *Lattice defects in plastic organic solids. Part 2.-Anomalous self-diffusion in succinonitrile.* Transactions of the Faraday Society, 1970. **66**: p. 1792–1798.

[118] Nazeeruddin, M.K., et al., *A swift dye uptake procedure for dye sensitized solar cells.* Chemical Communications, 2003(12): p. 1456–1457.

[119] K. Nazeeruddin, M., P. Pechy, and M. Gratzel, *Efficient panchromatic sensitization of nanocrystalline TiO2 films by a black dye based on a trithiocyanato-ruthenium complex.* Chemical Communications, 1997 (18): p. 1705–1706.

[120] Nazeeruddin, M.K., et al., *Combined Experimental and DFT-TDDFT Computational Study of Photoelectrochemical Cell Ruthenium Sensitizers.* Journal of the American Chemical Society, 2005. **127**(48): p. 16835–16847.

[121] Zedler, L., et al., *Ruthenium dye functionalized gold nanoparticles and their*

spectral responses. RSC Advances, 2012. **2** (10): p. 4463–4471.

[122] Li, L.-L. and E.W.-G. Diau, *Porphyrin-sensitized solar cells.* Chemical Society Reviews, 2013. **42**(1): p. 291–304.

[123] Higashino, T. and H. Imahori, *Porphyrins as excellent dyes for dye-sensitized solar cells: recent developments and insights.* Dalton Transactions, 2015. **44**(2): p. 448–463.

[124] Griffith, M.J., et al., *Porphyrins for dye-sensitised solar cells: new insights into efficiency-determining electron transfer steps.* Chemical Communications, 2012. **48**(35): p. 4145–4162.

[125] Campbell, W.M., et al., *Highly Efficient Porphyrin Sensitizers for Dye-Sensitized Solar Cells.* The Journal of Physical Chemistry C, 2007. **111**(32): p. 11760–11762.

[126] Yella, A., et al., *Porphyrin-Sensitized Solar Cells with Cobalt (II/III)–Based Redox Electrolyte Exceed 12 Percent Efficiency.* Science, 2011. **334**(6056): p. 629–634.

[127] Mathew, S., et al., *Dye-sensitized solar cells with 13% efficiency achieved through the molecular engineering of porphyrin sensitizers.* Nat Chem, 2014. **6** (3): p. 242–247.

[128] Wagner, K., et al., *Significant Performance Improvement of Porphyrin-Sensitized TiO2 Solar Cells under White Light Illumination.* The Journal of Physical Chemistry C, 2011. **115**(1): p. 317–326.

[129] Griffith, M.J., et al., *Cation Exchange at Semiconducting Oxide Surfaces: Origin of Light-Induced Performance Increases in Porphyrin Dye-Sensitized Solar Cells.* The Journal of Physical Chemistry C, 2013. **117**(23): p. 11885–11898.

[130] Mishra, A., M.K.R. Fischer, and P. Bäuerle, *Metal-Free Organic Dyes for Dye-Sensitized Solar Cells: From Structure: Property Relationships to Design Rules.* Angewandte Chemie International Edition, 2009. **48**(14): p. 2474–2499.

[131] Byung-Hong LEE, S.-Y.J., Seong Mu Jo, Dong Young Kim, Hyun-Ju Kim, *Dye-sensitized solar cell comprising metal oxide nanoball layer and preparation method thereof*. 2010: USA.

[132] Teng, C., et al., *Molecular Design of Anthracene-Bridged Metal-Free Organic Dyes for Efficient Dye-Sensitized Solar Cells.* The Journal of Physical Chemistry C, 2010. **114**(19): p. 9101–9110.

[133] Youn, J., et al., *Versatile α,ω-Disubstituted Tetrathienoacene Semiconductors for High Performance Organic Thin-Film Transistors.* Advanced Functional Materials, 2012. **22**(1): p. 48–60.

[134] Fong, H.H., et al., *Tetrathienoacene Copolymers As High Mobility, Soluble Organic Semiconductors.* Journal of the American Chemical Society, 2008. **130** (40): p. 13202–13203.

[135] Huang, J., et al., *Dibenzoannelated Tetrathienoacene: Synthesis, Characterization, and Applications in Organic Field-Effect Transistors.* Organic Letters, 2012. **14**(13): p. 3300–3303.

[136] Letizia, J.A., et al., *Variable Temperature Mobility Analysis of n-Channel, p-Channel, and Ambipolar Organic Field-Effect Transistors.* Advanced Functional Materials, 2010. **20**(1): p. 50–58.

[137] Wang, M., et al., *High-Performance Liquid and Solid Dye-Sensitized Solar Cells Based on a Novel Metal-Free Organic Sensitizer.* Advanced Materials, 2008. **20** (23): p. 4460–4463.

[138] Qin, H., et al., *An Organic Sensitizer with a Fused Dithienothiophene Unit for*

Efficient and Stable Dye-Sensitized Solar Cells. Journal of the American Chemical Society, 2008. **130**(29): p. 9202–9203.

[139] Lu, M., et al., *Organic Dyes Incorporating Bis-hexapropyltruxeneamino Moiety for Efficient Dye-Sensitized Solar Cells.* The Journal of Physical Chemistry C, 2011. **115**(1): p. 274–281.

[140] Chang, S.H., et al., *Broadband charge transfer dynamics in P3HT:PCBM blended film.* Optics Letters, 2013. **38** (24): p. 5342–5345.

[141] Zhou, N., et al., *Metal-Free Tetrathienoacene Sensitizers for High-Performance Dye-Sensitized Solar Cells.* Journal of the American Chemical Society, 2015. **137**(13): p. 4414–4423.

[142] Cid, J.-J., et al., *Molecular Cosensitization for Efficient Panchromatic Dye-Sensitized Solar Cells.* Angewandte Chemie, 2007. **119**(44): p. 8510–8514.

[143] Sayama, K., et al., *Efficient sensitization of nanocrystalline TiO2 films with cyanine and merocyanine organic dyes.* Solar Energy Materials and Solar Cells, 2003. **80**(1): p. 47–71.

[144] Lee, B., et al., *Cs2SnI6-Encapsulated Multidye-Sensitized All-Solid-State Solar Cells.* ACS Applied Materials & Interfaces, 2019. **11**(24): p. 21424–21434.

[145] Xie, Y., et al., *Porphyrin Cosensitization for a Photovoltaic Efficiency of 11.5%: A Record for Non-Ruthenium Solar Cells Based on Iodine Electrolyte.* Journal of the American Chemical Society, 2015. **137**(44): p. 14055–14058.

[146] Clifford, J.N., et al., *Sensitizer molecular structure-device efficiency relationship in dye sensitized solar cells.* Chemical Society Reviews, 2011. **40**(3): p. 1635–1646.

[147] Wu, H.-P., et al., *Molecular engineering of cocktail co-sensitization for efficient panchromatic porphyrin-sensitized solar cells.* Energy & Environmental Science, 2012. **5**(12): p. 9843–9848.

[148] Nguyen, L.H., et al., *A selective co-sensitization approach to increase photon conversion efficiency and electron lifetime in dye-sensitized solar cells.* Physical Chemistry Chemical Physics, 2012. **14** (47): p. 16182–16186.

[149] Pei, K., et al., *Cosensitization of D-A-π-A Quinoxaline Organic Dye: Efficiently Filling the Absorption Valley with High Photovoltaic Efficiency.* ACS Applied Materials & Interfaces, 2015. 7 (9): p. 5296–5304.

[150] Yum, J.-H., et al., *Efficient co-sensitization of nanocrystalline TiO2 films by organic sensitizers.* Chemical Communications, 2007(44): p. 4680–4682.

[151] Fan, S.-Q., et al., *Improved Efficiency of over 10% in Dye-Sensitized Solar Cells with a Ruthenium Complex and an Organic Dye Heterogeneously Positioning on a Single TiO2 Electrode.* The Journal of Physical Chemistry C, 2011. **115**(15): p. 7747–7754.

[152] Lan, C.-M., et al., *Enhanced photovoltaic performance with co-sensitization of porphyrin and an organic dye in dye-sensitized solar cells.* Energy & Environmental Science, 2012. **5**(4): p. 6460–6464.

[153] Yu, Z., et al., *Liquid electrolytes for dye-sensitized solar cells.* Dalton Transactions, 2011. **40**(40): p. 10289–10303.

[154] Wu, J., et al., *Electrolytes in Dye-Sensitized Solar Cells.* Chemical Reviews, 2015. **115**(5): p. 2136–2173.

[155] Zhou, N., et al., *Cross-Linkable Molecular Hole-Transporting*

Semiconductor for Solid-State Dye-Sensitized Solar Cells. The Journal of Physical Chemistry C, 2014. **118**(30): p. 16967–16975.

[156] Chung, I., et al., *All-solid-state dye-sensitized solar cells with high efficiency.* Nature, 2012. **485**(7399): p. 486–489.

[157] Law, C., R. Spence, and B. C. O'Regan, *Brief air heating of TiO2/dye films, to 120-250 [degree]C; the effect on resulting liquid junction dye sensitised solar cells (DSSCs) and melt-processed solid-state DSSCs.* Journal of Materials Chemistry A, 2013. **1**(45): p. 14154–14161.

[158] Liu, X., et al., *An Efficient Organic-Dye-Sensitized Solar Cell with in situ Polymerized Poly (3, 4-ethylenedioxythiophene) as a Hole-Transporting Material.* Advanced Materials, 2010. **22**(20): p. E150-E155.

[159] Johansson, E.M.J., et al., *Combining a Small Hole-Conductor Molecule for Efficient Dye Regeneration and a Hole-Conducting Polymer in a Solid-State Dye-Sensitized Solar Cell.* The Journal of Physical Chemistry C, 2012. **116**(34): p. 18070–18078.

[160] Ukaji, E., et al.,*The effect of surface modification with silane coupling agent on suppressing the photo-catalytic activity of fine TiO< sub> 2</sub> particles as inorganic UV filter.* Applied Surface Science, 2007. **254**(2): p. 563–569.

[161] Zhao, J., et al., *Surface modification of TiO< sub> 2</sub> nanoparticles with silane coupling agents.* Colloids and Surfaces A: Physicochemical and Engineering Aspects, 2012. **413**: p. 273–279.

[162] Scholin, R., et al., *Energy Level Shifts in Spiro-OMeTAD Molecular Thin Films When Adding Li-TFSI.* Journal of Physical Chemistry C, 2012. **116**(50): p. 26300–26305.

[163] Leijtens, T., et al., *Hole Transport Materials with Low Glass Transition Temperatures and High Solubility for Application in Solid-State Dye-Sensitized Solar Cells.* Acs Nano, 2012. **6**(2): p. 1455–1462.

[164] Papageorgiou, N., W.F. Maier, and M. Gratzel, *An Iodine/Triiodide Reduction Electrocatalyst for Aqueous and Organic Media.* Journal of The Electrochemical Society, 1997. **144**(3): p. 876–884.

[165] Papageorgiou, N., *Counter-electrode function in nanocrystalline photoelectrochemical cell configurations.* Coordination Chemistry Reviews, 2004. **248**(13–14): p. 1421–1446.

[166] Acharya, K.P., et al., *Pulsed laser deposition of graphite counter electrodes for dye-sensitized solar cells.* Applied Physics Letters, 2010. **97**(20): p. 201108–201103.

[167] Veerappan, G., K. Bojan, and S.-W. Rhee, *Sub-micrometer-sized Graphite As a Conducting and Catalytic Counter Electrode for Dye-sensitized Solar Cells.* ACS Applied Materials & Interfaces, 2011. **3**(3): p. 857–862.

[168] Murakami, T.N., et al., *Highly Efficient Dye-Sensitized Solar Cells Based on Carbon Black Counter Electrodes.* Journal of The Electrochemical Society, 2006. **153**(12): p. A2255-A2261.

[169] Murakami, T.N. and M. Grätzel, *Counter electrodes for DSC: Application of functional materials as catalysts.* Inorganica Chimica Acta, 2008. **361**(3): p. 572–580.

[170] Imoto, K., et al., *High-performance carbon counter electrode for dye-sensitized solar cells.* Solar Energy Materials and Solar Cells, 2003. **79**(4): p. 459–469.

[171] Huang, Z., et al., *Application of carbon materials as counter electrodes of dye-sensitized solar cells.*

Electrochemistry Communications, 2007. **9**(4): p. 596–598.

[172] Trancik, J.E., S.C. Barton, and J. Hone, *Transparent and Catalytic Carbon Nanotube Films.* Nano Letters, 2008. **8** (4): p. 982–987.

[173] Roy-Mayhew, J.D., et al., *Functionalized Graphene as a Catalytic Counter Electrode in Dye-Sensitized Solar Cells.* ACS Nano, 2010. **4**(10): p. 6203–6211.

[174] Saranya, K., M. Rameez, and A. Subramania, *Developments in conducting polymer based counter electrodes for dye-sensitized solar cells – An overview.* European Polymer Journal, 2015. **66**: p. 207–227.

[175] Gnanasekar, S., et al., *Pt-free, low-cost and efficient counter electrode with carbon wrapped VO2(M) nanofiber for dye-sensitized solar cells.* Scientific Reports, 2019. **9**(1): p. 5177.

[176] Lee, B., D.B. Buchholz, and R.P.H. Chang, *An all carbon counter electrode for dye sensitized solar cells.* Energy & Environmental Science, 2012. **5**(5): p. 6941–6952.

[177] Yuan, X.-Z., et al., *Electrical Fundamentals*

[178] *Electrochemical Impedance Spectroscopy in PEM Fuel Cells*. 2010, Springer London. p. 39-93.

[179] Kavan, L., J.H. Yum, and M. Grätzel, *Optically Transparent Cathode for Dye-Sensitized Solar Cells Based on Graphene Nanoplatelets.* ACS Nano, 2010. **5**(1): p. 165–172.

[180] Macagno, V.A., M.C. Giordano, and A.J. Arvía, *Kinetics and mechanisms of electrochemical reactions on platinum with solutions of iodine-sodium iodide in acetonitrile.* Electrochimica Acta, 1969. **14**(4): p. 335–357.

[181] Liberatore, M., et al., *Using EIS for diagnosis of dye-sensitized solar cells performance.* Journal of Applied Electrochemistry, 2009. **39**(11): p. 2291–2295.

[182] Zhu, G., et al., *Electrophoretic deposition of reduced graphene-carbon nanotubes composite films as counter electrodes of dye-sensitized solar cells.* Journal of Materials Chemistry, 2011. **21** (38): p. 14869–14875.

[183] Liu, Z., et al., *Revisiting thin silicon for photovoltaics: a technoeconomic perspective.* Energy & Environmental Science, 2020. **13**(1): p. 12–23.

[184] Kim, J.-H., H.-R. Kim, and J.-T. Kim, *Analysis of Photovoltaic Applications in Zero Energy Building Cases of IEA SHC/EBC Task 40/Annex 52.* Sustainability, 2015. 7(7): p. 8782–8800.

[185] Traverse, C.J., et al., *Emergence of highly transparent photovoltaics for distributed applications.* Nature Energy, 2017. **2**(11): p. 849–860.

Chapter 12

2D Organic-Inorganic Hybrid Perovskite Light-Absorbing Layer in Solar Cells

Meng Wang, Qunliang Song and Sam Zhang

Abstract

With the rapid development of high-performance perovskite solar cell, its instability has become an urgent problem to be solved. 2D perovskite is considered as a potential light absorbing material for perovskite solar cells due to its excellent stability. However, the preparation of high quality 2D perovskite films suitable for photovoltaic devices remains a challenge. In this chapter, based on the structural and photophysical properties of 2D perovskite thin film materials, the latest progress in 2D perovskite cells in recent years and the strategy of controlling the film quality of 2D perovskite are summarized, which is of great significance for the further development of 2D perovskite photovoltaic devices.

Keywords: perovskite solar cell, stability, 2D perovskite, material property, film quality

1. Introduction

To solve the energy crisis coming with the rapid development of the society, solar energy is a promising renewable energy source which can convert light to electricity in the photovoltaic devices [1]. Among all the photovoltaic devices, perovskite solar cell (PSC) attracts researchers' attention most for its power conversion efficiency (PCE) has been increased from 3.8 to 25.2% in just 10 years [2–6]. With strong optical absorption, high carrier mobility and small exciton binding energy, organic-inorganic perovskite materials are semiconductors with remarkable optical and electrical properties [7–9]. The advantages of solution manufacturing and processing are widely used by researchers in solar cells with various structures [10]. However, one of the major reasons why high-performance perovskite solar cells have not been applied to practical application is the instability of the materials.

The instability of organic-inorganic hybrid perovskite is caused by many intrinsic and extrinsic factors. The external environmental factors include moisture, heat, oxygen and many other factors. Moisture is considered to be an important factor for the instability of perovskite materials, while the presence of light and oxygen accelerates the degradation process [11–13]. In addition to the influence of these external environmental factors, some intrinsic properties of the perovskite material itself also directly lead to its instability, such as composition and ion migration [14]. The perovskite materials with excellent photoelectric properties have a strong ionic property, indicating that the activation energy of ion migration inside the crystal

IntechOpen

is low, and molecular dissociation and ion migration are prone to occur within the structure, which limits the structural stability of these materials [15].

In recent years, the two-dimensional (2D) perovskite structure formed by introducing large-size organic cations is proved to be more stable than its three-dimensional (3D) counterpart and it has become a potential light-absorber in the PSCs. There are many reasons for the 2D perovskite to exhibit higher stability. The 2D perovskite has higher formation energy and it is more difficult to be oxidized than the 3D structure [16]. Compared with 3D perovskite crystals, the bonding forces between organic ions and [PbI_6] octahedral units such as van der Waals forces and hydrogen bonds are stronger [17]. Due to the presence of large size organic cations, ion migration is blocked [18]. Meanwhile, the 2D perovskite layer can work as passivation layer and blocking layer of moisture and oxygen to enhance the stability of perovskite [19, 20].

Although 2D perovskite materials show great potential in terms of stability, the relatively lower PCE needs to be improved. In this chapter, based on the structural and photophysical properties of 2D perovskite, the latest progress made in 2D PSCs in recent years and strategies to improve the performance of 2D PSCs are summarized, which is of great significance for the further development of PSCs based on 2D perovskite materials. Finally, a brief conclusion and outlook is promoted.

2. Material properties of 2D perovskite materials

2.1 Structural properties of 2D perovskite

The stability of perovskite structure can be described by tolerance factor-t, whose calculation formula is: $t = \frac{Ra + Rx}{\sqrt{2}(Rb + Rx)}$, where R_a, R_b and R_x are the radius of the atom located at A, B, X site respectively [20]. The perovskite structure with tolerance factor in the range of 0.8–1.01 is stable and can form an ideal 3D cubic structure [21]. In the stable 3D structure, the octahedrons formed by lead ions and halogen ions are infinitely connected, and organic cations are located between the spaces formed by the octahedrons [22]. When the size of the organic cation at the A site increases, the tolerance factor will exceed the above range finally. Then the infinitely connected octahedral structure is broken, and a conductive inorganic layer and an insulating organic layer are alternately connected to each other [23]. The 2D layered perovskite film can be regarded as the infinitely connected octahedron structure separating by the large-size organic cations, and the thickness of the octahedron contained in each layer is n. The value of n is closely related to the ratio of large organic spacer cations, and represents the periodicity in the crystal structure. The perovskite crystal structure with n value ranging from n = 1 to n = ∞ is shown in **Figure 1**. The L-value there denotes the thickness of the inorganic layer in each compound [25].

All organic cations capable of forming a 2D perovskite structure have groups that can interact with the inorganic layer at their ends and can stably exist in the crystal structure. According to the structural characteristics of large-size organic cations, the obtained 2D perovskites can be divided into different types. And the corresponding PSCs have different performances accordingly. The 2D perovskite formed by organic cation similar to BA^+ and PEA^+, which have only one amino group at the end, is called the RP phase, which was first applied to 2D PSCs in 2014 [19]. According to the characteristics of the crystal structure and the chemical ratio of each ion, the chemical formula of the 2D RP phase perovskite is $AA_{n-1}B_nX_{3n+1}$. [26]

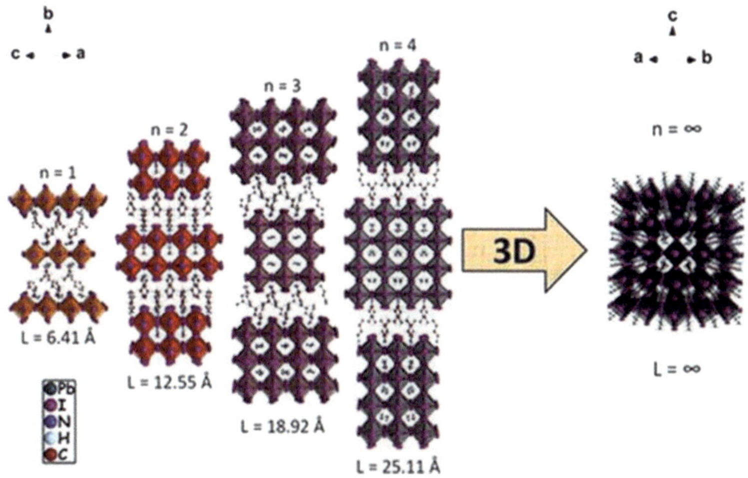

Figure 1.
Crystal structures of the 2D lead iodide perovskites, $(BA)_2(MA)_{n-1}Pb_nI_{3n+1}$, ranging from n = 1 to n = ∞. The L-value denotes the thickness of the inorganic layer in each compound [24]. Copyright 2016, American Chemical Society.

As shown in **Figure 2(a)**, the RP phase crystal is coupled by weak van der Waals forces. BA^+ and PEA^+ are the most widely studied organic spacer cations in 2D RP PSCs. In subsequent studies, organic cations such as AVA^+, PEI^+, PPA^+ have been extensively studied [27–29]. The development of new organic spacers cations is an important way to improve the performance of 2D perovskite solar cells. Recently, 2D RP PSCs with a record PCE more than 19% was prepared by using an organic-salt-assisted crystal growth (OACG) technique, which can induce the crystal growth and orientation, tune the surface energy levels, and suppress the losses caused by charge recombination [30]. The 2D perovskite formed by organic cation similar to EDA^{2+}, which have two amino groups that can interact with inorganic layers at both ends is called the DJ phase. Its chemical formula and crystal structure are shown in **Figure 2(b)** [31]. The DJ phase has better stability than RP perovskite (van der Waals interaction) because the spacer cations with two amino groups at both ends can form hydrogen bonds with inorganic plates without any gaps [32]. Ke et al. used 3-(amino methyl) piperidine ($3\text{-}AMP^{2+}$) as organic spacers. Compared with the single A-site cation, mixing cation $(3AMP)(MA_{0.75}FA_{0.25})_3Pb_4I_{13}$ perovskite has a narrower band gap, less inorganic skeleton distortion and the larger Pb-I-Pb angle [33]. Cohen used BDA^{2+} as an organic spacer cation and achieved 15.6% PCE

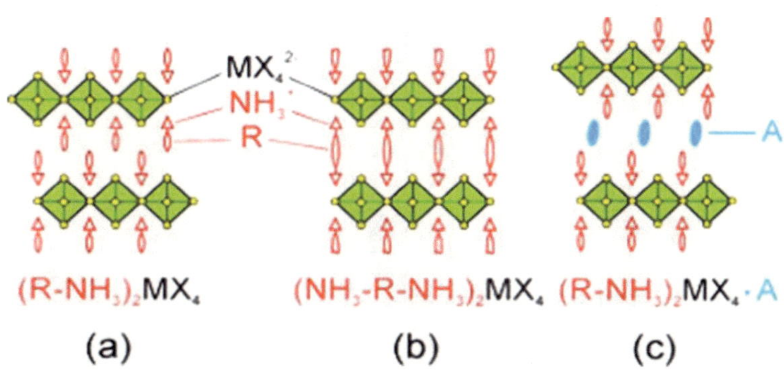

Figure 2.
Schematic representation of (a) RP phase (b) DJ phase (c) ACI phase [26, 31, 35]. Copyright 2000, springer. Copyright 2018, American Chemical Society. Copyright 2017, American Chemical Society.

without additives and any additional treatment [34]. The 2D perovskite formed by organic cation similar to GA^+, which can alternate interact with MA^+ in the organic layers, is called the ACI phase. Its chemical formula and crystal structure are shown in **Figure 2(c)** [35]. Due to its relatively small difference in ion size from MA^+ and FA^+, it has smaller exciton binding energy and weaker quantum confined effect. So, it is expected to achieve higher efficiency. The perovskite solar cells with BEA^{2+} ligand achieved high efficiency of 14.48 and 17.39% when doped with and without Cs^+ respectively [36]. Zhao's team achieved a high PCE of 18.48% by adding methyl ammonium chloride as an additive to effectively control the film quality of ACI 2D perovskite $(GA)(MA)_nPbnI_{3n+1}$ (n = 3), showing great potential of ACI perovskite with high stability and PCE [37].

2.2 Photophysical properties of 2D perovskite materials

Compared with 3D perovskites, 2D perovskite has a greater chemical and structural flexibility. The optical, electrical, and charge transfer properties can be regulated by controlling the width and composition of the potential well and barrier. In the 2D perovskite structure, a quantum well structure is formed between the insulating organic layer and the conductive inorganic layer, resulting in a quantum confinement effect [38]. The dielectric confinement effect is caused by the different dielectric constants of the potential well and the barrier, coupled with the quantum space limitation. The optical gap of 2D perovskites has a higher value than its 3D counterparts [39]. Zhang's work explored the inherent properties of 2D layered perovskites $(PEA)_2PbI_4(N)$ and $Cs_2PbI_4(N)$, and demonstrated that their structure and properties vary with N. The results reveal that both $(PEA)_2PbI_4(N)$ and $Cs_2PbI_4(N)$ are direct bandgap semiconductors. When $N \geq 3$, their band/optical gap and exciton binding energy vary linearly by 1/N. This work shows that ultra-thin 2D materials can become potential candidates for nano-optoelectronic devices, and nanoplates with $N \geq 3$ can have similar properties to bulk materials in terms of carrier migration and exciton separation, so they can be effectively applied for photovoltaic devices [40].

Figure 3 shows the absorption and emission spectra of a series of ultra-thin $(BA)_2(MA)_{n-1}Pb_nI_{3n+1}$ (n = 1–5) crystal layers mechanically peeled off from the pure phase (fixed n) single crystal by Blancon et al. In the exfoliated crystal, as n decreases from 5 to 1 (quantum well thickness varies from 3.139 to 0.641 nm), the band edge absorption and emission peaks monotonically increase from 1.85 to 2.42 eV. Due to the quantum and dielectric confinement effect, exciton binding

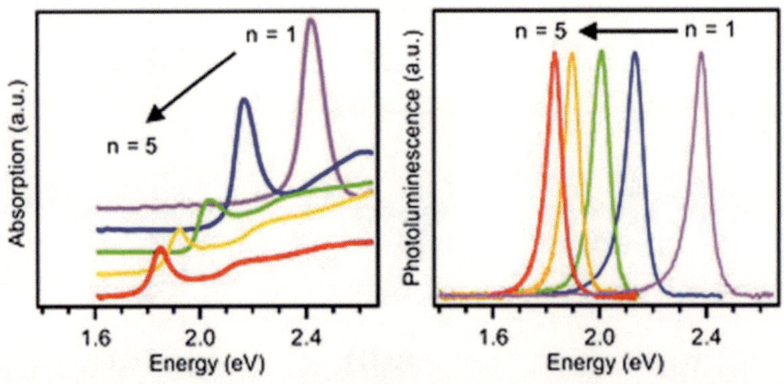

Figure 3.
(Left) absorption and (right) PL Spectrum of the exfoliated $(BA)_2(MA)_{n-1}Pb_nI_{3n+1}$ *crystals (from n = 1 to 5) [41]. Copyright 2017, American Association for the Advancement of Science (AAAS).*

energies are approximately one order of magnitude higher than the values found in 3D perovskites. Their results also show that in the film, the optical band gap is consistent with that of the exfoliated crystal at n = 1 and n = 2, but the red shifts about 200–300 meV at n = 3–5 [41].

3. Film quality control of 2D perovskite

Consistent with the situation in the 3D PSCs, it is of vital importance to control the film quality of absorption layer for a high-performance 2D PSC. For the synthesis of the light-absorption layer, the most-commonly used spin-coating method is applied. However, things become more difficult when it comes to the crystallization process of the 2D perovskite films. To improve the film quality of the 2D peroskite, many strategies are used to optimize the film quality, mainly focusing on adjusting the crystal orientation and the phase distribution, as summarized below.

3.1 Vertical orientation

The precursor solution of 2D perovskite is prepared by mixing and dissolving ammonium salt of the organic cation spacers, ammonium salt of the A-site organic cations and metal halide according to a certain proportion (depending on n value). During the crystallization process, competition was confirmed to occur between the large organic cations and A-site organic cations [18]. To be more specific, the large organic spacers tend to form a low-dimensional perovskite while the A-site cations tend to form a 3D structure. The low-n phases are prone to have a horizontal orientation instead of vertical orientation, which is unfavorable to the charge transfer. Since the presence of insulating large-size organic cations will hinder the charge transport out-of-plane, Tsai, H. et al. used a hot-casting technology in 2016 to prepare high-quality films, which means that the substrate is preheated before spin-coating the perovskite precursor solution. With this method, a more beneficial crystal growth along (111) and (202) planes is observed instead of random orientation [42]. The application of the hot-casting method has promoted the efficiency of 2D PSCs and confirmed the importance of the crystal orientation perpendicular to the substrate for the performance improvement. At present, hot-casting method is widely used for better performance of the 2D PSCs. However, it is hard to keep temperature accurate and uniform when transferring the substrate from the hotplate to the spin-coater. To solve this problem, Li et.al partially replaced the BA^+ with MA^+ in the BAMA quasi-2D perovskite, reducing the dependence on the preheating of the substrate. After being replaced, the quantum confinement effect of the perovskite film is weakened, the crystallization barrier is reduced, and higher quality perovskite film crystals and fewer defect states are obtained [43]. Beyond ion replacement, the morphology of the 2D DJ perovskite film with rigid piperidine ring was adjusted by MASCN additive at room temperature. By optimizing the amount of added MASCN, the perovskite film deposited on the substrate has good crystallinity, preferred orientation, reduced defects and better energy level alignment with the transport layer. The device had an inverted planar structure with a maximum PCE of 16.25%, which is the highest PCE of 2D DJ PVSCs without hot casting. After being exposed to air for 35 days, the unencapsulated device maintains about 80% of its initial efficiency (Hr 45 ± 5%). It provides a possiblely practical way for the development of high-performance 2D DJ PSCs [44].

Many other fabrication methods besides hot-casting have also been applied to obtain a vertical crystal orientation. Ke et al. used a two-step method of spin-coating a stoichiometric precursor containing PEAI and PbI_2, then performing FAI, and then dropping the solution and spin-coating it on the substrate to prepare 2D perovskite film [45]. Koh et al. also obtained preferential oriented growth of 2D perovskite perpendicular to the substrate through the immersion method, greatly improving the charge transport and extraction [46].

Adding additives was also proved to be an effective method to modify the crystal orientation. The role of additives in the 3D perovskite includes improving the morphology of the film, adjusting the energy level alignment, inhibiting the non-radiative recombination inside the film, eliminating the hysteresis of the device and so on [47–50]. Additives play the same roles in 2D PSCs. However, due to the different crystallization process, they have some other effects. Xinqian Zhang and his colleagues improved the PCE of $(PEA)_2(MA)_{n-1}Pb_nI_{3n+1}$ (n = 5) PSC from the initial 0.56% (without NH_4SCN) to 11.01% through optimized NH_4SCN addition. The performance improvement was attributed to the vertically oriented highly crystalline 2D perovskite film and balanced electron/hole transport [51]. NH_4SCN additives was also proved to be a simple and effective method to induce the growth direction of 2D DJ phase perovskite crystals perpendicular to the substrate, and at the same time the phase distribution in the perovskite crystals can be concentrated near the phases of n = 3 and n = 4. The quasi 2D DJ phase $(BDA)(MA))_4Pb_5I_{16}$ perovskite film based on NH_4SCN treatment has a PCE of 14.53%. In addition, after storing under an environmental condition of 50 ± 5% humidity for 900 hours, the device retained 85% of its initial PCE [52]. Xu Zhang and his colleagues demonstrated in their work that a 2D $BA_2(MA)_3Pb_4I_{13}$ PSC doped with cesium cation (Cs^+) had a PCE of up to 13.7%. The efficiency increased from 12.3% (without Cs^+) to 13.7% (with 5% Cs^+) due to perfectly controlled crystal orientation, increased grain size, excellent surface quality, reduced density of trap states, and enhanced carrier mobility [53].

3.2 Phase distribution

The coexistence of 2D perovskite phase and 3D perovskite phase lead to a phase impurity inside the low-dimensional films. Though hot-casting method has become an effective way to improve the efficiency of 2D PSCs, the impurity and phase distribution in 2D perovskite films were ignored at the early stage of studying 2D PSCs. Jin's work also shows that multiple phases of n = 2, 3, 4, and $n \approx \infty$ coexist in a 2D peovskite film with a nominal n value of 4. And they are naturally arranged perpendicular to the substrate. Through transient absorption spectroscopy analysis, they successfully observed continuous light-induced electron transfer from small n-phase to large n-phase driven by the band offset. And hole transfer in the opposite direction within hundreds of picoseconds was also observed. Exciton absorption peaks corresponding to different phases appear on the absorption spectrum, which is a strong evidence for the coexistence of mixed phases with different n values in the thin film. The strength of emission peaks corresponding to different n values was different when excited from the perovskite film side and the glass side, which confirmed the gradient distribution from small n value to large n value in the film [54]. Therefore, the mixed phases with multiple n values affect performance of 2D PSC from two aspects. One is phase purity, and the other is the arrangement of those phases. Besides gradient distribution in the vertical direction, other phase distributions may occur in low-dimensional perovskite films prepared by solution method. To realize high-performance low-dimensional perovskite cells, it is necessary to deliberately regulate the phase distribution inside the film.

In 2018, Liu's team used in situ time-resolved GIWAXS technology to track the transition process from precursor solution to solid film under different conditions of substrate temperatures and different solvents of precursor solutions. The results reveal that under lower temperature the intermediate phase formed by lead iodide crystal and solvate complexes can cause multiple RP phases with random grain orientations. It is better for the disordered solvate to transform to perovskite directly [55]. The mixed phases in the low-dimensional perovskite thin film crystals are prone to energy transfer, resulting in large V_{oc} loss. To get a more vertically phase distribution, later Liu's team transferred the device to a hot plate at a suitable temperature for several hours after the 2D perovskite device was prepared, which is called a slow post-annealing (SPA) method. Such a device obtains an open circuit voltage as high as 1.24 V, which proves that the quantum well effect in the perovskite film is reduced to greatly improve the charge transfer and extraction efficiency in the device. They compared the phase distribution of the SPA films with the film prepared by the hot casting method and at room temperature respectively. Inside the thin film prepared at room temperature, the phases with different n values are randomly distributed. There is a sudden and uneven phase gradient inside the thin film prepared by the hot-casting method, resulting in the presence of 2D phase and 3D phase at the bottom and top respectively. After SPA treatment, the phase distribution inside the thin film appears more orderly [56]. Tiefeng Liu and his colleagues reported that the phase distribution of different n values in a 2D perovskite film deposited on a hole transport layer is different from that on a glass substrate. Due to the colloidal characteristics of the perovskite precursor, the vertical distribution can be explained by the sedimentation equilibrium. The addition of acid changes the precursor from colloid to solution, thereby changing the phase distribution. The self-assembled layer was used to modify the acidic surface properties of the hole transport layer, which can cause the vertical distribution required for charge transport. The surface-modified 2D PSC had a higher open circuit voltage and a higher efficiency than the control device [57].

Zhou and his colleagues proved that by controlling the crystal growth direction and growth rate, the phase distribution and carrier transport of quasi-2D perovskite films can be controlled. They found that using ethyl acetate as an anti-solvent can change the growth direction of quasi-2D perovskites by accelerating the formation of surface crystals. In addition, through the introduction of MACl and DMSO in the preparation process, the film with the phases of $n = 3$ and $n = 4$ was successfully obtained. With the addition of MACl and DMSO in the precursor solution, an intermediate phase is formed, which slows down the rate of crystallization in the solution. In addition, by correlating the phase distribution with the device characteristics, it was shown that the performance of the solar cell is sensitive to the phase purity and phase distribution [58].

Zhang Jia et.al confirmed that the phase distribution obtained in the film prepared by the vacuum-poling method is different from the traditional films that the phases with n value from small to large are arranged in a gradient distribution from the bottom to the top. The research result showed that the phases with different n values show a uniform distribution inside the film. The uniform phase distribution was confirmed by the PL results. When excited from the perovskite film side and from the glass side, and after the film is peeled off by the tape, the measured PL spectrum showed negligible differences, proving the uniform distribution of the mixed phase. In this case, the excellent PCE up to 18% can be ascribed to the shortest charge transfer path [59].

4. Conclusions

The high stability and relatively poor capability to convert light to electricity of low-dimensional perovskite films are the two aspects that need to be balanced. Based on the above analysis, we conclude that crystal growth direction and phase distribution in the low-dimensional perovskite film determine the performance of the 2D PSCs. The crystal orientation in vertical direction and the proper phase distribution in favor of charge transfer are the targets for preparing high performance low-dimensional perovskite films. A few high-performance low-dimensional perovskite cells have emerged, showing high stability and efficiency comparable to 3D PSCs. Though the research on low-dimensional perovskite cells is far from enough and the crystal growth mechanism and carrier transport kinetics are still obscure, it is very promising to achieve efficient and stable photovoltaic devices using 2D perovskite materials as light absorbers.

Acknowledgements

This work was supported by the National Natural Science Foundation of China (Grant No.11774293), Chongqing Key Laboratory for Advanced Materials & Technologies of Clean Energies (JJNY201905). This work was also supported by Fundamental Research Funds for the Central Universities: SWU118105.

Conflict of interest

The authors declare no conflict of interest.

Author details

Meng Wang, Qunliang Song* and Sam Zhang*
Institute for Clean Energy and Advanced Materials, School of Materials and Energy, Southwest University, Chong Qing, China

*Address all correspondence to: qlsong@swu.edu.cn; samzhang@swu.edu.cn

References

[1] Barnham KWJ, Mazzer M, Clive B. Resolving the energy crisis: Nuclear or photovoltaics? Nature Materials. 2006;**5**(3):161-164. DOI: 10.1038/nmat1604

[2] Kojima A, Teshima K, Shirai Y. Organometal halide perovskites as visible-light sensitizers for photovoltaic cells. Journal of the American Chemical Society. 2009;**131**(17):6050-6051. DOI: 10.1021/ja809598r

[3] Kim H-S, Lee C-R, Im J-H. Lead iodide perovskite sensitized all-solid-state submicron thin film mesoscopic solar cell with efficiency exceeding 9%. Scientific Reports. 2012;**2**(1):591. DOI: 10.1038/srep00591

[4] Jiang Q, Zhao Y, Zhang X. Surface passivation of perovskite film for efficient solar cells. Nature Photonics. 2019;**13**(7):460-466. DOI: 10.1038/s41566-019-0398-2

[5] Min H. ,Kim M. ,Lee S. Efficient, stable solar cells by using inherent bandgap of α-phase formamidinium lead iodide. Science. 2019;**366**(6466):749. DOI: 10.1126/science.aay704

[6] National Renewable Energy Laboratory Best Research-Cell Effciency Chart. Available from: https://www.nrel.gov/pv/cellefficiency.html

[7] Lee MM, Teuscher J, Miyasaka T. Efficient hybrid solar cells based on meso-superstructured organometal halide perovskites. Science. 2012;**338**(6107):643-647. DOI: 10.112 6/science.1228604

[8] Wehrenfennig C, Eperon GE, Johnston MB. High charge carrier mobilities and lifetimes in organolead trihalide perovskites. Advanced Materials. 2014;**26**(10):1584-1589. DOI: 10.1002/adma.201305172

[9] D'innocenzo V, Grancini G, Alcocer M, et al. Excitons versus free charges in organo-lead tri-halide perovskites. Nature Communications. 2014;**5**(1):3586. DOI: 10.1038/ncomms4586

[10] Jung JW, Williams ST, Jen AKY. Low-temperature processed high-performance flexible perovskite solar cells via rationally optimized solvent washing treatments. RSC Advances. 2014;**4**(108):62971-62977. DOI: 10.1039/ C4RA13212B

[11] Leguy AMA, Hu Y. Campoy-Quiles M.l.Reversible hydration of CH3NH3PbI3 in films, single crystals, and solar cells. Chemistry of Materials. 2015;**27**(9):3397-3407. DOI: 10.1021/acs.chemmater.5b00660

[12] Divitini G, Cacovich S, Matteocci F. In situ observation of heat-induced degradation of perovskite solar cells. Nature Energy. 2016;**1**(2):15012. DOI: 10.1038/n energy.2 015.12

[13] Aristidou N, Sanchez-Molina I, Chotchuangchutchaval T. The role of oxygen in the degradation of methylammonium lead trihalide perovskite photoactive layers. Angewandte Chemie International Edition. 2015;**54**(28):8208-8212. DOI: 10.1002/anie. 20 1503153

[14] Misra RK, Aharon S, Li B. Temperature- and component-dependent degradation of perovskite photovoltaic materials under concentrated sunlight. The Journal of Physical Chemistry Letters. 2015;**6**(3):326-330. DOI: 10.1021/jz50264 2b

[15] Frost JM, Butler KT, Brivio F. Atomistic origins of high-performance in hybrid halide perovskite solar cells. Nano Letters. 2014;**14**(5):2584-2590. DOI: 10.1021/nl500 390f

[16] Wang R, Mujahid M, Duan Y. A review of perovskites solar cell stability. Advanced Functional Materials. 2019;**29**(47):1808843. DOI: 10.1002/adfm.201808843

[17] Quan LN, Yuan M, Comin R. Ligand-stabilized reduced-dimensionality perovskites. Journal of the American Chemical Society. 2016;**138**(8):2649-2655. DOI: 10.1021/jacs.5b11740

[18] Cao DH, Stoumpos CC, Farha O, et al. 2D homologous perovskites as light-absorbing materials for solar cell applications. Journal of the American Chemical Society. 2015;**137**(24):7843-7850. DOI: 10.1021/jacs.5b03796

[19] Smith IC, Hoke ET, Solis-Ibarra D. A layered hybrid perovskite solar-cell absorber with enhanced moisture stability. Angewandte Chemie, International Edition. 2014;**53**(42):11232-11235. DOI: 10.1002/anie.201406466

[20] Yang J, Siempelkamp BD, Liu D. Investigation of CH3NH3PbI3 degradation rates and mechanisms in controlled humidity environments using in situ techniques. ACS Nano. 2015;**9**(2):1955-1963. DOI: 10.1021/nn506864k

[21] Goldschmidt VM. Die Gesetze der krystallochemie. Naturwissenschaften. 1926;**14**(21):477-485. DOI: 10.1007/BF01507527

[22] Kieslich G, Sun S, Cheetham AK. Solid-state principles applied to organic–inorganic perovskites: New tricks for an old dog. Chemical Science. 2014;**5**(12):4712-4715. DOI: 10.1039/C4SC02211D

[23] Mitzi DB. Templating and structural engineering in organic-inorganic perovskites. Journal of the Chemical Society Dalton Transactions. 2001;(1):1-12. DOI: 10.1039/b007070j

[24] Stoumpos CC, Cao DH, Clark D, et al. Ruddlesden–popper hybrid lead iodide perovskite 2D homologous semiconductors. Chemistry of Materials. 2016;**28**(8):2852-2867. DOI: 10.1021/acs.chemmater.6b00847

[25] Gao P, Bin Mohd Yusoff AR, Nazeeruddin MK. Dimensionality engineering of hybrid halide perovskite light absorbers. Nature Communications. 2018;**9**(1):5028. DOI: 10.1038/s41467-018-07382-9

[26] Beznosikov BV, Aleksandrov KS. Perovskite-like crystals of the Ruddlesden-popper series. Crystallography Reports. 2000;**45**(5):792-798. DOI: 10.1134/1.1312923

[27] Grancini G, Roldán-Carmona C, Zimmermann I. One-year stable perovskite solar cells by 2D/3D interface engineering. Nature Communications. 2017;**8**(1):15684. DOI: 10.1038/ncomms15684

[28] Yao K, Wang X, Xu Y-x, Li F, Zhou L. Multilayered perovskite materials based on polymeric-ammonium cations for stable large-area solar cell. Chemistry of Materials. 2016;**28**(9):3131-3138. DOI: 10.1021/acs.chemmater.6b00711

[29] Cohen B-E, Wierzbowska M, Etgar L. High efficiency quasi 2D lead bromide perovskite solar cells using various barrier molecules. Sustainable Energy & Fuels. 2017;**1**(9):1935-1943. DOI: 10.1039/C7SE00311K

[30] Lai H, Lu D, Xu Z. Organic-salt-assisted crystal growth and orientation of quasi-2D Ruddlesden–Popper perovskites for solar cells with efficiency over 19%. Advanced Materials. 2020;**32**(33):2001470. DOI: 10.1002/adma.202001470

[31] Mao L, Ke W, Pedesseau L. Hybrid Dion–Jacobson 2D lead iodide

perovskites. Journal of the American Chemical Society. 2018;**140**(10):3775-3783. DOI: 10.1021/jacs.8 b00542

[32] Ahmad S, Fu P, Yu S. Dion-Jacobson phase 2D layered perovskites for solar cells with ultrahigh stability. Joule. 2019;**3**(3):794-806. DOI: 10.1016/j.joule.2018.11.026

[33] Ke W, Mao L, Stoumpos CC. Compositional and solvent engineering in Dion–Jacobson 2D perovskites boosts solar cell efficiency and stability. Advanced Energy Materials. 2019;**9**(10):1803384. DOI: 10.1002/aenm.201803384

[34] Cohen B-E, Li Y, Meng Q. Dion–Jacobson two-dimensional perovskite solar cells based on benzene dimethanammonium cation. Nano Letters. 2019;**19**(4):2588-2597. DOI: 10.1021/acs.nanolett.9b00387

[35] Soe CMM, Stoumpos CC, Kepenekian M. New type of 2D perovskites with alternating cations in the interlayer space, (C(NH2)3)(CH3NH3)nPbnI3n+1: Structure, properties, and photovoltaic performance. Journal of the American Chemical Society. 2017;**139**(45):16297-16309. DOI: 10.1021/jacs.7b09096

[36] Li P, Liang C, Liu X-L. Low-dimensional perovskites with diammonium and monoammonium alternant cations for high-performance photovoltaics. Advanced Materials. 2019;**31**(35):1901966. DOI: 10.1002/adma.201901966

[37] Luo T, Zhang Y, Xu Z. Compositional control in 2D perovskites with alternating cations in the interlayer space for photovoltaics with efficiency over 18%. Advanced Materials. 2019;**31**(44):1903848. DOI: 10.1002/adma.201903848

[38] Hattori T, Taira T, Era M. Highly efficient electroluminescence from a heterostructure device combined with emissive layered-perovskite and an electron-transporting organic compound. Chemical Physics Letters. 1996;**254**(1):103-108. DOI: 10.1016/0009-2614(96)00310-7

[39] Tanaka K, Takahashi T, Kondo T. Image charge effect on two-dimensional excitons in an inorganic-organic quantum-well crystal. Physical Review B. 2005;**71**(4):045312. DOI: 10.1103/PhysRevB.71.045312

[40] Zhang L, Liang W. How the structures and properties of two-dimensional layered perovskites MAPbI3 and CsPbI3 vary with the number of layers. The Journal of Physical Chemistry Letters. 2017;**8**(7):1517-1523. DOI: 10.1021/acs.jpclett.6b03005

[41] Blancon JC, Tsai H, Nie W. Extremely efficient internal exciton dissociation through edge states in layered 2D perovskites. Science. 2017;**355**(6331):1288. DOI: 10.1126/science.aal4211

[42] Tsai H, Nie W, Blancon J-C. High-efficiency two-dimensional Ruddlesden–Popper perovskite solar cells. Nature. 2016;**536**(7616):312-316. DOI: 10.1038/ nature18 306

[43] Li X, Wu G, Zhou J. Non-preheating processed quasi-2D perovskites for efficient and stable solar cells. Small. 2020;**16**(11):1906997. DOI: 10.1002/smll.201906997

[44] Wu H, Lian X, Tian S. Additive-assisted hot-casting free fabrication of Dion–Jacobson 2D perovskite solar cell with efficiency beyond 16%. Solar RRL. 2020;**4**(7):2000087. DOI: 10.1002/solr.202000087

[45] Ke L, Gan X, Zhao W. (C6H5C2H4NH3)2FAn-1PbnI3n+1: A quasi two-dimensional perovskite with high performance produced

via two-step solution method. Journal of Alloys and Compounds. 2019;**788**:954-960. DOI: 10.1016/j.jallcom.2019.02.280

[46] Koh T. M. ,Shanmugam V. ,Schlipf J. Nanostructuring mixed-dimensional perovskites: A route toward tunable, efficient photovoltaics. Advanced Materials. 2016;**28**(19):3653-3661. DOI: 10.1002/adma.201506141

[47] Li B, Li M, Fei C. Colloidal engineering for monolayer CH3NH3PbI3 films toward high performance perovskite solar cells. Journal of Materials Chemistry A. 2017;**5**(46):24168-24177. DOI: 10.1039/C7TA08761F

[48] Chen C, Song Z, Xiao C. Achieving a high open-circuit voltage in inverted wide-bandgap perovskite solar cells with a graded perovskite homojunction. Nano Energy. 2019;**61**:141-147. DOI: 10.1016/j.n anoen.2019.04.069

[49] Zou Y, Wang H-Y, Qin Y. Reduced defects of MAPbI3 thin films treated by FAI for high-performance planar perovskite solar cells. Advanced Functional Materials. 2019;**29**(7):1805810. DOI: 10.1002/adfm.201805810

[50] Son D-Y, Kim S-G, Seo JY. Universal approach toward hysteresis-free perovskite solar cell via defect engineering. Journal of the American Chemical Society. 2018;**140**(4):1358-1364. DOI: 10.1021/jacs.7b10430

[51] Zhang X, Wu G, Fu W. Orientation regulation of phenylethylammonium cation based 2D perovskite solar cell with efficiency higher Than 11%. Advanced Energy Materials. 2018;**8**(14):1702498. DOI: 10.1002/aenm.201702498

[52] Li F, Zhang J, Jo S. Vertical orientated Dion–Jacobson quasi-2D perovskite film with improved photovoltaic performance and Stability. Small Methods. 2020;**4**(5):1900831. DOI: 10.1002/smtd.201900831

[53] Zhang X, Ren X, Liu B. Stable high efficiency two-dimensional perovskite solar cells via cesium doping. Energy & Environmental Science. 2017;**10**(10):2095-2102. DOI: 10.1039/C7EE01145H

[54] Liu J, Leng J, Wu K. Observation of internal photoinduced electron and hole separation in hybrid two-dimentional perovskite films. Journal of the American Chemical Society. 2017;**139**(4):1432-1435. DOI: 10.1021/jacs.6b12581

[55] Zhang X, Munir R, Xu Z. L.Phase transition control for high performance Ruddlesden–Popper perovskite solar cells. Advanced Materials. 2018;**30**(21):1707166. DOI: 10.1002 / adma.201707166

[56] Wu G, Li X, Zhou J. Fine multi-phase alignments in 2D perovskite solar cells with efficiency over 17% via slow post-annealing. Advanced Materials. 2019;**31**(42):1903889. DOI: 10.1002/adma.201903889

[57] Liu T, Jiang Y, Qin M. Tailoring vertical phase distribution of quasi-two-dimensional perovskite films via surface modification of hole-transporting layer. Nature Communications. 2019;**10**(1):878. DOI: 10.1038/s41467-019-08843-5

[58] Zhou M, Fei C, Sarmiento JS. Manipulating the phase distributions and carrier transfers in hybrid quasi-two-dimensional perovskite films. Solar RRL. 2019;**3**(4):1800359. DOI: 10.1002/solr.201800359

[59] Zhang J, Qin J, Wang M. Uniform permutation of quasi-2D perovskites by vacuum poling for efficient, high-fill-factor solar cells. Joule. 2019;**3**(12):3061-3071. DOI: 10.1016/j.joule.2019.09.020

Chapter 13

Two-Dimensional Materials for Advanced Solar Cells

Manoj Kumar Singh, Pratik V. Shinde, Pratap Singh and Pawan Kumar Tyagi

Abstract

Inorganic crystalline silicon solar cells account for more than 90% of the market despite a recent surge in research efforts to develop new architectures and materials such as organics and perovskites. The reason why most commercial solar cells are using crystalline silicon as the absorber layer include long-term stability, the abundance of silicone, relatively low manufacturing costs, ability for doping by other elements, and native oxide passivation layer. However, the indirect band gap nature of crystalline silicon makes it a poor light emitter, limiting its solar conversion efficiency. For instance, compared to the extraordinary high light absorption coefficient of perovskites, silicon requires 1000 times more material to absorb the same amount of sunlight. In order to reduce the cost per watt and improve watt per gram utilization of future generations of solar cells, reducing the active absorber thickness is a key design requirement. This is where novel two-dimensional (2d) materials like graphene, MoS_2 come into play because they could lead to thinner, lightweight and flexible solar cells. In this chapter, we aim to follow up on the most important and novel developments that have been recently reported on solar cells. Section-2 is devoted to the properties, synthesis techniques of different 2d materials like graphene, TMDs, and perovskites. In the next section-3, various types of photovoltaic cells, 2d Schottky, 2d homojunction, and 2d heterojunction have been described. Systematic development to enhance the PCE with recent techniques has been discussed in section-4. Also, 2d Ruddlesden-Popper perovskite explained briefly. New developments in the field of the solar cell via upconversion and downconversion processes are illustrated and described in section-5. The next section is dedicated to the recent developments and challenges in the fabrication of 2d photovoltaic cells, additionally with various applications. Finally, we will also address future directions yet to be explored for enhancing the performance of solar cells.

Keywords: 2D materials, graphene, MoS_2, advanced solar cells, perovskites

1. Introduction

Because of excessive utilization and consumption, the conventional fuel sources started depleting rapidly. In this direction, there is an urgent need for reconstruction of energy infrastructure, which is based on environmentally sustainable energy technologies such as wind, water, and solar. The worldwide research attracted towards solar energy, converting light energy into electrical energy. Solar photovoltaic is a pollution-free, efficient, renewable, reliable, rich, and continual source of energy. The photovoltaic solar cell, well-known technique, provides the solution of energy source crises in the 21st century. The main mechanism for the conversion of light to electricity: photovoltaic effect, photoconductivity, and photovoltaic effect (bulk). There is the requirement of a p-n junction in which electron and holes (photo-induced) in p-type and n-type materials partitioned transport a gathered to an electrode for production of photocurrent. In 1839, Edmond Becquerel first of all showed the demonstration of photovoltaic effect [1–2]. In the absence of p-n junction the conductivity of the semiconductor sample rises (by the illumination), it happens when the number of free electrons is increased, this is famed as photoconductivity. The electricity generated through the photovoltaic solar cells is not so cost-efficient in comparison to the grid power which we are using today [3]. At the large scale the solar energy conversion which should be low cost, there is a need for such type semiconducting materials that will make the production processes easily measurable and economically feasible [4]. In this direction two-dimensional (2d) material is referred to as impediment in one dimension between the size range 0–100 nanometers (nm), while the rest of the two dimensions are of micrometer range [5]. Furthermore, the configuration of atom and bond strength in 2d is identic and much stronger than that of bulk materials [6]. Also, ultrathin 2d nanomaterials have uncommon properties from their alternative nanostructured materials, such as three-dimensional (3d) nanocubes, one-dimensional (1d) nanotubes, and zero-dimensional (0d) quantum dots. First, the ultra-thickness of 2d nanomaterials provides high charge carrier, high charge mobility both at low and 300 kelvin (K) temperature, and high thermal conductivity [7–9]. Second, quantum confinement of 2d nanomaterials especially single layer or atomic thick layer, displays a number of properties, such as conductivity, tunable bandgap, surface activity, and magnetic anisotropy [10–11]. Third, the quantum Hall Effect (QHE) is shown by defect-free 2d materials, even at 300 K. The defect-free 2d materials have the electrons with a concentric (scatter-less) motion that allows the high charge carrier [12–13]. Fourth, the large ultrahigh surface area, keeping atomic-sized thickness, shows them ultrahigh specific surface area [14–15]. Therefore, photovoltaic solar cell manufactured by two-dimensional materials is a well-versed method in between of scientific community.

In the present chapter, we aim to follow up on the most important and novel developments that have been recently reported on solar cells. Section-2 is devoted to the properties, synthesis techniques of different 2d materials like graphene, transition metal dichalcogenides (TMDs), and perovskites. In the next section-3, various types of photovoltaic cells, 2d Schottky, 2d homojunction, and 2d heterojunction have been described. Systematic development to enhance the power conversion efficiency (PCE) with recent techniques has been discussed in section-4. Also, 2d Ruddlesden-Popper perovskite explained briefly. New developments in the field of the solar cell via upconversion and downconversion processes are illustrated and described in section-5. The next section is dedicated to the recent developments and challenges in the fabrication of 2d photovoltaic cells, additionally with various applications. Finally, we will also address future directions yet to be explored for enhancing the performance of solar cells.

2. Photovoltaic materials

2.1 Graphene

The dimension is the key factor to classify carbon allotropes/nanostructures into four groups, 0d (quantum dots, fullerenes), 1d (nanohorns, nanoribbons, carbon nanotubes), 2d (graphene) and 3d (diamond, graphite) structures [16–17]. A new area of research started with the groundbreaking discovery of graphene in 2004 by Novoselov and his co-authors in his famous publication "Electric field effect in atomically thin carbon films" and awarded jointly Nobel prize for it [18]. Graphene is a single layer structure with sp^2 hybridization in which carbon atoms are arranged in a hexagonal honeycomb lattice. It is a semi-metal with zero-bandgap, large specific surface area (2630 m^2g^{-1}), high Young's modulus (1.1 TPa), and high thermal conductivity (3×10^3 W m^{-1} K^{-1} at 300 K) [6, 19–22]. Graphene also provides the optical and electrical properties as excellent transparency (97.7% in the visible spectrum) and electrical conductivity ($\approx 10^4$ Ω^{-1} cm^{-1}) [23–24]. These exotic properties of graphene make it special in several optoelectronic applications. In solar cells, instead of indium doped tin oxide (ITO) and fluorine-doped tin oxide (FTO), graphene attracted attention due to flexibility, chemical stability, and high transmittance [20, 25–26]. These excellent dimensional, structural, optical, and electrical properties depict the graphene as a suitable aspirant for photovoltaic cells.

One of the well-known methods to synthesis the graphene is thermal chemical vapor deposition. In the thermal chemical vapor deposition (CVD), copper substrate placed into the quartz tube and then precursor gases (in the specific ratio) are allowed to flow at very high temperatures in the furnace [27]. After some time, single layer, bilayer, or multilayer deposition of graphene revels, this depends upon the internal conditions of experiments like temperature, pressure, reaction time, and gas flow rate [28]. The more advancement in the synthesis of graphene on Ni was achieved by Somani *et al.* [29]. In this, the camphor ($C_{10}H_{16}O$) has been taken as the precursor. Moreover, the large-scale monolayer graphene was produced by Obraztsov and co-others via a CVD method [30]. Another attempt has been performed to manufacture graphene on Cu foil (industrial base) via thermal CVD of methane with 1000°C temperature by Lia and co-workers in 2009 [31].

2.2 Transition metal dichalcogenides

Although graphene has various excellent properties, due to zero-bandgap, work-function, and toxic nature, the research on new atomically thin 2d materials gained attention. These necessities have been fulfilled by TMDs. These2d materials attracted more attention as they have grown on a flexible surface and can be bears the stress and deformation [32–34]. Generally, TMDs are formulized as $MX_2$where M expresses the transition metal from group IV-VIII, (M = Ti, Zr, Hf, V, Nb, Cr Ta, Mo, W, etc.) and X is a chalcogen atom (X = S, Se, Te) [35–36]. TMDs have opened the new pipeline of research as having tunable bandgap (1–2 eV) and explore an excellent picture of electrical, optical, and mechanical properties [37–39]. Various combinations of TMDs such as MoS_2, CrS_2, WS_2, TiS_2, $MoSe_2$, $CrSe_2$, WSe_2, $TiSe_2$ etc. found in metallic, semiconductor and insulator phase [40]. TMDs are a collection of big crystal family, found in different phases such as 1 T, 2H, and 3R., having two-third materials with layered structure [41]. In particular, MoS_2 shows mechanically 30% more strength than steel and can be ruptured after warping 1%. It generates the most distensible and strongest semiconducting materials [36, 42]. Counter electrodes manufactured by platinum (Pt) were replaced by MoS_2 in photovoltaic devices [43].

Typically, the synthesis approaches like exfoliation, hydrothermal, CVD, molecular beam epitaxy (MBE), and atomic layer deposition (ALD) are used to prepare the desired size of TMDs [44–48].

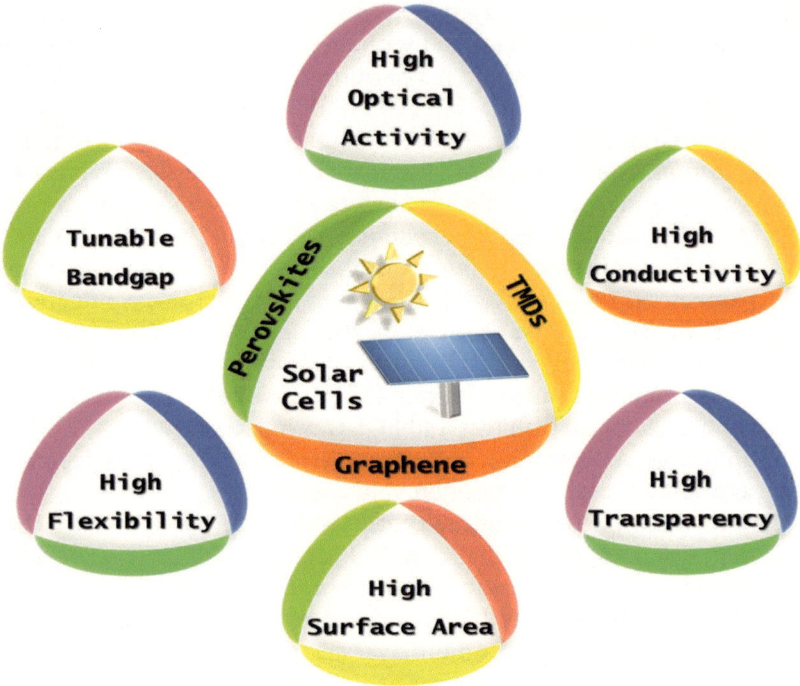

Figure 1.
Schematic illustration of exotic properties of 2d materials used for solar cell devices.

2.3 Perovskites

Perovskites are a mixture of organic–inorganic materials, which offer high absorption coefficients, direct bandgap, high charge carrier mobility, and long charge carrier diffusion length [49–50]. This is why the research groups attracted more and more attention by 2d perovskites for a long time. There are three types of halide perovskite (2d) (i) organic–inorganic mixed halide perovskite, (ii) 2d Ruddlesden-Popper perovskites, (iii) inorganic halide perovskite [51]. The typically Perovskite structure is given by ABX_3, where A indicates monovalent cation such as methyl rubidium (Rb), ammonium, and formamidinium; B represents heavy materials like tin (Sn) and lead (Pb); and X shows a halogen anion (i. e. chlorine, bromine, iodine). A unique type of properties provides highly defected bulk structures, indicate chemical compound through which the device operation power has been smoothed. The performance of 2d perovskite solar cells can be improved by obtaining a very high output voltage (under the circumstance of open circuit V_{oc}). The photovoltaic solar cells should be free from all recombination losses and this can be achieved by suppressing losses up to unity while quantum yield must be highest. [52–53].

The synthesis of 2d organic–inorganic mixed halide perovskite fabricated in two steps: (i) formation of lead halide (nano-platelets) on muscovite mica using van der Waals epitaxy in vapor transport CVD system, (ii) Ag as-solid heterophase reaction (using methylammonium halide molecules) used to obtain perovskite from platelets. However, the structure fabricated via this method is a 3d perovskite but using the universal scotch tape-based mechanical exfoliation method 2d perovskites is obtained [54–57]. **Figure 1** shows schematic illustration of exotic properties of 2d materials useful for solar cell devices.

3. Photovoltaic in domain of 2d materials

3.1 Photovoltaic based on 2d Schottky junction

During the photovoltaic processes (under illumination), electron–hole pairs are formed. These pairs are also termed as photogenerated carriers and they can be equal and more energetic (with incident photons) by the bandgap of the semiconductor. The conjunction of electron–hole pairs accorded on the electrodes and they are isolated through the junction internal field (electric) [58]. When the difference between the Fermi level of semiconductor and metal work function is generated, a Schottky junction enters in the pictures and photocurrent starts to develop. Net photocurrent has been maintained in asymmetric Schottky barriers (metal having different work function), whereas symmetric metal contact structure produces no net photocurrent. The important characteristics terms associated with the photovoltaic device illustrated in **Table 1**. Fontana *et al.* [58] synthesized a MoS_2 based (50 nm thick) phototransistor with palladium (Pd) and gold (Au) for drain contact and source, respectively. When two different materials are used for the source and drain contacts, such as hole-doping Pd and electron-doping Au, the Schottky junctions formed at the MoS_2 contacts generates a photovoltaic effect. **Figure 2a** displays the optical image of the device. **Figure 2b** shows the current vs. voltage curve at zero gate voltage, corresponding to the branch of the hysteresis with higher current, where the Fermi energy is shifted into the MoS_2 conduction band. Shin *et al.* [59] reported the graphene/porous silicon Schottky-junction solar cells by employing graphene transparent conductive electrodes doped with silver nanowires. The Ag nanowires-doped graphene/PSi solar cells show a maximum PCE of 4.03%. Yi and his co-workers developed Schottky junction photovoltaic cells based on multilayer $Mo_{1-x}W_xSe_2$ with x = 0, 0.5, and 1 [60]. To generate built-in potentials, Pd and Al were used as the source and drain electrodes in a lateral structure, while Pd and graphene were used as the bottom and top electrodes in a vertical structure.

3.2 Photovoltaic based on 2d homojunction

Due to the very low efficiency of the Schottky junction, more research efforts are required to improve photovoltaic processes in the semiconducting p-n junction.

Term	Description
Short-circuit current (I_{sc})	It is defined as the current flowing through the device (under illumination) and at zero external bias having contact shorted.
Power conversion efficiency (PCE)	It is defined as the ratio of electrical power generated to the incident light power.
External quantum efficiency (EQE)	The ratio defines by the amount of charge carriers moving through the device (under short-circuit current) to the all number of colliding photons on it.
Internal quantum efficiency (IQE)	Shows the ratio of the amount of charge carriers moving through the device (under short-circuit current) to all numbers of absorbed photons.
Open-circuit voltage (V_{oc})	The voltage produced by the device having no current flow (under illumination)
Fill factor (FF)	It is describing the ratio of maximum electric power generated to the product of its open-circuit voltages and its short-circuit current.

Table 1.
Main terms to demonstrate the photovoltaic device.

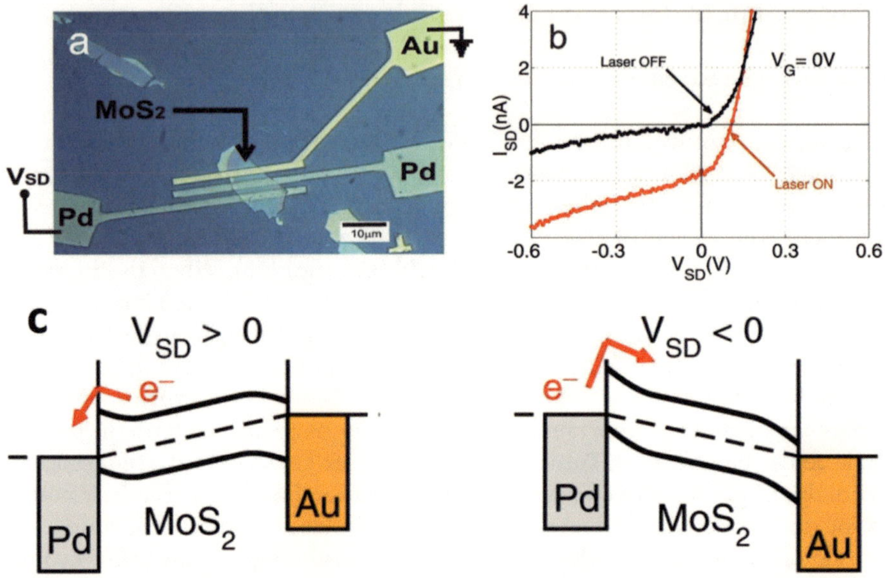

Figure 2.
Photovoltaic effect with Pd-Au bias configuration. (a) Optical image of the device. The spacing between the electrodes is 2 μm. (b) Current vs. source-drain voltage (at V_G = 0) showing strong asymmetry and photoresponse with diode-like behavior for the Pd-Au bias configuration indicated in (a). (c) Effect of source-drain bias on a device with one Au and one Pd electrode. Reprinted with permission from [58]. Copyright (2013) springer nature.

Using a splitting gate on monolayer WSe_2, Pospischil *et al.* [61], Baugher *et al.* [62], and Ross *et al.* [63] effectuated experimentally p-n junctions. This demonstration of WSe_2 monolayer flake has been achieved with mechanical exfoliation after that shifted onto a pair of split gates slipover with previously formed gate dielectric materials (SiN, HfO_2). The charge density and the conduction type (into the monolayer thick channel) can be modulated by electrostatic control after dissimilar voltages have been applied on the two local gates and the automatically thin p-n junction was formed. Due to this remarkable rectification in the diode behavior occur, finally able to photovoltaic generation. Taking the gap within two gates as a photoactive area the power conversion of 0.5% was demonstrated by Pospischil *et al.* with V_{oc} of 0.64 V and illumination of 140 mW/cm^2 (halogen source). The remarkable results are in the picture with very high efficiency of photovoltaic energy conversion, assuming the monolayer WSe_2 (95% transparency), which opens a pipeline of single-layer TMDs for semi-transparent solar cells. Memaran *et al.* [64] successfully demonstrated the composition of electrostatically generated $MoSe_2$ multilayer p-n junction to achieve high photovoltaic performance.

In addition to modifying the photovoltaic parameters, the 2d black phosphorous (BP) has attracted more attention of researchers due to its remarkable optical and electrical properties, keeping in mind its unique bandgap (≈0.3–2.0 eV), in-plane anisotropy and high carrier mobility i.e. 1000 cm^2/Vs, hence BP shows the possibility for broadband optoelectronic applications [65–67]. Choi *et al.* [68] develop a technique to form a lateral homogeneous 2d MoS_2 p-n junction by partially stacking 2d h-BN as a mask to p-dope MoS_2. The fabricated lateral MoS_2 p-n junction with asymmetric electrodes of Pd and Cr/Au displayed a highly efficient photo response such as maximum external quantum efficiency of ~7000%, specific directivity of ~5 x10^{10} Jones, and light switching ratio of ~10^3. **Figure 3** shows the fabrication of the MoS_2 p-n diode. **Figure 3b** shows the optical microscopy image of MoS_2 p-n diode with hetero electrodes.

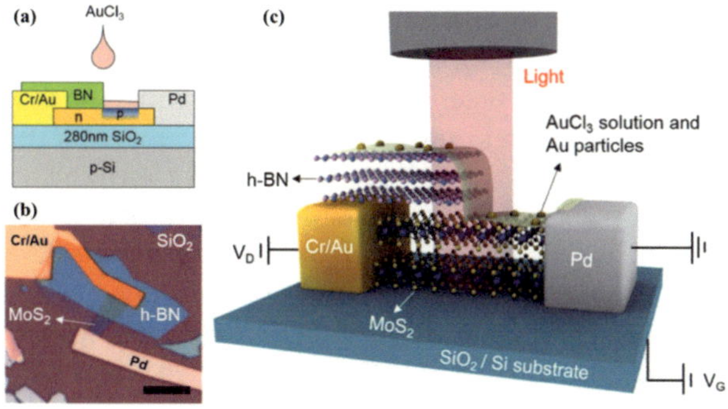

Figure 3.
Fabrication of the MoS_2 p-n diode. (a) Cross-sectional diagram and (b) optical microscopy image of MoS_2 p-n diode with hetero electrodes. The scale bar indicates 10 μm. (c) Three-dimensional schematic and circuit diagrams of the fabricated MoS_2 p-n diodes under light illumination. Reprinted with permission from [68]. Copyright (2014) American Chemical Society.

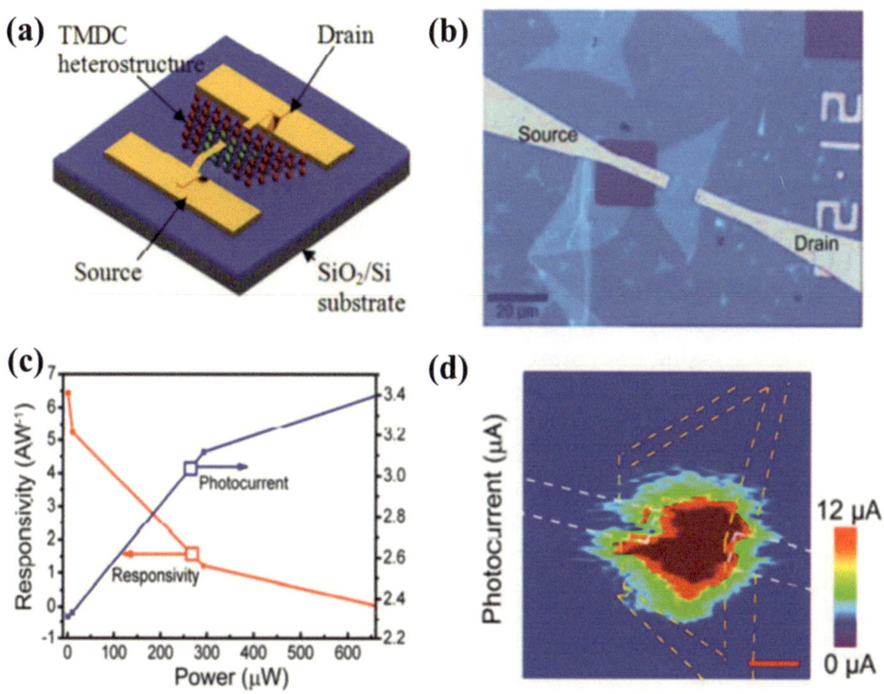

Figure 4.
(a) Schematic diagram of the photodiode based on the WSe_2/WS_2 heterojunction. (b) Optical image of the photodiode device. (c) Photocurrent and responsivity as a function of the incident power. d) the photocurrent image is taken from the device in (b). Reprinted with permission from [71]. Copyright (2017) American Chemical Society.

3.3 Photovoltaic based on 2d heterojunction

The p-n heterojunctions work as a basic backbone of various optoelectronic devices and applications due to various theoretical and experimental restrictions, there is the need for manufacturing designed heterostructures. Duan *et al.* [69] demonstrate the modulated MoS_2-$MoSe_2$ and WS_2-WSe_2 lateral heterostructures by thermal chemical vapor deposition (CVD) technique. The well-known n-type (WS_2) and p-type (WSe_2) builds natured heterojunction p-n diode with current rectification. When such heterojunction p-n diodes are characterized. The photovoltaic effect with a V_{oc} of ≈ 0.47 V and I_{sc} of ≈1.2 nA was established with laser illumination conditions (514 nm and 30nW). The internal quantum efficiency (IQE) and external quantum efficiency (EQE) were found to be 43% and 9.9% respectively The active regime was selected as lightly doped WS_2 and WS_2-WSe_2 interface region through photocurrent mapping, implies with the fact "A large fraction of the depletion layer is localized to the lightly doped WS_2 of the diode" [69]. Atomically sharp in-plane WSe_2/MoS_2 interface automatically comes into the picture with a high-magnification scanning transmission electron microscope (STEM), these force the photovoltaic effect of the intrinsic single layer p-n heterojunction was indorsed [70]. Li *et al.* [71] successfully fabricated a composition graded doped lateralWSe_2/WS_2 heterostructure by ambient pressure CVD in a single heat-cycle. The optoelectronic device (**Figure 4a-b**) based on the lateral WSe_2/WS_2 heterostructure shows improved photodetection performance in terms of a reasonable responsivity and a large photoactive area. The photocurrent and photo-responsivity are also depicted in **Figure 4c**. The photocurrent appears to increase nonlinearly, whereas the photoresponsivity decreases as the light power increases, with the highest obtained photoresponsivity of 6.5 $A.W^{-1}$. The reduction of the photoresponsivity at higher light powers may be ascribed to the reduction of unoccupied states in the conduction bands of WSe_2 and WS_2.

4. Perovskite 2d materials for photovoltaic cells

4.1 Transport layers in regular (n-i-p) photovoltaic

The layer-by-layer deposition technique is used to manufacture photovoltaic solar cell devices (PSCs). In these types of constructions, the order of charge selective layers in the manner can prosecute subdivide the devise configuration in two ways, regular PSCs (n-i-p) and inverted PSCs (p-i-n). The PSCs have two parts, (a) metal contact, (b) transparent conductive glass (TCO), while a slice of the observer has been arranged between hole transporting layer (HTL) and electron transporting layer (ETL). When the perovskite absorbs the light, an exciton i. e. the carriers are partitioned and moved towards the adequate layer, HTL, and ETL. Hence the charge carriers are shifted to the different electrodes. Moreover, ETL and HTL are performing two main roles, control the perovskite crystal growth, and extract and move the charge carriers. It is well versed that the hysteresis phenomenon is chiefly linked with the characteristics and interface of the charge selective layers to the perovskite [72–73]. Some of the remarkable features of ideal ETL and HTL materials are high transparency, high charge mobility, inherent stability, low-cost manufacturing, and appropriate energy alignment.

2,2,7,7-tetrakis (N,Npdimethoxyphenylamino)-9,9-spirobifluorene (Spiro-OMeTAD) and TiO_2 are the wall known materials that can be used as hole transport material (HTM) and electron transport material (ETM) respectively in the formation of n-i-p PSCs. On the other hand poly(3,4 ethylenedioxythiophene)–polystyrene sulfonate (PEDOT:PSS) and fullerene derivatives (e.g., 6,6-phenyl-C61-butyric acid methyl ester (PCBM)) has been taken as the HTM and ETM to manufactured the p-i-n PSCs [74–76]. Singh *et al.* [77] described the direct synthesis of MoS_2 (transparent, thin, and uniform) films on FTO- coated glass by the use of microwave irradiation and utilize this ETL in perovskite solar cells. TiO_2-coated FTO, SnO_2, MoS_2, and FTO-only substrates were prepared and their photovoltaic performances were checked and they have maximum PCEs with 17.15%, 15.80%, 13.14%, and 6.11% respectively [77]. When the article examination of the lifetime of charge carriers and charge recombination dynamics of ETL/Perovskite seems to be MoS_2 as shortest charge carrier lifetime as with other ETLs, which shows the charge extraction.

Yin *et al.* [78] reported the PCE of 17.37% by taking TiS_2 nanosheets (8–9 layers) suspended in isopropanol alcohol (IPA) in the perovskite solar cells. Uniquely, encapsulated perovskite solar cells with TiS_2 ETL demonstrate the highest stable under the conditions (UV irradiation 10 mW cm^{-2} and in ambient atmosphere for 50 h), with the result of 90% PCE. A double of SnO_2 and 2d TiS_2 synchronously, used for the PCE of 21.73% performed by Huang *et al.*, as every potential ETLs [79].

Applying the self-assembly stacking deposition method, the PCE of SnS_2 ETL based device, up to 20.12% has reported by Zhao *et al.* [80]. The Ti_3C_2 integration into a SnO_2 ETL for low-temperature planer perovskite solar cells by Yang and his co-workers by varying the loading of MXene from 0 to 2.5 wt% [81]. The device formed has the value of PCEs modified from the value of 17.23% to 18.34% at 1 wt% (increasing the loading of MXene). In the addition, the device fabricated through a SnO_2-$Ti_3C_2$is highly stable (with, RH-20%), and shows the PCE up to 80% of their initial performance after 700 h (identical with SnO_2-only devices) [81].

To increase the charge transfer efficiency, the perovskite crystal size, and lower the defect density, Guo *et al.* [82] explained $Ti_3C_2T_X$ MXene as an external material for the perovskite precursor solution. The results reveal that the incorporated device experimentally verified a PCE of 17.41% with short-circuit current (J_{SC}) of 22.26 mA cm^{-1} which is comparably high with the pristine perovskite device having the PCE of 15.54% and J_{SC} of 20.67 mA cm^{-1}.

Wang and his research group [83] experimentally verified that whenever a tiny amount of black phosphorus added to the $MAPbI_3$ starting solution (precursor), the photostability and efficiency of perovskite solar cells have been critically enhanced. The few-layer black phosphorus is proved to obtain ample perovskite to the size greater than 500 nm with the comparison bare $MAPbI_3$ film size (<400 nm). Taking the complex structure FTO/c-TiO_2/SnO_2/perovskite/Spiro-OMeTAD/Ag and $MAPbI_3$/BP for PSCs, the unique PCE of 20.65% was achieved, having less hysteresis and high reproducibility. Under the continuous white light-emitting diode (illumination of 100 mW cm^{-2} within the N_2 glove box) the $MAPbI_3$/BP-based PSCs show an excellent PCE limit of 94% (1000 h) [83–84].

The spiro-OMeTAD HTL and perovskite on active buffer layer (liquid phase exfoliated few-layer MoS_2 nanosheets) instead by Cappaso and co-workers and tried to solve the problem [85]. The above arrangement completes two necessary conditions i.e. prominent layer to ease the injection process and hole collection and performing like a barrier so that metal electrode migration can be curbed. The N-methyl-2-pyrrolidone solvent is very famous for the experimentalist to efficient MoS_2 [86–87]. On the other hand, some study proves this solvent not suitable for perovskite, to make it perovskite favorable solvent (IPA), a solvent exchange process has to be required.

Najaf*et al.* [88] developed "graphene interface engineering" strategy based on van der Waals MoS_2 QD/graphene hybrids that enable $MAPbI_3$-based PSCs to achieve a PCE up to 20.12% (average PCE of 18.8%).The van der Waals hybridization of MoS_2 quantum dots (QDs) with functionalized reduced graphene oxide (f-RGO), obtained by chemical silanization induced linkage between RGO and (3-mercaptopropyl)trimethoxysilane. This results in homogenize the deposition of the hole transport layer (HTL) or active buffer layer (ABL) onto the perovskite film since the two-dimensional nature of RGO effectively plugs the pinholes of theMoS_2 QD films. **Figure 5a** represents the sketch of mesoscopic $MAPbI_3$-based PSC exploiting MoS_2 QDs:f-RGO hybrids as both HTL and ABL. The normalized

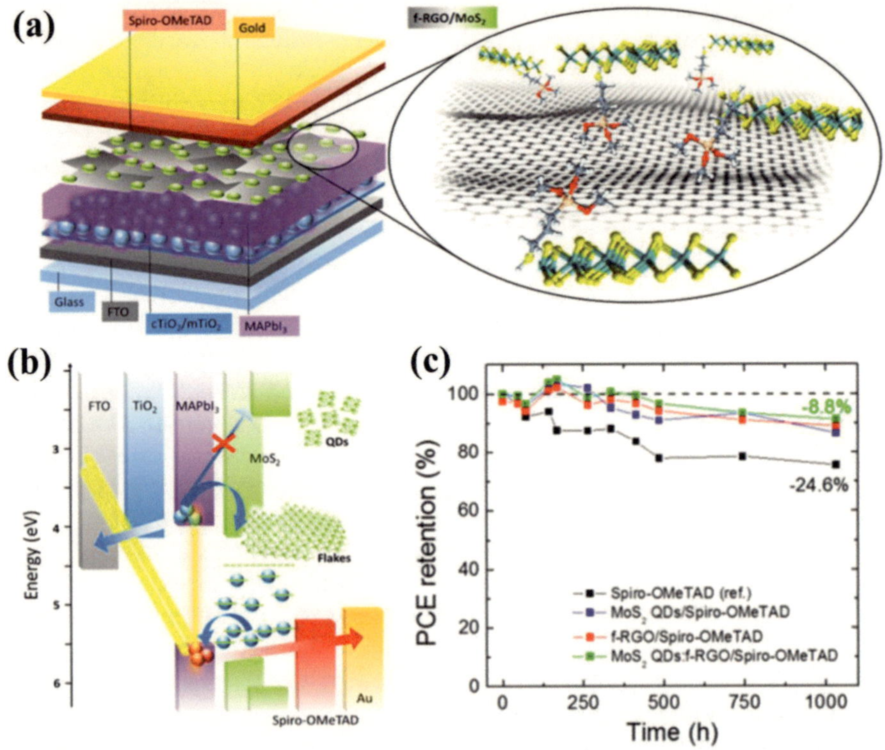

Figure 5.
(a) Sketch of mesoscopic $MAPbI_3$-based PSC exploiting MoS_2 QDs:f-RGO hybrids as both HTL and ABL. (b) Scheme of the energy band edge positions of the materials used in the different components of the assembled mesoscopic $MAPbI_3$-based PSC. (c) Normalized PCE trends vs. time extracted by I – V characteristics under 1 sun illumination. Reprinted with permission from [88]. Copyright (2018) American Chemical Society.

PCE trends vs. time extracted by I – V characteristics under 1 SUN illumination, periodically acquired during the shelf life test (ISOS-D-1), shows in **Figure 5c**.

4.2 Transport layers in inverted (p-i-n) photovoltaic

The organic solar cell is the key to fabricate the p-i-n PSC structures [89]. Huang *et al.* [90] successfully showed that with coverage optimization, a planar p-i-n^{++} device with a PCE of over 11% was achieved. This also suggests that the ETL may not be necessary for an efficient device as long as the perovskite coverage is approaching 100%. **Figure 6a-b** presents the device architecture of the perovskite solar cells with (a) and without (b) a TiO_2 ETL. **Figure 6c** shows the current density-voltage curves of two typical $CH_3NH_3PbI_{3-x}Cl_x$-based solar cells grown on FTO substrates with and without UVO treatment under simulated AM1.5G solar irradiation (100 mW/cm^2). Jeng and co-workers reported that perovskites have the ability to transport the holes [91]. To achieve a PCE of 15.1%, Hu *et al.* proposed a surface-modification technique in which the ITO surface/optimize energy level by using the cesium salt solution [92].

Liu and his co-workers reported layer free PSC with an efficiency of 13.5% to obtain this, perovskite layer directly put together with the ITO surface by using a sequential layer deposition method. The above arrangement proves that to enhance device efficiency ETL is not required always [93]. Ke *et al.* [94] manufactured ETL free PSC with efficiency 14.14% and V_{oc} 1.4 V deposited directly on FTO substrate (via a one-step solution process) in which no hole blocking layers

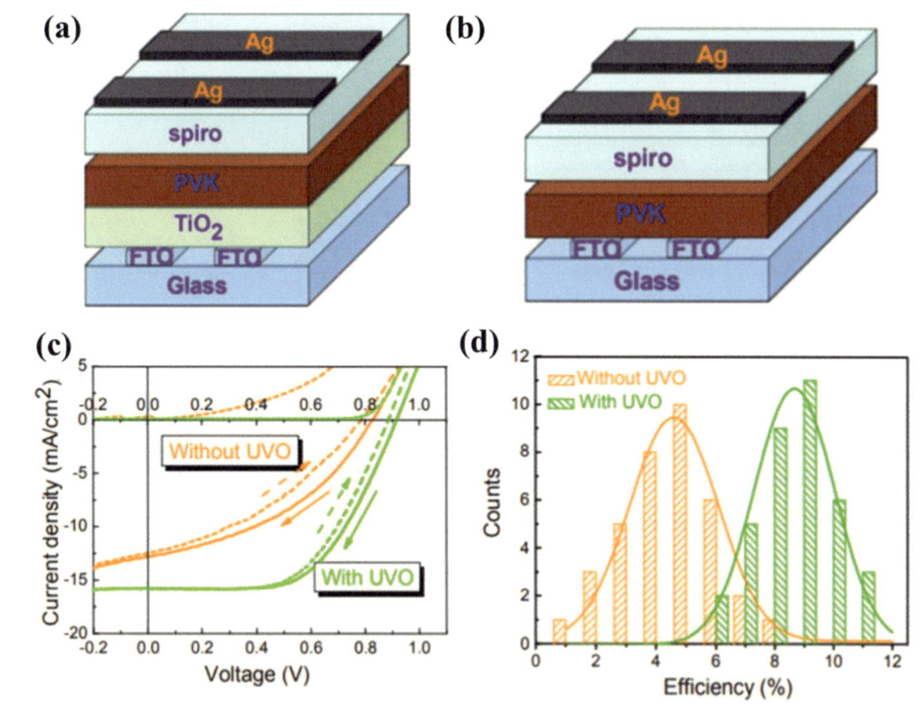

Figure 6.
Device architecture of the perovskite solar cells (a) with TiO_2 ETL and (b) without TiO_2 ETL (c) J-V curves of $CH_3NH_3PbI_{3-x}Cl_x$-based solar cells with and without UVO treatment (d) histograms of their corresponding PCE values measured for 36 separate devices (c). Reprinted with permission from [90]. Copyright (2016) American Chemical Society.

are required. He further described that TiO_2 (electron-transporting material) is not a perfect interfacial material. It is also described Liu *et al.* [93], Chen *et al.* [95], and Prochowicz *et al.* [96] that the efficiency of compact layer-free devices can be higher when various film-forming techniques are to be used.

Various quantities of black phosphorus quantum dots (BPQDs) mixed with $MAPbI_3$ precursor solution to form p-i-n inverted devices [97]. These BPQDs based perovskite films revealed less non-reactive defects, larger grain size, and higher crystallinity, with a comparison of no BPQDs perovskite films. Further, it is clear from some experimental facts that BPQDs also work as heterogeneous nucleation sites. This leads to the growth of perovskite crystals with homogeneity. The PCE of 20% was obtained for additive-assisted perovskite film. Adding the BPQDs on the lower surface of $MAPbI_3$-the enhanced crystalline of $MAPbI_3$-BPQDs film has been achieved.

4.3 2d Ruddlesden-popper perovskite

The hybrid organic–inorganic halide perovskite (OIHPs) are very unsuitable in the industrial fabrication of solar device as they lose the stability on heating, moisture, and light. These major issues have been resolved by S. N. Ruddlesden and P. Popper to fabricate the perfect candidates known as Ruddlesden-Popper perovskite (RPPS). This is the mixture of 2D/3D materials and can be used in LEDs as it has an intense photoluminescence feature [98]. The 2D Ruddlesden-Popper (2DRP) perovskites are the topic of great interest and research because highly stable PSCs can be fabricated by them [99]. The general form of 2DRP is $(A^{'})_2\ (A)_{n-1}B_n\ X_{3n+1}$, where $A^{'}$ indicates $R\text{-}NH_3$ or $H_3N\text{-}R$ (R an aromatic ligand or large aliphatic alkyl chain) and works as an insulating layer to partitioned the various inorganic layers. A describes small cation

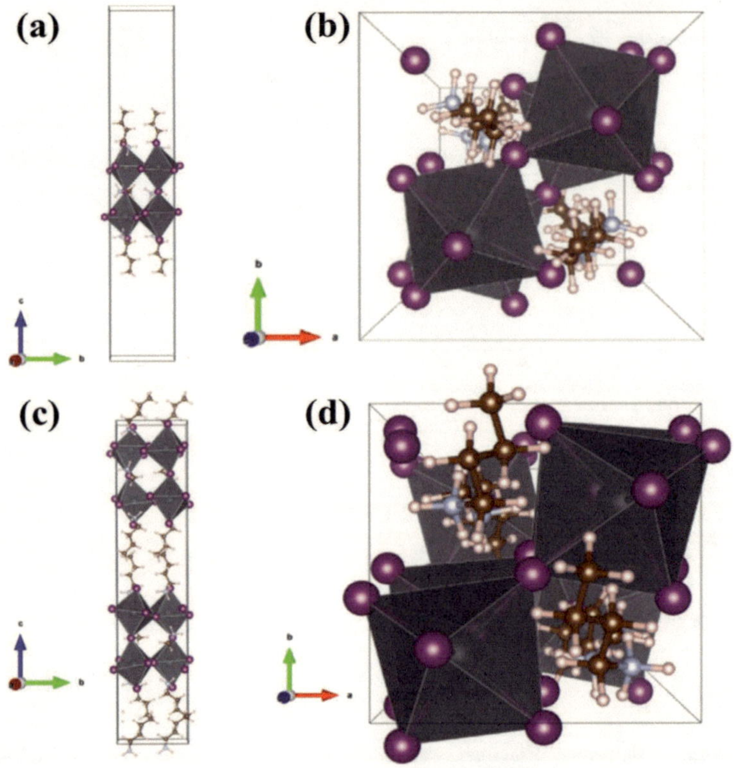

Figure 7.
(a) Lateral view and (b) top view of the n = 2 sheet (NS-RPP) optimized structure. Same (c, d) for the bulk QW (QW-RPP). [gray: Pb; purple: I; Brown: C; light blue: N; white: H atoms]. Reprinted with permission from [98]. Copyright (2018) American Chemical Society.

such as $CH_3NH_3^+$ and Cs^+. B represents Pb^{2+} and Sn^{2+}, divalent metal cation, while X describes the halides. Various values of small n provide us strict 2D structure (n = 1), quasi-2D structure (n = 2–5), conventional 3D structure (n = ∞), and represents the number of metal halide, monolayer sheets [100]. These 2DRP excellently perform thermal stability, humidity stability, and structure stability [101–106]. Giorgi *et al.* [98] showed a lateral and top view of the nanosheets Ruddlesden–Popper organic–inorganic halide perovskites (NS-RPPs) optimized structure in **Figure 7a-b**. Also, lateral and top view of the quantum-well Ruddlesden–Popper organic–inorganic halide perovskites (QW-RPPs) structures in **Figure 7c-d**. The solution base synthesis, colloidal base method, liquid, and vapor-based epitaxy, exfoliation method, and single crystal growth are the well-known growing technique to fabricate 2DRP perovskites [107]. Niu *et al.* [108] prepared mono and few layers $(C_6H_9C_2H_4NH_3)_2$ PbI_4 flakes via the method of micromechanical exfoliation.

5. Conversion treatment for photovoltaic cells

5.1 Downconversion of perovskite photovoltaic cell

In the previous investigations, various techniques and methods have been adopted to improve the efficiency of solar cells, higher by the Shockley and Queisser limit (32%). The phenomenon of splitting, low energy photons by high energy photons (single) is known as downconversion (quantum cutting). An ample work has been done on downconversion for photovoltaic devices. This was performed with lanthanide ions because of its excellent optical properties. Various experiments also proved the use of nanomaterials as down converters. If we chose the lanthanide ions and design of the solar cell, there will be a maximum benefit to using the down-conversion materials. The host materials should possess properties like low scattering, absorption strength, thermal and chemical strength, high transmittance, photo-stability, and excitation energy [109–110]. The necessary conditions to pick the lanthanide ion are good electrical and chemical stability and high emission lifetime. Downconversion Tb^{3+}-Yb^{3+} has also been demonstrated in $GdAl_3(BO_3)_4$ [111], $GdBO_3$ [112], Y_2O_3 [113], CaF_2 nanocrystals [114] and lanthanum borogermanate glass [115]. Tsai *et al.* [116] explored the graphene quantum dots (GQDs) as a down conversion in then-type Si heterojunction (SHJ) solar cells.

The photographic image and the cross-sectional schematic of the SHJ solar cell are shown in **Figure 8a-b**, respectively. **Figure 8c-d** shows the low-magnification and high-magnification top-view SEM images of the micro pyramids of SHJ solar cells. The device with 0.3 wt % of GQDs shows the highest short-circuit current (J_{SC}) and fill factor (FF) of 37.47 mA/cm^2 and 72.51%, respectively, which leads to the highest PCE of 16.55% (**Figure 8f**). The external quantum efficiency (EQE) of the devices with 0.3 wt % and without GQDs and the EQE enhancement are shown in **Figure 8g**. The efficiency enhancement is due to the photon down conversion phenomenon of GQDs to make more photons absorbed in the depletion region for effective carrier separation, leading to the enhanced photovoltaic effect. Various down conversion materials were described and synthesized to enhance UV stability and UV photon harvesting [117–118].

5.2 Upconversion of perovskite photovoltaic cell

The conversion (nonlinear optical process) in which minimum two low energy photons, present in the near-infrared region into high energy photon with the visible region known as upconversion [119–120]. The upconversion materials contain large bandgap, seem to be most favorable for solar cell applications. Various uses of upconversion materials are optical data storage, medical therapy, display technology, light-harvesting, temperature sensors, and solid-state lighting. Trupke *et al.* [121] theoretically investigated that if a perfect upconvertor is used with conventional single-junction bifacial solar cells (bandgap 2 eV), we can obtain PCE of 47.6% (non-concentrated sunlight) and 63.2% concentrated sunlight. Lanthanide based upconverters and organic upconverters are very common to improve the efficiency of photovoltaic devices. The elements lanthanum to lutetium are used as the upconverters. In addition, enhancement in the photocurrent has been achieved by the use of two commercial upconverters on both sides of Si solar cells (Pan *et al.*) [122].

The nano precursor upconversion materials Er^{3+}/Yb^{3+} co-doped with TiO_2 and LaF_3 have been explored by Shan *et al.* [123]. Shang *et al.* [124] also explored the various techniques to enhance efficiency via upconversion. The utilization of light beyond the visible region is not possible by PSCs ($CH_3NH_3PBI_3$), due to their intrinsic bandgap. The upconversion is a specific way, to harvest this regime and convert it in the visible regime, so that the PSCs IR response can be increased. Chen and co-workers reported that the efficiency will be increased if $LiYF_4$: Yb^{3+}, Er^{3+} single-crystal attached in the front part of PSCs [125]. Taking nano prisms NaYF4: Yb^{3+}, Er^{3+}, which is hydrothermally formed to the TiO_2 layer in PSCs, the efficiency enhancement has been demonstrated by Roh *et al.* [126]. In another study, Wang *et al.* introduced efficiency increment via hydrothermally grown 3% Er^{3+} and 6% Yb^{3+} co-doped TiO_2 nanorod in PSCs [127]. In 2016, He *et al.* integrated $NaYF_4$:Yb^{3+}, Er^{3+} nanoparticles as mesoporous electrode with PSCs (CH_3NH3PbI_3 based), this leads a short circuit current density 0.74 mAcm^{-2} excite with 980 nm laser with

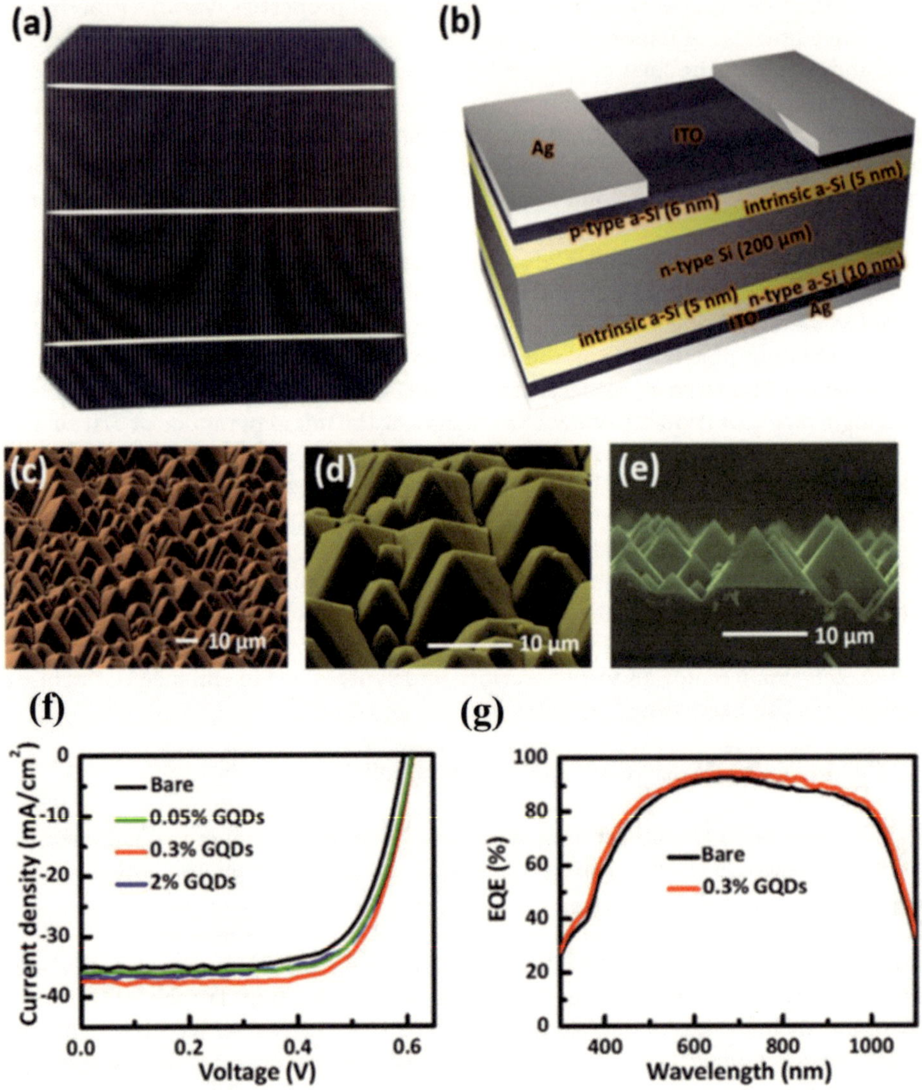

Figure 8.
(a) Photographic image, (b) schematic, (c) low-resolution, (d) high-resolution, and (e) cross-sectional SEM images of SHJ solar cells. (f) J – V characteristics of SHJ solar cells with various concentrations of GQDs. (g) EQE spectra of SHJ solar cells without and with 0.3 wt % of GQDs. Reprinted with permission from [116]. Copyright (2016) American Chemical Society.

28 W cm^{-2} [128]. The sample (semiconductor plasmon-sensitized nanocomposites i.e. $mCu_{2-x}S@SiO_2@\ Er_2O_3$), which changes the broadband infrared light to visible light by the localized surface plasmon resonance (LSPR) of $Cu_{2-x}S$ integrated with TiO_2 paste [129]. This arrangement is used in PSCs as an electron extraction layer with an enhanced response of 800–1000 nm. With the dependency of the power density of the 980 nm extraction layer (>0.85 $mAcm^{-2}$), the short circuit current density was noticed to rise 45 Wcm^{-2}.

6. Recent developments on 2d perovskite photovoltaic cells

The 2d layered perovskites are more advanced and useful than their 3d structures. The unique electroluminescent property of 2d perovskites makes them more superior to their 3d counterparts. 2d perovskites structures have large exciton binding energy than 3d structures [130–131]. Due to this fact, 2d perovskite has a high photoluminescence quantum yield (PLQY) with enhanced radiative recombination. The appearance of cascaded energy structures in 2d perovskites films (mixed n layer thickness) leads to a fast and efficient energy transfer from lower-n quantum wells to higher-n quantum wells. These results in the decreased exciton quenching effect: occurring the enhanced van der Walls interactions in organic molecules and hydrophobic organic ligands, 2d perovskites show enhanced and ambient and thermal stability in the comparison with 3d perovskites. In 2d perovskites, the electrical and optical properties can be tuned more and advanced applications like circular-polarized emission and broadband emission due to their excellent chemical tenability [132–135].

You *et al.* [136] developed a novel annealing approach formethylammonium lead iodide ($MAPbI_3$) and $(CsPbI_3)_{0.05}(FAPbI_3)_{0.95}(MAPbBr_3)_{0.05}$ mixed perovskite

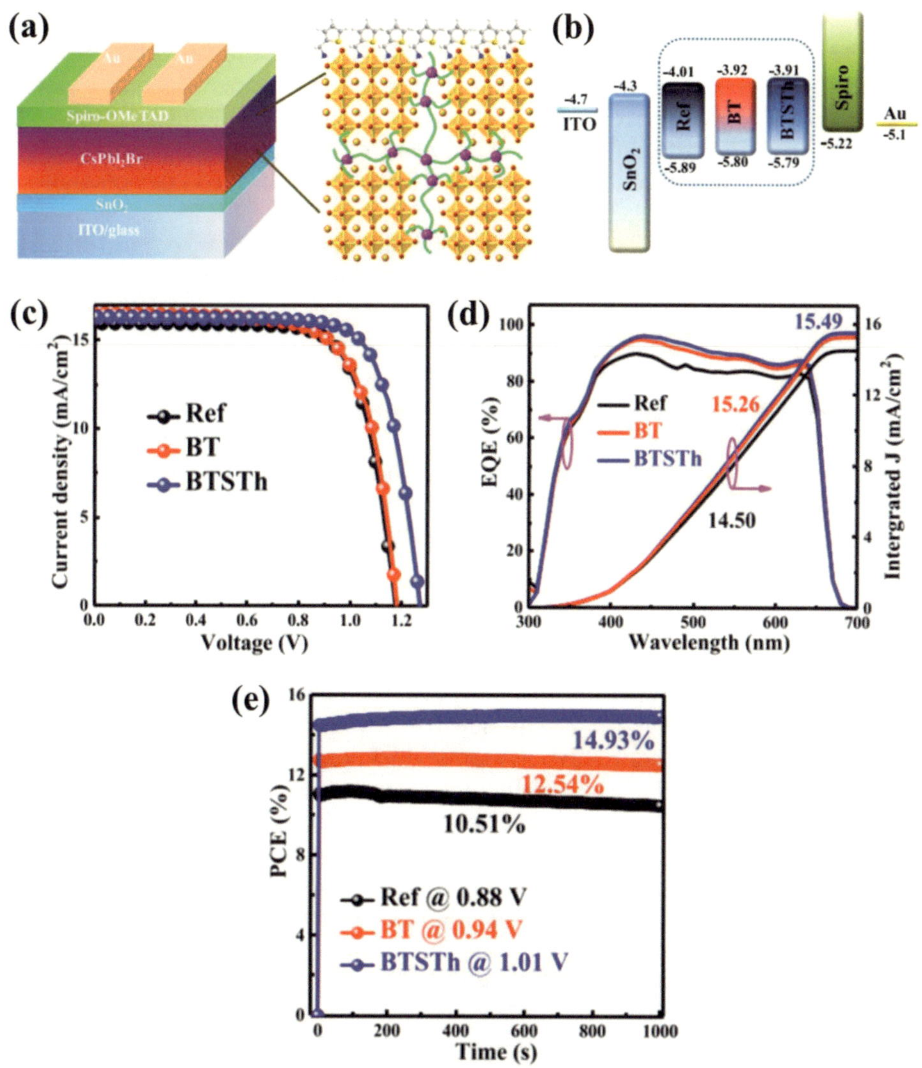

Figure 9.
(a) Schematic illustration of the device and the dual-protection $CsPbI_2Br$ film. (b) Band alignment between various layers of a complete device. (c) the J – V curves of the devices fabricated from the ref., BT, and BTSTh films. (d) the EQE spectra of the various devices. (e) the stabilized maximum power output of the devices. Reprinted with permission from [140]. Copyright (2020) American Chemical Society.

films by fast laser beam scanning. Under optimum conditions, high-quality perovskite films with good crystallinity, preferred orientation, and low density of defects device offers PCE nearly about 20%. Wang *et al.* [137] showed an efficient strategy to tune the band structure and electron mobility of the ETL by adding NH_4Cl to the sol–gel-derived ZnO precursor. Low temperature (160°C) fabricated $CsPbIBr_2$solar cells recorded high efficiency of 10.16%. Li *et al.* [138] fabricated the $MAPbI_3$perovskite solar cells and showed the device performance is strongly influenced by the TiO_2 electron transport layer. Oz *et al.* [139] studied the effect of lead(II)propionate additive on the stabilization of $CsPbI_2Br$all-inorganic perovskite, and the use of a novel dopant-freepolymer hole transport material (synthesized by us) for photovoltaic performance assessment of $CsPbI_2Br$ solar cells.

Fu and his co-workers report a dual-protection strategy via incorporating monomer trimethylolpropane triacrylate (TMTA) into$CsPbI_2Br$ perovskite bulk and capping the surface with 2-thiophenemethylammonium iodide (Th – NI) [140]. The fabricated devices show a greatly improved efficiency from 12.17 to 15.58% with an opening circuit voltage (V_{oc}) of 1.286 V. **Figure 9a** presents a schematic illustration of the device and the dual-protection $CsPbI_2Br$ film. The UPS measurements are conducted to reveal the electronic structure changes the calculated results are drawn in **Figure 9b**. The photocurrent density-voltage (J-V) curves of the optimal devices under AM 1.5 illumination are presented in **Figure 9c.** The ref. device shows the best efficiency of 12.17% with a V_{oc} of 1.151 V, and TMTA doped film (BT) devices show an improved efficiency of 13.88% after incorporating 1 mg/mL of TMTA. In **Figure 9e**, the Th – NI modified BT film (BTSTh) device exhibits the improved output efficiency with 14.93% for 1000 s, while the ref. shows the attenuated efficiency and remains 10.51% under the same operation, which indicates the better operational stability for the BTSTh devices.

Li *et al.* [141] reported vertically aligned 2D/3D Pb – Sn perovskites with enhanced charge extraction and suppressed phase segregation for efficient printable solar cells. Wang *et al.* [142] successfully fabricated high-quality $CsPbBr_3$ films via additive engineering with NH_4SCN. The incorporation of NH_4^+ and pseudo-halide ion SCN^- into the precursor solution, a smooth and dense $CsPbBr_3$ film with good crystallinity and low trap state density can be obtained.

7. Neoteric challenges in development of 2d photovoltaic cells

It is crucial that the entire energy of any absorbed photon is harvested for next-generation photovoltaics. The unique features of 2d layered materials such as high crystalline quality, transparency, atomic thickness, make them special to use in photovoltaic solar cells. Thus, it is important to synthesized and demonstrate 2d flexible photovoltaic devices on industrial scale. While some different issues and problems will be faced by the experimentalist regarding an efficiency performance. The specific requirements in the fabrication of photovoltaic devices are the large-area synthesis, highly controllable, low cost, atomically thin, and recyclable fabrication of materials and their devices. The above features are specified not only for all potential electronic devices but also for optoelectronic devices. Increased number of layers of 2d materials, the conductivity of the material improves at the cost of reduced transparency. The application of 2dmaterials like TMDs and perovskites in photovoltaic devices has also been investigated over the last few years. The advantages of using such materials for solar cells have been explored based on the high absorption coefficient of these materials in the visible to the near-infrared part of the solar spectrum. So there is a need for more investigation of the heterostructure based on these materials, which can synergize the performance of the device. Although the efficiency of perovskite solar cells has been boosted to over 25%, further improving the efficiency towards their theoretical Shockley–Queisser efficiency limit of more than 30% and improving the stability towards commercial application deserve more intensive research. The micromechanical cleavage method provides us structural study and device performance but this cannot be used as industry level as low production rate and low yield. This lack noted by the scientific community and remarkable advancement has been achieved by fabricating 2d materials at the industry level within the past few years. Moreover, TMDs, hBN, graphene, and perovskite are critically fabricated with different methods like physical vapor transport, CVD, layer by layer conversion; etc. The samples obtained in this way have properties like controllable thickness, electronic properties scalable sizes, and high crystal quality. Also, a liquid-based wet chemical method provides two unique requirements high control power and low-cost manufacturing of atomically thin 2D nanomaterials. Moreover, it is necessary to achieve excellent crystal quality and film uniformity before using the LBWC technique in large scale manufacturing.

8. Conclusion

In this chapter, we have enlightened specific properties and synthesis techniques of graphene, TMDs, and perovskite. We have described recent progress made with graphene, graphene-based 2D materials for solar photovoltaics. In addition, 2d Schottky junction, homojunction, and heterojunction are explained briefly. The unique account of the charge carrier transport layer as ETL and HTL has been done fairly. Moreover, regular n-i-p and inverted p-i-n structure are the key features. Furthermore, 2DRP perovskite, upconversion/downconversion of perovskite cells are also discussed briefly. In the last of the section very recent developments on 2d perovskite photovoltaic cells have been critically explained with the facts. Various challenges in the fabrication and development of today's devices are pointed out. Therefore, the outlook towards 2d materials should be optimistic and needs more attention in the near future.

Author details

Manoj Kumar Singh[1]*, Pratik V. Shinde[2], Pratap Singh[1] and Pawan Kumar Tyagi[3]

1 Department of Physics under School of Engineering and Technology (SOET), Central University of Haryana (CUH), Mahendergarh, Haryana, India

2 Centre for Nano and Material Sciences, Jain University, Bangalore, India

3 Department of Applied Physics, Delhi Technological University, Delhi, India

*Address all correspondence to: manojksingh@cuh.ac.in

References

[1] Bequerel ER. C.R.Acad.Sci. 1839;9 145-149.

[2] Park NG. Research Direction toward Scalable, Stable, and High Efficiency Perovskite Solar Cells. Advanced Energy Materials 2020;10 1903106(1-14). DIO: 10.1002/aenm.201903106.

[3] Solanki CS and Beaucarne G. Advanced solar cell concepts. Energy for Sustainable Development 2007;11 17-23. DOI: 10.1016/S0973-0826(08)60573-6.

[4] Wang L, Huang L, Wee Tan WC, Feng X, Chen L, Huang X, et al. 2D Photovoltaic Devices: Progress and Prospects. Small method 2018;2 1700294 (1-20). DOI: 10.1002/smtd.201700294

[5] Edvinsson T. Optical quantum confinement and photocatalytic properties in two-, one- and zero-dimensional nanostructures. Royal Society Open Science 2018;5 180387-180403. DOI: 10.1098/rsos.180387.

[6] Butler SZ, Hollen SM, Cao L, Cui Y, Gupta JA, Gutierrez HR, et al. Progress, Challenges, and Opportunities in Two-Dimensional Materials Beyond Graphene. ACS Nano 2013;7 2898-2926. DOI: 10.1021/nn400280c.

[7] Gupta A, Sakthivel T, Seal S. Recent development in 2D materials beyond graphene. Progress in Materials Science 2015;73 44-126. DOI: 10.1016/j.pmatsci.2015.02.002.

[8] Chen L, Zhang C, Liu. The Optimal Electronic Structure for High-Mobility 2D Semiconductors: Exceptionally High Hole Mobility in 2D Antimony. Journal of American Chemical Society 2019;141 16296-16302. DOI: 10.1021/jacs.9b05923.

[9] Song H, Liu J, Liu B, Wu J, Cheng HM, Kang F. Two-Dimensional Materials for Thermal Management Applications. Joule 2018;2 442-463. DOI: 10.1016/j.joule.2018.01.006.

[10] Zhang H, Ultrathin Two-Dimensional Nanomaterials, ACS Nano 2015;9 9451-9469. DOI: 10.1021/acsnano.5b05040.

[11] Guo Y. Surface chemical-modification for engineering the intrinsic physical properties of inorganic two-dimensional nanomaterials. Chemical Society Review 2015;44 637-646. DOI: 10.1039/C4CS00302K.

[12] Zhang Y, Tan YW, Stormer HL, Kim P. Experimental observation of the quantum Hall effect and Berry's phase in graphene. Nature Letters 2005;438 201-204. DOI: 10.1038/nature04235.

[13] Novoselov KS, Jiang Z, Zhang Y, Morozov SV, Stormer HL, Zeitler U et al. Room-Temperature Quantum Hall Effect in Graphene. Science 2007;315 1379. DOI: 10.1126/science.1137201.

[14] Stoller MD, Park S, Zhu Y, An J, Ruoff RS. Graphene-Based Ultracapacitors 2008;8 3498-3502. DOI: 10.1021/nl802558y.

[15] Pomerantseva E, Gogotsi Y. Two-dimensional heterostructures for energy storage. Nature Energy 2017;2(7) 1-6. DOI: 10.1038/nenergy.2017.89.

[16] Tucek J, Btonski P, Ugolotti P, Swain AK, Enokib T, Zboril R. Emerging chemical strategies for imprinting magnetism in graphene and related 2D materials for spintronic and biomedical applications. Chemical Society Reviews 2018;47 3899-3990. DOI: 10.1039/C7CS00288B.

[17] Georgakilas V, Perman JA, Tucek J, Zboril R. Broad family of carbon nanoallotropes: classification, chemistry, and applications of

fullerenes, carbon dots, nanotubes, graphene, nanodiamonds, and combined superstructures. Chemical Reviews 2015;115 4744-4822. DOI: 10.1021/cr500304f.

[18] Novoselov KS, Geim AK, Morozov SV, Jiang D, Zhang D, Dubonos SV et al. Electric Field Effect in Atomically Thin Carbon Films. Science 2004;306(5696) 666-669. DOI: 10.1126/science.1102896.

[19] Allen MJ, Tung VC, Kaner RB. Honeycomb Carbon: A Review of Graphene. Chemical Reviews 2010;110 132-145. DOI: 10.1021/cr900070d.

[20] Mahmoodi T, Wang Y, Hahn YB. Graphene and its derivatives for solar cells application. Nano Energy 2018;47 51-65. DOI: 10.1016/j.nanoen.2018.02.047.

[21] Akinwande D, Brennan CJ, Bunch JS, Egberts P, Felts JR, Gao H et al. A review on mechanics and mechanical properties of 2D materials-Graphene and beyond. Extreme Mechanics Letters 2017;13 42-77. DOI: 10.1016/j.eml.2017.01.008.

[22] Yu1 X, Cheng H, Zhang M, Zhao Y, Qu L. Graphene-based smart materials, Nature Reviews Materials 2017;2 1-13. DOI: 10.1038/natrevmats.2017.46.

[23] Das S, Pandey D, Thomas J, Roy. The Role of Graphene and Other 2D Materials in Solar Photovoltaics. Advanced Materials 2018;31(1) 1802722(1-35). DOI: 10.1002/adma.201802722.

[24] Cao M, Xiong DB, Yang L, Li S, Xie Y, Guo Q et al. Ultrahigh Electrical Conductivity of Graphene Embedded in Metals, Advanced Functional Materials 2019;29(17) 1806792(1-8). DOI: 10.1002/adfm.201806792.

[25] Ma Y, Zhi L. Graphene-Based Transparent Conductive Films: Material Systems, Preparation and Applications. Small Methods 2018;3(1) 1800199(1-32). DOI: 10.1002/smtd.201800199.

[26] Liu N, Chortos A, Lei T, Jin L, KimTR, Bae WG et al., Ultratransparent and stretchable graphene electrodes, Science Advances 2017;3 1700159(1-10). DOI: 10.1126/sciadv.1700159.

[27] Adetayo A, Runsewe D. Synthesis and Fabrication of Graphene and Graphene Oxide: A Review. Open Journal of Composite Materials 2019;9 207-229. DOI: 10.4236/ojcm.2019.92012.

[28] Das S, Sudhagar P, Kang YS, Choi W. Synthesis and Characterization of Graphene. In: Lu W Baek JB, Dai L (ed.) Carbon Nanomaterials for Advanced Energy Systems: Advances in Materials Synthesis and Device Applications. John Wiley & Sons; 2015. p1-2.

[29] Somani PR, Somani SP, Umeno M. Planer nano-graphenes from camphor by CVD. Chemical Physics Letters 2006;430 56-59. DOI: 10.1016/j.cplett.2006.06.081.

[30] Obraztsov AN, Obraztsova EA, Tyurnina AV, Zolotukhin AA. Chemical vapor deposition of thin graphite films of nanometer thickness. Carbon 2007;45 2017-2021. DOI: 10.1016/j.carbon.2007.05.028.

[31] Li X, Cai1 W, An1 J, Kim, Nah J, Yang D et al. Large-Area Synthesis of High-Quality and Uniform Graphene Films on Copper Foils. Science 2009;324 1312-1314. DOI: 10.1126/science.1171245.

[32] Cheng R, Jiang S, Chen Y, Liu Y, Weiss, Cheng HC et al. Few-layer molybdenum disulfide transistors and circuits for high-speed flexible electronics. Nature Communications 2014;5 5143(1-9). DOI: 10.1038/ncomms6143.

[33] Akinwande D, Petrone N, Hone J. Two-dimensional flexible

nanoelectronics. Nature Communications 2014;5 5678(1-12). DOI: 10.1038/ncomms6678.

[34] Choi W, Choudhary N, Han GH, Park J, Akinwande D, Lee YH. Recent development of two-dimensional transition metal dichalcogenides and their applications. Materialstoday 2017;20(3) 116-130. DOI: 10.1016/j.mattod.2016.10.002.

[35] Li C, Cao Q, Wang F, Xiao Y, Li Y, Delaunay JJ et al. Engineering graphene and TMDs based van der Waals heterostructures for photovoltaic and photoelectrochemical solar energy conversion. Chemical Society Reviews 2018;47 4981-5037. DOI: 10.1039/C8CS00067K.

[36] Wang QH, Zadeh KK, Kis A, Coleman JN, Strano MS. Electronics and optoelectronics of two-dimensional transition metal dichalcogenides. Nature nanotechnology 2012;7 699-712. DOI: 10.1038/nnano.2012.193

[37] Mak KF, Lee C, Hone J, Shan J, Heinz TF. Atomically thin MoS2: a new direct-gap semiconductor. Physical Review Letters 2010;105 136805(1-4). DOI: 10.1103/PhysRevLett.105.136805.

[38] Mak KF, He K, Lee C, Lee GH, Hone J, Heinz TF et al. Tightly bound trions in monolayer MoS_2. Nature Materials 2013;12 207-211. DOI: 10.1038/nmat3505.

[39] Cao T, Wang G, Han W, Ye H, Zhu C, Shi J et al. Valley-selective circular dichroism of monolayer molybdenum disulphide. Nature Communications 2012;3 1-5. DOI: 10.1038/ncomms1882.

[40] Wilson JA, Yoffe AD. The transition metal dichalcogenides discussion and interpretation of the observed optical, electrical and structural properties. Advances in Physics 1969;18(73) 193-335. DOI: 10.1080/00018736900101307.

[41] Velicky M, Toth PS. From two-dimensional materials to their heterostructures: An electrochemist's perspective. Applied Materials Today 2017;8 68-103. DOI: 10.1016/j.apmt.2017.05.003.

[42] Simone Bertolazzi, Jacopo Brivio, Kis A. Stretching and Breaking of Ultrathin MoS_2. ACS Nano 2011; 5 9703-9709. DOI: 10.1021/nn203879f.

[43] Iqbal MZ, Alam S, Faisal MM, Khan S. Recent advancement in the performance of solar cells by incorporating transition metal dichalcogenides as counter electrode and photoabsorber. International Journal of Energy Research 2019; 43 3058-3079. DOI: 10.1002/er.4375.

[44] Ansari SA, Fouad H, Ansari SG, Sk MP, Cho MH. Mechanically exfoliated MoS_2 sheet coupled with conductive polyaniline as a superior supercapacitor electrode material. Journal of colloid and interface science 2017;504 276-282. DOI: 10.1016/j.jcis.2017.05.064.

[45] Ding W, Hu L, Dai J, Tang X, Wei R, Sheng Z, et al. Highly ambient-stable 1T-MoS_2 and 1T-WS_2 by hydrothermal synthesis under high magnetic fields. ACS nano 2019;13(2) 1694-1702. DOI: 10.1021/acsnano.8b07744.

[46] Chae WH, Cain JD, Hanson ED, Murthy AA, Dravid VP. Substrate-induced strain and charge doping in CVD-grown monolayer MoS_2. Applied Physics Letters 2017;111(14) 143106(1-5). DOI: 10.1063/1.4998284.

[47] Zhan L, Wan W, Zhu Z, Shih TM, Cai W. MoS_2 materials synthesized on SiO_2/Si substrates via MBE. Journal of Physics: Conference Series 2017;864(1) 012037(1-4). DOI: 10.1088/1742-6596/864/1/012037.

[48] Yang J, Liu L. Trickle Flow Aided Atomic Layer Deposition (ALD)

Strategy for Ultrathin Molybdenum Disulfide (MoS_2) Synthesis. ACS applied materials & interfaces 2019;11(39) 36270. DOI: 10.1021/acsami.9b12023.

[49] Xiao M, Huang F, W Huang, Dkhissi Y, Zhu Y, Etheridge J, et al. A Fast Deposition Crystallization Procedure for Highly Efficient Lead Iodide Perovskite Thin Film Solar Cells. Angewandte Chemie 2014; 126(37)10056-10061. DOI: 10.1002/ange.201405334.

[50] Cai B, Xing Y, Yang Z, Zhang WH, Qiu J, High performance hybrid solar cells sensitized by organolead halide perovskites. Energy Environmental Science 2013;**6** 1480-1485. DOI: 10.1039/C3EE40343B.

[51] Lan C, Zhou Z, Wei R, Ho JC, Two-dimensional perovskite materials: From synthesis to energy-related applications. Materials Today Energy 2019;11 61-82. DOI: 10.1016/j.mtener.2018.10.008.

[52] Kulkarni SA, Mhaisalkar SG, Mathews N, Boix PP. Perovskite Nanoparticles: Synthesis, Properties, and Novel Applications in Photovoltaics and LEDs. Small Methods 2019;3 1800231(1-16). DOI: 10.1002/smtd.201800231.

[53] Ha ST, Liu X, Zhang Q, Giovanni D, Sum TC, Xiong Q. Synthesis of Organic–Inorganic Lead Halide Perovskite Nanoplatelets: Towards High-Performance Perovskite Solar Cells and Optoelectronic Devices. Advanced Optical Materials 2014;2 838-844. DOI: 10.1002/adom.201400106.

[54] Utama MIB, Mata M, Magen C, Arbiol J, Xiong Q. Twinning-, Polytypism-, and Polarity-Induced Morphological Modulation in Nonplanar Nanostructures with van de Waals Epitaxy. Advanced Functional Materials 2013;23 1636-1646. DOI: 10.1002/adfm.201202027.

[55] Utama MIB, Peng ZP, Chen R, Peng B, Xu XL, Dong YJ, et al. Vertically Aligned Cadmium Chalcogenide Nanowire Arrays on Muscovite Mica: A Demonstration of Epitaxial Growth Strategy. Nano Letters 2011;11 3051-3057. DOI: 10.1021/nl1034495.

[56] Gillet M, Delamare R, Gillet E. Growth of epitaxial tungsten oxide nanorods. Journal of Crystal Growth 2005;279 93-99. DOI: 10.1016/j.jcrysgro.2005.01.089.

[57] Tan C, Cao X, Wu XJ, He Q, Yang J, Zhang, et al. Recent advances in ultrathin two-dimensional nanomaterials. Chemical Reviews 2017;117(9) 6225-6331. DOI: 10.1021/acs.chemrev.6b00558.

[58] Fontana M, Deppe T, Boyd AK, Rinzan M, Liu AY, Paranjape M. Electron-hole transport and photovoltaic effect in gated MoS_2 Schottky junctions. Scientific Reports 2013;3 1634(1-5). DOI: 10.1038/srep01634.

[59] Shin DH, Kim JH, Kim JH, Jang CW, Seo SW, Lee HS, Kim S, Choi SH. Graphene/porous silicon Schottky-junction solar cells. Journal of Alloys and Compounds 2017;715:291-296. DOI: 10.1016/j.jallcom.2017.05.001.

[60] Yi SG, Kim SH, Park S, Oh D, Choi HY, Lee N, et al. $Mo_{1-X} W_X Se_2$-Based Schottky Junction Photovoltaic Cells. ACS applied materials & interfaces 2016;8(49) 33811-33820. DOI: 10.1021/acsami.6b11768.

[61] Pospischil A, Furchi M M, Mueller T. Solar-energy conversion and light emission in an atomic monolayer p-n diode. Nature Nanotechnology 2014;9(4), 257-261. DOI: 10.1038/nnano.2014.14

[62] Baugher BW, Churchill HO, Yang Y, Herrero PJ. Optoelectronic devices based on electrically tunable p-n diodes

in a monolayer dichalcogenide. Nature Nanotechnology 2014;9(4) 262(1-6). DOI: 10.1038/nnano.2014.25.

[63] Ross JS, Klement P, Jones AM, Ghimire NJ, Yan J. Electrically tunable excitonic light-emitting diodes based on monolayer WSe2 p-n junctions. Nature Nanotechnology 2014;9(4) 268-272. DOI: 10.1038/nnano.2014.26.

[64] Memaran S, Pradhan NR, Lu Z, Rhodes D, Ludwig J, Zhou Q, et al. Pronounced Photovoltaic Response from Multilayered Transition-Metal Dichalcogenides PN-Junctions. Nano Letters 2015;15(11) 7532-7538. DOI: 10.1021/acs.nanolett.5b03265.

[65] Li L, Yu Y, Ye GJ, Ge Q, Ou X, Wu H et al. Black phosphorus field-effect transistors. Nature Nanotechnology 2014;9(5) 372-377. DOI: 10.1038/nnano.2014.35.

[66] Liu H, Neal AT, Zhu Z, Luo Z, Xu X, Tomanek D et al, Phosphorene: An Unexplored 2D Semiconductor with a High Hole Mobility. ACS Nano 2014;8(4) 4033-4041. DOI: 10.1021/nn501226z.

[67] Xia F, Wang H, Jia Y. Rediscovering black phosphorus as an anisotropic layered material for optoelectronics and electronics. Nature Communications 2014;5 4458(1-6). DOI: 10.1038/ncomms5458.

[68] Choi MS, Qu D, Lee D, Liu X, Watanabe K, Taniguchi T et al. Lateral MoS_2 p–n junction formed by chemical doping for use in high-performance optoelectronics. ACS nano 2014;8(9) 9332-9340. DOI: 10.1021/nn503284n.

[69] Duan X, Wang C, Shaw JC, Cheng R, Chen Y, Li H et al. Lateral epitaxial growth of two-dimensional layered semiconductor heterojunctions. Nature nanotechnology 2014;9(12) 1024-1030. DOI: 10.1038/nnano.2014.222.

[70] Gong Y, Lin J, Wang X, Shi G, Lei S, Lin Z, et al. Vertical and in-plane heterostructures from WS_2/MoS_2 monolayers. Nature materials 2014;13(12) 1135-1142. DOI: 10.1038/nmat4091.

[71] Li Z, Zheng J, Zhang Y, Zheng C, Woon WY, Chuang MC. et al. Synthesis of ultrathin composition graded doped lateral WSe_2/WS_2 heterostructures. ACS Applied Materials & Interfaces 2017;9(39) 34204-34212. DOI: 10.1021/acsami.7b08668.

[72] Snaith HJ, Abate A, Ball JM, Eperon GE, Leijtens T, Noel NK. et al. Anomalous Hysteresis in Perovskite Solar Cells. The Journal of Physical Chemistry Letters 2014;5 1511-1515. DOI: 10.1021/jz500113x.

[73] Baena JPC, Steier L, Tress W, Saliba M, Neutzner S, Matsui T, Highly efficient planar perovskite solar cells through band alignment engineering. Energy Environmental Science 2015;5 2928-2934. DOI: 10.1039/C5EE02608C.

[74] Guo Z, Zhuang J, Ma Z, Xia H, Yi J, Zhou W, et al. Improving the performance of lead acetate-based perovskite solar cells via solvent vapor annealing, Adv. Funct. Mater. 2019;21 4753-4762. DOI: 10.1039/C9CE00724E.

[75] Roose B, Wang Q, Abate A. The Role of Charge Selective Contacts in Perovskite Solar Cell Stability, Advanced Energy Materials 9 (2018) 1803140-1803159. DOI: 10.1002/aenm.201803140.

[76] Posudievsky OY, Konoshchuk NV, Shkavro AG, Koshechko VG, Pokhodenko VD. Structure and electronic properties of poly(3,4-ethylenedioxythiophene) poly(styrene sulfonate) prepared under ultrasonic irradiation. Synthetic Metals 2014;195 335-339. DOI: 10.1016/j.synthmet.2014.07.001.

[77] Singh R, Giri A, Pal M, Thiyagarajan K, Kwak J, Lee JJ, Perovskite solar cells with an MoS_2 electron transport layer. Journal of Materials Chemistry A 2019;7 7151-7158. DOI: 10.1039/C8TA12254G.

[78] Yin G, Zhao H, Feng J, Sun J, Yan J, Liu Z, et al. Low-temperature and facile solution-processed two-dimensional TiS 2 as an effective electron transport layer for UV-stable planar perovskite solar cells. Journal of Materials Chemistry A. 2018;6(19):9132-9138. DOI: 10.1039/C8TA01143E.

[79] Huang P, Chen Q, Zhang K, Yuan L, Zhou Y, Song B, et al. 21.7% efficiency achieved in planar n–i–p perovskite solar cells via interface engineering with water-soluble 2D TiS_2. Journal of Materials Chemistry A 2019;7(11) 6213-6219. DOI: 10.1039/C8TA11841H.

[80] Zhao X, Liu S, Zhang H, Chang SY, Huang W, Zhu B, et al. 20% efficient perovskite solar cells with 2D electron transporting layer. Advanced Functional Materials 2019;29(4) 1805168(1-8). DIO: 10.1002/adfm.201805168.

[81] Yang L, Dall'Agnese Y, Hantanasirisakul K, Shuck CE, Maleski K, Alhabeb M, SnO_2–Ti_3C_2 MXene electron transport layers for perovskite solar cells. Journal of materials chemistry A 2019;7(10) 5635-5642. DOI: 10.1039/C8TA12140K.

[82] Guo Z, Gao L, Xu Z, Teo S, Zhang C, Kamata Y et al. High Electrical Conductivity 2D MXene Serves as Additive of Perovskite for Efficient Solar Cells. Small 2018;14 1802738(1-8). DOI: 10.1002/smll.201802738.

[83] Wang Y, Zhang H, Zhang T, Shi W, Kan M, Chen J et al. Photostability of $MAPbI_3$ perovskite solar cells by incorporating black phosphorus. Solar RRL 2019;3(9) 1900197(1-7). DOI: 10.1002/solr.201900197.

[84] Long R, Fang W, Akimov AV. Nonradiative electron–hole recombination rate is greatly reduced by defects in monolayer black phosphorus: ab initio time domain study. The Journal of Physical Chemistry Letters 2016;7(4) 653-659. DOI: 10.1021/acs.jpclett.6b00001.

[85] Capasso A, Matteocci F, Najafi L, Prato M, Buha J, Cinà L et al. Few Layer MoS_2 Flakes as Active Buffer Layer for Stable Perovskite Solar Cells. Advanced Energy Materials 2016;6(16) 1600920(1-12). DOI: 10.1002/aenm.201600920.

[86] Jawaid A, Nepal D, Park K, Jespersen M, Qualley A, Mirau P, et al. Mechanism for liquid phase exfoliation of MoS_2. Chemistry of Materials 2016;28(1) 337-348. DOI: 10.1021/acs.chemmater.5b04224.

[87] Gupta A, Arunachalam V, Vasudevan S. Liquid-phase exfoliation of MoS_2 nanosheets: the critical role of trace water. The journal of physical chemistry letters 2016;7(23) 4884-4890. DOI: 10.1021/acs.jpclett.6b02405.

[88] Najafi L, Taheri B, Martín-García B, Bellani S, Girolamo DD, Agresti A, et al. MoS_2 quantum dot/graphene hybrids for advanced interface engineering of a $CH_3NH_3PbI_3$ perovskite solar cell with an efficiency of over 20%. ACS nano 2018;12(11) 10736-10754. DOI: 10.1021/acsnano.8b05514.

[89] Meng L, You J, Guo TF, Yang Y. Recent advances in the inverted planar structure of perovskite solar cells. Accounts of chemical research 2016;49(1) 155-165. DOI: 10.1021/acs.accounts.5b00404.

[90] Huang L, Xu J, Sun X, Du Y, Cai H, Ni J, et al. Toward revealing the critical role of perovskite coverage in highly efficient electron-transport layer-free perovskite solar cells: an energy band and equivalent circuit model

perspective. ACS Applied Materials & Interfaces 2016;8(15) 9811-9820. DOI: 10.1021/acsami.6b00544.

[91] Jeng JY, Chiang YF, Lee MH, Peng SR, Guo TF, Chen P, et al. $CH_3NH_3PbI_3$ perovskite/fullerene planar heterojunction hybrid solar cells. Advanced Materials 2013;25(27) 3727-3732. DOI: 10.1002/adma.201301327.

[92] Hu Q, Wu J, Jiang C, Liu T, Que X, Zhu R, et al. Engineering of electron-selective contact for perovskite solar cells with efficiency exceeding 15%. ACS nano. 2014;8(10) 10161-10167. DOI: doi.org/10.1021/nn5029828.

[93] Liu D, Yang J, Kelly TL. Compact layer free perovskite solar cells with 13.5% efficiency. Journal of the American Chemical Society 2014;136(49) 17116-17122. DOI: 10.1021/ja508758k.

[94] Ke W, Fang G, Wan J, Tao H, Liu Q, Xiong L, et al. Efficient hole-blocking layer-free planar halide perovskite thin-film solar cells. Nature communications 2015;6(1) 1-7. DOI: 10.1038/ncomms7700.

[95] Chen Z, Dong Q, Liu Y, Bao C, Fang Y, Lin Y, et al. Thin single crystal perovskite solar cells to harvest below-bandgap light absorption. Nature communications. 2017;8(1) 1-7. DOI: 10.1038/s41467-017-02039-5.

[96] Prochowicz D, Yadav P, Saliba M, Saski M, Zakeeruddin SM, Lewiński J, et al. Mechanosynthesis of pure phase mixed-cation $MA_XFA_{1-X}PbI_3$ hybrid perovskites: photovoltaic performance and electrochemical properties. Sustainable Energy & Fuels 2017;1(4) 689-693. DOI: 10.1039/C7SE00094D.

[97] Yang W, Chen J, Lian X, Li J, Yao F, Wu G, et al. Black Phosphorus Quantum Dots Induced High Quality Perovskite Film for Efficient and Thermally Stable Planar Perovskite Solar Cells. Solar Rrl. 2019;3(8) 1900132(1-8). DOI: 10.1002/solr.201900132.

[98] Giorgi G, Yamashita K, Palummo M. Nature of the electronic and optical excitations of Ruddlesden–Popper hybrid organic–inorganic perovskites: The role of the many-body interactions. The journal of physical chemistry letters 2018;9(19) 5891-5896. DOI: 10.1021/acs.jpclett.8b02653.

[99] Zheng Y, Niu T, Ran X, Qiu J, Li B, Xia Y, et al. Unique characteristics of 2D Ruddlesden–Popper (2DRP) perovskite for future photovoltaic application. Journal of Materials Chemistry A.2019;7(23):13860-13872. DOI: 10.1039/C9TA03217G.

[100] Gao X, Zhang X, Yin W, Wang H, Hu Y, Zhang Q, et al. Ruddlesden–Popper Perovskites: Synthesis and Optical Properties for Optoelectronic Applications. Advanced Science 2019;6(22) 1900941(1-22). DOI: 10.1002/advs.201900941.

[101] Smith IC, Hoke ET, Solis-Ibarra D, McGehee MD, Karunadasa HI. A Layered Hybrid Perovskite Solar-Cell Absorber with Enhanced Moisture Stability. Angewandte International Edition Chemie 2014;53 11232-11235. DOI: 10.1002/anie.201406466.

[102] Mitzi DB, Templating and structural engineering in organic–inorganic perovskites. J. Chem. Soc., Dalton Transaction 2001 1-12. DOI: 10.1039/B007070J.

[103] Quan LN, Yuan M, Comin R, Voznyy O, Beauregard EM, Hoogland S et al. Ligand-Stabilized Reduced-Dimensionality Perovskites Journal of American Chemical Society 2016;138 2649-2655. DOI: 10.1021/jacs.5b11740.

[104] Eperon GE, Paterno GM, Sutton RJ, Zampetti A, Haghighirad AA, Cacialli F et al. Inorganic caesium lead

iodide perovskite solar cells. Journal of Materials Chemistry A 2015;3 19688-19695. DOI: 10.1039/C5TA06398A.

[105] Protesescu L, Yakunin S, Bodnarchuk MI, Krieg F, Caputo R, Hendon CH et al. Nanocrystals of Cesium Lead Halide Perovskites ($CsPbX_3$, X = Cl, Br, and I): Novel Optoelectronic Materials Showing Bright Emission with Wide Color Gamut. Nano Letters 2015;5 3692-3696. DOI: 10.1021/nl5048779.

[106] Zhang T, Dar MI, Li G, Xu F, Guo, Gratzel M et al. Bication lead iodide 2D perovskite component to stabilize inorganic a-$CsPbI_3$ perovskite phase for high-efficiency solar cells. Science Advances 2017;3 1700841(1-6). DOI: 10.1126/sciadv.1700841.

[107] Tian X, Zhang Y, Zheng R, Wei D, Liu J. Two-dimensional organic–inorganic hybrid Ruddlesden–Popper perovskite materials: preparation, enhanced stability, and applications in photodetection. Sustainable Energy & Fuels 2020;4(5):2087-2113. DOI: 10.1039/C9SE01181A.

[108] Niu W, Eiden A, Prakash VG, Baumberg JJ. Exfoliation of self-assembled 2D organic-inorganic perovskite semiconductors. Applied Physics Letters 2014;104(17) 171111(1-4). DOI: 10.1063/1.4874846.

[109] Bünzli JCG, Chauvin AS: Lanthanides in solar energy conversion. In Jean-Claude GB, Vitalij KP, (ed.) Handbook on the physics and chemistry of rare earths. Elsevier;2014. p169-281.

[110] Huang X, Han S, Huang W, Liu X. Enhancing solar cell efficiency: the search for luminescent materials as spectral converters. Chemical Society Reviews. 2013;42(1) 173-201. DOI: 10.1039/c2cs35288e.

[111] Zhang QY, Yang GF, Jiang ZH. Cooperative downconversion in Gd $Al_3(BO_3)_4$: RE^{3+}, Yb^{3+}(RE= Pr, Tb, and Tm). Applied Physics Letters 2007;91(5) 051903(1-3). DOI: 10.1063/1.2757595.

[112] Li Z, Zeng Y, Qian H, Long R, Xiong Y. Facile synthesis of $GdBO_3$ spindle assemblies and and microdisks as versatile host matrices for lanthanide doping. CrystEngComm. 2012;14(11):3959-3964. DOI: 10.1039/C2CE06596G.

[113] Yuan JL, Zeng XY, Zhao JT, Zhang ZJ, Chen HH, Yang XX. Energy transfer mechanisms in Tb^{3+}, Yb^{3+} codoped Y_2O_3 downconversion phosphor. Journal of Physics D. Applied Physics 2008;41(10) 105406(1-6). DOI: 10.1088/0022-3727/41/10/105406.

[114] Ye S, Zhu B, Chen J, Luo J, Qiu JR. Infrared quantum cutting in Tb^{3+}, Yb^{3+} codoped transparent glass ceramics containing Ca F_2 nanocrystals. Applied Physics Letters 2008;92(14) 141112(1-3). DOI: 10.1063/1.2907496.

[115] Liu X, Ye S, Qiao Y, Dong G, Zhu B, Chen D et al. Cooperative downconversion and near-infrared luminescence of Tb^{3+}–Yb^{3+} codoped lanthanum borogermanate glasses. Applied Physics B 2009;96(1) 51-55. DOI: 10.1007/s00340-009-3478-z.

[116] Tsai ML, Tu WC, Tang L, Wei TC, Wei WR, Lau SP et al. Efficiency enhancement of silicon heterojunction solar cells via photon management using graphene quantum dot as downconverters. Nano letters 2016;16(1) 309-313. DOI: 10.1021/acs.nanolett.5b03814.

[117] Hou X, Xuan T, Sun H, Chen X, Li H, Pan L. High-performance perovskite solar cells by incorporating a $ZnGa_2O_4$: Eu^{3+} nanophosphor in the mesoporous TiO_2 layer. Solar Energy Materials and Solar Cells 2016;149 121-127. DOI: 10.1016/j.solmat.2016.01.021.

[118] Chen W, Luo Q, Zhang C, Shi J, Deng X, Yue L et al. Effects of down-conversion CeO_2: Eu^{3+} nanophosphors in perovskite solar cells. Journal of Materials Science: Materials in Electronics 2017;28(15) 11346-11357. DOI: 10.1007/s10854-017-6928-0.

[119] Wang HQ, Batentschuk M, Osvet A, Pinna L, Brabec CJ. Rare-earth ion doped up-conversion materials for photovoltaic applications. Advanced Materials 2011:2675-2680. DOI: 10.1002/adma.201100511.

[120] Badescu V, Badescu AM. Improved model for solar cells with up-conversion of low-energy photons. Renewable Energy 2009;34(6) 1538-1544. DOI: 10.1016/j.renene.2008.11.006.

[121] Trupke T, Green MA, Würfel P. Improving solar cell efficiencies by up-conversion of sub-band-gap light. Journal of applied physics. 2002;92(7) 4117-4122. DOI: 10.1063/1.1505677.

[122] Pan AC, Canizo CD, Luque A. Characterization of up-converter layers on bifacial silicon solar cells. Materials Science and Engineering: B 2009;159 212-215. DOI: 10.1016/j.mseb.2008.10.058.

[123] Shan GB, Demopoulos GP. Near-infrared sunlight harvesting in dye-sensitized solar cells via the insertion of an upconverter-TiO_2 nanocomposite layer. Advanced materials 2010;22(39) 4373-4377. DOI: 10.1002/adma.201001816.

[124] Shang Y, Hao S, Yang C, Chen G. Enhancing solar cell efficiency using photon upconversion materials. Nanomaterials 2015;5(4) 1782-1809. DOI: 10.3390/nano5041782.

[125] Chen X, Xu W, Song H, Chen C, Xia H, Zhu Y et al. Highly Efficient $LiYF_4$: Yb^{3+}, Er^{3+} Upconversion single crystal under solar cell spectrum excitation and photovoltaic application. ACS Applied Materials & Interfaces 2016;8(14) 9071-9079. DOI: 10.1021/acsami.5b12528.

[126] Roh J, Yu H, Jang J. Hexagonal β-$NaYF_4$: Yb^{3+}, Er^{3+} nanoprism-incorporated upconverting layer in perovskite solar cells for near-infrared sunlight harvesting. ACS Applied Materials & Interfaces 2016;8(31) 19847-19852. DOI: 10.1021/acsami.6b04760.

[127] Wang X, Zhang Z, Qin J, Shi W, Liu Y, Gao H et al. Enhanced photovoltaic performance of perovskite solar cells based on Er-Yb co-doped TiO_2 nanorod arrays. Electrochimica Acta 2017;245 839-845. DOI: 10.1016/J.ELECTACTA.2017.06.032.

[128] He M, Pang X, Liu X, Jiang B, He Y, Snaith H, Lin Z. Monodisperse Dual-functional upconversion nanoparticles enabled nearinfrared organolead halide perovskite solar cells. Angewandte International Edition Chemie 2016;128(13):4352-4356. DOI: 10.1002/anie.201600702.

[129] Zhou D, Liu D, Jin J, Chen X, Xu W, Yin Z et al. Semiconductor plasmon-sensitized broadband upconversion and its enhancement effect on the power conversion efficiency of perovskite solar cells. Journal of Materials Chemistry A 2017;5(32) 16559-16567. DOI: 10.1039/C7TA04943A.

[130] Straus DB, Kagan CR. Electrons, excitons, and phonons in two-dimensional hybrid perovskites: connecting structural, optical, and electronic properties. The journal of physical chemistry letters 2018;9(6) 1434-1447. DOI: 10.1021/acs.jpclett.8b00201.

[131] Katan C, Mercier N, Even J. Quantum and dielectric confinement effects in lower-dimensional hybrid perovskite semiconductors. Chemical

reviews 2019;119(5) 3140-3192. DOI: 10.1021/acs.chemrev.8b00417.

[132] Zhang L, Liu Y, Yang Z, Liu SF. Two dimensional metal halide perovskites: Promising candidates for light-emitting diodes. Journal of Energy Chemistry 2019;37 97-110. DOI: 10.1016/j.jechem.2018.12.005.

[133] Sutherland BR, Sargent EH. Perovskite photonic sources. Nature Photonics 2016;10(5) 295-305. DOI: 10.1038/nphoton.2016.62. DOI: 10.1016/j.jechem.2018.12.005.

[134] Kim YH, Cho H, Lee TW. Metal halide perovskite light emitters. Proceedings of the National Academy of Sciences. 2016 Oct 18;113(42):11694-702. DOI: 10.1073/pnas.1607471113.

[135] Kim YH, Cho H, Heo JH, Kim TS, Myoung N, Lee CL et al. Multicolored organic/inorganic hybrid perovskite light-emitting diodes. Advanced materials 2015;27(7) 1248-1254. DOI: 10.1002/adma.201403751.

[136] You P, Li G, Tang G, Cao J, Yan F. Ultrafast laser-annealing of perovskite films for efficient perovskite solar cells. Energy & Environmental Science 2020;13(4) 1187-1196. DOI: 10.1039/C9EE02324K.

[137] Wang H, Cao S, Yang B, Li H, Wang M, Hu X et al. NH_4Cl-Modified ZnO for High-Performance $CsPbIBr_2$ Perovskite Solar Cells via Low-Temperature Process. Solar Rrl 2020;4(1) 1900363. DOI: 10.1002/solr.201900363.

[138] Li Y, Hoye RL, Gao HH, Yan L, Zhang X, Zhou Y et al. Over 20% Efficiency in Methylammonium Lead Iodide Perovskite Solar Cells with Enhanced Stability via "in Situ Solidification" of the TiO_2 Compact Layer. ACS Applied Materials & Interfaces 2020;12(6) 7135-7143. DOI: 10.1021/acsami.9b19153.

[139] Oz S, Jena AK, Kulkarni A, Mouri K, Yokoyama T, Takei I et al. Lead (II) Propionate Additive and a Dopant-Free Polymer Hole Transport Material for $CsPbI_2Br$ Perovskite Solar Cells. ACS Energy Letters 2020;5(4) 1292-1299. DOI: 10.1021/acsenergylett.0c00244.

[140] Fu S, Zhang W, Li X, Wan L, Wu Y, Chen L et al. Dual-Protection Strategy for High-Efficiency and Stable $CsPbI_2Br$ Inorganic Perovskite Solar Cells. ACS Energy Letters 2020;5(2) 676-684. DOI: 10.1021/acsenergylett.9b02716.

[141] Li C, Pan Y, Hu J, Qiu S, Zhang C, Yang Y et al. Vertically Aligned 2D/3D Pb–Sn Perovskites with Enhanced Charge Extraction and Suppressed Phase Segregation for Efficient Printable Solar Cells. ACS Energy Letters 2020;5(5) 1386-1395. DOI: 10.1021/acsenergylett.0c00634.

[142] Wang D, Li W, Du Z, Li G, Sun W, Wu J et al. Highly efficient $CsPbBr_3$ planar perovskite solar cells via additive engineering with NH_4SCN. ACS Applied Materials & Interfaces 2020;12(9) 10579-10587. DOI: 10.1021/acsami.9b23384.

Chapter 14

Graphene-Based Material for Fabrication of Electrodes in Dye-Sensitized Solar Cells

Nguyen Huu Hieu

Abstract

Graphene-based materials have been widely studied for the fabrication of electrodes in dye-sensitized solar cells (DSSCs). The use of graphene in the cathode is to reduce the amount of platinum (Pt), which in turn is expected to reduce the production cost of DSSCs. Additionally, in the structure of cathode, graphene acts as a supporting material to reduce the particle sizes of Pt and helps to maintain the high efficiency of DSSCs. For anodes, graphene can provide a more effective electron transfer process, resulting in the improvement of efficiency of DSSCs. In this chapter, the use of graphene-based materials for fabrication of cathodes and anodes in DSSCs, including platinum/reduced graphene oxide composite (Pt/rGO) and zinc oxide/reduced graphene oxide composite (ZnO/rGO) is discussed. The fabricated DSSCs were tested using current density-voltage (J-V) curves to evaluate the efficiency. The results of efficiency demonstrate that Pt/rGO is the potential material for fabrication of cathode in DSSCs, which helps to reduce the amount of Pt and maintain the high efficiency. The efficiency values of DSSCs fabricated from ZnO/rGO anodes show that the incorporation of reduced graphene oxide in the ZnO could improve the performance of DSSCs.

Keywords: dye-sensitized solar cell, cathode, anode, graphene, zinc oxide, platinum, composite

1. Introduction

The human demand for energy has rapidly risen due to technological and economic developments. The energy problems and environmental pollutions are alarming concerns that have to be addressed urgently; therefore, researchers have explored the availability of renewable energies, such as wind energy, biomass energy, solar cells, etc. Among these types of energy, solar cells have become a potential solution to the energy shortage in terms of cost, geographical distribution, and sustainability. Since the breakthrough achievement by Grätzel and O'Regan in 1991, dye-sensitized solar cells (DSSCs) have been drawing significant attention as an alternative to fossil fuels due to the low cost, low energy consumption, simple fabrication process, and environmental friendliness [1, 2]. DSSCs are third-generation photovoltaic cells consisting of four primary components: photoanode, dye-sensitizer, electrolyte solution, and cathode.

 IntechOpen

Recently, cathodes of DSSCs are fabricated from platinum (Pt). Because Pt is a noble and expensive metal, the usage of Pt for fabrication of cathodes could lead to the increase in the production cost of DSSCs. Hence, many efforts had been made for the reduction of cathode fabrication costs to reduce the production costs of DSSCs [3].

Many research studies have been conducted for replacing Pt in cathodes by using other materials in combination with Pt. According to previous studies, copious low-cost materials were studied for fabrication of cathodes in DSSCs, such as carbonaceous materials, conductive polymers, alloys, metal oxide, transition metal-based materials including metal sulfides, metal carbides, metal nitrides, etc. [4]. Among these materials, carbon-based materials such as carbon vulcan, carbon black, activated carbon, carbon nanofibers, carbon nanotubes, and graphene have attracted more attention from researchers due to the relatively low cost, high stability, high chemical inertness, and high catalytic behavior [5]. In comparison with other carbon-based materials, graphene showed better properties, such as: having the highest electrical conductivity, fast charged carrier mobility, good chemical stability, and high surface area. These properties make graphene one of the most potential materials for fabrication of cathodes in DSSCs [6, 7]. Graphene can be synthesized from graphene oxide (GO) using the chemical reduction method, in which the synthesized graphene product is known as reduced graphene oxide (rGO). Numerous studies combined Pt and graphene for fabricating solar cells, fuel cells, and for other catalytic applications [8, 9]. In DSSCs, platinum/reduced graphene oxide (Pt/rGO) composite has been widely used for fabrication of cathode.

For anodes, the electron recombination processes in anode material (ZnO and TiO_2) were a phenomenon that decreased the efficiency of DSSCs. Due to the high electron mobility, lower recombination rate, electron lifetime is considerably higher in ZnO as compared to TiO_2, good transparency to visible light, high photo activity and nanocrystalline ZnO of varying morphologies, ZnO is considered as a potential material for fabrication of anodes in DSSCs. In order to increase the efficiency of ZnO-based DSSCs, graphene was studied for combination with ZnO, which could reduce the electron recombination processes of anodes. Numerous efforts have been made to investigate the performance of DSSCs fabricated from zinc oxide/reduced graphene oxide (ZnO/rGO) anodes.

In order to emphasize the potential of graphene as a promising material for fabrication of cathodes and anodes in DSSCs, this chapter is aiming to provide an overview on the current issues of DSSCs that need to be improved and the recently studied materials for fabrication of electrodes in DSSCs, especially carbonaceous materials. Subsequently, the synthesis of Pt/rGO and ZnO/rGO composite materials and the effect of synthesized materials on the performance of fabricated DSSCs are discussed. Additionally, the characterization results of Pt/rGO and ZnO which were synthesized by our group were also presented to illustrate the morphologies and structure of these materials.

2. Dye-sensitized solar cells

The structure and working principle of DSSCs are presented in **Figure 1**.

An electrode is a solid electrical conductor through which an electrical current enters or leaves an electrochemical cell or other mediums like metallic solids, liquids, gases, plasmas, or vacuums. Electrodes are usually made of good electrical conducting materials. In an electrochemical cell, an electrode is referred to as either an anode or a cathode. The anode is defined as the electrode at which electrons leave the cell and the oxidation occurs, while the cathode is defined as the electrode at

DOI: http://dx.doi.org/10.5772/intechopen.93637

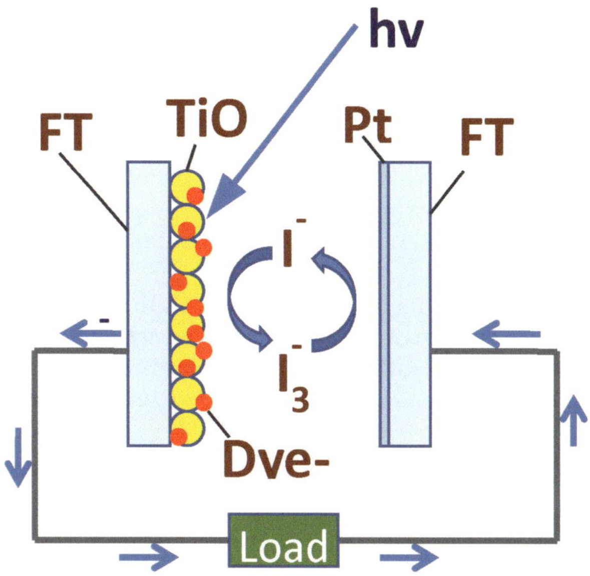

Figure 1.
Structure of DSSCs.

which electrons enter the cell and reduction occurs. For DSSCs, the anode is the light collector, thus it is also called photoanode. The anode had a deposited layer of a metal oxide semiconductor. The cathode is the electrode on which Pt and other conducting materials are deposited. The cathode in DSSCs is often called the counter electrode. In 1991, Grätzel first introduced the term "counter electrode" in his pioneering publication about dye-sensitized solar cells [10, 11].

A photoanode is typically a layer of nanocrystalline titanium dioxide (TiO_2), with a thickness of about 10 μm, coated on a transparent conductive oxide (TCO) glass substrate, such as an indium-doped tin oxide (ITO) or fluorine-doped tin oxide (FTO). Morphologies of TiO_2 materials (rod, spherical, hierarchical, and tubular) significantly affect the light-harvesting, charge injection, and charge-collecting properties of DSSCs [12, 13].

Dye-sensitizer plays a crucial role in improving light absorption within the visible region. Dye-sensitizer is described as an electron pump with inputting power from light photons. When the light irradiates the photoanode of DSSCs, electrons from the highest occupied molecular orbital (HOMO) of dye-sensitizer transfer to the lowest unoccupied molecular orbital (LUMO), then the electrons flow to the photoanode. Many types of dye-sensitizer have been used in DSSCs, such as ruthenium dyes and natural dyes. Dye N719 is a ruthenium-based dye that is used widely in DSSCs [10, 11].

After producing electrons, dye-sensitizer becomes oxidized states, and electrons need a medium to transfer back to dye-sensitizer and complete the external circuit of DSSCs. In DSSCs, electrolyte plays a role as a shuttle, which transfers electrons from cathode back to dye-sensitizer. A good electrolyte system used in DSSCs must have excellent electrical conductivity, low viscosity for easier and faster diffusion of electrons, and good interfacial contact with the nanocrystalline semiconductor of photoanode and cathode. The most popular electrolyte system used in DSSCs is iodide/triiodide (I^-/I_3^-) [10, 11].

The last main component of a DSSC is the cathode. A typical cathode of DSSCs is thin Pt layer coated on the TCO (typically FTO). At the interface between the Pt layer and the electrolyte, a reduction reaction of electrolyte occurs. This phenomenon happens when electrons move from external loads to the cathode of DSSCs. The Pt film has three main roles in a DSSC: a catalyst for redox reaction of electrolyte, an electrolyte regenerator, and a conducting material [10, 11].

The working principle of DSSCs is different from that of conventional silicon-based solar cells. In silicon-based solar cells, the semiconductor p-n junction carried out many processes, both absorbing light and sending out current. In DSSCs, those tasks are separate. The main role of the semiconductor is to transfer electron from dye-sensitizer to FTO glass, whereas electron suppliers are the dye-sensitizer [10, 11].

The light illuminates the photoanode of the DSSCs then dye-sensitizer will absorb appropriate wavelengths and turn into excited states S^*. In excited states, the dye releases electrons that will diffuse into the conduction band of semiconductor TiO_2 and reach FTO glass. The exciting state of sensitizers is shown in Eq. (1).

$$h\nu + S \rightarrow S* + e \tag{1}$$

After that, electrons move to external loads and get back to the DSSCs at the cathode. Dye-sensitizer molecules turn into oxidized states S^+ after providing electron. Subsequently, those molecules are reduced by I^- in the electrolyte to restore to initial states and electrolyte changes from the oxidized state I^- to the reduced state I_3^- as a result of reduction reactions of dye-sensitizers, as shown in Eq. (2).

$$3I^- + 2S^+ \rightarrow I_3^- + 2S \tag{2}$$

At the cathode, the reduced state I_3^- receives electrons coming back to the DSSCs from the external load and restores to I^-. Redox mediators move back and forth between cathode and anode mainly relying on the diffusion process. The process is shown in Eq. (3).

$$I_3^- + 2e \rightarrow 3I^- \tag{3}$$

The above processes take place when DSSCs are illuminated continuously. Then a current is generated and flow in the external load. However, electrons in the conduction band of TiO_2 may follow three other routes to join in recombination reactions, as shown in Eqs. (4)–(6).

$$e_{TiO_2} + S^+ \rightarrow S \tag{4}$$

$$e_{TiO_2} + I_3^- \rightarrow 3I^- \tag{5}$$

$$e_{SnO_2} + I_3^- \rightarrow 3I^- \tag{6}$$

Eq. (6) takes place on the interface between FTO and electrolyte, where TiO_2 does not cover. The electron transfer from FTO to the external load is significantly faster than that of Eq. (6). Eq. (5) occurs more often than other ones because the concentration of I_3^- is much higher than that of S^+. These recombination reactions make the external current deteriorate and cause declines in voltage of DSSCs, which lead to low power conversion efficiency [10, 11].

Many efforts have been made for commercialization of DSSCs, including the investigation to enhance the efficiency and reduce the production cost of DSSCs, especially the research on cathodes and anodes. Various materials have been studied for the fabrication of electrodes in DSSCs: carbonaceous materials, conductive polymers, alloys, metal oxide, transition metal-based materials, etc. Among these materials, graphene is one of the most prospective materials for the synthesis of composites for fabrication of electrodes in DSSCs [14, 15].

3. Graphene for fabrication of electrodes in DSSCs

3.1 Carbonaceous materials

Carbon has long been an intriguing material because it had two allotropes which were widely known: diamond and graphite. Although these materials have the same elemental composition, the properties of graphite are very different from those of diamond. Although researchers working on carbon have long been aware of other forms, usually they were regarded as a nuisance and discarded if their formation could not be avoided. The importance of the recent "discoveries" of fullerenes and carbon nanotubes resides in the fact that their structures were elucidated for the first time. The most recent "discovery" is that of graphene, it is simply one of the many parallel sheets constituting graphite. The ingenuity resided in the preparation of a single isolated sheet, which opened the possibility of examining experimentally what was already a well-studied material theoretically [16].

Carbon-based materials have shown great versatility due to the ability to chemically combine with other carbon-based materials and with a range of different elements to form strong covalent bonds. As a result, these materials exhibit excellent characteristics such as high strength, high density, and high hardness. Their research, development and innovation are taking place in various fields, and studies employing the development of carbon-based materials have shown many positive results for a wide variety of structures, which has allowed the development of several materials with different applications [17].

Graphite is one of the most common allotropes of carbon, and the most stable form of carbon under standard conditions. Graphite is another promising electrode electrocatalyst and conducting layer material due to the abundant defect sites and good electronic properties. Particularly, the defect sites from edge planes are preferred to those of basal planes, as the former exhibits the faster electrons transport and charge transfer [18].

The three new materials, graphene, carbon nanotubes and fullerenes, can be called "nanocarbon" materials. Like graphite, the structures of these carbonaceous materials consist of sp^2 orbitals. Fullerenes contain 12 pentagons and have some sp^3 character. Fullerenes are spheroidal molecules and are made exclusively of carbon atoms (e.g. C60, C70). Their unique hollow cage-like shape and structural analogy with clathrin-coated vesicles in cells support the idea of the potential use of fullerenes as drug or gene delivery agents. Fullerenes display a diverse range of biological activity, which arises from their reactivity, due to the presence of double bonds and bending of sp^2 hybridized carbon atoms, which produces angle strain. Fullerenes can act either as acceptors or donors of electrons. When irradiated with ultraviolet or visible light, fullerenes can convert molecular oxygen into highly reactive singlet oxygen. Thus, fullerenes have the potential to inflict photodynamic damage on biological systems, including damage to cellular membranes, inhibition of various enzymes [18].

Carbon nanotubes (CNTs) are formed by a single cylindrically shaped graphene sheet called single-wall carbon nanotubes (SWCNTs) or several graphene sheets arranged concentrically called multi-wall carbon nanotubes (MWCNTs). CNTs have been proposed as the prospective substitutes for the conventional Pt in DSSCs due to their outstanding advantages of large surface area, high electrical conductivity, and chemical stability [18, 19]. Additionally, CNTs could also be used for synthesis of composite materials of anodes in DSSCs, including the ZnO nanowires/CNTs and TiO2/CNTs, in order to offer a potential platform to enhancement surface area and decrease of carrier recombination in DSSCs [20, 21].

As mentioned, carbonaceous materials are quite attractive for replacement of Pt in DSSCs due to the high electronic conductivity, corrosion resistance toward I_2 electrolyte, high reactivity for I_3^- reduction, and low cost. The lower catalytic activity of carbon compared to Pt can be compensated by increasing the active surface area of the electrode by using a porous electrode structure. For example, porous carbon electrodes are easily prepared from graphite powder, which consists of plate-like crystals that, on deposition from a liquid dispersion and drying, will preferentially align in the plane of the counter electrodes, resulting in a high conductivity in this plane. Numerous carbonaceous materials were studied for the fabrication of electrodes in DSSCs, using carbon vulcan, carbon black, activated carbon, carbon nanofibers, carbon nanotubes, graphene or the combination of these materials to fabricate the high-performance electrodes of DSSCs, like graphite-activated carbon [22], carbon black-graphite [23]. Among these materials, graphene has attracted the most attention of researchers due to its outstanding properties [5, 24].

3.2 Graphene

Andre Konstantin Geim and Konstantin Sergeevich Novoselov of the University of Manchester received the 2010 Nobel Prize in Physics for their pioneering research on graphene. Graphene is a flat monolayer of carbon atoms arranging like the structure of honeycombs with one atom thickness. Due to the special thickness, graphene is considered as a 2D material, as shown in **Figure 2**. Carbon atoms in graphene lattice are hybridized sp^2 with the C-C bond length of 1.42 Å. Graphene is one of the basic carbon allotropes, including graphite, carbon nanotube, and fullerene. Graphene possesses not only all properties of graphite but also other extraordinary characteristics. Graphene has high carrier mobility at room

Figure 2.
Structure of graphene.

temperature (~200,000 $cm^2 V^{-1} s^{-1}$), high specific area (2630 $m^2 g^{-1}$), excellent thermal conductivity (~3000 $Wm^{-1} K^{-1}$), high Young modulus (~1 TPa), and high optical transparency (97.7%). Moreover, although graphene has low-density, it is up to 50 times stronger than steel [24, 25].

Because graphene is an atom-thick layer, it is a perfect nanoscale material and, therefore, has great potential in a very wide range of applications in the field of nanotechnology. Graphene has been extensively studied for nano-technological applications in field-effect transistors, solar cells, fuel cells, supercapacitors, rechargeable batteries, optical modulators, chemical sensors, drug delivery, and biomedical applications, in addition to other areas [26, 27].

There are numerous methods for synthesis of graphene: mechanical exfoliation, chemical vapor deposition, epitaxial growth, chemical reduction, etc. [28, 29]. Among these methods, chemical reduction is a commonly used route for graphene synthesis. In this method, graphene is synthesized by reduction of graphene oxide (GO), which has the similar structure to graphene, with oxygen-containing functional groups introduced into the hexagon structure of graphene sheets. The synthesized graphene by this method is called rGO. Chemical reduction method is a cost-effective and widely available method for the mass production of rGO compared with thermal reduction and other methods. Otherwise, the synthesis procedure of graphene with bottom-up methods has revealed difficulty for applications in industry [30]. In this study, GO was synthesized from graphite (Gi) using the Hummers' method and GO was reduced to create rGO.

3.3 Graphene for fabrication of cathodes in DSSCs

In DSSCs, cathode carries out three functions. As a catalyst, the cathode facilitates the regeneration of redox couple i.e., the oxidized state of redox couple is reduced by accepting electrons at the surface of the electrode. As a positive electrode of primary cells, the cathode collects electrons from the external circuit and transfers electrons into the cell. As a mirror, unabsorbed light came through the photoanode and the electrolyte, which was partially reflected in the cell to enhance the utilization of the light [5, 31].

The cathode in DSSCs conventionally includes two parts: the substrate and the catalytic layer. The type of substrate mostly used in DSSCs is conductive glass substrates made by coating a layer of TCO on transparent glass. In DSSCs, the TCO substrate plays an important role in transmitting the incident light and leading the electrical current. Both transmittance and conductivity are crucial for the electrode of DSSCs. The conductive glass that is widely used in the fabrication of DSSCs is FTO [5].

For promoting the commercialization of DSSCs, the production cost of DSSCs needs to be reduced. Moreover, the efficiency of DSSCs is expected to maintain at an acceptable level. At present, Pt is still the most appropriate material for the fabrication of cathode in DSSCs. However, this noble metal has limited availability and relatively high cost that hinder the large-scale production of DSSCs. Moreover, Pt cathodes show poor resistance toward corrosion in iodide solution, which may result in the formation of PtI_4. The use of Pt in cathodes is considered as one of many reasons that prevent the commercialization of DSSCs. On the other hand, carbon is the material that can be found everywhere on the planet. Carbon-based materials had wide applications in technology including solar technology [5]. Recently, graphene has been explored as a novel material with many outstanding characteristics, which makes these materials become one of the most promising alternatives of Pt in DSSCs. In recent years, graphene and graphene-based materials have been demonstrated to be an adequate substitute for Pt, to cut off the use of the noble metal and maintain the performance of DSSCs [32, 33]. Many research studies have been conducted to synthesized Pt/graphene or Pt/rGO with different methods. Yen et al. mixed GO and H_2PtCl_6 precursor salt in an ethylene glycol/ H_2O mixture and heated the mixture at 120°C to form the Pt/rGO nanocomposite [34]. With the similar method, Khoa et al. mixed GO and H_2PtCl_6 with different amount of GO and then the mixture was heated at 350°C to obtain the Pt/thermally reduced graphene oxide composite [35]. Instead of thermal reduction method, Wan et al. used $NaBH_4$ as reducing agent for synthesis of Pt/rGO composite for cathode of DSSCs [36]. Recently, Yu et al. synthesized the Pt nanoparticles-loaded holey reduced graphene oxide framework materials with the addition of aqueous hydrofluoric acid, resulting in the high electrocatalytic activity and efficient electron/ ion transport material for cathode fabrication of DSSCs [33]. After that, Suriani et al. combined graphene, SWCNT, and Pt to create the rGO/SWCNT hybrid film with low Pt loading for fabrication of cathode in DSSCs [37]. Beside these studies, there were various advanced methods for synthesis of Pt/graphene composite and other graphene-based materials: chemical vapor deposition [38, 39], hydrothermal, coating with different layers [40], etc. All of the fabricated DSSCs from mentioned studies exhibited high conversion efficiency that can be compared with DSSCs fabricated from the pure Pt cathodes.

Our group conducted the experiments to investigate the performance of Pt/rGO cathodes in DSSCs. Accordingly, rGO was synthesized from GO, which was synthesized from Gi using the improved Hummers' method. The Pt/rGO composite pastes were fabricated from rGO and H_2PtCl_6 with different weight percents of rGO: 0, 10, 20, 30, 40, 50, and 100 wt%, marked as PG0, PG10, PG20, PG30, PG40, PG50, and PG100, respectively. These composite pastes were used for fabrication of cathodes in DSSCs using the screen-printing technology. DSSCs were assembled with fabricated cathodes, Dyesol TiO2 anodes, N719 dye, and Dyesol High Stability Electrolyte (HSE). The efficiency of fabricated DSSCs was measured using the current density-voltage (J-V) curves. The J-V curves and the photovoltaic parameters of fabricated DSSCs are presented in **Figure 3** and **Table 1**.

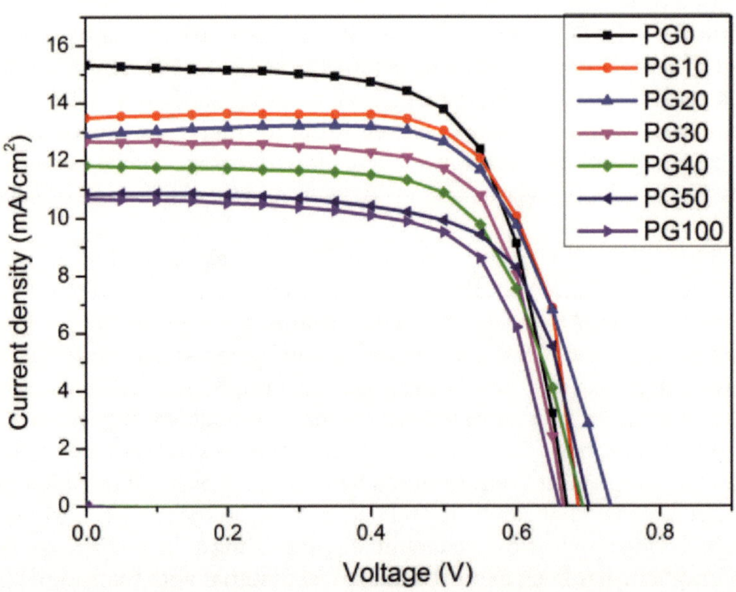

Figure 3.
J-V curves of DSSCs fabricated from Pt/rGO cathodes with different weight percents of rGO. Reproduced with permission Ref. [32]. Copyright 2020, Elsevier.

DSSCs	VOC (V)	J_{SC} (mA cm^{-2})	ff	η (%)
PG0	0.67	15.33	0.67	6.89
PG10	0.73	13.49	0.68	6.64
PG20	0.73	12.86	0.68	6.42
PG30	0.66	12.67	0.70	5.94
PG40	0.69	11.82	0.66	5.44
PG50	0.70	10.84	0.68	5.18
PG100	0.66	10.67	0.68	4.76

Table 1.
Photovoltaic parameters of DSSCs fabricated from Pt/rGO cathodes with different weight percents of rGO. Reproduced with permission Ref. [32]. Copyright 2020, Elsevier.

The J-V curve results showed that the conversion efficiency values of fabricated DSSCs was reduced when the amount of H_2PtCl_6 was reduced. In the structure of cathode, Pt nanoparticles played the catalytic and charge transferring roles. Since the intrinsic conductivity of rGO was relatively lower than that of Pt, the catalytic activities of Pt/rGO cathodes were lowered, leading to the decrease in efficiency [32]. However, rGO with the high surface area could play the role of supporting material to improve the electrochemical activity of Pt/rGO cathode. Therefore, Pt nanoparticles could be uniformly decorated on the surface of rGO lattice [41]. The high surface area of the Pt/rGO helps to maintain the high efficiencies of DSSCs in spite of the low conductivity of rGO, compared with Pt [42]. These results exhibited that the PG10 and PG20 DSSCs had better performance compared to other composite DSSCs, with efficiency values higher than 93% compared to PG0. The J-V results showed that PG20 was the appropriate material for fabrication of cathodes in DSSCs with high amount of Pt replacement and high conversion efficiency.

The Pt/rGO material in our study was investigated using different characterization technique like Fourier-transform infrared spectroscopy (FTIR), Raman spectroscopy, X-ray diffraction (XRD), and transmission electron microscopy (TEM). As shown in **Figure 4a**, the FTIR spectrum of GO revealed the vibrations peaks of the epoxide (C–O–C), alkene (C=C), and carboxyl functional groups (–COOH), at about 1050, 1628, and 1738 cm^{-1}, respectively. The broad peak at 3386 cm^{-1} originated from vibrations of hydroxyl (–OH) groups. The FTIR result demonstrated that during the oxidation process of Gi by improved Hummers' method, the oxygen-containing functional groups were introduced to the carbon framework of Gi to obtain GO [43]. For rGO, characteristic peaks could not be obviously determined, proving that the functional groups in the structure of GO were reduced. Similarly, the characteristic peaks of functional groups in the spectrum of PG20 were decreased, compared with GO, due to the functional groups remaining in the carbon lattice of PG20 after the reduction process [44].

As shown in **Figure 4b**, the Raman spectra of Gi, GO, rGO, and PG20 showed two characteristic peaks at around 1350 and 1580 cm^{-1}, corresponding to D and G-band peak. While G band relates to the vibration of sp^2 carbon network, the D band attributed to the structural defects and partial distortion in the structure of sp^2 carbon network. The ratios of D-band to G-band (I_D/I_G) represent the levels of defect in graphene-based materials [45]. The I_D/I_G value of GO is higher than that of Gi and lower than rGO or PG20, showing that the oxidation and reduction process increased the levels of defect in the structure of graphene sheets. The I_D/I_G value of PG20 was measured to be 1.12, where that of rGO accounts for 1.07.

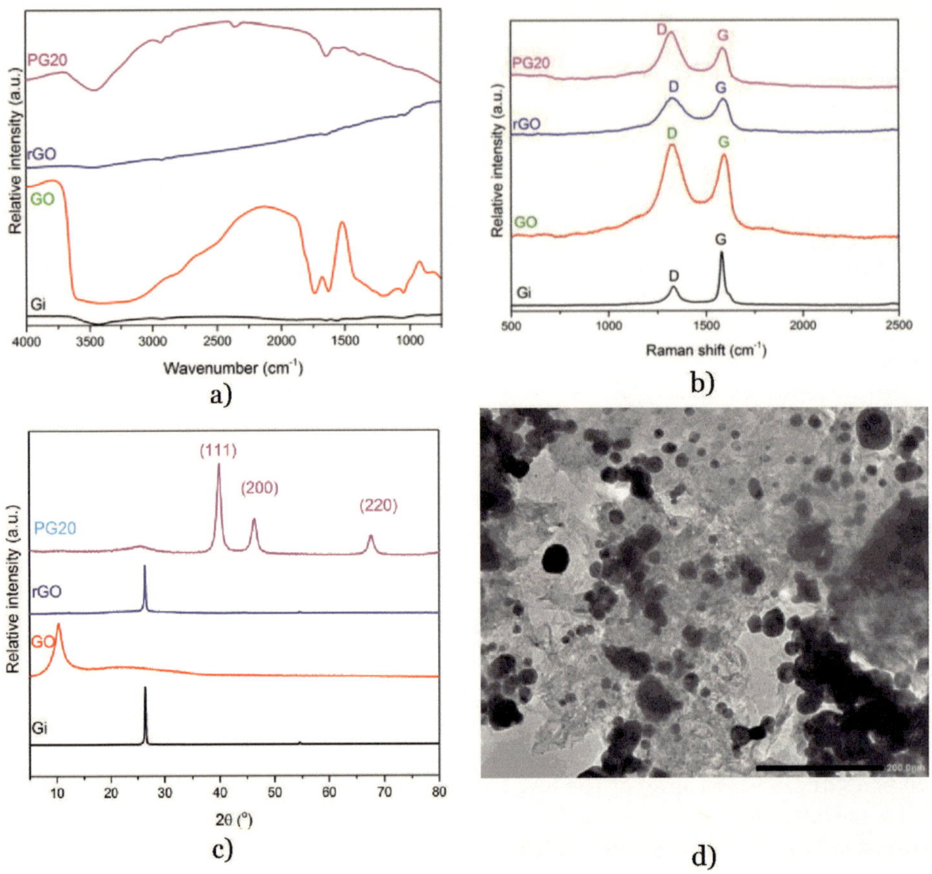

Figure 4.
Characterization of PG20 and precursor materials: (a) FTIR spectra, (b) Raman spectra, (c) XRD patterns, and (d) TEM image. Reproduced with permission Ref. [32]. Copyright 2020, Elsevier.

This means that the level of defect in the structure of PG20 is higher than that of rGO, this was attributed to the incorporation of Pt nanoparticles into the rGO sheets [46].

The XRD patterns of Gi, GO, rGO, and PG20 are presented in **Figure 4c**. In the pattern of GO, there is a characteristic peak (002) at 2θ = 10.42°, corresponding to the interlayer distance of 0.85 nm. For rGO, the (002) peak appears at 2θ = 26.87°; the interlayer distance between the rGO sheets was determined to be 0.33 nm. The diffraction peak of GO pattern is not observed in the pattern of rGO, indicating the removal of functional groups of GO during the reduction process. In the pattern of PG20, the diffraction peaks are determined at 2θ = 39.97, 46.36, and 67.69 °, corresponding to the (111), (200), and (220) crystalline planes of Pt nanoparticles, respectively. Additionally, there is a diffraction peak at 2θ = 25.32° in the pattern of PG20, which was similar to that of rGO. The XRD result proved the formation of Pt particles from H_2PtCl_6 and the reduction of GO to create the rGO sheets, proving the successful synthesis of Pt/rGO [46].

The morphology of PG20 composite was investigated using the TEM images. As shown in **Figure 4d**, rGO is observed to be the semi-transparent thin layer, indicating that the 2D structure of rGO was maintained after the annealing process. The Pt nanoparticles are observed as the black spheres which were decorated on the rGO sheets. From the TEM images, the sizes of the Pt nanoparticles were estimated to be in range of 10–30 nm. Besides, the Pt nanoparticles were eventually decorated on the rGO sheets. However, the Pt nanoparticles tended to agglomerate to form Pt clusters. The TEM images showed the high degree of agglomeration of Pt nanoparticles and the role of rGO as an effective supporting material in order to keep the distribution of Pt particles. The role as a supporting material of rGO helped to maintain the efficiency of PG20 DSSCs by increasing the surface areas of the materials.

The investigation of Pt/rGO cathodes in DSSCs proved that rGO was an excellent replacement for Pt in cathodes of DSSCs. By using the Pt/rGO composite for fabrication cathodes of DSSCs, the amount of Pt in DSSCs could be reduced and the efficiency could be maintained.

3.4 Graphene for the fabrication of anodes in DSSCs

Three key factors that affect the DSSC efficiency have been extensively studied: photo-electron generation, charge carrier transfer, and surface reaction. The unique and outstanding properties of graphene are ideal for addressing these factors. Graphene offers a 2D conductive support path for electron transfer, which can improve the electron transfer in photoanode materials and reduce the electron-hole recombination rate. For an example of TiO_2, without carbonaceous material supporter, electrons that are injected into TiO_2 nanoparticles may transfer around and need a much longer transfer distance. Graphene provides a faster electron transfer path and significantly reduces the electron-hole recombination rate in the TiO_2 material layer. Additionally, the large surface area creates more occasions for reactive group decoration and enhances the chemical reaction and reduction processes [11, 47].

ZnO materials have been made into composite/hybrid materials along with other metallic as well as graphitic structures to enhance their mechanical and electrochemical properties [48]. ZnO has superior optical and electrical properties that include high electron mobility in the order of 1500 $cm^2 V^{-1} s^{-1}$ at room temperature, wide band gap energy of 3.3 eV, and a high excitation (electron-hole) binding energy of 60 meV. However, to overcome the drawback of poor catalytic activity in ZnO due to its photoelectron recombination, many studies had incorporated graphene into the ZnO matrix to its efficiency and reduce the electron recombination process [49, 50]. There are two main method for fabrication of ZnO/rGO composite: the in-situ method, which used the Zn(II) precursor salts and mixed with GO or rGO [51, 52], and the ex-situ method, which used the ZnO nanomaterials (nanoparticles, nano wires, nanorods, etc.) and mixed with rGO [53, 54].

Our group conducted the experiments to investigate the performance of ZnO/rGO anodes in DSSCs. Accordingly, ZnO/rGO composite materials were synthesized from $Zn(O_2CCH_2)_3$ and GO. The ZnO/rGO anodes were fabricated from ZnO/rGO composite materials with different weight percent of rGO: 0, 0.1, 0.5, 1, and 5 wt% corresponding to ZnO, ZnO/rGO1, ZnO/rGO2, ZnO/rGO3, and ZnO/rGO4, respectively. DSSCs were assembled with fabricated anodes, cathodes from Dyesol Platinum Paste, N719 dye, and HSE electrolyte. The efficiency values of fabricated DSSCs were measured using the J-V curves. The results showed that the DSSCs fabricated from ZnO/rGO anodes demonstrated the high efficiencies, compared with DSSCs using the ZnO anode. The J-V curves and the photovoltaic parameters of fabricated DSSCs are presented in **Figure 5** and **Table 2**.

As shown in **Figure 5**, the efficiencies of ZnO/rGO1, ZnO/rGO2, and ZnO/rGO3 DSSCs were determined to be 1.25, 1.47, and 1.55%, respectively, which were higher than that of ZnO DSSC (1.08%). During the transfer process of the excited electron from N719 to FTO, the electron-hole recombination could significantly deteriorate the performance of DSSCs, leading to the decrease in the conversion

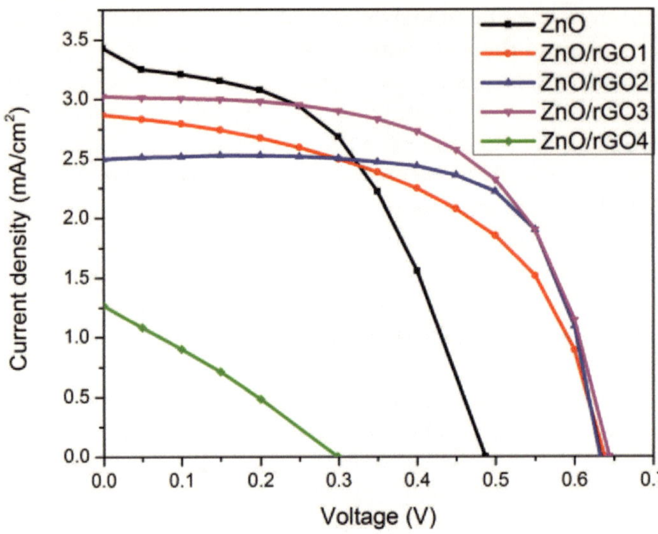

Figure 5.
J-V curves of DSSCs fabricated from ZnO/rGO anodes with different weight percents of rGO. Reproduced with permission Ref. [50]. Copyright 2020, Chemical Engineering Transaction.

DSSCs	VOC (V)	J_{SC} (mA cm^{-2})	ff	η (%)
ZnO	0.48	3.43	0.49	1.08
ZnO/rGO1	0.64	2.87	0.51	1.25
ZnO/rGO2	0.63	2.50	0.70	1.47
ZnO/rGO3	0.64	3.02	0.60	1.55
ZnO/rGO4	0.30	1.27	0.28	0.14

Table 2.
Photovoltaic parameters of fabricated from ZnO/rGO anodes with different weight percents of rGO. Reproduced with permission Ref. [50] Copyright 2020, Chemical Engineering Transaction.

efficiency values. The addition of rGO, a material with high conductivity and high electron mobility, could improve the transfer pathway of the excited electron. The introduction of rGO in the structure of the anode material could help to prevent the recombination reactions and enhanced the efficiency of DSSC. ZnO/rGO4 DSSC demonstrated a low efficiency, because transmittance of anode was drastically decreased, due to the high amount of rGO in ZnO/rGO4 composite. Besides, the excessive amount of rGO could become the recombination center that increased the recombination reactions [55, 56]. Therefore, the appropriate rGO weight percent in the composite was 1%, corresponding to the ZnO/rGO3 DSSC.

The ZnO/rGO material in our study was characterized using various methods including: FTIR spectroscopy, Raman spectroscopy, XRD patterns, and TEM images. **Figure 6a** demonstrates the FTIR spectra of ZnO/rGO3 and precursor materials. As mentioned above, the FTIR results showed that Gi was oxidized to obtain GO. The characteristic peaks of GO were reduced or disappeared on the spectra of GO and ZnO/rGO, indicating that the functional group of GO was reduced to create rGO and ZnO/rGO. The FTIR spectrum of ZnO/rGO3 revealed two typical peaks at 507.14 and 433.23 cm^{-1}, corresponding to the Zn-O vibration of hexagonal wurtzite structure of ZnO [49].

As shown in **Figure 6b**, Raman spectra of ZnO/rGO3 and precursor materials had the D and G-band peak. In the Raman spectra of ZnO and ZnO/rGO3, the

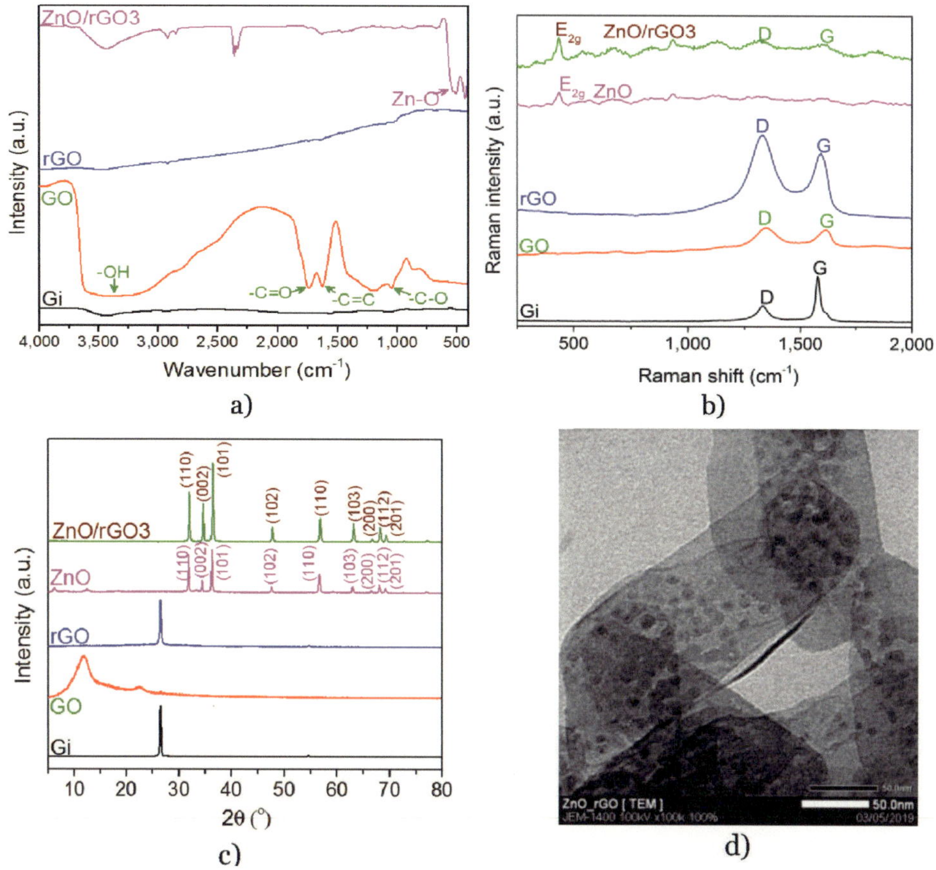

Figure 6.
Characterization of ZnO/rGO3 and precursor materials: (a) FTIR spectra, (b) Raman spectra, (c) XRD patterns, and (d) TEM image. Reproduced with permission Ref. [50]. Copyright 2020, Chemical Engineering Transaction.

vibration peak at 436 cm^{-1} was assigned to the E_2 non-polar phonon modes, corresponding to the hexagonal crystal structure of ZnO nanoparticles. The D-band and G-band peaks of ZnO/rGO3 spectrum were unclear, due to the low amount of rGO in the structure of ZnO/rGO3 [57].

The XRD patterns of ZnO/rGO3 and precursor materials are demonstrated in **Figure 6c**. The characteristic diffraction peaks of ZnO/rGO3 were observed at 2θ = 31.8, 34.5, 36.3, 47.5, 56.6, 62.8, 67.9, and 69.0°, corresponding to the (110), (002), (101), (102), (110), (103), (112), and (201) crystalline plane of wurtzite structures; similar with the JCPDS File No.36-1451 of ZnO [55, 58]. The diffraction peaks of GO and rGO were not observed in the pattern of ZnO/rGO3, because the weight percent of rGO in ZnO/rGO3 composite was low (1%).

As shown in **Figure 6d**, the TEM images of the ZnO/rGO3 demonstrated the decoration of ZnO nanoparticles on rGO sheets. As mentioned above, rGO is observed to be the semi-transparent thin layer, while the ZnO nanoparticles are the black spheres. The average size of ZnO nanoparticles was estimated to be about 10 nm. As can be seen, the ZnO nanoparticles were evenly distributed on rGO sheets. These results exhibited the good incorporation of rGO sheets and ZnO nanoparticles in the structure of ZnO/rGO3.

The investigation of ZnO/rGO anodes in DSSCs proved that the incorporation of rGO into the structure of ZnO could enhance the efficiencies of DSSCs. Therefore, the ZnO/rGO composite is the appropriate material for the fabrication of anodes in DSSCs.

4. Conclusion

In this chapter, the use of graphene as a supporting material for electrodes in DSSCs was discussed. Graphene-based DSSCs have emerged as an important class of photovoltaic material for many recent years. Graphene could be synthesized by a chemical reduction method and called rGO. This chapter demonstrates that graphene (or rGO) is the prospective material for the synthesis of composite materials, including Pt/rGO and ZnO/rGO, which were applied for the fabrication of cathodes and anodes in DSSCs. Graphene has shown great potential to replace Pt as cathodes for fabricating DSSCs. The mentioned experiments of PG20 demonstrated that Pt nanoparticles supported on rGO could help to reduce the amount of Pt in cathodes of DSSCs, while the high efficiencies were maintained. For anodes, the incorporation of rGO in ZnO/rGO composites enhanced the efficiencies of DSSCs, by reducing the electron recombination processes in the structure of anodes in DSSCs.

Based on the reported results, it is clear that graphene materials will have wide and promising applications in DSSCs. Various graphene-based materials with distinct properties should be extensively studied before they are practically applied for producing DSSCs.

Conflict of interest

The authors declare no conflict of interest.

Author details

Nguyen Huu Hieu[1,2]

1 Key Laboratory of Chemical Engineering and Petroleum Processing (CEPP Lab), Ho Chi Minh City University of Technology (HCMUT), VNU-HCMC, Vietnam

2 Faculty of Chemical Engineering, HCMUT, VNU-HCMC, Vietnam

*Address all correspondence to: nhhieubk@hcmut.edu.vn

References

[1] Pashaei B, Shahroosvand H. Molecularly engineered ruthenium polypyridyl complexes for using in dye-sensitized solar cell. Inorganic Chemistry Communications. 2020;**112**:107737. DOI: 10.1016/j.inoche.2019.107737

[2] Subramaniam K, Athanas AB, Kalaiyar S. Dual anchored ruthenium (II) sensitizer containing 4-nitro-phenylenediamine Schiff base ligand for dye sensitized solar cell application. Inorganic Chemistry Communications. 2019;**104**:88-92. DOI: 10.1016/j.inoche.2019.03.043

[3] Arjunan AT, Senthil TS. Dye sensitised solar cells. Materials Technology. 2013;**28**(1-2):9-14. DOI: 10.1179/1753555712Y.0000000040

[4] Ye M, Wen X, Wang M, Iocozzia J, Zhang N, Lin C, et al. Recent advances in dye-sensitized solar cells: From photoanodes, sensitizers and electrolytes to counter electrodes. Materials Today. 2015;**18**(3):155-162. DOI: 10.1016/j.cej.2016.07.001

[5] Chen M, Shao LL. Review on the recent progress of carbon counter electrodes for dye-sensitized solar cells. Chemical Engineering Journal. 2016;**304**:629-645. DOI: 10.1016/j.cej.2016.07.001

[6] Kumar S, Mahajan M, Singh R, Mahajan A. Silver nanoparticles anchored reduced graphene oxide for enhanced electrocatalytic activity towards methanol oxidation. Chemical Physics Letters. 2018;**693**:23-27. DOI: 10.1016/j.cplett.2018.01.003. DOI: 10.1016/j.cplett.2019.136797

[7] Cai L, Hou B, Shang Y, Xu L, Zhou B, Jiang X, et al. Synthesis of Fe3O4/graphene oxide/pristine graphene ternary composite and fabrication electrochemical sensor to detect dopamine and hydrogen peroxide. Chemical Physics Letters. 2019;**736**:136797. DOI: 10.1016/j.cplett.2019.136797

[8] Sun L, Li Y, Dan Y, Lu X, Li X, Wang F, et al. Self-assembled composite thin film counter electrode of cobalt sulfide/functionalized graphene for dye-sensitized solar cells. Thin Solid Films. 2019;**679**:8-14. DOI: 10.1016/j.tsf.2019.04.012

[9] Wei L, Wang P, Yang Y, Dong Y, Fan R, Song W, et al. Enhanced performance of dye sensitized solar cells by using a reduced graphene oxide/TiO2 blocking layer in the photoanode. Thin Solid Films. 2017;**639**:12-21. DOI: 10.1016/j.tsf.2017.08.011

[10] Gong J, Liang J, Sumathy K. Review on dye-sensitized solar cells (DSSCs): Fundamental concepts and. Renewable and Sustainable Energy Reviews. 2012;**16**(8):5848-5860. DOI: 10.1016/j.rser.2012.04.044

[11] Singh E, Nalwa HS. Graphene-based dye-sensitized solar cells: A review. Science of Advanced Materials. 2015;**7**(10):1863-1912. DOI: 10.1166/sam.2015.2438

[12] Low FW, Lai CW. Recent developments of graphene-TiO2 composite nanomaterials as efficient photoelectrodes in dye-sensitized solar cells: A review. Renewable and Sustainable Energy Reviews. 2018;**82**:103-125. DOI: 10.1016/j.compositesb.2018.01.013

[13] Yeoh ME, Chan KY. Recent advances in photo-anode for dye-sensitized solar cells: A review. International Journal of Energy Research. 2017;**41**(15):2446-2467. DOI: 10.1002/er.3764

[14] Tasis D. Recent progress on the synthesis of graphene-based

nanostructures as counter electrodes in DSSCs based on iodine/iodide electrolytes. Catalysts. 2017;**7**(8):234. DOI: 10.3390/catal7080234

[15] Ngidi NP, Ollengo AM, Nyamori VO. Heteroatom-doped graphene and its application as a counter electrode in dye-sensitized solar cells. International Journal of Energy Research. 2019;**43**(5):1702-1734. DOI: 10.1002/er.4326

[16] Ramsden JJ. Carbon-based nanomaterials and devices. In: Nanotechnology. United Kingdom: Elsevier Inc; 2011. pp. 189-197. DOI: 10.1016/B978-0-08-096447-8.00009-0

[17] Santos MC, Maynart C, Aveiro LR, Paz EC, Pinheiro VS. Carbon-based materials: Recent advances, challenges, and perspectives. In: Reference Module in Materials Science and Materials Engineering. [book on the Internet]. Elsevier Inc. DOI: 10.1016/B978-0-12-803581-8.09262-6

[18] Slepička P, Hubáček T, Kolská Z, Trostová S, Kasálková N, et al. The properties and application of carbon nanostructures. In: F Y. Polymer Science. London: InTech; 2013. pp. 175-201. DOI: 10.5772/51062

[19] Chen H, Liu T, Ren J, He H, Cao Y, Wang N, et al. Synergistic carbon nanotube aerogel–Pt nanocomposites toward enhanced energy conversion in dye-sensitized solar cells. Journal of Materials Chemistry A. 2016;**4**(9):3238-3244. DOI: 10.1039/C5TA10185A

[20] Kilic B. Produce of carbon nanotube/ZnO nanowires hybrid photoelectrode for efficient dye-sensitized solar cells. Journal of Materials Science: Materials in Electronics. 2019;**30**(4):3482-3487. DOI: 10.1007/s10854-018-00624-y

[21] Kilic B, Turkdogan S, Astam A, Ozer OC, Asgin M, Cebeci H, et al. Preparation of carbon nanotube/TiO2 mesoporous hybrid photoanode with iron pyrite (FeS2) thin films counter electrodes for dye-sensitized solar cell. Scientific Reports. 2016;**6**:27052. DOI: 10.1038/srep27052

[22] Sun KC, Memon AA, Arbab AA, Sahito IA, Kim MS, Yeo SY, et al. Electrocatalytic porous nanocomposite of graphite nanoplatelets anchored with exfoliated activated carbon filler as counter electrode for dye sensitized solar cells. Solar Energy. 2018;**167**:95-101. DOI: 10.1016/j.solener.2018.04.002

[23] Don MF, Ekanayake P, Nakajima H, Mahadi AH, Lim CM, Atod A. Acetylene carbon black-graphite composite as low-cost and efficient counter electrode for dye-sensitized solar cells (DSSCs). Ionics. 2019;**25**(11):5585-5593. DOI: 10.1007/s11581-019-03071-9

[24] Roy-Mayhew JD, Aksay IA. Graphene materials and their use in dye-sensitized solar cells. Chemical Reviews. 2014;**114**(12):6323-6348. DOI: 10.1021/cr400412a

[25] Mohan VB, Lau KT, Hui D, Bhattacharyya D. Graphene-based materials and their composites: A review on production, applications and product limitations. Composites Part B: Engineering. 2018;**142**:200-220. DOI: 10.1016/j.compositesb.2018.01.013

[26] Zhu Y, Murali S, Cai W, Li X, Suk JW, Potts JR, et al. Graphene and graphene oxide: Synthesis, properties, and applications. Advanced Materials. 2010;**22**(35):3906-3924. DOI: 10.1002/adma.201001068

[27] Avouris P, Dimitrakopoulos C. Graphene: Synthesis and applications. Materials Today. 2012;**15**(3):86-97. DOI: 10.1016/S1369-7021(12)70044-5

[28] Lee HC, Liu WW, Chai SP, Mohamed AR, Aziz A, Khe CS, et al. Review of the synthesis, transfer,

characterization and growth mechanisms of single and multilayer graphene. RSC Advances. 2017;**7**(26):15644-15693

[29] Sattar T. Current review on synthesis, composites and multifunctional properties of graphene. Topics in Current Chemistry. 2019;**377**(2):10. DOI: 10.1007/s41061-019-0235-6

[30] Smith AT, LaChance AM, Zeng S, Liu B, Sun L. Synthesis, properties, and applications of graphene oxide/reduced graphene oxide and their nanocomposites. Nano Materials Science. 2019;**1**(1):31-47. DOI: 10.1016/j.nanoms.2019.02.004

[31] Samantaray MR, Mondal AK, Murugadoss G, Pitchaimuthu S, Das S, Bahru R, et al. Synergetic effects of hybrid carbon nanostructured counter electrodes for dye-sensitized solar cells: A review. Materials. 2020;**13**(12):2779. DOI: 10.3390/ma13122779

[32] Cuong LV, Thinh ND, Nghia LTT, Khoa ND, Hung LK, Dat HH, et al. Synthesis of platinum/reduced graphene oxide composite pastes for fabrication of cathodes in dye-sensitized solar cells with screen-printing technology. Inorganic Chemistry Communications. 2020;**118**:108033. DOI: 10.1016/j.inoche.2020.108033

[33] Yu M, Wu X, Zhang J, Meng Y, Ma Y, Liu J, et al. Platinum nanoparticles-loaded holey reduced graphene oxide framework as freestanding counter electrodes of dye sensitized solar cells and methanol oxidation catalysts. Electrochimica Acta. 2017;**258**:485-494. DOI: 10.1016/j.electacta.2017.11.086

[34] Yen MY, Teng CC, Hsiao MC, Liu PI, Chuang WP, Ma CCM, et al. Platinum nanoparticles/graphene composite catalyst as a novel composite counter electrode for high performance dye-sensitized solar cells. Journal of Materials Chemistry. 2011;**21**(34):12880-12888. DOI: 10.1039/C1JM11850A

[35] Khoa NT, Van Thuan D, Kim SW, Park S, Van Tam T, Choi WM, et al. Facile fabrication of thermally reduced graphene oxide–platinum nanohybrids and their application in catalytic reduction and dye-sensitized solar cells. RSC Advances. 2016;**6**(2):1535-1541. DOI: 10.1039/C5RA21896A

[36] Wan L, Luo T, Wang S, Wang X, Guo Z, Xiong H, et al. Pt/graphene nanocomposites with low Pt-loadings: Synthesis through one-and two-step chemical reduction methods and their use as promising counter electrodes for DSSCs. Composites Science and Technology. 2015;**113**:46-53. DOI: 10.1016/j.compscitech.2015.03.015

[37] Suriani AB, Mohamed A, Othman MHD, Mamat MH, Hashim N, Ahmad MK, et al. Reduced graphene oxide-multiwalled carbon nanotubes hybrid film with low Pt loading as counter electrode for improved photovoltaic performance of dye-sensitised solar cells. Journal of Materials Science: Materials in Electronics. 2018;**29**(13):10723-10743. DOI: 10.1007/s10854-018-9139-4

[38] Capasso ABS, Palma AL, Najafi L, Castillo AEDR, Curreli N, Pellegrini V. CVD-graphene/graphene flakes dual-films as advanced DSSC counter electrodes. 2D Materials. 2019;**6**(3):035007. DOI: 10.1088/2053-1583/ab117e

[39] Cheng CE, Lin CY, Shan CH, Tsai SY, Lin KW, Chang CS, et al. Platinum-graphene counter electrodes for dye-sensitized solar cells. Journal of Applied Physics. 2013;**114**(1):014503. DOI: 10.1063/1.4812498

[40] Lin CA, Lee CP, Ho ST, W TC, Chi YW, Huang KP, et al. Nitrogen-doped Graphene/platinum counter

electrodes for dye-sensitized solar cells. ACS Photonics. 2014;**1**(12):1264-1269. DOI: 10.1021/ph500219r

[41] Shih PT, Dong RX, Shen SY, Vittal R, Lin JJ, Ho KC. Transparent graphene–platinum nanohybrid films for counter electrodes in high efficiency dye-sensitized solar cells. Journal of Materials Chemistry A. 2014;**2**(23):8742-8748. DOI: 10.1039/C3TA12931D

[42] Wan L, Zhang Q, Wang S, Wang X, Guo Z, Dong B, et al. A two-step reduction method for synthesizing graphene nanocomposites with a low loading of well-dispersed platinum nanoparticles for use as counter electrodes in dye-sensitized solar cells. Journal of Materials Science. 2015;**50**(12):4412-4421. DOI: 10.1007/s10853-015-8998-9

[43] Chen W, Yan L, Bangal PR. Chemical reduction of graphene oxide to graphene by sulfur-containing compounds. The Journal of Physical Chemistry C. 2010;**114**:19885-19890. DOI: 10.1021/jp107131v

[44] Kempegowda R, Antony D, Malingappa P. Graphene–platinum nanocomposite as a sensitive and selective voltammetric sensor for trace level arsenic quantification. International Journal of Smart and Nano Materials. 2014;**5**(1):17-32. DOI: 10.1080/19475411.2014.898710

[45] Yang M, Zhao X, Yao C, Kong Y, Ma L, Shen X. Nanostructured cation disordered Li2FeTiO4/graphene composite as high capacity cathode for lithium-ion batteries. Materials Technology. 2016;**31**(9):537-543. DOI: 10.1080/10667857.2016.1192372

[46] Qiao J, Zhang L, Gao S, Li N. Facile fabrication of graphene-supported Pt electrochemical sensor for determination of caffeine. Applied Biochemistry and Biotechnology. 2020;**190**(2):529-539. DOI: 10.1007/s12010-019-03104-z

[47] Guo X, Lu G, Chen J. Graphene-based materials for photoanodes in dye-sensitized solar cells. Frontiers in Energy Research. 2015;**3**:50. DOI: 10.3389/fenrg.2015.00050

[48] Vittal R, Ho KC. Zinc oxide based dye-sensitized solar cells: A review. Renewable and Sustainable Energy Reviews. 2017;**70**:920-935. DOI: 10.1016/j.rser.2016.11.273

[49] Philip MR, Nguyen H, Babu R, Krishnakumar V, Bui TH. Polyol synthesis of zinc oxide-graphene composites: Enhanced dye-sensitized solar cell efficiency. Current Nanomaterials. 2018;**31**(5):52-60. DOI: 10.2174/2405461503666180507124310

[50] Nghia LTT, Phat LT, NTT M, Dat HH, Hung LK, Hien TM, et al. Synthesis of zinc oxide/reduced graphene oxide composites for fabrication of anodes in dye-sensitized solar cells. Chemical Engineering Transactions. 2019;**78**:61-66. DOI: 10.3303/CET2078011. DOI: 10.3303/CET2078011

[51] Jayabal P, Gayathri S, Sasirekha V, Mayandi J. Preparation and characterization of ZnO/graphene nanocomposite for improved photovoltaic performance. Journal of Nanoparticle Research. 2014;**16**(11):2640. DOI: 10.1007/s11051-014-2640-7

[52] Abdullah H, Atiqah NA, Omar A, Asshaari I, Mahalingam S, Razali Z, et al. Structural, morphological, electrical and electron transport studies in ZnO–rGO (wt%= 0.01, 0.05 and 0.1) based dye-sensitized solar cell. Journal of Materials Science: Materials in Electronics. 2015;**26**(4):2263-2270. DOI: 10.1007/s10854-015-2679-y

[53] Song JL, Wang X. Effect of incorporation of reduced graphene

oxide on ZnO-based dye-sensitized solar cells. Physica E: Low-Dimensional Systems and Nanostructures. 2016;**81**:14-18. DOI: 10.1016/j.physe.2016.02.005

[54] Chang WC. Graphene/ZnO nanoparticle composite photoelectrodes for dye-sensitized solar cells with enhanced photovoltaic performance. Journal of Nanoscience and Nanotechnology. 2016;**16**(9):9160-9165. DOI: 10.1166/jnn.2016.12900

[55] Bykkam S, Kalagadda VR, Kalagadda B, Selvam KP, Hayashi Y. Ultrasonic-assisted synthesis of ZnO nano particles decked with few layered graphene nanocomposite as photoanode in dye-sensitized solar cell. Journal of Materials Science: Materials in Electronics. 2017;**28**(8):6217-6225. DOI: 10.1007/s10854-016-6301-8. DOI: 10.1007/s10854-016-6301-8

[56] Siwach B, Mohan D, Jyoti D. To investigate opulence of graphene in ZnO/graphene nanocomposites based dye sensitized solar cells. Journal of Materials Science: Materials in Electronics. 2017;**28**(15):11500-11508. DOI: 10.1007/s10854-017-6946-y

[57] Khurana G, Sahoo S, Barik SK, Katiyar RS. Improved photovoltaic performance of dye sensitized solar cell using ZnO–graphene nano-composites. Journal of Alloys and Compounds. 2013;**578**:257-260. DOI: 10.1016/j.jallcom.2013.05.080. DOI: 10.1016/j.jallcom.2013.05.080

[58] Siwach B, Mohan D, Singh KK, Kumar A, Barala M. Effect of carbonaceous counter electrodes on the performance of ZnO-graphene nanocomposites based dye sensitized solar cells. Ceramics International. 2018;**44**(17):21120-21126

Chapter 15

Effect of Combination of Natural Dyes and the Blocking Layer on the Performance of DSSC

Md. Mosharraf Hossain Bhuiyan, Fahmid Kabir, Md. Serajum Manir, Md. Saifur Rahaman, Md. Robiul Hossain, Prosenjit Barua, Bikram Ghosh, Fumiaki Mitsugi, Tomoaki Ikegami, Saiful Huque and Mubarak Ahmad Khan

Abstract

Over the years, researchers have been working on replacing sensitized dye for dye sensitized solar cells (DSSC), because of its low production cost, biodegradability, and non-toxicity. However, the overall performance of natural dye-based DSSCs is low compared to the DSSCs sensitized with Ruthenium based dyes. The combination of natural dyes with an optimized choice of the extracting solvents and the proper volume ratio of mixture of the dyes, enhances inherent properties, such as absorption and adsorption of the dyes. It also allows the device to utilize photon energy more efficiently over the entire visible wavelength. As a result, DSSC sensitized with the dye mixture shows higher absorbance, and cumulative absorption properties over the whole visible region than the DSSC fabricated with individual dyes and showed higher photocurrent. Another effective way to improve cell efficiency is by using a blocking layer. The blocking layer increases the photocurrent, is mainly due to the improvement of the electron recombination at the transparent conducting oxide/electrolyte interfaces. Also, the blocking layer's compact structure creates an effective pathway for electron transportation; thus, the device's photocurrent increases. Additionally, a slight improvement in the open-circuit voltage and fill factor was observed, thus cell efficiency enhances significantly. By both the proper ratio of dye mixture and the blocking layer improves cell performance of DSSC and opens a new pathway for future studies.

Keywords: DSSC, natural dye, natural dye based DSSC, dye combination, blocking layer

1. Introduction

The global energy demand has been continuously increasing due to the continuous growth of the world population, economic development, standard of living, and drive of modern technologies. In the current situation, about 87% of the primary energy needs are mostly supplied through fossil fuels (coal, oil, and gas) [1, 2].

However, the sources of these fossil fuel reserves are depleting very fast. The existing sources of energy are inadequate, and if the fuel consumption continued at current usage rates, it would last only about 50 years [3]. On the other hand, burning fossil fuels releases carbon dioxide and other greenhouse gases (e.g., water vapor, methane, nitrous oxide, sulfur dioxide, and other ozone-depleting substance) in the atmosphere, making them the primary contributors to global warming and climate change. Because of the depletion of fossil fuels and global warming, in recent years, researchers are endeavoring severe attempts to find out various ways to meet energy demand around the world. Renewable energy could be an eco-friendly, alternative, sustainable energy resource because they are inexhaustible and will not pollute the environment for us or those of future generations by emitting harmful gases. Many alternative renewable energy sources have already been available, such as solar, hydro, wind, biogas, biomass, geothermal, wave, and tidal energy. Among all renewable resources, solar energy can be the solution to the problem of dwindling fossil-fuel reserves [4].

Solar energy is the cleanest and most abundant renewable energy source available. The solar cell or photovoltaic (PV) device is used for converting the energy of sunlight into useable electrical energy. The generated energy from solar does not produce any harmful emissions, consumes no fossil fuels, has no moving parts, and requires little maintenance. The development of PV technology is growing, and intensive research works are undertaken worldwide to improve cell performance and reduce the cost of the cell. With a history dating back over 60 years, since the very first silicon bipolar solar cell, the last three decades silicon solar cell has seen immeasurable advancement in both the performance of experimental and commercial cells. First-generation silicon solar cells showed their value in the market with the advantages, including high efficiency (26.6%), high reliability, low cost, ease of fabrication, and environmentally friendly traits [5–8]. Second generation thin-film (e.g., amorphous silicon (a-Si), cadmium telluride (CdTe), and copper indium gallium selenide (CIGS)) solar cells are cheaper than the mature Si solar cells; additionally, thin-films are easier to handle and more flexible. However, the shortage of Tellurium and Indium makes it hard to manufacture solar cells commercially. Also, Cadmium is extremely poisonous and medical problems with environmental impact [9]. This significant concern opened the method of exploration of finding other elective materials and further innovation for solar cells. Several new thinner-films have surfaced through concentrated research with higher potential, including dye-sensitized solar cell (DSSC), perovskite solar cell (PSC), copper zinc tin sulfide (CZTS) solar cell, organic solar cell (OSC), and quantum dot solar cell (QDSC) [10].

The DSSC belongs to the group of thin-films, functions on a semiconductor generated into an electrolyte and a light-sensitive anode [11]. In 1988, Brian O'Regan and Michael Grätzel at UC Berkeley, USA initially co-invented the modern version of DSSC and later they further developed this work at the ÉcolePolytechniqueFédérale de Lausanne, Switzerland [12]. Brian O'Regan and Michael Grätzel reported the first modern version of DSSC in 1991 with an efficiency of 7.1–7.9% [13, 14]. DSSC can convert the solar energy to electrical energy by using photosensitive dye. DSSC is fabricated by using conventional roll-printing systems. The semi- transparency and semi-flexibility of DSSC offer a diversity of usages not appropriate for glass-based construction and most of the materials used are low-cost. However, practical elimination of several expensive elements has proven to be difficult, notably Pt and ruthenium (Ru). The energy conversion efficiency of the most recent laboratory-developed module is approximately 14.7% [15]. This chapter is focused on the improvement of efficiency of DSSC by the combination of natural dyes and the blocking layer. In this work, structure and operation principle of the third generation dye-sensitized solar cell (DSSC) has been discussed in the second section. Section 3 explains the combination of natural dyes with an optimized choice of the extracting solvents and the dye mixture's at proper volume ratio, enhancing the dye sensitizer's inherent properties, such as absorption and adsorption, thus improving the cell efficiency. Section 4 explains the comprehensive study of the blocking layer and its effect on the cell efficiency, and finally in section five, overall conclusions and accomplishments of this study have been mentioned.

2. Basics of DSSC

As shown in **Figure 1**, a typical DSSC consists of five different parts, such as, (1) transparent conducting oxide (TCO), glass substrate, (2) anode (wide band-gap semiconductor material layer on TCO), (3) photosensitizer, (4) electrolyte and (5) cathode (platinum/carbon layer on TCO). The components of a DSSC are: two transparent conductive oxide [indium tin oxide (ITO), fluorine-doped tin oxide (FTO), Indium Zinc Oxide (IZO) and Aluminum Zinc Oxide (AZO)] glass electrodes. One of the electrodes is the anode, the working anode, which is printed with semiconductor material [TiO_2, ZnO, SnO_2, $SrTiO_3$, Zn_2SnO_4, Nb_2O_5, etc.] nanoparticles (particle size around 20–50 nm). The semiconductor oxides are sensitized with a photosensitizer (metal complex sensitizer, metal-free organic sensitizer or natural dye sensitizer), which absorbs the photons. The other electrode is the counter electrode [platinum or carbon coated TCO] and in between the two working electrodes is the electrolyte containing the redox couple [I^-/I_3^-, Br^-/Br_3^-, $SCN^-/(SCN)_3^-$ and $SeCN^-/(SeCN)_3^-$, etc.].

Figure 2 illustrates the schematic diagram of the basic working principle of a typical TiO_2 based DSSC. All other semiconductor-based DSSC such as ZnO, SnO_2, etc. works under the same principle. Under illumination, a photo-excited electron is injected from the excited state of the dye (D*) from the highest occupied molecule orbital (HOMO) to lowest un-occupied molecular orbital (LUMO) (Eq. (1)). The excited electron is injected to the conduction band of the semiconductor material. The injected electron percolates through the semiconductor material by a driving chemical diffusion gradient and is collected at the TCO glass substrate (Eq. (2)). After passing through an external circuit, and reaches the counter electrode, thus,

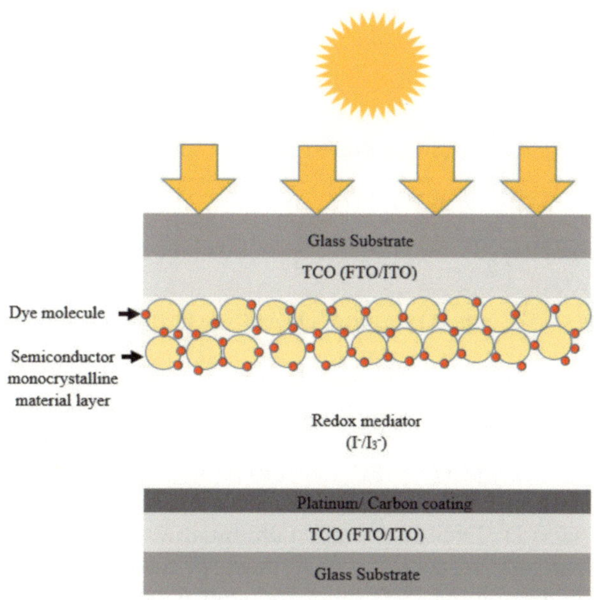

Figure 1.
Schematic diagram of basic structure of DSSC.

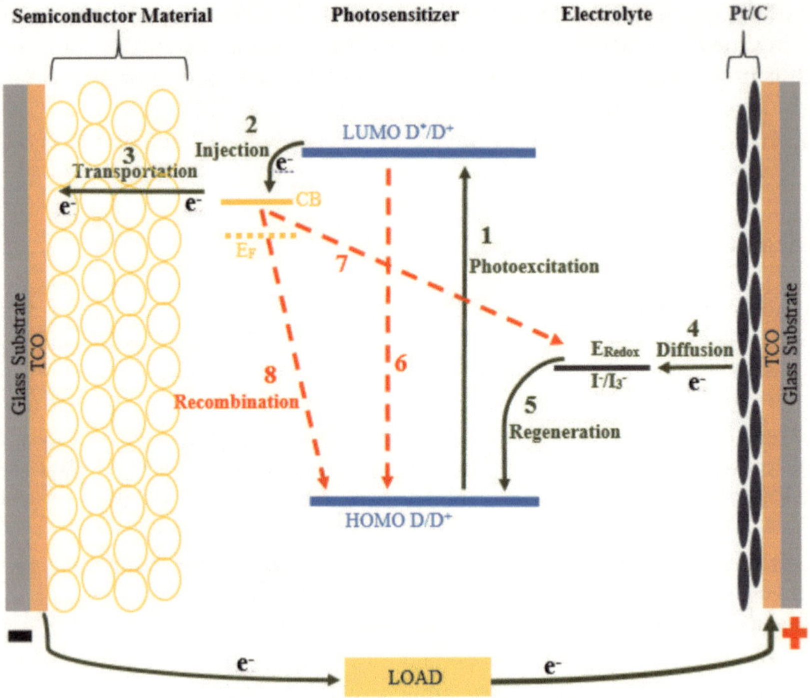

Figure 2.
Operation principal of typical DSSC.

dye regeneration takes place due to the acceptance of electrons from I^- ion redox mediator, and I^- gets oxidized to I_3^- (Eq. (3)). To complete the circle, by electron donation, I^- ions regenerated by the reduction of I_3^- ions at the cathode (Eq. (4)). However, some undesirable reactions are simultaneously taking place, such as non-radiation relaxation (Eqs. (5) and (6) no. red arrow in **Figure 2**), recombination of injected electrons with the oxidized dye (Eqs. (6) and (7 no. red arrow in **Figure 2**) and recombination of injected electrons with I_3^- (Eqs. (7) and (8) no. red arrow in **Figure 2**). In brief, the sequence of events in a DSSC is as follows [16]:

$$TiO_2|D + h\nu \rightarrow TiO_2|D^* \text{ Excitation of dye upon illumination} \quad (1)$$

$$TiO_2|D^* \rightarrow TiO_2|D^+ + e^- \text{ Oxidation of dye due to injection of electrons in } TiO_2 \text{ photoanode} \quad (2)$$

$$TiO_2|D^+ + \frac{3}{2}I^- \rightarrow TiO_2|D + \frac{1}{2}I_3^- \text{ Oxidation of electrolyte} \quad (3)$$

$$\frac{1}{2}I_3^- + 2e^-|CE \rightarrow \frac{3}{2}I^- \text{ Restoration of electrolyte at the counter electrode} \quad (4)$$

$$TiO_2|D^* \rightarrow TiO_2|D \text{ Recombination of dye} \quad (5)$$

$$TiO_2|D^+ + e^- \rightarrow TiO_2|D \text{ Dye recover to ground state} \quad (6)$$

$$I_3^- + 2e^-|TiO_2 \rightarrow 3I^- \text{ Recombination of electrolyte} \quad (7)$$

D: Dye sensitizer; D^*: Excited dye upon illumination; D^+: Oxidized dye.

Nemours researchers are working on to improve cell performance by different means, such as modifying the TCO/semiconductor material interface by blocking layer; modifying semiconductor material by doping, annealing time, radiation, carbon nanotubes, etc.; modifying the absorption properties of the dyes by Ru dye, organic dye, dye mixture, etc. to enhance cell efficiency [17–22].

3. Effect of combination of natural dyes

The dye in DSSCs has a vital role in harnessing solar energy from the sun and converts it into useable electrical energy. The primary charges in the dyes separate through photo-excitation, and photo-excited dyes inject electrons into the conduction band of semiconductor material. A dye should fulfill some pre-requisites to be considered efficient dye: (1) binding firmly with the semiconductor material; (2) higher molar absorption capabilities for maximum absorption from visible to IR-region; (3) fast electron transfer; (4) LUMO of the dye should be higher than the conduction band of semiconductor for efficient electron injection into the semiconductor material; (5) HOMO of the dye should be lower than the redox couple for efficient regeneration of oxidized dye; and (6) slow degradation (or do not degrade at all) [16, 23–25]. The dyes used in DSSC are divided into three types: metal complexes dye sensitizer, metal-free organic dye sensitizer, and natural dye sensitizer. Metal complexes dye sensitizers, such as polypyridyl complexes of Ruthenium (Ru), Osmium (Os), metal porphyrin, phthalocyanine are the most efficient and durable dye for DSSC application. However, these dyes have a complex synthesis process, release chemicals as a by-product, and require rear-earth material for the synthesis process. As a result, the overall fabrication process highly depended on the rear earth material that is neither sustainable nor economical. On the other hand, metal-free organic dye sensitizer has advantages over metal complex dye sensitizer, reducing the use of rear-earth material, higher molar absorption co-efficient, and preprocessing color. However, these advantages are offset by their instability, tedious manufacturing process, tendency to undergo degradation, and toxicity. These significant limitations influenced scientists to work on possible replacements for metal complexes or metal-free organic dye sensitizers [16].

Over the years, significant research has been done to determine the possibility of replacing sensitized dye. Natural dye has several advantages over sensitized dyes. These include low production cost, high availability, easy access, simple fabrication technique, biodegradable, environment friendly, purity grade, non-toxic, and reducing the use of rear-earth material. Natural dye-based DSSCs have attracted considerable attraction as an alternative way to produce low-cost dyes to a large extent by extracting dyes from natural resources. In nature, some vegetables, fruits, flowers, leaves, seeds, roots, stems, bacteria, and algae exhibit various colors due to plant pigmentation [16]. The natural dyes are four major families which are chlorophyll, anthocyanin, carotenoids, and flavonoids [26, 27].

Chlorophyll, which is the most widespread pigment occurring naturally in plants, fungi, bryophytes and algae. The molecular structure of a chlorophyll consists of a Magnesium-containing tetrapyrrolic ring, encircled by other side chains. The chlorophylls are classed mainly as chlorophyll-a, chlorophyll-b, chlorophyll-c1, chlorophyll-c2, chlorophyll-d, and chlorophyll-f. They absorb light from red, blue, and violet in the visible wavelengths with an absorptionmaximumof $\sim$670 nm while reflecting green wavelengths. Chlorophyll dye molecule create an electronic coupling with the conduction band of semiconductor material through the carboxylic groups, which helps to anchor the dye molecules and transfer injected electron efficiently from the dye sensitizer to the conduction band of semiconductor material [16, 28]. **Figure 3** shows the basic molecular structure of chlorophyll and the binding chlorophyll and semiconductor material (e.g., TiO_2).

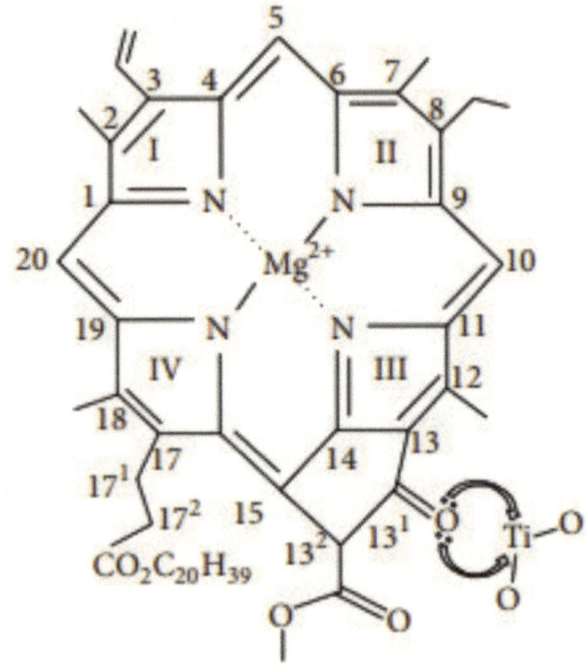

Figure 3.
Chlorophyll-semiconductor material (i.e., TiO_2) interaction [29].

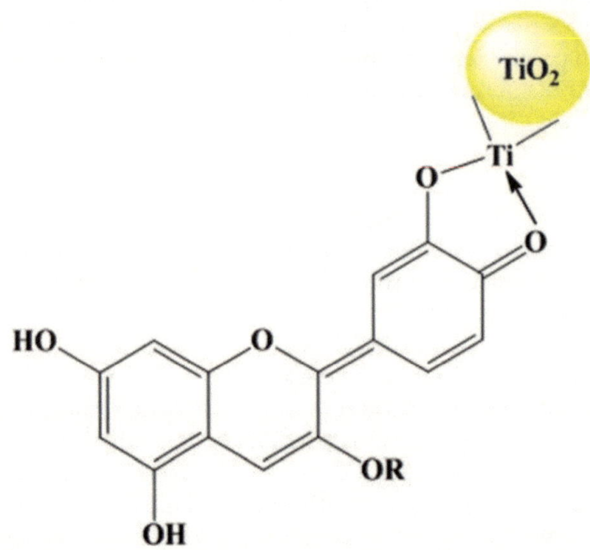

Figure 4.
Basic structure of anthocyanin and anthocyanin-semiconductor material (TiO_2) intaraction [31].

Anthocyanins are also an abundant and widespread group of water-soluble pigments in plants. They absorb light at the longest wavelengths. Depending on the pH value, anthocyanins are responsible for the existence of attractive colors, such as red, orange, magenta, pink, blue, blue-black and purple floral [16, 30]. Generally, the carbonyl and hydroxyl functional groups in the anthocyanin dye sensitizers create an electronic coupling with the semiconductor material's conduction band, which helps transfer the excited electron efficiently to the conduction band of semiconductor material [16]. **Figure 4** shows the basic interaction between anthocyanin and semiconductor material (e.g., TiO_2).

Carotenoids occur in many plants and algae, as well as several bacteria, and fungi. It contributes to yellow, orange, and red colors and allows them to absorb short-wave visible light [32]. Carotenoids can be divided into two major types: xanthophylls (with oxygen) and carotenes (purely hydrocarbons and without oxygen) [16, 33]. **Figure 5** illustrates the interaction between carotenoids- semiconductor material (i.e., TiO_2).

Flavonoids are essential floral pigments. The development of a specific color depends on the accumulation of flavonoid chromophores and other intrinsic and extrinsic factors. Chemically, the flavonoids have a C6- C3- C6 carbon framework with two connected two phenyl rings (A and B) and a heterocyclic ring (C). Depending on the oxidation potential of the C-ring, the particular flavonoids absorb light in the visible wavelength. Till now, over 5000 flavonoids have been identified from different plants. Most of the flavonoid pigment has loosely or unbound electrons. Thus less energy is required for excitation of such electrons is lower compared to the others. As a result, those pigment molecules can be energized by the light within the visible range [16].

The overall cell efficiency of natural dye-based DSSCs is comparably low compared to DSSCs sensitized with sensitized dyes. Due to the inadequate interaction between dyes and semiconductor surface, a significant reduction of the cell's short-circuit current. The pigment's long structure obstructs the dye molecules to form a bond with the oxide surface of the semiconductor materials effectively. Those are the field of works that are yet to be developed in natural dye DSSCs to achieve high-efficiency devices and device stability. To further raise the efficiency of the DSSC combination of dyes has been explored and reported DSSC or to broaden the absorption spectrum [35–39]. A combination of natural dyes with an optimized choice of the extracting solvent enhances the absorption of solar light and allowed utilization of the photon energy more efficiently. As a result, DSSC sensitized with the dye mixture shows higher absorbance, and cumulative absorption properties over the entire visible region than the DSSC fabricated with single individual dyes [35, 36].

Kabir et al. studied the effect of chlorophyll and anthocyanin dye mixture on the cell performance of natural dye-based DSSC. They also mixed the dyes at five different volume ratios to find the optimized dye mixture. The cell conversion efficiency of DSSC fabricated with individual chlorophyll, and anthocyanin dyes

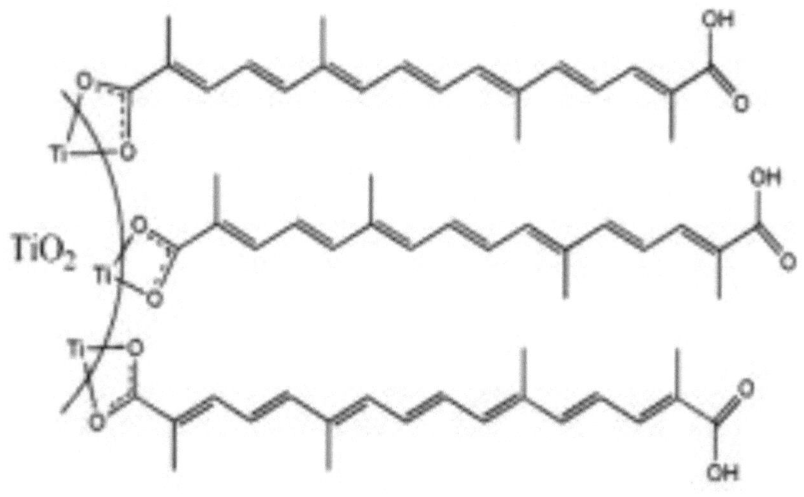

Figure 5.
Carotenoids -semiconductor material (i.e., TiO_2) interaction [34].

were 0.466% and 0.531%, respectively. DSSC co-sensitized with the optimized dye mixture (20% chlorophyll +80% anthocyanin) showed cell conversion efficiency of 0.847%, which is almost 1.82 and 1.6 times higher than the cell efficiency of the individual chlorophyll and anthocyanin dye-sensitized DSSC's (shown in **Figure 6**). The chemical characteristics study of the dye showed that no new bond except has formed; however, few shifts in the adsorption peak was observed (Shown in **Figure 7** and **Table 1**). Similar characteristics were seen when dyes were adsorbed the TiO_2 semiconductor material (shown in **Figure 8**. and **Table 2**, [36].

Figure 9 illustrates the UV–visible absorption spectra of natural chlorophyll (green), anthocyanin (red), and the optimum combination of dyes (green + red) diluted in ethanol. The dye mixture has demonstrated the cumulative absorption properties of both individual green and red dye.

Kabir et al. also studied the effect of betalain and curcumin dye combination on the cell performance of natural dye-based DSSC. They also mixed the dyes at three different volume ratios to find the optimized dye combination. The optimized dye mixture demonstrated the cumulative absorption properties of both individual betalain and curcumin dye (shown in **Figure 10**). The DSSC fabricate with the combination of betalain and curcumin dye also showed superior cell performance than DSSC manufactured with individual betalain and curcumin dye (shown in **Figure 11** and **Table 3**) [35].

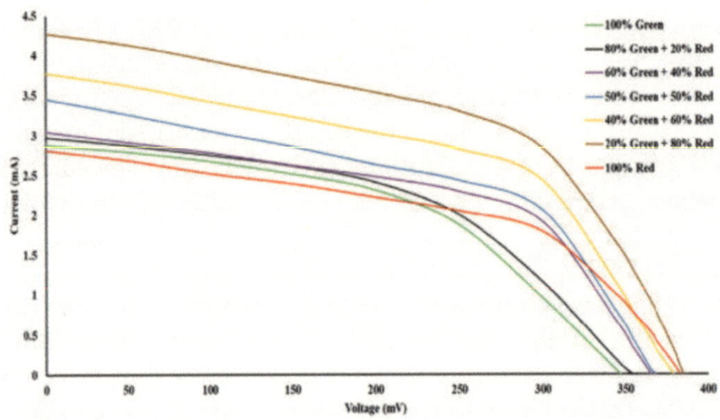

Figure 6.
I-V characteristics of DSSC fabricated with chlorophyll, anthocyanin and different combinations [36].

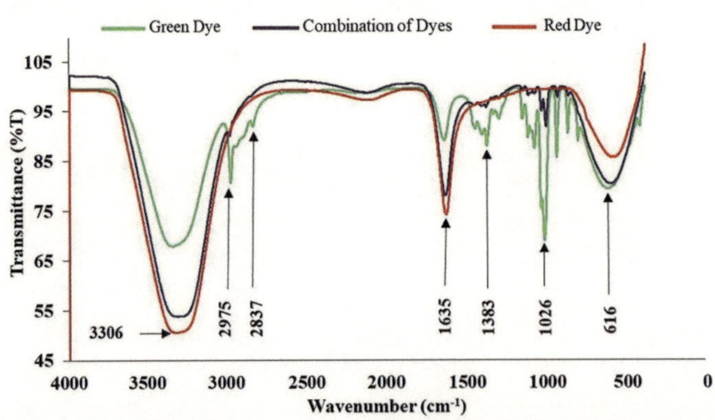

Figure 7.
FT-IR adsorption spectra of natural chlorophyll (green), anthocyanin (red), and combination (20% green +80% red) of dyes (without TiO2) [36].

Functional group	Absorption range (in cm^{-1})	Type of vibration	Intensity	Absorption peak of green dye (in cm^{-1})	Absorption peak of combination of dyes (in cm^{-1})	Absorption peak of red dye (in cm^{-1})
Alkyl Halide (C—Cl)	600–800	Stretch	Strong	616	610	606
Alkene (=C—H)	675–1000	Bending	Strong	944	950	—
Ether (C—O)	1000–1300	Stretch	Strong	1017	1026	—
Amine (C—N)	1080–1360	Stretch	Weak	1338	1354	—
Aromatic (C=C)	1400–1600	Stretch	Medium weak	1422	1404	—
Alkene (C=C)	1620–1680	Stretch	Variable	1652	1619	1635
Alkane (C—H)	2820–2850	Stretch (symmetric)	Strong	2837	—	—
Alkane (C—H)	2850–3000	Stretch (asymmetric)	Strong	2975	—	—
Alcohol (O—H)	3200–3600	Stretch	Broad and strong	3346	3320	3329

Table 1.
IR absorption of organic functional groups of natural green, red, and combination of dyes (20% green +80% red) without TiO_2 [36].

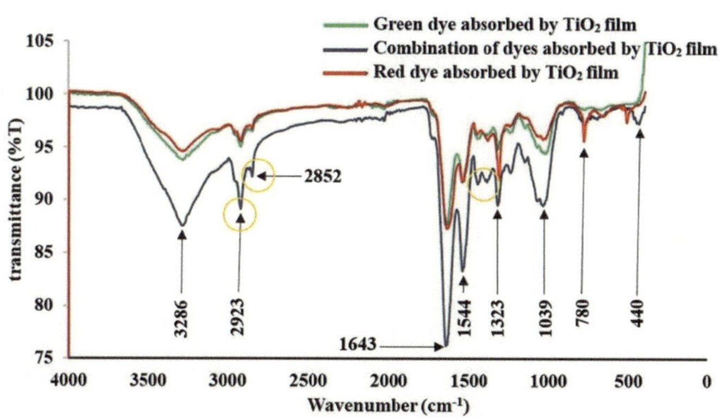

Figure 8.
FT-IR adsorption spectra of natural chlorophyll (green), anthocyanin (red), and combination (20% green +80% red) of dyes (with TiO_2) [36].

Nonetheless, to the best of our knowledge, combination of dyes have a positive impact on the cell performance of natural based DSSC.

4. Effect of blocking layer in DSSC

In DSSCs, a porous layer of nanostructuredsemiconductor materials such as TiO_2 [40–45], ZnO [46–48], SnO_2 [49, 50], $SrTiO_3$ [51, 52] Zn_2SnO_4 [53, 54] and Nb_2O_5 [55] called a photo anode, covered with photosynthetic dye. The photo anode of

Functional group	Absorption range (in cm^{-1})	Type of vibration	Intensity	Absorption peak of green dye by TiO_2 film (in cm^{-1})	Absorption peak of combination of dyes by TiO_2 film (in cm^{-1})	Absorption peak of red dye by TiO_2 film (in cm^{-1})
Ti—O—Ti	400–800	Stretch	Strong	438	440	515
Alkene (=C—H)	675–1000	Bending	Strong	817	782	780
Ether (C—O)	1000–1300	Stretch	Strong	1042	1039	1040
Amine (C—N)	1080–1360	Stretch	Weak	1324	1323	1315
Aromatic (C=C)	1400–1600	Stretch	Medium weak	1546	1544	1538
Alkene (C=C)	1620–1680	Stretch	Variable	1639	1643	1636
Alkane (C—H)	2820–2850	Stretch (symmetric)	Strong	2848	2852	2856
Alkane (C—H)	2850–3000	Stretch (asymmetric)	Strong	2924	2923	2924
Alcohol (O—H)	3200–3600	Stretch	Broad and strong	3384	3286	3281

Table 2.
IR absorption of organic functional groups of natural green, red, and combination of dyes (20% green +80% red) with TiO_2 [36].

DSSC influences the photo generated current. Highly porous structures and large surface areas of the nanostructured semiconductor materials increase the dye absorption and move the photo-induced electron towards the load [56]. Intensive research has been undertaken by the DSSC research community to increase photo-induced current and understand the mechanisms responsible for losses in the cell. Radiation less relaxation of energized dye, electron recombination with the oxidized dye; and electron recombination with the tri-iodide in the electrolyte are the main reasons for limiting the photocurrent in the cell. Generally, the first two have a negligible impact, while the last one shows a significant impact [56, 57].

Electron recombination occurs when electron transfer to I_3^- in the electrolytes via semiconductor material and the TCO. Electron recombination through both routes needs to be reduced to prevent loss. In I^-/I_3^- redox couple, the electron transfer via the TCO is negligible due to small exchange current density between I_3^--I^-. Generally, the losses via the FTO under short-circuit condition is insignificant, because the Fermi level of the TCO (i.e., FTO) is close to the redox Fermi level. However, under illumination, the quasi-Fermi level of the semiconductor material (i.e., TiO_2) rises rapidly with distance from the TCO (as shown in **Figure 12a**). As a result, a higher driving force is observed when electron transfer from the semiconductor material to I_3^-, which is much higher than in the bulk of the sensitized layer that is close to the TCO glass substrate. Thus, I_3^- electrons are anticipated to recombine with the semiconductor material at short-circuit conditions [57].

However, under illumination, the open-circuit condition is entirely different (shown in **Figure 12b**). Due to the rise ($\sim$0.7 eV) of the Fermi level of TCO glass substrate, a much higher driving force is observed when electron transfer from the TCO glass substrate to I_3^-. Thus, the electron recombination with I_3^- via the TCO glass substrate and the back reaction by these two routes causes a photo stationary state in the cell [57].

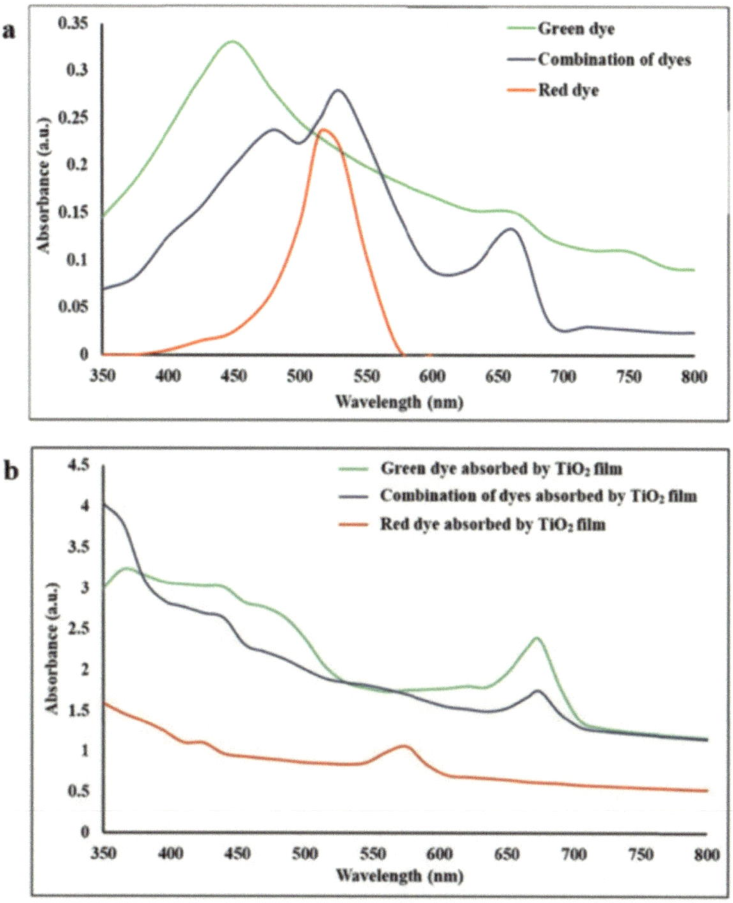

Figure 9.
(a) Absorption spectra of diluted natural chlorophyll (green), anthocyanin (red), and the optimum combination of dyes without TiO_2, and (b) absorption spectra of diluted natural chlorophyll (green), anthocyanin (red), and the optimum combination of dyes with TiO_2 [36].

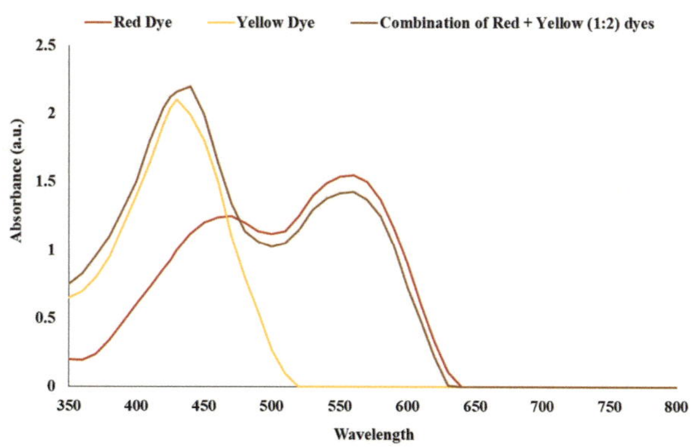

Figure 10.
Absorption properties of betalain, curcumin, and combination of dyes [35].

The blocking layer works as a barrier layer at the TCO/semiconductor material interface to improve cell performance. Studies had shown that a significant improvement in photo induced current observed when the blocking layer was introduced in the cell. Park and colleagues found that due to the blocking layer's

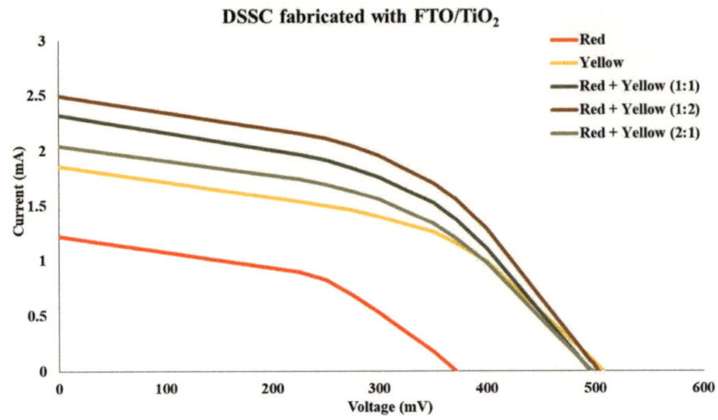

Figure 11.
I-V characteristics of DSSC fabricated with betalain, curcumin, and combination of dyes [35].

Dye/ Combination of dyes	Dye ratio	V_{oc} (mV)	I_{sc} (mA)	FF	η%	Dye loading (mol mm^{-3} X 10^7)
Red		371.6 ± 09.5	1.218 ± 0.039	0.487 ± 0.008	0.220 ± 0.016	1.05
Yellow		507.2 ± 10.5	1.857 ± 0.026	0.503 ± 0.004	0.473 ± 0.020	1.09
Red + Yellow	1:1	495.5 ± 09.4	2.319 ± 0.015	0.508 ± 0.003	0.583 ± 0.018	1.09
Red + Yellow	1:2	502.7 ± 11.5	2.494 ± 0.022	0.518 ± 0.002	0.649 ± 0.020	1.08
Red + Yellow	2:1	497.1 ± 14.3	2.041 ± 0.025	0.508 ± 0.003	0.515 ± 0.024	1.09

Table 3.
Photovoltaic performance of DSSC fabricated with FTO/TiO$_2$ [35].

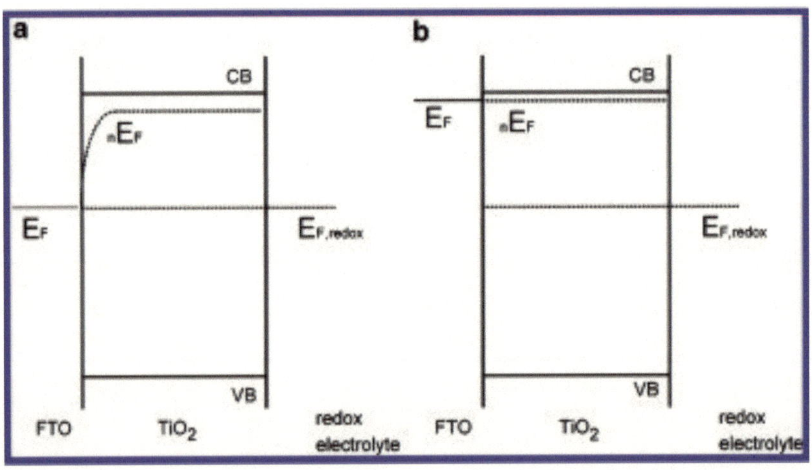

Figure 12.
Schematic of DSSC in the absence of a blocking layer. (a) under short circuit conditions, the Fermi level in the FTO is close to the redox Fermi level results in rapid electron-transfer kinetics to I3 -. (b) under open-circuit conditions, the Fermi level in the FTO moves up as the electron quasi-Fermi level rises and results in a photo stationary state [57].

presence, total transfer resistance at the blocking layer/electrolyte interface increased that increased cell performances by preventing electron recombination near the TCO glass substrate [58, 59]. Fabregat and co-workers found that BL

DOI: http://dx.doi.org/10.5772/intechopen.94760

improved physical contact between the TCO and semiconductor material that produce higher photo conversion efficiency. However, the advantage obtained by utilizing blocking layer is lost if the layer is too thick, and, generally, generates a series of resistance and an electron barrier that reduces the charge collection efficiency [59, 60].

A significant amount of photo-induced electron recombined and results in lower photocurrent. Recombination of the electron at the interfaces reduces the photocurrent and affects the fill factor; thus, cell performance decreases [60]. The complete photo anode is constructed layer-by-layer stack of suitably designed structures to maximize different cell functionalities. The recombination losses in DSSCs occurred primarily at the interface between the glass substrate of TCO and the electrolyte. The compact blocking layer acts as a physical barrier and physically separates and reduces the contact surface area between the TCO glass substrate from the electrolyte [59]. By employing the blocking layer with suitable thickness, the recombination can be reduced; and photo induced current and fill factor increase, leading to the DSSC efficiency improvement. Studies also showed that the blocking layer also improved the open-circuit-photo voltage of the cell [61]. The schematic on the effect of blocking layer is shown in **Figure 13**.

There are many kinds of preparation methods for blocking layers in DSSCs, including spin coating, deep coating, spray coating, sol–gel, sputtering, hydrothermal technique, etc. Spin-coating is a simple method for preparing uniform thin films onto flat substrates. Generally, the spin coating method includes deposition, spinup, spinoff, and evaporation [62]. Usually, a small amount of coating material is applied to the center of the substrate then rotated at speed up to 10,000 rpm to spread the coating material by centrifugal force. Rotation is continued while the fluid spins off the substrate's edges until the desired thickness of the film is

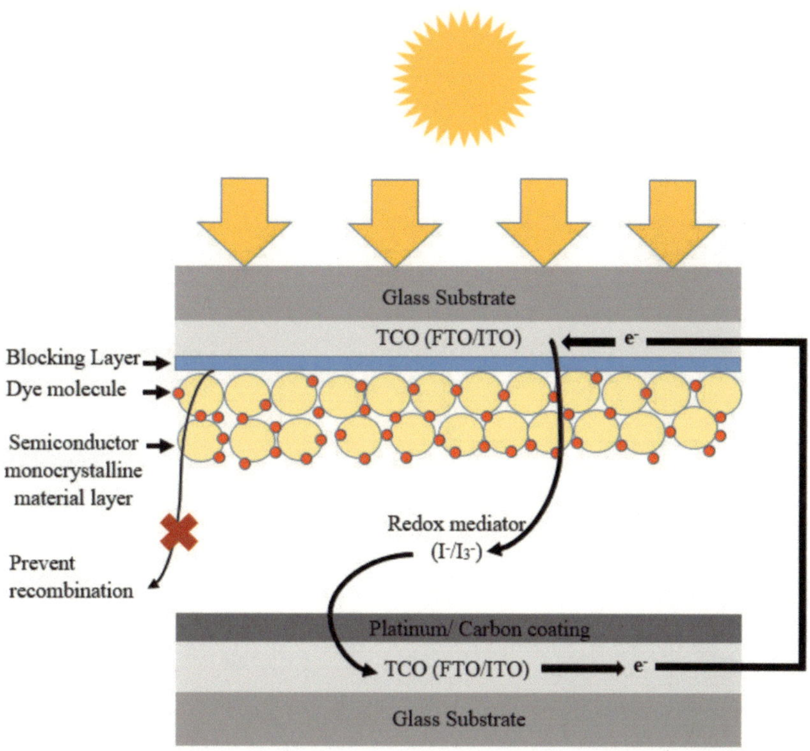

Figure 13.
Schematic diagram of the DSSC including a blocking layer for preventing recombination.

achieved. The film's thickness also depends on the solvent and solvents concentration [63].

Yeol et al. prepared a ZnO precursor on FTO substrates for the blocking layer. For ZnO precursor, a homogeneous mixture of 2.195 g zinc acetate dehydrate, 20 mL isopropanol, and 0.605 mL monoethanolamine (MEA) was prepared. The concentration was 0.5 M, with MEA: zinc acetate molar ratio of 1: 1. The prepared solution was stirred for 2 hrs at 200 rpm at 60°C, then stirred at the same rpm at ambient temperature for 22 hrs. For the spin-coated film, rotation speed and duration were held at 3000 rpm and 20 s, respectively. They annealed the spin-coated films at 500° C for 1 h to form a blocking layer of ZnO (55 nm to 310 nm). The ZnO blocking layer thickness is a function of the number of deposition cycles in the spin-coating process. ZnO blocking layer thickness increased linearly with the number of deposition cycles, a typical feature of the spin-coating technique [64]. **Figure 14a** illustrates the morphology of FTO. **Figure 14b** and **c** show that ZnO nanoparticles are distributed uniformly across the FTO substrate's surface to form a compact layer. Comparing both **Figure 14a** and **c**, when the thickness of the ZnO blocking layer increased, the size of the ZnO nanoparticles also slightly increased.

Yeol et al. showed that the effect of ZnO blocking layer and increasing its thickness on the cell performance of TiO_2 based DSSC. **Table 4** lists photovoltaic performance and **Figure 15** illustrates the J–V characteristics of the cell, including ZnO blocking layers of different thicknesses. The value of open-circuit voltage (V_{oc}) and fill factor (FF) of the DSSC improves, though the short-circuit current decreased. The increase of open-circuit voltage is due to the blocking of electron injection from the TiO_2 conduction band to the FTO [64, 65]. Due to the increased electron density in the TiO_2, the Fermi level rises. However, further an increase in the thickness of the ZnO blocking layer, the value of short circuit current decreased

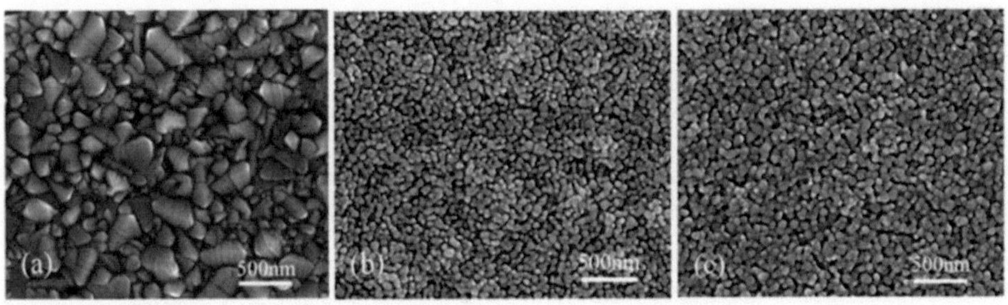

Figure 14.
FESEM images of (a) bare FTO, (b) FTO/ ZnO blocking layer (120 nm), (c) FTO/ ZnO blocking layer (310 nm) [64].

Sample	V_{oc}(mV)	J_{sc} (mA/cm^2)	Efficiency (%η)	Fill Factor (FF)
FTO	695	8.48	3.86	0.66
FTO/ZnO (55 nm)	708	8.30	3.96	0.67
FTO/ZnO (120 nm)	728	8.18	4.34	0.73
FTO/ZnO (220 nm)	744	6.64	3.63	0.73
FTO/ZnO (275 nm)	745	4.83	2.69	0.75
FTO/ZnO (310 nm)	781	3.05	1.66	0.70

Table 4.
Photovoltaic properties of TiO_2 based DSSCs including ZnO blocking layer of different thicknesses (for N3 dye) [64].

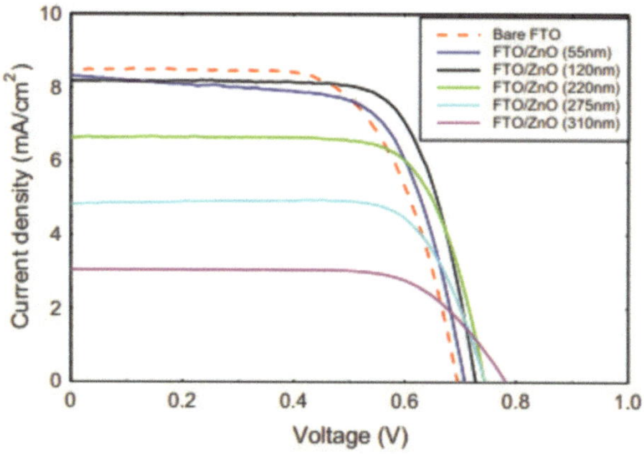

Figure 15.
J–V characteristics of DSSCs including ZnO blocking layer of different thicknesses [64].

Revolution per minute (rpm)	Thickness (nm)	Rughness (nm)
0	0	21.14
2000	10–30	10.91
1000	40–60	11.68
500	120–150	14.05

Table 5.
Thickness and root-mean-square roughness of TiO_2 thin films [66].

rapidly. As a result, cell performance decrease despite the slight improvement in the V_{oc} and FF values because the excessively thick ZnO layer blocks the electron injection from the conduction band of TiO_2 to the FTO substrate [64].

Lee et al. introduced an additional spin-coated TiO_2 thin film between the FTO and TiO_2 (semiconductor material) as a blocking layer for the electron injected from the exited photosensitizer. A homogeneous mixture of 29.0 mg titanium tetraisopropoxide [Ti $(OC_3H_7)_4$], and 100 ml isopropanol [$(CH_3)_2CHOH$] was prepared. Then the solution of 7.5 ml HCl in 100 ml of isopropanol was added drop by drop to the [Ti $(OC_3H_7)_4$]-[$(CH_3)_2CHOH$] solution at 0°C under continuous stirring, and afterward the solution was allowed to stand for less than 1 h at the same temperature. The solution was smeared on FTO substrates and rotated at 500, 1000, and 2000 rpm for 40 s to ensure uniformity. The samples were heated for 1 h at 100°C; they were sintered for 30 min at 450°C [66].

Lee et al. prepared several TiO_2 gel films with the spin coating method with different thicknesses. The thickness and roughness of the TiO_2 layers are among the most critical factors in the cell performance of DSSC [66]. **Table 5** lists the thickness and root-mean-square roughness of TiO_2 thin films. SEM images of 10 μm thin films (surface and cross-sections) are shown in **Figure 16**. **Table 6** lists photovoltaic performance and **Figure 17** illustrates the J–V characteristics of the cell, including ZnO blocking layers of different thicknesses.

TiO_2 layers also enhance the contact property between the FTO and TiO_2 electrode. **Figure 16c** and **d** illustrate the photovoltaic performance and the J-V characteristics curve of DSSC with different blocking layer thicknesses. The thickness of the TiO_2 blocking layer affects the efficiency of DSSC. As thin films' rpm increased, the thickness and roughness of the TiO_2 blocking layer also decreased, and the film becomes smooth and uniform. This increase in the smoothness and uniformity of

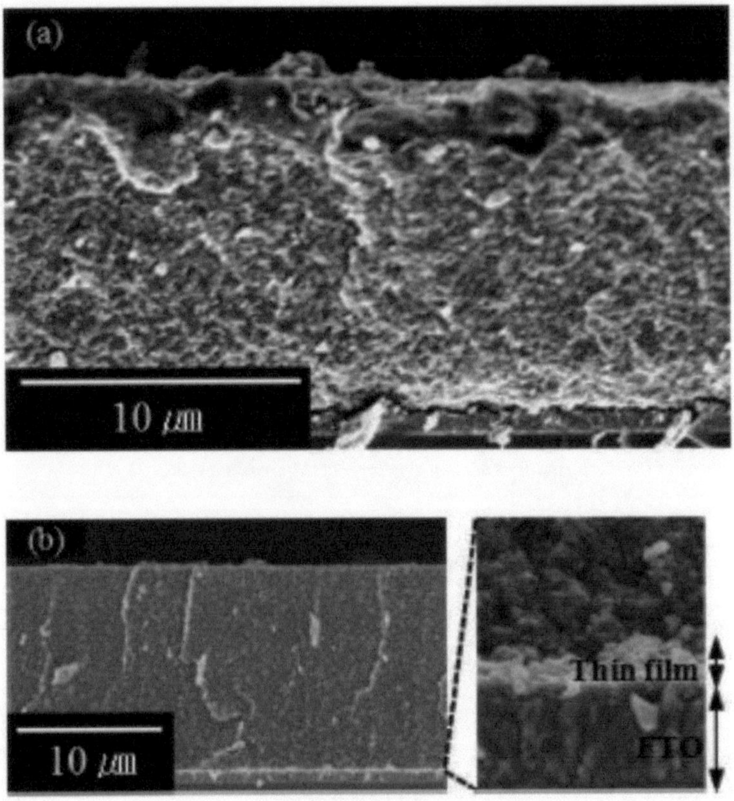

Figure 16.
Cross-sectional SEM images of the main-TiO2/FTO (a), main- TiO2/TiO2 thin film/FTO applied to a DSSC (b) [66].

Thickness (nm)	Open-circuit voltage, V_{oc} (V)	Short-circuit current density, J_{sc} (mA/cm^2)	Fill factor, FF (%)	Efficiency, η (%)	Electron lifetime, T_e (ms)	Resistance at Pt. counter electrode, R_{Ct1} (Ω)	Charge transfer resistances at the TiO_2/dye/electrolyte interface R_{Ct2} (Ω)
0	0.65	11.09	62	4.43	14.1	5.3	28.8
10–30	0.74	11.92	64	5.62	20.1	4.3	19.1
40–60	0.72	11.58	65	5.39	18.2	4.7	19.7
120–150	0.70	11.21	60	4.68	16.6	7.6	21.9

Table 6.
The cell performance of DSSCs based on TiO2 layers (10.5 μm) compressed at different thickness of thin films during the preparation for ruthenium 535 (Solaronix Co. N3) dye [66].

the TiO_2 blocking layer results in improved cell performance. The increased number of efficiently transferred photo generated electrons to the TiO_2 electrode results in an improvement in short-circuit current [67]. By suppressing the recombination of electrons injected from excited photosensitizers in the TiO_2 and electrolyte interface, a higher value of open-circuit voltage was obtained [57]. Their study also showed that when the resistance at the FTO/TiO_2 layer interface was decreased, the electron lifetime in DSSCs was increased [66].

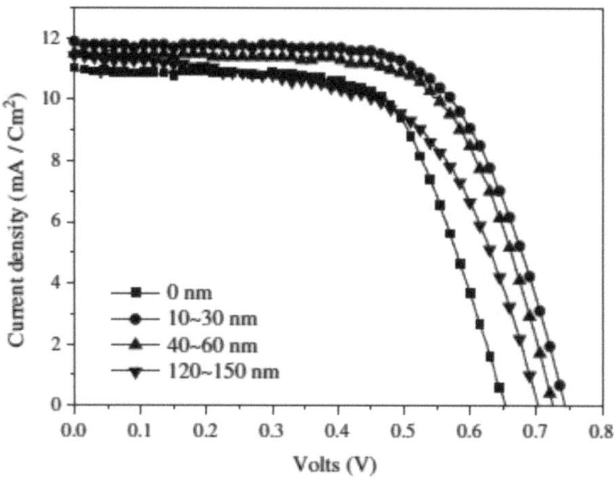

Figure 17.
I–V curves for DSSCs with TiO_2 blocking layers at different thickness [66].

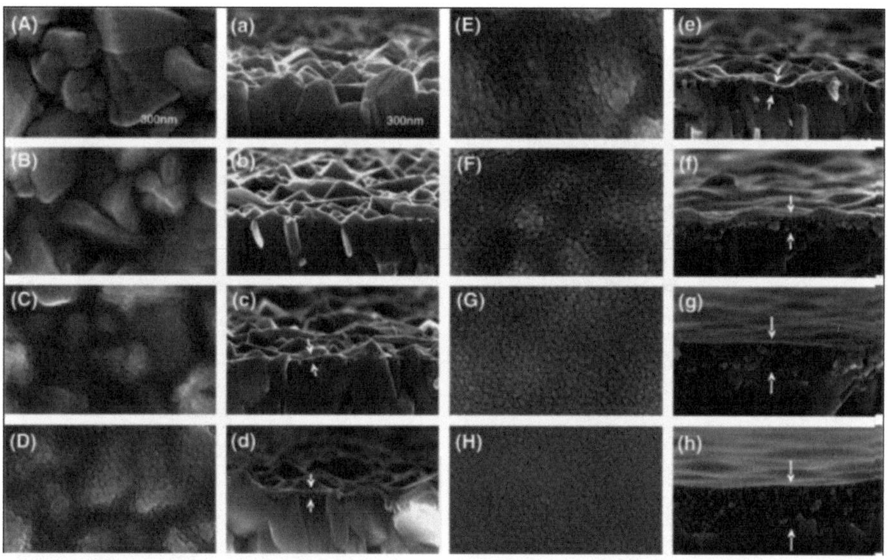

Figure 18.
Surface (A–H) and cross-section (a–h) SEM micrographs for the bare FTO, blocking layer-deposited FTO substrates from the Ti precursor solutions with the concentration of 0.05, 0.1, 0.15, 0.2, 0.4, 0.8, and 1.2 M, respectively [58].

Yoo et al. showed the impact of precursor concentration in the cell conversation efficiency of DSSC. For the blocking layer, a 1-butanol solution contained titanium (IV) bis(ethylacetoacetato) di-isopropoxide precursor was spin-coated on an FTO glass, followed by annealing at 500°C in air for 30 min. The concentration of the solution was varied from 0.05 M to 1.2 M [58]. **Figure 18** illustrates the SEM of bare FTO and blocking layer-deposited FTO glasses (surface and cross-sections). **Table 7** lists the photovoltaic property of DSSC with a blocking layer, where short-circuit current density increases with increasing the precursor concentration (and increased blocking layer thickness).

Zou et al. studied the effect of the $TiCl_4$ blocking layer (or pre-treatment) in ZnO based DSSC. **Figure 19** shows the fabricated ZnO films, with and without $TiCl_4$ pre-treatment on the FTO glass substrate. From **Figure 19a-d**, it can be seen that fabricated ZnO films have porous flakes, both with and without blocking

Ti precursor concentration (M)	Short-circuit current, J_{sc} (mA/cm^2)	Open-circuit voltage, V_{oc} (V)	Fill factor, FF	Efficiency, η (%)	Area (cm^2)
Without blocking layer	0.01	0.588	0.356	0.002	0.99
0.05	0.05	0.861	0.475	0.020	0.99
0.10	0.08	0.865	0.482	0.033	1.02
0.15	0.10	0.869	0.530	0.046	1.02
0.20	0.14	0.871	0.564	0.069	1.02
0.40	0.21	0.881	0.618	0.114	1.02
0.80	0.38	0.884	0.648	0.218	1.02
1.20	0.56	0.883	0.615	0.304	1.02

Table 7.
Photocurrent-voltage characteristics of DSSC comprising only blocking layers for N719 dye [58].

Figure 19.
SEM images of (a) FTO/ZnO with $TiCl_4$ pretreatment. (b) FTO/ZnO without $TiCl_4$ pretreatment. (c) and (d) the amplification figure of (a) and (b), respectively [68].

Samples	Open-circuit voltage, V_{oc} (V)	Short-circuit current, J_{sc} (mA/cm^2)	Fill factor, FF	Efficiency, η (%)
Without $TiCl_4$	0.3977	1.07	0.2829	0.12
$TiCl_4$	0.4759	2.86	0.3967	0.54

Table 8.
Photovoltaic performance of ZnO based DSSCs with $TiCl_4$ pretreatment and without $TiCl_4$ pretreatment [68].

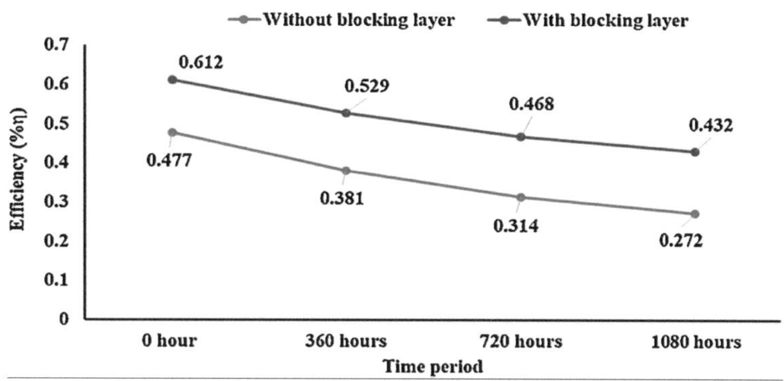

Figure 20.
Effect of post TiCl4 treatment in the cell performance of TiO2 based DSSC [69].

layers. However, $TiCl_4$ treated ZnO anode has a larger flake, which can offer a large surface area to absorb much more dyes. **Table 8** lists the photovoltaic property of DSSC with $TiCl_4$ blocking layer [68].

Kabir et al. studied the effect of post-treatment of $TiCl_4$ in TiO_2 based DSSC. Post $TiCl_4$ treatment not only increases the overall cell conversion efficiency of DSSC but also enhances cell stability. $TiCl_4$ treated TiO_2 anode based DSSC's degradation rate is much lower than the $TiCl_4$ untreated TiO_2 anode based DSSC. Studies showed that $TiCl_4$ treated TiO_2 anode based DSSC's cell stability of the increase around 38–44.5%. **Figure 20** shows the effect of post $TiCl_4$ treatment in the cell conversion efficiency of TiO_2 based DSSC [69].

Cameron et al. [57], Heo et al. [70], Yu et al. [71] used spray coating method to prepare TiO_2 blocking layer. Introducing the blocking layer into the device decreases charge carrier trapping and recombination. Subsequently, short-circuit current increases significantly. Additionally, a slight improvement in the open-circuit voltage and fill factor is observed, thus cell efficiency enhances significantly.

5. Summary

In conclusion, natural dye is a promising alternative to replace the metal complexes or organic dyes in the DSSC application. They are low-cost, abundant, eco-friendly, simple extraction procedures, and non-toxic. The combination of natural dyes with an optimized choice of the mixture of the volume ratio of the extracting dye extracting solvent accounts for many possible interactions that promise to provide more charge injection upon sensitization and allowed utilization of the photon energy more efficiently. DSSC co-sensitized with the dye mixture shows higher absorbance, and cumulative absorption properties over the entire visible region than the DSSC fabricated with individual dyes.

A blocking layer in DSSC provides good adhesion between the transparent conducting oxide (e.g., ITO, FTO, etc.) and an active semiconductor layer, TCO (e.g., TiO_2, ZnO, etc.). It also represses the electron back transport between electrolyte and TCO by blocking direct contact. Also, it offers a more uniform layer than bare TCO glass substrate. The conventional blocking suppresses electron leakage, recombination, and trapping; thus, the photovoltaic performance of the DSSC improves. Introducing a blocking layer in the DSSCs show lower dark current and operates efficiently under high-intensity sunlight and ambient light conditions.

Acknowledgements

This work is supported by the by the Ministry of Science & Technology, Government of the People's Republic of Bangladesh under the Special Allocation Research project for Science & Technology, FY 2019-2020, entitled, "Development of nanotechnology based solar cells using low cost natural dyes extracted from vegetables and the effect of gamma radiation on its performance", (Ref. No.: 39.00.0000.009.06.024.19/ID-478-494).

Author details

Md. Mosharraf Hossain Bhuiyan[1,2]*, Fahmid Kabir[3], Md. Serajum Manir[4], Md. Saifur Rahaman[1], Md. Robiul Hossain[3], Prosenjit Barua[3], Bikram Ghosh[3], Fumiaki Mitsugi[5], Tomoaki Ikegami[5], Saiful Huque[3] and Mubarak Ahmad Khan[4]

1 Institute of Nuclear Science and Technology, Atomic Energy Research Establishment, Bangladesh Atomic Energy Commission, Dhaka, Bangladesh

2 Department of Computer Science and Engineering, Central University of Science and Technology, Mirpur, Dhaka, Bangladesh

3 Institute of Energy, University of Dhaka, Dhaka, Bangladesh

4 Institute of Radiation and Polymer Technology, Atomic Energy Research Establishment, Bangladesh Atomic Energy Commission, Dhaka, Bangladesh

5 Faculty of Advanced Science and Technology, Kumamoto University, Kumamoto, Japan

*Address all correspondence to: mosharraf22003@yahoo.com; mosharraf22003@baec.gov.bd

References

[1] "OPEC: Organization of the Petroleum Exporting Countries, 2011. World Oil Outlook.," [Online]. Available: http://www.opec.org/opec_web/static_files_project/media/downloads/publications/WOO_2011.pdf.

[2] "OPEC: Organization of the Petroleum Exporting Countries, 2012. World Oil Outlook.," [Online]. Available: http://www.opec.org/opec_web/static_files_project/media/downloads/publications/WOO2012.pdf.

[3] R. R. Baadhe, R. Potumarthi and V. Gupta, "Lipase-Catalyzed Biodiesel Production: Technical Challenges," Elsevier, 2014, pp. 119–129.

[4] N. Thejo Kalyani and S. J. Dhoble, "Empowering the Future With Organic Solar Cell Devices," Nanomaterials for Green Energy, 2018, p. 325–350.

[5] J. Liu, Y. Yao, S. Xiao and X. and Gu, "Review of status developments of high-efficiency crystalline silicon solar cells," *Journal of Physics D: Applied Physics,* vol. 51, no. 12, p. 123001., 2018.

[6] T. D. Lee and A. and U. Ebong, "A review of thin film solar cell technologies and challenges," *Renewable and Sustainable Energy Reviews,* vol. 70, pp. 1286–1297, 2017.

[7] F. Kabir and e. al., "Stability study of natural green dye based DSSC," *Optik,* vol. 181, pp. 458–464, 2019.

[8] M. M. H. Bhuiyan and e. al., "Economic evaluation of a stand-alone residential photovoltaic power system in Bangladesh," *Renewable Energy,* vol. 21, no. 3–4, pp. 403–410, 2000.

[9] V. Fthenakis and K. and Zweibel, "Fthenakis, V. and Zweibel, K., 2003. CdTe PV: Real and perceived EHS risks (No. NREL/CP-520-33561)," in *National Renewable Energy Lab.(NREL)*, Golden, CO (United States)., 2003.

[10] "Third-generation photovoltaic cell," [Online]. Available: https://en.wikipedia.org/wiki/Third-generation_photovoltaic_cell. [Accessed 20 7 2020].

[11] M. Fitra, I. Daut, N. Gomesh, M. Irwanto and Y. and M. Irwan, "Dye solar cell using Syzigium Oleina organic dye," *Energy Procedia,* Vols. 341–348., pp. 341–348., 2013.

[12] "Dye-sensitized solar cell," [Online]. Available: https://en.wikipedia.org/wiki/Dye-sensitized_solar_cell. [Accessed 20 07 2020].

[13] B. O'Regan and M. and Grätzel, "A Low-Cost, High-Efficiency Solar Cell Based on Dye-Sensitized Colloidal TiO2 Films," *Nature,* vol. 353, pp. 737–740, , 1991 .

[14] M. Grätzel, "Dye-sensitized solar cells," *Journal of Photochemistry and Photobiology C: Photochemistry Reviews,* vol. 4, p. 145–153. , 2003.

[15] K. Kakiage, Y. Aoyama, T. Yano, K. Oya, J. I. Fujisawa and M. and Hanaya, "Highly-efficient dye-sensitized solar cells with collaborative sensitization by silyl-anchor and carboxy-anchor dyes," *Chemical Communications,,* vol. 51, no. 88, pp. 15894–15897, 2015.

[16] N. T.R.N. Kumara and e. al., "Recent progress and utilization of natural pigments in dye sensitized solar cells: A review," *Renewable and Sustainable Energy Reviews,* vol. 78, pp. 301–317, 2017.

[17] K. Balachandran and e. al., "Enhancing power conversion efficiency of DSSC by doping SiO2 in TiO2 photo anodes," *Materials Science in Semiconductor Processing,* vol. 35, pp. 59–65, 2015.

[18] B. Roose and e. al., "Doping of TiO 2 for sensitized solar cells," *Chemical Society Reviews,* vol. 44, no. 22, pp. 8326–8349, 2015.

[19] F. Kabir and e. al., "Effect of MWCNT's concentration in TiO2 based DSSC and degradation study of the cell," *Journal of Renewable and Sustainable Energy,* vol. 11, no. 2, p. 023502, 2019.

[20] A. Islam and e. al., "Enhanced photovoltaic performances of dye-sensitized solar cells by co-sensitization of benzothiadiazole and squaraine-based dyes," *ACS applied materials & interfaces,* vol. 8, no. 7, pp. 4616–4623, 2016.

[21] T. A. Ruhane and e. al., "Photo current enhancement of natural dye sensitized solar cell by optimizing dye extraction and its loading period," *Optik,* vol. 149, pp. 174–183, 2017.

[22] T. A. Ruhane and e. al., "Impact of photo electrode thickness and annealing temperature on natural dye sensitized solar cell," *Sustainable energy technologies and assessments,* vol. 20, pp. 72–77, 2017.

[23] G. Calogero and e. al., "Synthetic analogues of anthocyanins as sensitizers for dye-sensitized solar cells," *Photochemical & Photobiological Sciences,* vol. 12, no. 5, pp. 883–894, 2013.

[24] A. Hagfeldt and e. al., "Dye-Sensitized Solar Cells," *Chem. Rev.,* vol. 10, no. 11, p. 6595–6663, 2010.

[25] J. A. Mikroyannidis and e. al., "Low band gap dyes based on 2-styryl-5-phenylazo-pyrrole: Synthesis and application for efficient dye-sensitized solar cells," *Journal of Power Sources,* vol. 196, no. 8, pp. 4152–4161, 2011.

[26] S. Shalini and e. al., "Review on natural dye sensitized solar cells: Operation, materials and methods," *Renewable and Sustainable Energy Reviews,* vol. 51, p. 1306–1325, 2015.

[27] M. R. Narayan, "Review: Dye sensitized solar cells based on natural photosensitizers," *Renewable and Sustainable Energy Reviews,* vol. 16, no. 1, pp. 208–215, 2012.

[28] Y. Amao and e. al., "Preparation and properties of dye-sensitized solar cell using chlorophyll derivative immobilized TiO2 film electrode," *Journal of Photochemistry and Photobiology A: Chemistry,* vol. 164, no. 1–3, pp. 47–51, 2004.

[29] H. C. HAssan and e. al., "A High Efficiency Chlorophyll Sensitized Solar Cell with Quasi Solid PVA Based Electrolyte," *International Journal of Photoenergy,* p. Article ID 3685210, 2016.

[30] K. Davies, Plant pigments and their manipulation., Blackwell publishing, 2004.

[31] I. C. Maurya and e. al., "Natural dye extracted from saraca asoca flowers as sensitizer for TIO2-based dye-sensitized solar cell," *Journal of Solar Energy Engineering,* vol. 138, no. 5, 2016.

[32] C. Zho and e. al., "The regulation of carotenoid pigmentation in flowers," *Archives of Biochemistry and Biophysics,* vol. 504, no. 1, pp. 132–141, 2010.

[33] E. Murillo, "Native carotenoids composition of some tropical fruits," *Food Chemistry,* vol. 140, no. 4, pp. 825–836, 2013.

[34] E. C. Prima, "Performance of natural carotenoids from Musa aromatica and *Citrus medica* var Lemon as photosensitizers for dye-sensitized solar cells with TiO2 nanoparticle," *Advanced Materials Research,* vol. 789, pp. 167–170, 2013.

[35] F. Kabir and e. al., "Enhance cell performance of DSSC by dye mixture, carbon nanotube and post TiCl4 treatment along with degradation study," *Sustainable Energy Technologies*

and Assessments, vol. 35, pp. 298–307, 2019.

[36] F. Kabir and e. al., "Development of dye-sensitized solar cell based on combination of natural dyes extracted from Malabar spinach and red spinach," *Results in Physics,* vol. 14, p. 102474, 2019.

[37] F. Kabir and e. al., "Effect of combination of natural dyes and post-TiCl4 treatment in improving the photovoltaic performance of dye-sensitized solar cells," *Comptes Rendus Chimie,* vol. 22, no. 9–10, pp. 659–666, 2019.

[38] D. D. Pratiwi and e. al., "Performance improvement of dye-sensitized solar cells (DSSC) by using dyes mixture from chlorophyll and anthocyanin," *Journal of Physics: Conference Series,* vol. 909, p. 012025, 2017.

[39] H. Bashar and e. al., "Study on combination of natural red and green dyes to improve the power conversion efficiency of dye sensitized solar cells," *Optik,* vol. 185, pp. 620–625, 2019.

[40] F. Kabir, M. M. H. Bhuiyan, M. R. Hossain, H. Bashar, M. S. Rahaman, M. S. Manir, S. M. Ullah, S. S. Uddin, M. Z. I. Mollah, R. A. Khan and S. and Huque, "Improvement of efficiency of Dye Sensitized Solar Cells by optimizing the combination ratio of Natural Red and Yellow Dyes," *Optik,* vol. 179, pp. 252–258, 2019.

[41] k. Usha, B. Mondal, D. Sengupta and e. al., "Development of multilayered nanocrystalline TiO2 thin films for photovoltaic application," *Optical Materials,* vol. 36, no. 6, pp. 1070–1075, 2014.

[42] W. Q. Wu, Y. F. Xu, C. Y. Su and D. B. Kuang, "Ultra-long anatase TiO2 nanowire arrays with multi-layered configuration on FTO glass for high-efficiency dye-sensitized solar cells," *Energy & Environmental Science,* vol. 7, no. 2, pp. 644–649, 2014.

[43] P. Sun, X. Zhang, L. Wang, F. Li, Y. Wei, C. Wang and Y. and Liu, "Bilayer TiO 2 photoanode consisting of a nanowire–nanoparticle bottom layer and a spherical voids scattering layer for dye-sensitized solar cells," *New Journal of Chemistry,* vol. 39, no. 6, pp. 4845–4851, 2015.

[44] H. H.G. Tsai, C. J. Tan and W. and H. Tseng, "Electron transfer of squaraine-derived dyes adsorbed on TiO2 clusters in dye-sensitized solar cells: a density functional theory investigation," *The Journal of Physical Chemistry C,* vol. 119, no. 9, pp. 4431–4443, 2015.

[45] A. K.Chandiran, F. Sauvage, M. Casas-Cabanas, P. Comte, S. M. Zakeeruddin and M. Graetzel, "Doping a TiO2 photoanode with Nb5+ to enhance transparency and charge collection efficiency in dye-sensitized solar cells," *The Journal of Physical Chemistry C,* vol. 114, no. 37, pp. 15849–15856, 2010.

[46] D. Sengupta, P. Das, U. Kasinadhuni, B. Mondal and K. Mukherjee, "Morphology induced light scattering by zinc oxide polydisperse particles: promising for dye sensitized solar cell application," *Journal of Renewable and Sustainable Energy,* vol. 6, no. 6, p. 063114, 2014.

[47] X. Chen, Z. Bai, X. Yan, H. Yuan, G. Zhang, P. Lin, Z. Zhang, Y. Liu and Y. Zhang, "Design of efficient dye-sensitized solar cells with patterned ZnO–ZnS core–shell nanowire array photoanodes," *Nanoscale,* vol. 6, no. 9, pp. 4691–4697., 2014.

[48] L. Lin, X. Peng, S. Chen, B. Zhang and Y. Feng, "Preparation of diverse flower-like ZnO nanoaggregates for dye-sensitized solar cells," *RSC Advances,* vol. 5, no. 32, pp. 25215–25221, 2015.

[49] K. Manseki, T. Sugiura and T. Yoshida, "Microwave synthesis of size-controllable SnO 2 nanocrystals for dye-sensitized solar cells," *New Journal of Chemistry,* vol. 38, no. 2, pp. 598–603, 2014.

[50] H. J. Snaith and C. Ducati, " SnO2-based dye-sensitized hybrid solar cells exhibiting near unity absorbed photon-to-electron conversion efficiency," *Nano letters,* vol. 10, no. 4, pp. 1259–1265, 2010.

[51] E. Guo and L. Yin, "Tailored SrTiO3/TiO2 heterostructures for dye-sensitized solar cells with enhanced photoelectric conversion performance," *Journal of Materials Chemistry A,* vol. 3, no. 25, pp. 13390–13401, 2015.

[52] C. W. Kim, S. P. Suh, M. J. Choi, Y. S. Kang and Y. S. Kang, " Fabrication of SrTiO 3–TiO 2 heterojunction photoanode with enlarged pore diameter for dye-sensitized solar cells," *Journal of Materials Chemistry A,* vol. 1, no. 38, pp. 11820–11827, 2013.

[53] Y. F. Wang, K. N. Li, Y. F. Xu, H. S. Rao, C. Y. Su and D. B. Kuang, "Hydrothermal fabrication of hierarchically macroporous Zn 2 SnO 4 for highly efficient dye-sensitized solar cells," *Nanoscale,* vol. 5, no. 13, pp. 5940–5948., 2013.

[54] C. Chen, Y. Li, X. Sun, F. Xie and M. Wei, "Efficiency enhanced dye-sensitized Zn 2 SnO 4 solar cells using a facile chemical-bath deposition method," *New Journal of Chemistry,* vol. 38, no. 9, pp. 4465–4470, 2014.

[55] D. Hwang, J. S. Jin, H. Lee, H. J. Kim, H. Chung, D. Y. Kim and S. K. D. Y. Jang, "Hierarchically structured Zn 2 SnO 4 nanobeads for high-efficiency dye-sensitized solar cells," *Scientific reports,* vol. 4, p. 7353, 2014.

[56] T. S. Bramhankar and e. al., "Effect of Nickel–Zinc Co-doped TiO2 blocking layer on performance of DSSCs," *Journal of Alloys and Compounds,* vol. 817, p. 152810, 2020.

[57] P. J. Cameron and L. M. Peter, "Characterization of titanium dioxide blocking layers in dye-sensitized nanocrystalline solar cells," *The Journal of Physical Chemistry B,,* vol. 107, no. 51, pp. 14394–14400, 2003.

[58] B. Yoo and e. al., "Chemically Deposited Blocking Layers on Fto Substrates: Effect ofPrecursor Concentration on Photovoltaic Performance of Dye-Sensitized Solar Cells," *Journal of Electroanalytical Chemistry,* vol. 638, no. 1, pp. 161–166, 2010.

[59] G. S. Selopal and e. al., "Effect of blocking layer to boost photoconversion efficiency in ZnO dye-sensitized solar cells," *ACS applied materials & interfaces,* vol. 6, no. 14, pp. 11236–11244, 2014.

[60] M. S. Goes and e. al., "Impedance spectroscopy analysis of the effect of TiO2 blocking layers on the efficiency of dye sensitized solar cells," *The Journal of Physical Chemistry C,* vol. 116, no. 23, pp. 12415–12421, 2012.

[61] Z. Shaban and e. al., "Optimization of ZnO thin film through spray pyrolysis technique and its application as a blocking layer to improving dye sensitized solar cell efficiency," *Current Applied Physics,* vol. 16, no. 2, pp. 131–134, 2016.

[62] S. Ahmadi and e. al., "The role of physical techniques on the preparation of photoanodes for dye sensitized solar cells," *International Journal of Photoenergy,* vol. Article ID 198734, 2014.

[63] "Spin coating," [Online]. Available: https://en.wikipedia.org/wiki/Spin_coa ting. [Accessed 20 07 2020].

[64] M. E. Yeoh and K. and Y. Chan, "Efficiency enhancement in

dye-sensitized solar cells with ZnO and TiO 2 blocking layers," *Journal of Electronic Materials,* vol. 48, no. 7, pp. 4342–4350., 2019.

[65] Y. Liu and e. al., "Influences on photovoltage performance by interfacial modification of FTO/mesoporous TiO2 using ZnO and TiO2 as the compact film," *Journal of alloys and compounds,* vol. 509, no. 37, pp. 9264–9270., 2011.

[66] J. G. Lee and e. al., "Enhancement of photovoltaic performance in dye-sensitized solar cells with the spin-coated TiO2 blocking layer," *Journal of nanoscience and nanotechnology,* vol. 12, pp. 6026–6030, 2012.

[67] B. Pradhan and e. al., "Vertically aligned ZnO nanowire arrays in Rose Bengal-based dye-sensitized solar cells," *Solar energy materials and solar cells,* vol. 91, no. 9, pp. 769–773, 2007.

[68] X. Zou and e. al., "TiCl4 Pretreatment and Electrodeposition Time Investigations of ZnO Photoelectrodes Preparation for Dye Sensitized Solar Cells," *International Journal of Photoenergy*\, p. Article ID 890563, 2014.

[69] F. Kabir and S. N. Sakib, "Various impacts of blocking layer on the cell stability in natural dye based dye-synthesized solar cell," *Optik,* vol. 180, pp. 684–690, 2019.

[70] J. Heo and e. al., "Room temperature synthesis of highly compact TiO 2 coatings by vacuum kinetic spraying to serve as a blocking layer in polymer electrolyte-based dye-sensitized solar cells," *Journal of Thermal Spray Technology,* vol. 24, no. 3, pp. 328–337, 2015.

[71] H. Yu and e. al., "An efficient and low-cost TiO2 compact layer for performance improvement of dye-sensitized solar cells," *Electrochimica Acta, 54(4), pp.1319–1324.,* vol. 54, no. 4, pp. 1319–1324, 2009.

Chapter 16

IIIrd Generation Solar Cell

Paweł Kwaśnicki

Abstract

Light harvesting for generation of electric energy is one of the most important research topics in applied sciences. First, for an efficient harvesting one needs a material with a broad light absorption window having a strong overlap with the sunlight spectrum. Second, one needs an efficient conversion of photoexcited carriers into produced current or voltage which can be used for applied purposes. The maximum light conversion coefficient in semiconductor systems is designated by so called Shockley-Queisser law, which is around 32% for an optimal bandgap value of 1,2–1,3 eV. However the efficiency may be increased using a solutions based on semiconductor nano materials such as quantum dots. Solar cells based on such a structures are included in the group of 3rd generation solar sell. 3rd generation solar cell encompasses multiple materials as a base of cell, such as: perovskite, organic, polymers and biomimetics. The most promising and in the same time most discussed are quantum dots and perovskite. Both material has a potential to revolutionize the solar cell industry due to their wide absorption range and high conversion coefficient. Nonetheless before the most common used material in photovoltaic namely silicon is replace one must overcome few major issues such as: stability and lifetime for at least 5 to 10 years or more, manufacturing process for a large surfaces and low production cost as well as recycling after the time of optimal use.

Keywords: DSSC, QDSC, transparent PV

1. Introduction

Light harvesting for generation of electric energy is one of the most important research topics in applied sciences. First, for an efficient harvesting one needs a material with a broad light absorption window having a strong overlap with the sunlight spectrum. Second, one needs an efficient conversion of photoexcited carriers into produced current or voltage which can be used for applied purposes. The maximum light conversion coefficient in semiconductor systems is designated by so called Shockley-Queisser law, which is around 32% for an optimal bandgap value of 1.2–1.3 eV. However the efficiency may be increased using a solutions based on semiconductor nano materials such as quantum dots. Solar cells based on such a structures are included in the group of 3rd generation solar cell. 3rd generation solar cell encompasses multiple materials as a base of cell, such as: perovskite, organic, polymers and biomimetics. The most promising and in the same time most discussed are quantum dots and perovskite. Both material has a potential to revolutionize the solar cell industry due to their wide absorption range and high conversion coefficient. Nonetheless before the most common used material in photovoltaic namely silicon is replace one must overcome few major issues such as: stability and lifetime for at least 5 to 10 years or more, manufacturing process for a large surfaces and low production cost as well as recycling after the time of optimal use.

IntechOpen

In this chapter we focus on two most promising material for photovoltaic application. The basic overview of organometallic properties of perovskites and quantum dots from the point of view of photovoltaics and formulation description of the electronic structure in the form of a simplified effective Hamiltonian as an approximation of a tight tie will be presented. The electronic structure plays a key role in the photovoltaic effect and is responsible for the high efficiency of the effect. Additionally perovskites or quantum dots show the spin-orbit coupling in the general form, this coupling can increase the carrier's lifetime - the quantity important for solar cell applications.

2. Perovskites for photovoltaics

Some perovskite-structured oxides have an internal electrical field, which plays an important role as it leads to the separation of electrons and holes generated in the process of light absorption. These oxides have the general structure of the ABO_3 type. In general, there are quite a few different materials called perovskites, but the crystalline structure for all perovskites is similar. Perovskite oxides and, above all, organometallic halogen perovskites play an important role for photoelectronics and photovoltaics. Nonetheless perovskite oxides turned out to be inefficient in terms of photovoltaics. The interest in perovskite materials increased significantly towards the end of the last year a decade, when a series of works appeared showing the possibility of increasing efficiency in organometallic perovskites [1]. It turned out that there was a fairly broad class organometallic halide perovskites of the type $CH_3NH_3PbX_3$ (X = I, Br, Cl), which show promising properties from the photovoltaic point of view. Although the first results gave relatively low photovoltaic efficiency, however this efficiency is quite fast it grew with new research. Besides, the conducted research did not show any significant restriction on the upper limit of the photovoltaic efficiency organometallic perovskites, which now reaches over 20%, which in turn gives hope for its further growth. The main advantages of organometallic halide perovskites are their relatively low levels price and relatively simple technology, which makes these materials competitive. Recent research results show that the efficiency of the laudatory prototypes of perovskite solar cell are already equalled and even exceeded the silicon based solar cell. Hence the great interest these materials from the point of view of application in photovoltaic cells [2, 3]. Of course, these materials also have weaknesses. One of these weaknesses is the lead toxicity they contain. The second is quite rapid degradation resulting from the sensitivity of photovoltaic cells based on them on humidity and the effect of ultraviolet radiation to which they are exposed. Therefore, the main lines of current research are not only aimed at further increasing efficiency photovoltaic, but also removing these undesirable weaknesses. As mentioned for photovoltaic the most interesting and promising are halide perovskites, the crystal structure of these materials has the general form ABX_3, where A is the cation of the methylammonium group CH3NH3 for organometallic halide perovskites (metal cation for oxides), B is the metal cation Pb or Sn (the smaller metal cation in the case of oxides), while X is a Cl, Br or I anion for halide perovskites (O for oxides). The unit cell of the ABX3 perovskite crystal in the cubic phase is shown in **Figure 1**.

One of the most promising materials is a perovskite with the chemical composition $CH_3NH_3PbI_3$, because in this case the photovoltaic efficiency turned out to be the highest in this class of materials. It is worth noting, however, that the class of organometallic perovskites is in fact quite rich and contains many elements, which allows the use not only of single perovskites, but their more complex structures, e.g. double perovskites or systems composed of various materials [4]. The high photoelectric efficiency of organometallic perovskites is related to their electronic properties.

This material is a semiconductor with a band gap width of about 1.6 eV. The light absorption coefficient is very high while energy losses associated with the possibility of non-radiative electron processes (e.g., electron–hole recombination by phonons) are relatively low. Moreover, the mobility of the carriers (electrons and holes) in these perovskite materials is quite high due to the low effective mass of the carriers. All these properties underlie high photovoltaic efficiency. On the other hand, the physical mechanisms underlying these properties are not yet fully researched and elucidated.

The excellent photovoltaic properties of perovskites are related to their electronic structure, in particular to the quantum states of electrons and holes in the conduction and valence bands, respectively. In the case of organometallic halide perovskites these properties are related to the organic CH_3NH_3 positive ion and its orientation with respect to the crystallographic axes [5].

Even better results using perovskite material from energy harvesting point of view may be achieved using hybrid structure. Recent discovery by the group of Prof. Miyasaka of a highly efficient light-to-voltage conversion in hybrid organic–inorganic perovskites [6] made these material promising elements for photovoltaics, especially taking into account simple low-cost fabrication technology. The basic structure of the perovskite-based solar cell is presented in **Figure 2**.

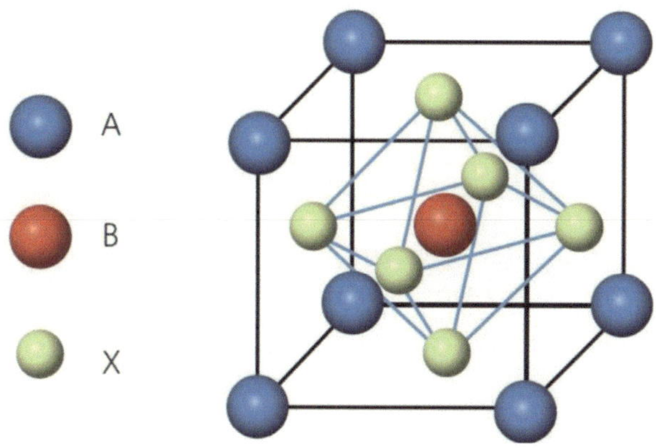

Figure 1.
Perovskite crystal unit cell, a - large cation (methylammonium group CH3NH3), B - smaller cation (Pb or Sn), X - anion (I, Br or Cl).

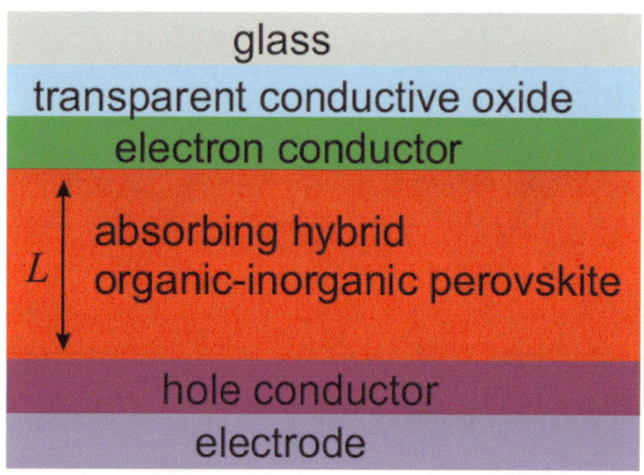

Figure 2.
Schematic picture of a hybrid organic–inorganic perovskite solar cell. (figure source: USA). Department of Energy website.

The first two compounds of this family investigated by Prof. Miyasaka group, that is $CH_3NH_3PbBr_3$ and $CH_3NH_3PbI_3$, deposited on the TiO_2 surface, demonstrated a good power conversion efficiency η higher than 3% (now reaching as high as 20%, similar to that in best conventional semiconductor-based solar cell) accompanied by the open-circle voltage Voc higher that 0.6 V and generated current density Jsc higher than 5.5 mA/cm^2. These results of Prof. Miyasaka group attracted a great deal of attention and caused a tide of research in this field.

These compounds belong to the family of perovskite structures, similar to the high- temperature superconductors, where the main element is represented by Cu-O octahedrons. Although some structure elements of these groups of materials are similar, their physical properties are very different. In general, all the perovskites are known for formation of different structures and a variety of temperature-induced structural transitions.

Due to a large variety of the organic cations, the entire family of hybrid organic-inorganic perovskites potentially contains more than 1000 members [7], all different in their properties. Structure-wise, the main element of these compounds as presented in two-dimensional projection in **Figure 3** is an octahedron built by metal and halogen ions, these elements are surrounded by organic layers.

Despite several years of extensive research efforts, many microscopic properties, which can help in the understanding of the high photovoltaic efficiency in these compounds, remain unknown. This holds true even for $CH_3NH_3PbI_3$ and

Figure 3.
Typical crystal structure of hybrid organic–inorganic perovskite compounds.

$HC(NH_2)_2PbI_3$ - the leaders of the race for the low-cost high photovoltaic efficiency. Electrical properties such as conductivity of these compounds can strongly depend on the temperature since due to a relatively soft lattice, various structural phase transitions occur in the temperature range of the order of 100 K - the property common for all perovskite structures, as mentioned above. Here we review and analyze properties of these materials in the context of their applicability for photovoltaics and connect these properties to the processes related to their possibly high efficiency.

Typical hybrid perovskite structure has the known form of a vertex-sharing networks of BX6 octahedrons as shown in **Figure 3**, which can be modeled as negatively charged $PbI^{-}{}_{3}$ elements. The bandstructure and optical properties are due mainly to the metal and halogen orbitals. Mutual overlap of these orbitals determines the matrix elements of interatomic hopping and, in turn, the band structure, corresponding to direct-band semiconductors with the bandgap Eg. This gap can be evaluated by different techniques. The overlap of the orbitals forms a relatively small effective mass of the carriers as well as the optical properties of these systems.

The sunlight has the following physical properties of our interest. Its spectrum corresponds to maximum in the green light region at photon energies $\hbar\omega$ close to 2.2 eV. This implies that the main contribution to the absorption in hybrid perovskites starts from the infrared part of the sunlight spectrum and its intense visible and ultraviolet parts can be used efficiently. The specific intensity of sunlight at the Earth's surface at a sunny day is of the order of 1000 W/m^2 corresponding to a relatively strong electric field of the order of 10^2–10^3 V/m. To understand the performance of a photovoltaic cell, one can establish that the energy flux of 1000 W/m^2 corresponds to the photon flux of the order of $10^{18}\ s^{-1}\ cm^{-2}$. Being totally absorbed in a layer of the thickness of $L \sim 10$ μm (10^{-3} cm), this photon flux corresponds to the free carrier bulk (3D) generation rate of the order of $10^{21}\ s^{-1}\ cm^{-3}$; corresponding to two-dimensional (2D) concentration injection rate $10^{18}\ s^{-1}\ cm^{-2}$. Although the exact electrodynamics of solar cells is very complex [8], by using the Gauss theorem for the electric field produced by charge distribution, one can obtain that this concentration corresponds to the rate of the voltage generation. The resulting bias V is of the order of $(dV/dt) \times G^{-1}$, where G^{-1} is the characteristic time corresponding to the relaxation of the optical injection, e.g. by electron–hole recombination or trapping carriers by lattice defects, leading to $V \sim 10^8) \times G^{-1}$. With the reasonable estimation of $G^{-1} \sim 10^{-9}$ s the value of $V \sim 0{,}1$ V. Simultaneously, current J produced in such a sample of $S = 1\ cm^2$ area, is of the order of 0,1 A. These numbers demonstrate that the first perovskite solar cells produced voltage close the maximal one, although the generated current was somewhat less than this basic estimate.

The light-induced transitions produce electron–hole pairs in the energy interval E > Eg, where Eg is the fundamental gap at the R – point of the Brillouin zone. The fundamental gap can be understood from the Coulomb energy arguments for the energy necessary to transfer electron from halogen to the IV-group heavy ion.

A qualitative plot of injected distribution of electrons in the energy representation is presented in **Figure 4**. This strongly nonequilibrium distribution then relaxes to the quasi equilibrium which, as we will see below, determines the performance of the photovoltaic devices. The relaxation process is mainly attributed to the multiple emissions of phonons.

The energy relaxation processes are understood even less sufficiently than the origin of the carrier's finite mobility. Indeed, due to a complex unit cell structure, crystals possess a large variety of phonon branches (of the order of 100) of different nature and symmetry. Here we propose a simple picture of the relaxation due to electron–phonon coupling. The analysis done in [9] shows that coupling to acoustic

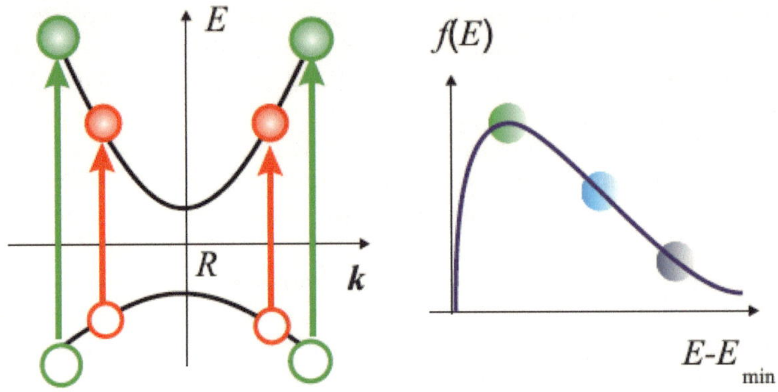

Figure 4.
Interband transitions caused by different photons, and electron distribution over the energy states, as injected. The behavior of the distribution at energies close to the minimum of the conduction band Ec is due to small density of states ~ $\sqrt{E - E_c}$ while the high-energy behavior is mainly determined by decrease in the spectral density of the incident light. Emin corresponds to the bottom of the conduction band.

phonons (with the frequency linear in the momentum) would lead to high mobilities of the order of 10^3 cm^2 V^{-1} s^{-1} and, therefore, this coupling is not the limiting factor for the observed low mobilities. Therefore, we concentrate on the relevant coupling to optical phonons. The coupling is due to the asymmetry of the field and change of the hopping integrals due to change in the interatomic distances. The value of the deformation potential constant is attributed to two main effects. First effect is the change in the site energy, corresponding to atomic displacement in the crystal field formed by its interaction with surrounding ions. Second effect is the changing in the overlap transfer integrals between the iodine and the lead orbitals, contributing to the electron energy as well.

The energy relaxation of the photoexcited electrons due to electron–phonon coupling with optical phonons, is relatively fast and occurs on the time scale of the order of 10 ps. This fast relaxation demonstrates that a thermalized room-temperature energy distribution is quickly produced. As a result, the performance of the photovoltaic elements with typical involved time scales of the order of 1–10 ns, is determined by the thermalized distributions, where the static local defects, either charged or not, structural disorder, and low-frequency optical phonons can play a role for the kinetics of the carriers distributions. The relation of these energy relaxation processes to the photovoltaic performance of real solar cells needs experimental studies and remains to be investigated.

The light absorption is efficient due to the band structure of perovskite materials having a direct bandgap close to 1.5 eV in the vertex point of the Brillouin zone. As a result, almost the entire sunlight spectrum can be absorbed. The efficiency of the absorption, in addition, is enhanced by relatively large momentum matrix elements between the group-IV heavy metal and halogen atoms resulting from their spatial overlap what makes perovskite material promising material for III generation of photovoltaic.

3. Quantum dots for photovoltaics

Quantum dots are one of the most interesting objects in modern fundamental and applied solid state physics, including applications for photovoltaics systems [10]. Typical sizes of the quantum dots of the order of 10 nanometers determine the majority of their physical properties, including the spectra of light absorption and properties of light-injected carries. These spectra determine the application of quantum dots in photogalvanics and photovoltaics systems. In contrast to bulk materials, where free electron–hole pairs can be produced optically, a strong confinement of carriers in quantum dots and resulting interaction between them leads to formation of exciton-like states. This effect qualitatively modifies optical properties of quantum dots and can make them useful for solar cell elements applications [10].

Various types of quantum dots can be used in photovoltaics: semiconductor polycrystalline and granular materials, quantum dots obtained by epitaxial methods or from colloidal solutions, nanoparticles of organic dyes. There are also a number of possibilities for the architecture of photovoltaic cells. Their common feature is that the phenomenon of multiple exciton excitation in dots a is used the energized charges (electrons and holes) are conducted to the electrodes in various ways, ensuring, however, their spatial separation. One possibility is to use scatter dots.

in a conductive material (e.g. in organic polymers). With the appropriate concentration of the dots, the discharge of the charge from the dots to the electrodes can be accomplished by coupling between the quantum dots. For regular networks of dots (single, double, or three-dimensional), discrete states of dots are transformed into mini-electron bands, ensuring charge transport. Photovoltaic cells with the use of regular quantum dot networks and their electronic mini-band structure (also called intermediate bands) have become one of the important directions in the development of photovoltaics. The essence of this type of solution is the fact that in the area between the electrodes in the p-n junction there is a layer containing dots quantum between which the distance is so small that an intermediate band is created in this area during the energy gap. This allows the use of low-energy photons (with energy lower than the width of the output semiconductor gap) to generate electrons in the conduction band and holes in the valence band. This is due to optical transitions from the valence band to the intermediate band and from the intermediate band to the conduction band. An important element is also that the recombination processes are in the case of the intermediate band much less likely than in the case of isolated quantum dots. In this case, it is enough for the wave functions of the dots to be quite delocalized. This can be achieved in systems with complexes quantum suppositories instead of regular networks.

Semiconductor quantum dots are usually produced of two types of binary materials. First type is usually referred to as III-V semiconductors, where one of the elements belongs to the III group of the periodic table of elements and the other belongs to the V group such as GaAs, InAs, InSb, and similar ones. The other group, named in the same way, is the II-VI semiconductors such as ZnS, ZnSe, CdS, and similar ones. In addition, coated quantum dots, where the core and the coating layer are made of different materials of the same (usually II-VI) type can be produced and used for different applications. Conventionally produced quantum dots show a variety of sizes and shapes. On one hand, this variety of quantum dot geometries extends the ability to use them for various applications, including photovoltaics. On the other hand, this variety hampers controllability of their applications. This circumstance should be taken into account in the analysis of all applications quantum dots.

Direct calculations of properties of quantum dots are very difficult, simple, but still highly efficient, approach relies on employing of the effective mass approximation with the Hamiltonian

$$H_0 = \frac{\hbar^2 k^2}{2m^*} + U(r)$$

where $\hbar^2 k^2 / 2\,m^*$ is the kinetic energy, k is the electron momentum, m_ is the electron effective mass, and U (r) is the effective confining potential of the quantum dot. For model calculations the anisotropic oscillator form of the confinement

$$U(R)=\frac{m^*}{2}\left(\Omega_x^2 x^2+\Omega_y^2 y^2+\Omega_z^2 z^2\right)$$

where Ω i are the corresponding frequencies, is assumed. One-dimensional representation of this potential and corresponding wave functions are shown in **Figure 5**. This form, being useful for basic understanding, especially of the ground state of the system, has strong limitations for the analysis of applications of quantum dots in photovoltaics and photogalvanics, where highly excited states are involved.

Another form of the potential is given by:

$$\mathrm{U(r)=0\ for\ r\ inside\ the\ quantum\ dot}$$

$$\mathrm{U(r)=\infty\ for\ r\ outside\ the\ quantum\ dot}$$

and determines the quantum dot shape. Usual model shapes of quantum dot are ellipsoidal (in simple realizations, spherical) or coin-like cylindrical with the radius much larger than the width, as shown in **Figure 6**.

It is well-known that in semiconductors, although the band electron velocities, being of the order of $v \sim 10^8$ cm/s, are much smaller than the speed of light $c = 3 \times 10^{10}$ cm/s, the relativistic effects, dependent on the v/c ratio, should be taken into account. In both types of semiconductors these relativistic effects lead to a coupling between electron momentum and electron spin, which appears due to the effect of the electric field in the unit cell of a binary semiconductor without inversion symmetry.

The Hamiltonian describing the spin-orbit coupling in bulk III-V materials has the form

$$H_{SO}=\alpha_D(\sigma\kappa)$$

presenting the Dresselhaus realization of the spin-orbit coupling [11]. In this Hamiltonian σ is the vector of Pauli matrices, and κ is defined by.

$$\kappa_x=k_x(k_y^2-k_z^2).$$

with the cyclic permutation of the indices defining the other two components. The coupling constant α_D being of the order of 10 eVA^3 leads to spin splitting of the band electron states of the order of 1 meV at electron concentrations of the order of 10^{18} cm^{-3}.

For II-VI compounds the bulk Hamiltonian has the form

$$H_{SO}=\alpha\left(\sigma_y k_x-\sigma_x k_y\right),$$

usually referred to as the" Rashba Hamiltonian".

The same form of the Hamiltonian describes the spin-orbit coupling in two-dimensional systems with a structural inversion asymmetry. Nonzero values of acan be achieved, in addition, by applying an electric field across the two-dimensional structure or a quantum dot. Usually both (Rashba and Dresselhaus) terms are present in two-dimensional electron systems and quantum dots.

The corresponding spectrum of the Rashba Hamiltonian is given by:

$$E(k) = \frac{\hbar^2 k^2}{2m^*} \pm \alpha k$$

as shown in **Figure 7**.

In the absence of an external magnetic field the presented states are double-degenerate as dictated by the time-reversal symmetry of the spin-orbit coupled Hamiltonian. In the presence of such a field, the spin-orbit coupling terms should be augmented by the Zeeman coupling for the interaction between electron spin and external magnetic field in the form:

$$Hz = \frac{g}{2}(\sigma B)$$

Here B is the magnetic field applied to the quantum dot, and g is the g-(Lande) factor.

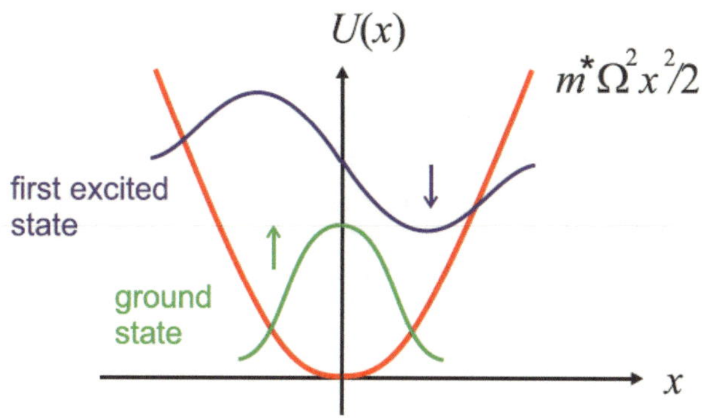

Figure 5.
One-dimensional parabolic potential modeling a simple quantum dot. Schematic plots of the ground and first excited states with opposite spins (marked by up- and down-arrows) are presented. Spin-orbit coupling couples these two spatial states with opposite spins and, as result, leads to the spin-position entanglement and modification of the charge densities.

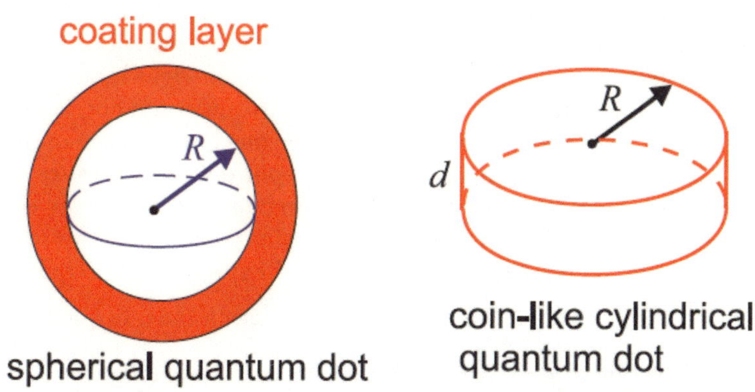

Figure 6.
*(a) Typical spherical quantum dot with a coating layer. Typical value of the radius R is of the order of 10 nm.
(b) Typical coin-like cylindrical model of a quantum dot. The width d is of the order or less than 10 nm.*

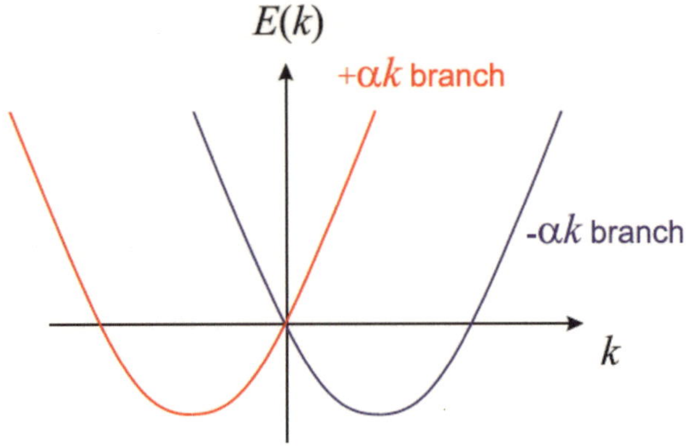

Figure 7.
Two branches of spectrum of two-dimensional carriers in the presence of the Rashba spin-orbit coupling with $E(k)=\frac{\hbar^2k^2}{2m^*}\pm\alpha k$. *This shape of the spectrum corresponds to the anomalous velocity in the presence of the spin-orbit coupling: The zeromomentum states have nonzero velocities while finite-momentum states at* $k=\frac{m^*\alpha}{\hbar}$ *have zero velocities.*

Being a relativistic effect, the g – factor is strongly related to the spin-orbit coupling and the Zeeman term plays an important role in the physics of the quantum dot. The values of g- factor can vary by two orders of magnitude, e.g. between −0.45 for GaAs and approximately −50 in InSb. The corresponding spin splitting can reach the values of the order of 10 meV at realistic fields of the order of 10 T. We mention here that while the spin-orbit coupling for a given quantum dot is fixed by its material and shape [12], the Zeeman coupling can easily be modified by applying a magnetic field. Thus, the absorption spectra and optics-related properties of the quantum dot can be modified as well.

The geometry of a quantum dot plays the crucial role for its spectrum and spin-orbit coupling, and, therefore in the light absorption and photovoltaics effects. A qualitative effect of the spin-orbit coupling in quantum dots is the quantum entanglement of particle spin and its position, where the particle wavefunction cannot be presented as a product of the spin and coordinate states. This entanglement determines several processes in quantum dots.

Since in the optical absorption electrons and holes are produced, similar spin-orbit coupling and Zeeman Hamiltonians should be defined for the holes as well. The Coulomb interaction between electrons and holes plays the crucial role in the light absorption in quantum dots. The spectrum of holes is described by 4 × 4 matrices due to degeneracy of the spectrum consisting of "light" and "heavy" hole branches. Dependent on the material and shape of the quantum dot, spin-orbit coupling and the Zeeman coupling for holes can be much stronger than that for electrons. In the presence of spin-orbit coupling, spin-defined states are prone to relaxation and loosing well-defined spin values. In bulk systems, where electron momentum is a well-defined quantum number, spins relax mainly due to the celebrated Dyakonov-Perel mechanism [13] with the spin relaxation rate of the order of $\left(\frac{\alpha k}{\hbar}\right)^2 \tau_k$, where τ_k is the momentum relaxation time. In quantum dots the situation is qualitatively different since the localization suppresses the free motion of carriers. Here the relaxation is mainly due to the so-called" admixture mechanism" caused by the spin-orbit coupling between different orbital states with opposite spins and interaction of these orbital states with lattice vibrations. This mechanism strongly depends on the applied magnetic field. In relatively week magnetic fields, the spin relaxation rate is small, and spin states are conserved for long times of the order of 1 ms or more. If the spin-orbit coupling is sufficient, the resulting spin-position entanglement influences the spin states, and, therefore, the orbital wave functions and the exchange interaction between the injected and original carriers in the quantum dots. In addition, spin states of the injected carriers can be controlled by polarization of the incident light [14]. These interactions should be clearly seen in the absorption spectra and, as a result, influence the photovoltaics and photogalvanic application.

We mention here that a new very large class of very promising for the photovoltaics materials such as perovskites or quantum dots show the spin-orbit coupling in the general form of the Rashba and Dresselhaus terms. Therefore, the spin-orbit coupling-related aspects of their applications can be treated in general terms similarly to semiconductors. This coupling can increase the carrier's lifetime - the quantity important for solar cell applications. However, these rather complex compounds, demonstrating a great variety of different properties, are not yet reliably functionalized in the form of quantum dots. At the same time, whether the combined effects of the spin-orbit coupling lead to an increase or to a decrease in the efficiency of the light-to-voltage conversion in solar cells, is not yet clear even on the qualitative scale.

Acknowledgements

This work was supported by the National Centre for Research and Development under the project No. POIR.01.02.00-00-0265/17-00.

Author details

Paweł Kwaśnicki[1,2]

1 Department of Physical Chemistry and Physicochemical Basis of Environmental Engineering, Institute of Environmental Engineering in Stalowa Wola, John Paul II Catholic University of Lublin, Kwiatkowskiego 3A, 37-450 Stalowa Wola, Poland

2 Research and Development Centre for Photovoltaics, ML System S.A. Zaczernie 190G, 36-062 Zaczernie, Poland

*Address all correspondence to: pawel.kwasnicki@mlsystem.pl

References

[1] A. Kojima et al, „Organometalic halide perovskites as visible-light sensitize for photovoltaic," *J. Am. Chem. Soc*, p. 131, 2009.

[2] J. C. H. M. A. Loi, „Hybrid solar cells – perovskites under the Sun," *Nature Materials 12*, p. 1087, 2013.

[3] P. M. H. J. Li, „Circular photovoltaic effect in organometal halide perovskite CH3NH3PbI3," *arXiv:*, p. 1607, 2016.

[4] M. Nakamura et al, „Spontaneous polarization and bulk photovoltaic effect driven by polar," *Phys. Rev. Lett*, p. 156801, 2016.

[5] C. M. e. al, „Revealing the role of organic cations in hybrid halide perovskite CH3NH3PbI3," *Nature Communications*, p. 6:7026, 2015.

[6] K. T. Y. S. a. T. M. Akihiro Kojima, „Organometal halide perovskites as visible-light sensitizers for photovoltaic cells," *J. Am. Chem. Soc.*, p. 131, 2009.

[7] T. D. H. S. K. e. a. Chiho Kim, „A hybrid organic-inorganic perovskite dataset," *Scient. Data 4,*, p. 170057 , 2017.

[8] J. O.-M. a. W. R. Calderon-Munoz, „Thermal influence on charge carrier transport in solar cells based on GaAs PN junctions," *Journ. of Appl. Phys.*, p. 154502, 2014.

[9] C. C. S. M. G. K. e. a. Pierre-Adrien Mante, „Directional Negative Thermal Expansion and Large Poisson Ratio," *J. Phys. Chem. Lett*, p. 14398, 2017.

[10] M. C. B. J. M. L. M. L. R. J. E. a. J. C. J. A. J. Nozik, *Chem. Rev. 110*, p. 6873, 2010.

[11] G. Dresselhaus, „Spin-Orbit Coupling Effects in Zinc Blende Structures," *Phys.Rev.* , p. 580, 1955.

[12] E. L. I. a. U. R. A. A. Kiselev, „Electron g factor in one- and zero-dimensional semiconductor nanostructures," *Phys. Rev. B* , p. 6353 , 1998.

[13] M. Dyakonov and V. Perel, „SPIN ORIENTATION OF ELECTRONS ASSOCIATED WITH THE INTERBAND," *Sov. Phys. JETP 33*, p. 1053, 1971.

[14] A. V. K. a. Y. V. Nazarov, „Spin-flip transitions between Zeeman sublevels in semiconductor quantum dots," *Phys. Rev. B 64*, p. 125316, 2001.

Chapter 17

Nanostructured Transition Metal Compounds as Highly Efficient Electrocatalysts for Dye-Sensitized Solar Cells

Yi-June Huang and Chuan-Pei Lee

Abstract

Nowadays, the requirement of energy increases every year, however, the major energy resource is fossil fuel, a limiting source. Dye-sensitized solar cells (DSSCs) are a promising renewable energy source, which could be the major power supply for the future. Recently, the transition metal component has been demonstrated as potential material for counter electrode of platinum (Pt)-free DSSCs owing to their excellent electrocatalytic ability and their abundance on earth. Furthermore, the transition metal components exist different special nanostructures, which provide high surface area and various electron transport routs during electrocatalytic reaction. In this chapter, transition metal components with different nanostructures used for the application of electrocatalyst in DSSCs will be introduced; the performance of electrocatalyst between intrinsic heterogeneous rate constant and effective electrocatalytic surface area are also be clarified. Final, the advantages of the electrocatalyst with different dimensions (i.e., one to three dimension structures) used in DSSCs are also summarized in the conclusion.

Keywords: counter electrode, dye-sensitized solar cells, nanostructures, and transition metal components

1. Introduction

In this century, the energy requisition and environment caring arrive at the highest point in history. The clean and economical renewable energy resource is urgently needed for us. Photovoltaics, named solar cells, tremendous progress has been achieved in efficiency (η), reproducibility, and stability [1–3]. It has been considered as one of the most promising renewable energy sources. Photovoltaics are classified to three generations, as shown in **Figure 1** [1, 4–7]. The first-generation solar cells, named silicon-based solar cells or the traditional solar cells, made up of crystalline silicon. These solar cells demonstrate high efficiency and significant demand in the market, but the production cost of crystalline silicon materials limited the large-scale industrial applications. The second-generation is cadmium telluride (CdTe)/cadmium indium gallium diselenide (CIGS) based solar cells. The solar cells could be produced with large-scale and well efficiency (14–22%). The first and second generations are the most widely solar cells at present. However, they are scarcity, the toxicity of materials, high-temperature, and high-vacuum processes

IntechOpen

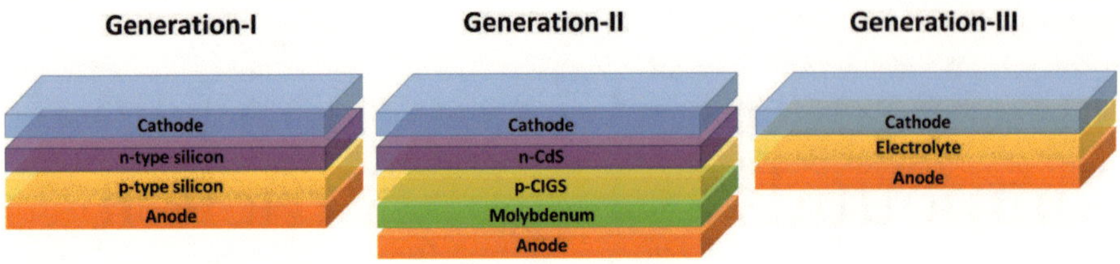

Figure 1.
The scheme of three generation photovoltaic solar cells.

that restrict further applications. Dye-sensitized solar cells (DSSCs), classed third-generation solar cells, have gained attention and be regarded as prospective solar cells for the photovoltaic technologies in recent years as potential cost-effective alternatives to the first and second generations solar cells [8–11]. Furthermore, the DSSCs have outstanding performance in an indoor, dim light environment [12–14].

Typically, DSSCs are consist of three sections, including photoanode, electrolyte, and counter electrode (CE), that respond to different functions, as shown in **Figure 2** [1, 4, 5, 8–10]. The photoanode converts the photon into the electron by the dye. The electrolyte keeps the function of the photoanode by iodine ion. The CE catalyzes the redox reduction in the electrolyte, which is an obvious influence on the photovoltaic performance, long-term stability, and cost of the device. In other words, the CE is a crucial component of DSSCs.

The CE is classified into three components, that are electrocatalyst, transparent conducting oxide, and substrate, as shown in **Figure 2**. Among them, the electrocatalyst is the key factor to promise the function of CE [1, 7–9, 15, 16]. As shown in **Figure 3**, between electrolyte and CE, the reaction of reduction iodide/triiodide (I^-/I_3^-) redox couple is that: The first stage, diffusion, triiodide diffuses from electrolyte bulk to near the CE for regenerating electrolyte. The second stage, decomposition, triiodide decomposes to iodide and iodine. The iodide is used to renew the dye and iodine will go to the next step. The third stage, adsorption, the CE adsorbs iodine near the CE. The fourth stage, electrocatalysis, electrocatalyst catalyzes reduction reaction, transferring iodine to iodide. The final stage, desorption, the CE desorbs iodide to complete regenerate the electrolyte. According to this mechanism, the electrocatalytic ability, it also represents the reaction rate in here, and the specific structure are the major affections for the reduction reaction.

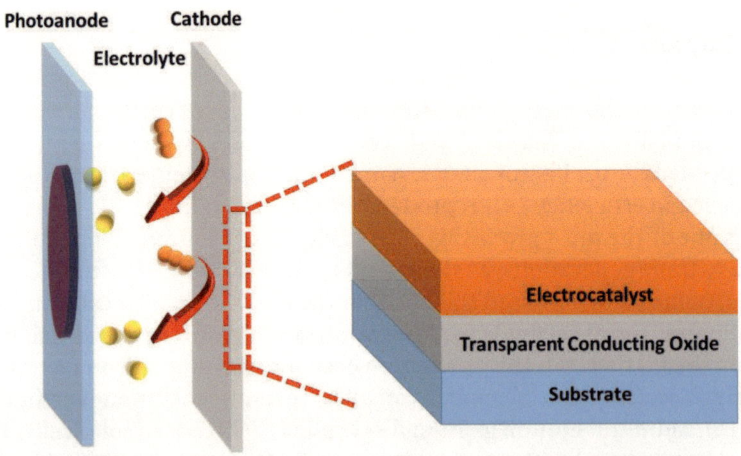

Figure 2.
The scheme of dye-sensitized solar cells and counter electrode (cathode).

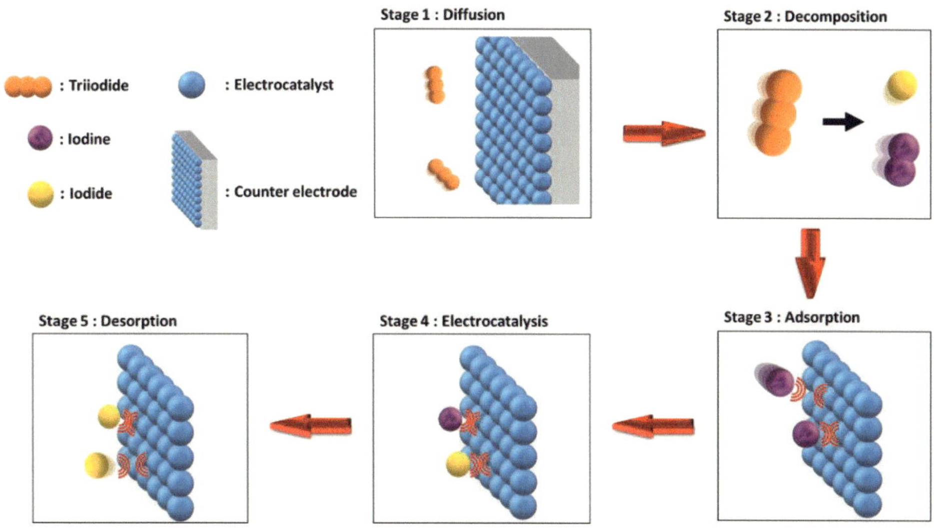

Figure 3.
The scheme of reduction iodide/triiodide (I^-/I_3^-) redox couple in counter electrode.

The traditional electrocatalyst of DSSCs is Platinum (Pt), which has an outstanding electrocatalytic ability [10, 15–20]. However, Pt, noble metal, is rare on earth that present expensive prices and difficult shapes the specific structure. Up to date, there are a few non-Pt nanomaterials that could have comparable electrocatalytic ability to that of Pt. There have two ways to raise the electrocatalytic reduction reaction. The intrinsic electrocatalytic ability of the electrocatalyst is directly related the electrocatalytic ability. In other words, the choice of material is very important. The other way is to design the nanostructure of the electrocatalyst for I_3^- reduction regarding with the charge transfer route and the surface area.

Transition metal compounds (TMCs) possess d-electron filling in e_g orbitals, which promote excellent electrocatalytic performance in partially filled condition [4, 19, 21–24]. So, they are interested to replace Pt. But most of TMCs still show poorer electrocatalytic ability than Pt. To overcome the challenge, TMCs are synthesized with various nanostructure, which is an important factor for increasing electrocatalytic ability [20–22, 25]. A nanostructure is defined if any dimension of the structure is lower than 100 nm, the structure is the nanostructure. Basically, nanostructure divides into four groups: zero-dimensional (0D, *e.g.* nanoparticle, nanocube, *etc.*), one-dimensional (1D, *e.g.* nanorod, nanotube, nanoneedle, *etc.*), two-dimensional (2D, *e.g.* nanosheet, nanopental *etc.*) and hierarchical nanostructures, as shown in **Figure 4**. In view of 1D, 2D, and hierarchical nanostructures have

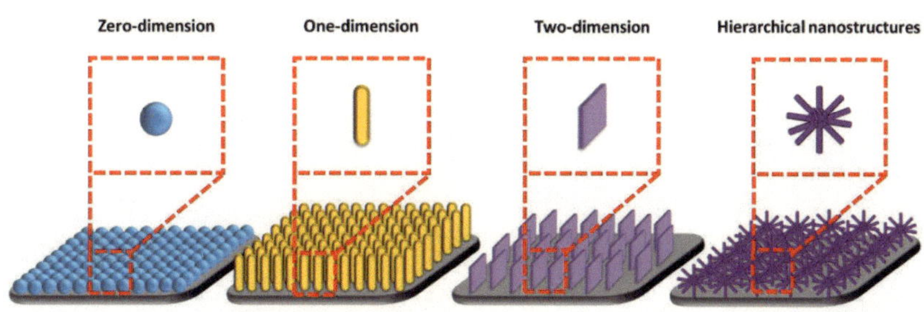

Figure 4.
The scheme of zero-dimensional (0D), one-dimensional (1D), two-dimensional (2D), and hierarchical nanostructure.

complex structure. In this chapter, we will systematically discuss their plural strategies (including high electrochemical surface area, directional electron transferring pathways, decrease diffusion control, *etc.*) to promote the electrocatalytic ability for DSSCs performance.

2. One-dimensional nanostructure (1D)

One-dimensional TMCs nanostructure is expected that it provides the 1D electron transfer pathways, promoting electrolyte penetration, and more reaction area [26–34]. However, the vertical 1D structure is rarely obtained because it is difficult to synthesize. Herein, we focus on that the 1D structure has been directly obtained without the template method, in **Figures 5** and **6**. Their corresponding efficiencies are listed in **Table 1**. In **Figure 5**, it shows horizontal 1D TMCs nanostructure SEM images of MoN nanorod, $W_{18}O_{49}$ nanowire, NiS nanorod, $CoSe_2$ nanorod, $Co_{0.85}Se$ nanotubes, $CoSe_2/CoSeO_3$ nanorod, and Ni_3S_4 nanorod that were synthesized by Song et al., Zhou et al., Yang et al., Sun et al., Yuan et al., Huang et al., and Huang et al., respectively [27–33]. Song et al. reported that MoN nanorod morphology reveals enhancement of diffusion kinetics for the active electrochemical process, as shown in **Figure 5a** [27]. So that the MoN nanorod has higher V_{OC} and J_{SC} than MoN nanoparticle. Zhou synthesized $W_{18}O_{49}$ nanowire (**Figure 5b**), having oxygen vacancies within the range of $WO_{2.625}$ to WO_3, *via* the solvothermal method [28]. Their efficiency is 4.85% for Co ion electrolyte. Yang et al. obtained NiS nanorod (**Figure 5c**), which is α type, through chemical bath method [29]. It has η of 5.20%,

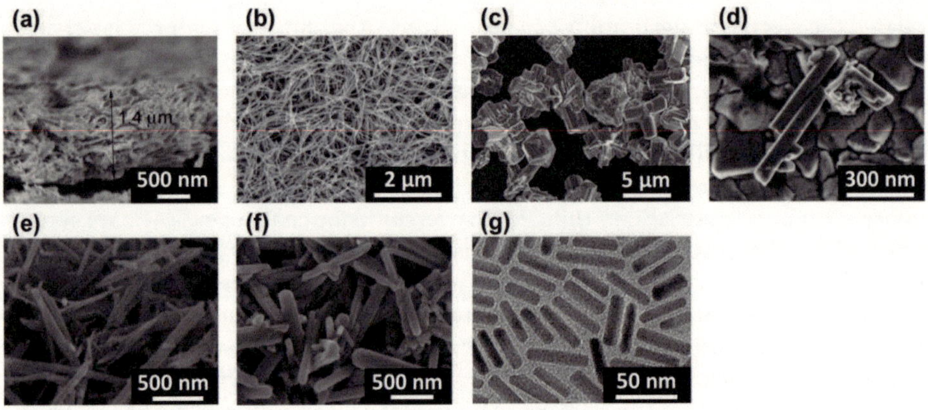

Figure 5.
The SEM of horizontal 1D nanostructure with (a) MoN, (b) $W_{18}O_{49}$, (c) NiS, (d) $CoSe_2$, (e) $Co_{0.85}Se$, (f) $CoSe_2/CoSeO_3$, (g) Ni_3S_4 [27–33].

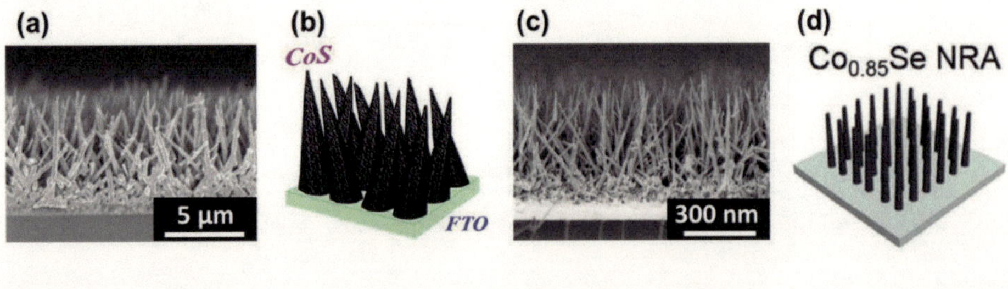

Figure 6.
The pseudo-vertical 1D nanostructure with (a) and (b) CoS and (c) and (d) $Co_{0.85}Se$ [26, 34].

Materials	η (%)	V_{OC} (V)	J_{SC} (mA cm^{-2})	FF	η/η_{Pt}	Ref
CoS	7.67	0.71	16.31	0.66	1.00	[26]
MoN	7.29	0.74	15.26	0.65	0.98	[27]
$W_{18}O_{49}$	4.85	0.80	9.26	0.67	1.08	[28]
NiS	5.20	0.68	11.42	0.67	0.83	[29]
$CoSe_2$	10.20	0.75	18.55	0.73	1.25	[30]
$Co_{0.85}Se$	5.34	0.71	14.51	0.52	0.71	[31]
$CoSe_2/CoSeO_3$	7.54	0.82	14.32	0.64	0.95	[32]
Ni_3S_4	7.31	0.75	15.53	0.63	0.93	[33]
$Co_{0.85}Se$	8.35	0.74	15.76	0.71	1.08	[34]

Table 1.
A partial list of literature on the DSSCs with 1D TMCs nanostructure based CEs.

which is better than the nanoparticle NiS (4.20%). The reason is that the nanorod affords lower charge transfer resistance than the nanoparticle. Sun et al. acquired $CoSe_2$ nanorod (**Figure 5d**), possessing a single orthorhombic crystal structure, by hydrothermal method [30]. The $CoSe_2$ nanorod exits the excellent performance (10.20%), even better than the Pt. They remind that single $CoSe_2$ nanorod has great electrocatalytic ability, lower charge resistance, and higher adsorption capacity for electrolyte. Yuan et al. prepared $Co_{0.85}Se$ nanotubes (**Figure 5e**) by a simple hydrothermal method [31]. It shows η of 5.34%, which lower than Pt, obviously. Huang et al. obtained $CoSe_2/CoSeO_3$ nanorod (**Figure 5f**) through a microemulsion-assisted hydrothermal synthesis [32]. It reveals η of 7.54%, which is approach Pt performance. This result contributes to the 1D electron transfer pathways. Huang et al. synthesized the Ni_3S_4 nanorod (**Figure 5g**) *via* a one-pot colloidal synthesis [33]. And it has η of 7.31%, which is quite close Pt. As listed in **Table 1**, there have a few of the 1D TMCs nanostructures existing the better performance than the Pt.

Most of them are vertical 1D TMCs nanostructures. The horizontal 1D TMCs nanostructures could not support the vertical electron transfer pathways and promote the electrolyte penetration. So most of them display lower performance than the Pt.

The vertical 1D TMCs nanostructure is an ideal condition, as shown in **Figure 4**. Kung et al. and Jin et al. directly synthesized pseudo-vertical 1D nanostructure array with CoS and $Co_{0.85}Se$, respectively, as shown in **Figure 6** [26, 34]. This structure sufficiently acts the 1D TMCs nanostructure advantages, including favorable for fast diffusion of redox species within the CE film, 1D direction electron channel, enhance electrolyte penetration, and more reaction area. Both of them exhibit higher value of η than Pt. In other words, they could straightly replace the Pt function for CE in DSSCs.

3. Two-dimensional (2D)

Geim and Grigorieva classified 2D materials into three groups [35]. First group, graphene type contains graphene, fluorographene, graphene oxide, hBN, *etc.*; second group, 2D chalcogenides (transition metal) type includes MoS_2, NbS_2, $NbSe_2$, $CoSe_2$, $MoSe_2$, $ZrSe_2$, GaSe, GaTe, InSe, Bi_2Se_3, Bi_2Te_3, *etc.*; final group, 2D oxides type involves TiO_2, MnO_2, V_2O_5, RuO_2, perovskite-based materials ($LaNb_2O_7$, $Ba_4Ti_3O_{12}$, $Ca_2Ta_2TiO_{10}$ *etc.*), hydroxides ($Ni(OH)_2$, $Eu(OH)_2$, *etc.*), *etc.* Research of 2D TMCs nanostructure is intensified in recently [16, 36]. The bandgap energy is reduced by decreasing the layer of the TMCs [16, 35, 37–40]. In other words, a single or a few layers of the 2D TMCs nanostructure presents excellent electrocatalytic ability. Besides that, the 2D TMCs nanostructure has advantages including enhancing the diffusion of electrolyte, vertical electron channel, and special optical property.

In this section, the partial works of literature are chosen depending on the electrocatalytic performance and structure. Their corresponding SEM images and efficiency parameters are shown in **Figures 7** and **8**, and **Table 2**, respectively. In **Figure 7**, Ibrahem et al., Huang et al., and Mohammadnezhad et al. applied the horizontal 2D nanostructure with $NbSe_2$, $MoSe_2$, and $Cu_2ZnSnS_xSe_{4-x}$ in CE for DSSCs [41–43]. Ibrahem et al. reported that the $NbSe_2$ nanosheet (**Figure 7a**) has the best performance among nanosheet, nanorod, and nanoparticle [41]. They mention that nanosheet could provide high surface area and coverage. And the $NbSe_2$ nanosheet existed η of 7.73%, which is better than the Pt CE. The result indicates that $NbSe_2$ nanosheet substitutes to the noble metal Pt in DSSCs. Following the idea, $MoSe_2$ and $Cu_2ZnSnS_xSe_{4-x}$ nanosheet show the η of 7.58% and 5.73%, respectively. However, both of their values of η are lower than the Pt. To increase the performance of 2D TMCs nanostructure, the pseudo-vertical 2D nanostructure was synthesized and provide the vertical electron channel. The pseudo-vertical 2D nanostructure with MoS_2, $CoSe_2$, MoS_2, $Cu_xZn_ySn_zS$, and $CoNi_2S_4$ were obtained by Antonelou et al., Chiu et al., Raj et al., Chiu et al., and Patil et al., respectively [44–48]. Antonelou et al. obtained the MoS2 nanosheet with η of 8.40%, which has thicknesses down to the 1-2 nm scale. Chiu et al. acquired the nanoclimbing-wall-like $CoSe_2$ (**Figure 8a**) through an electrodeposition process, by using bathes with different pH values.

Its performance is 8.92%. They mentioned that vertical nanowall provides conducting charge for electrocatalytic reduction, as shown in **Figure 9a**. Raj et al. synthesized reflectivity of MoS_2 nanosheet (**Figure 8b**), which has η of 7.50%, through chemical vapor deposition (CVD). The reflectivity of MoS_2 nanosheet is

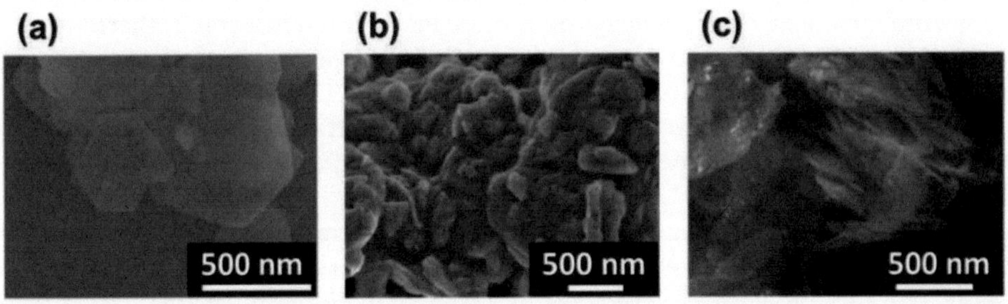

Figure 7.
The SEM of 2D nanostructure with (a) $NbSe_2$, (b) $MoSe_2$, (c) $Cu_2ZnSnS_xSe_{4-x}$ [41–43].

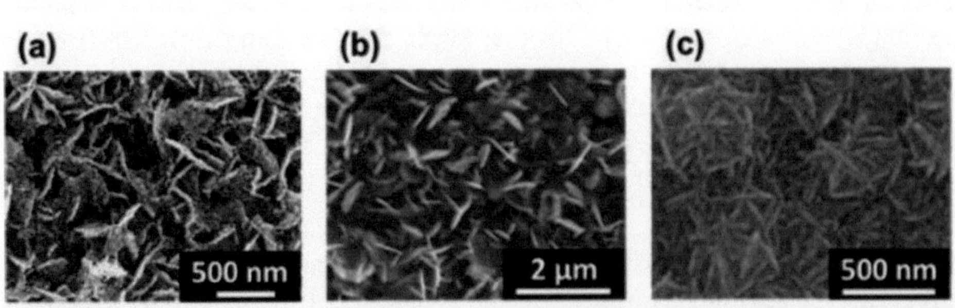

Figure 8.
The pseudo-vertical 2D nanostructure with (a) $CoSe_2$, (b) MoS_2, and (c) $CoNi_2S_4$ [45–48].

Materials	η (%)	V_{OC} (V)	J_{SC} (mA cm^{-2})	FF	η/η_{Pt}	Ref
$NbSe_2$	7.73	0.74	16.85	0.62	1.10	[41]
MoS_2	8.40	0.74	22.60	0.50	0.97	[44]
$CoSe_2$	8.92	0.73	18.03	0.67	1.08	[45]
$MoSe_2$	7.58	0.70	15.97	0.67	0.97	[42]
MoS_2	7.50	0.71	15.20	0.70	1.03	[46]
$Cu_xZn_ySn_zS$	7.44	0.67	16.57	0.66	1.03	[47]
$CoNi_2S_4$	8.86	0.66	19.21	0.70	0.98	[48]
$Cu_2ZnSnS_xSe_{4-x}$	5.73	0.69	12.60	0.66	0.99	[43]

Table 2.
A partial list of literature on the DSSCs with 2D TMCs nanostructure based CEs.

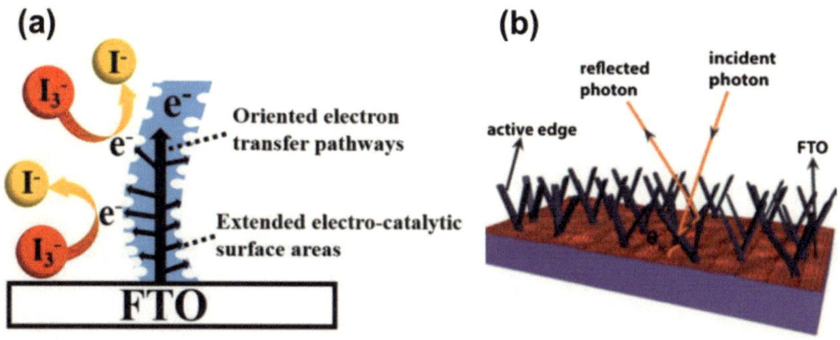

Figure 9.
The mechanism of 2D nanostructure with (a) $CoSe_2$ and (b) MoS_2 [45, 46].

raised their high reflectivity facilitates the absorbance of more photons, and more active edge sites exposed to redox couple in the electrolyte, as shown in **Figure 9b**. Chiu et al. gained $Cu_xZn_ySn_zS$ nanowall-structure by thermal solvent method, and it shows performance 7.44%. The performance is attributed to improves the carrier transport pathway and effectively reduces the interface resistance. Patil et al. utilized a simple one-step solution-based fabrication method for $CoNi_2S_4$ interconnected nanosheet (**Figure 8c**). The $CoNi_2S_4$ exhibits η of 8.86%, which attributes to a larger active surface area with favorable charge transport. The pseudo-vertical 2D nanostructure has obviously improvement of electrocatalytic ability than the normal 2D nanostructure. It is not only providing vertical transport pathways and active surface area but also contributes to reflection photon. Those properties make 2D TMCs nanostructure have the potential to alternative Pt as an electrocatalyst.

4. Hierarchical nanostructure

Basically, 0D nanostructure possesses a high reaction area; 1D and 2D nanostructure offers directional electron pathways and enhance electrolyte penetration. But they have their own weakness. For example, 0D nanostructure is easy aggregation and has larger heterogeneous resistance; 1D and 2D nanostructure have lower reaction area. A hierarchical nanostructure consists of the nanostructure with multidimensional subunits (0D, 1D, and 2D). It merges various subunits, so it has multidimensional nanostructure advantages, including high reaction area, benefit electron transfer, avoiding aggregation, enhance electrolyte diffusion, and offer directional electron pathways.

Herein, we list partial literature with hierarchical TMCs nanostructure. **Figure 10** shows SEM of Ni_3Se_4 with sea urchins-like structure, $TiO_{1.1}Se_{0.9}$ with nanospheres and 1D nanorods, $NiCo_{0.2}$ with hollow structure and nanoclusters, $NiCo_2S_4$ with ball-in-ball structure, $NiS@MoS_2$ with feather duster-like hierarchical structure, $CoSe_2/CoSeO_3$ with hierarchical urchin-like structure, CuO/Co_3O_4 with core-shell structure and $CoS_2/NC@Co\text{-}WS_2$ with yolk-shell structure by Lee et al., Li et al., Jiang et al., Jiang et al., Su et al., Huang et al., Liao et al., and Huang et al., respectively [49–56]. And their efficiency parameters are listed in **Table 3**. Lee et al. synthesized the Ni_3Se_4 sea urchins-like structure (**Figure 10a**) through one-step and low temperature hydrothermal process [49]. It reveals η of 8.31%, which attribute to the high active electrocatalytic surface area. Li et al. obtained $TiO_{1.1}Se_{0.9}$ with nanospheres and 1D nanorods (**Figure 10b**) *via* a simple dip-coating process and rapid thermal annealing (RTA) process [50]. The $TiO_{1.1}Se_{0.9}$ exhibits η of 9.47%, which is better than the Pt. The result is established that the nanospheres can work as electro-catalytic active sites, and the nanorods can function not only as electro-catalytic active sites but also as fast electron transport channels, as shown in **Figure 11a**. Jiang et al. synthesized $NiCo_{0.2}$ hollow structure and nanoclusters, having uniform spherical particles with an average diameter of about 2 μm and shell thickness of around 200 nm (**Figure 10c**), *via* a thermal method [51]. It shows η of 9.30% and displays that the novel spherical structures can efficiently promote the transfer of electrons from the conductive carbon frameworks to metal nanoparticles, thus resulting in high electrocatalytic activity for the reduction. Jiang et al. acquired $NiCo_2S_4$ ball-in-ball structure (**Figure 10d**) by a thermal method [52]. Its efficiency is 9.49%, which is attributed to the rougher surface, higher surface area, and high diffusion coefficient for redox. Su et al. obtained $NiS@MoS_2$ feather duster-like hierarchical structure, which has η of 8.58% [53]. They propose that feather duster-like hierarchical structure array can support the fast electron transfer and electrolyte diffusion channels, moreover, it also can render abundant active catalytic sites and large electron injection efficiency from CE to the electrolyte. Huang et al. gained $CoSe_2/CoSeO_3$ hierarchical urchin-like structure (**Figure 10g**), the nanoparticle-composed sphere is the central core with a diameter of about 50 nm surrounded by several hexagonal prisms, through a one-step hydrothermal method [54]. The $CoSe_2/CoSeO_3$ reveals η of 9.29% and is mention that the

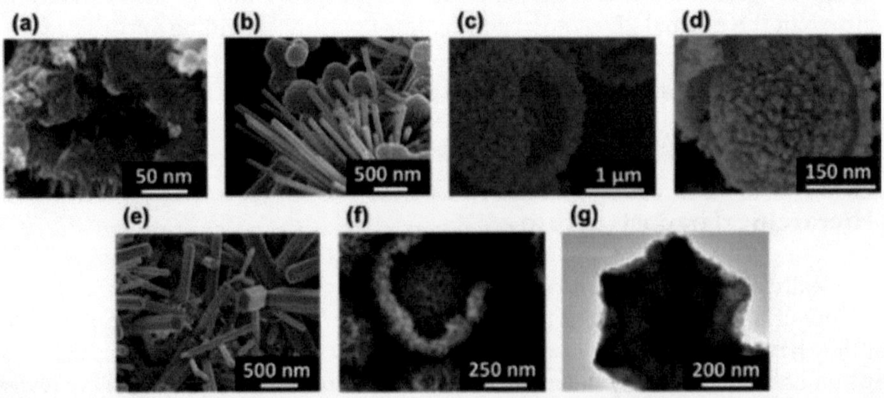

Figure 10.
The SEM of hierarchical nanostructure with (a) Ni_3Se_4, (b) $TiO_{1.1}Se_{0.9}$, (c) $NiCo_{0.2}$, (d) $NiCo_2S_4$, (e) $CoSe_2/CoSeO_3$, (f) CuO/Co_3O_4, and (g) $CoS_2/NC@Co\text{-}WS_2$ [49–56].

Materials	η (%)	V_{OC} (V)	J_{SC} (mA cm^{-2})	FF	η/η_{Pt}	Ref
Ni_3Se_4	8.31	0.75	16.27	0.69	1.03	[49]
$TiO_{1.1}Se_{0.9}$	9.47	0.79	17.22	0.70	1.22	[50]
$NiCo_{0.2}$	9.30	0.78	17.80	0.67	1.16	[51]
$NiCo_2S_4$	9.49	0.84	17.40	0.647	1.14	[52]
$NiS@MoS_2$	8.58	0.77	16.64	0.67	1.05	[53]
$CoSe_2/CoSeO_3$	9.29	0.82	16.09	0.70	1.12	[54]
CuO/Co_3O_4	8.34	0.73	18.13	0.63	1.06	[55]
$CoS_2/NC@Co\text{-}WS_2$	9.21	0.82	16.50	0.67	1.13	[56]

Table 3.
A partial list of literature on the DSSCs with hierarchical TMCs nanostructure based CEs.

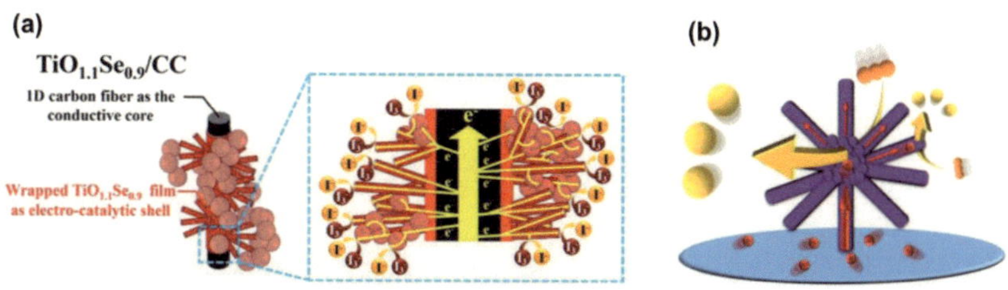

Figure 11.
The mechanism of hierarchical nanostructure with (a) $TiO_{1.1}Se_{0.9}$ and (b) $CoSe_2/CoSeO_3$ [50, 54].

urchin-like structure possessing the hexagonal prism structure and nanoparticles to provide both rapid electron transport routes and a reasonably high surface area for electro-catalytic reactions, as shown in **Figure 11b**. Liao et al. obtained CuO/Co_3O_4 core-shell structure (**Figure 10f**) *via* a facile self-templated method [55]. The CuO/Co_3O_4 has η of 8.34% and an excellent electronic transmission channel and more adsorption sites for the redox couple, which greatly enhances the subsequent redox process. Huang et al. acquired $CoS_2/NC@Co\text{-}WS_2$ with yolk-shell structure (**Figure 10g**) [56]. By virtue of larger surface area and more effective active sites, the $CoS_2/NC@Co\text{-}WS_2$ (η of 9.21%) has better performance than the Pt.

In this section, it can be found that the hierarchical TMCs nanostructure has better performance than the Pt in CE. In other words, they can efficiently raise the TMCs performance, so the hierarchical TMCs nanostructure could replace Pt directly.

5. Conclusion

The electrocatalytic ability of catalysts is usually determined by below two points: one is the intrinsic electrocatalytic activity, and another is the nanostructure. The nanostructure of TMCs can briefly be classified into 0D, 1D, 2D, and hierarchical nanostructures; those have different properties and could obviously affect the electrocatalytic ability. Herein, the partial reports about DSSCs with the electrocatalysts having 1D, 2D, or hierarchical nanostructures are selected for introduction and discussion. 1D nanostructure possesses several advantages, including the 1D electron transfer pathways, promoting electrolyte penetration, avoiding stack problem, and high reaction area. However, not all the electrocatalysts with 1D nanostructure show better performance than the Pt in DSSC application. Some of them lied down on substrate; so, the advantage on vertical electron transport rout is not given. Furthermore, as the stacking problem comes out, it will lose surface are for reaction. 2D nanostructures possess the active site on edges or defects, and their 2D structure could provide the benefits below, such as directional electron and diffusion channels; these properties boost their DSSC performances obviously. However, the stacking problem and poor activity on basal plane of 2D materials also retarding their practical performance in DSSCs. Hierarchical nanostructure incorporates the profits of subunits, so it displays high reaction area, benefit electron transport rout, avoiding aggregation, enhanced electrolyte diffusion, *etc*. Several reports already demonstrated that the TMCs with hierarchical nanostructures show excellent electrocatalytic ability in DSSCs; they even exhibit better electrocatalytic performance than that of Pt.

Acknowledgements

This work was supported by the Ministry of Science and Technology (MOST) of Taiwan, under grant numbers 107-2113-M-845-001-MY3.

Author details

Yi-June Huang[1] and Chuan-Pei Lee[2*]

1 Department of Chemical Engineering, National Taiwan University, Taiwan

2 Department of Applied Physics and Chemistry, University of Taipei, Taiwan

*Address all correspondence to: cplee@utaipei.edu.tw

References

[1] Mrinalini Madoori, Islavath Nanaji, Prasanthkumar Seelam, Giribabu Lingamallu, Stipulating low production cost solar cells all set to retail...! Chem Rec. 2019; 19; 661-674, DOI: 10.1002/tcr.201800106.

[2] Zhou Liang, Zhuang Zechao, Zhao Huihui, Lin Mengting, Zhao Dongyuan, Mai Liqiang, Intricate hollow structures: controlled synthesis and applications in energy storage and conversion. Advanced Materials. 2017; 29; 1602914-1602942, DOI: 10.1002/adma.201602914.

[3] Park Nam-Gyu, Segawa Hiroshi, Research direction toward theoretical efficiency in perovskite solar cells. ACS Photonics. 2018; 5; 2970-2977, DOI: 10.1021/acsphotonics.8b00124.

[4] Guo J., Liang S., Shi Y., Hao C., Wang X., Ma T., Transition metal selenides as efficient counter-electrode materials for dye-sensitized solar cells. Physical Chemistry Chemical Physics. 2015; 17; 28985-28992, DOI: 10.1039/c5cp04862a.

[5] Yun Sining, Qin Yong, Uhl Alexander R., Vlachopoulos Nick, Yin Min, Li Dongdong, Han Xiaogang, Hagfeldt Anders, New-generation integrated devices based on dye-sensitized and perovskite solar cells. Energy & Environmental Science. 2018; 121; 476-526, DOI: 10.1039/c7ee03165c.

[6] Hisham A. Maddah Vikas Berry, Sanjay K. Behura, Biomolecular photosensitizers for dye-sensitized solar cells: Recent developments and critical insights. Renewable and Sustainable Energy Reviews. 2020; 121; 109678-109702, DOI: 10.1016/j.rser.2019.109678.

[7] Kouhnavard Mojgan, Ludin Norasikin Ahmad, Ghaffari Babak V, Sopian Kamarozzaman, Ikeda Shoichiro, Carbonaceous materials and their advances as a counter electrode in dye-sensitized solar cells: challenges and prospects. ChemSusChem. 2015; 8; 1510-1533, DOI: 10.1002/cssc.201500004.

[8] Yun Sining, Hagfeldt Anders, Ma Tingli, Pt-free counter electrode for dye-sensitized solar cells with high efficiency. Advanced Materials. 2014; 26; 6210-6237, DOI: 10.1002/adma.201402056.

[9] Wu Jihuai, Lan Zhang, Lin Jianming, Huang Miaoliang, Huang Yunfang, Fan Leqing, Luo Genggeng, Lin Yu, Xie Yimin, Wei Yuelin, Counter electrodes in dye-sensitized solar cells. Chemical Society Reviews. 2017; 46; 5975-6023, DOI: 10.1039/c6cs00752j.

[10] Fakharuddin Azhar, Jose Rajan, Brown Thomas M., Fabregat-Santiago Francisco, Bisquert Juan, A perspective on the production of dye-sensitized solar modules. Energy & Environmental Science. 2014; 7; 3952-3981, DOI: 10.1039/c4ee01724b.

[11] Yuan Huihui, Wang Wei, Xu Di, Xu Quan, Xie Junjie, Chen Xinyu, Zhang Tao, Xiong Changjun, He Yunlong, Zhang Yumei, Liu Yan, Shen Hujiang, Outdoor testing and ageing of dye-sensitized solar cells for building integrated photovoltaics. Solar Energy. 2018; 165; 233-239, DOI: 10.1016/j.solener.2018.03.017.

[12] Cao Yiming, Liu Yuhang, Zakeeruddin Shaik Mohammed, Hagfeldt Anders, Grätzel Michael, Direct contact of selective charge extraction layers enables high-efficiency molecular photovoltaics. Joule. 2018; 2; 1108-1117, DOI: 10.1016/j.joule.2018.03.017.

[13] Cheng Rui, Chung Chih-Chun, Zhang Hong, Liu Fangzhou, Wang Wei-Ting, Zhou Zhiwen, Wang Sijia,

Djurišić Aleksandra B., Feng Shien-Ping, Tailoring triple-anion perovskite material for indoor light harvesting with restrained halide segregation and record high efficiency beyond 36%. Advanced Energy Materials. 2019; 1901980-1901987, DOI: 10.1002/aenm.201901980.

[14] Chen Han-Ting, Huang Yi-June, Li Chun-Ting, Lee Chuan-Pei, Lin Jiann T., Ho Kuo-Chuan, Boron nitride/sulfonated polythiophene composite electrocatalyst as the TCO and pt-free counter electrode for dye-sensitized solar cells: 21% at dim light. ACS Sustainable Chemistry & Engineering. 2020 8 5251-5259, DOI: 10.1021/acssuschemeng.0c00097.

[15] Wu Mingxing, Lin Xiao, Wang Yudi, Wang Liang, Guo Wei, Qi Daidi, Peng Xiaojun, Hagfeldt Anders, Gratzel Michael, Ma Tingli, Economical Pt-free catalysts for counter electrodes of dye-sensitized solar cells. Journal of the American Chemical Society. 2012; 134; 3419-3428, DOI: 10.1021/ja209657v.

[16] Singh Eric, Kim Ki Seok, Yeom Geun Young, Nalwa Hari Singh, Two-dimensional transition metal dichalcogenide-based counter electrodes for dye-sensitized solar cells. RSC Advances. 2017; 7; 28234-28290, DOI: 10.1039/c7ra03599c.

[17] Gao Min-Rui, Jiang Jun, Yu Shu-Hong, Solution-based synthesis and design of late transition metal chalcogenide materials for oxygen reduction reaction (ORR). Small. 2012; 8; 13-27, DOI: 10.1002/smll.201101573.

[18] Chen Wei-Fu, Iyer Shilpa, Iyer Shweta, Sasaki Kotaro, Wang Chiu-Hui, Zhu Yimei, Muckerman James T., Fujita Etsuko, Biomass-derived electrocatalytic composites for hydrogen evolution. Energy & Environmental Science. 2013; 6; 1818-1826, DOI: 10.1039/c3ee40596f.

[19] Kong Desheng, Cha Judy J., Wang Haotian, Lee Hye Ryoung, Cui Yi, First-row transition metal dichalcogenide catalysts for hydrogen evolution reaction. Energy & Environmental Science. 2013; 6; 3553-3558, DOI: 10.1039/c3ee42413h.

[20] Prasad Saradh, Durai G., Devaraj D., AlSalhi Mohamad Saleh, Theerthagiri J., Arunachalam Prabhakarn, Gurulakshmi M., Raghavender M., Kuppusami P., 3D nanorhombus nickel nitride as stable and cost-effective counter electrodes for dye-sensitized solar cells and supercapacitor applications. RSC Advances. 2018; 8; 8828-8835, DOI: 10.1039/c8ra00347e.

[21] Cui Xiaodan, Xie Zhiqiang, Wang Ying, Novel CoS_2 embedded carbon nanocages by direct sulfurizing metal-organic frameworks for dye-sensitized solar cells. Nanoscale. 2016; 8; 11984-11992, DOI: 10.1039/c6nr03052a.

[22] Li Chun-Ting, Tsai Yu-Lin, Ho Kuo-Chuan, Earth abundant silicon composites as the electrocatalytic counter electrodes for dye-sensitized solar cells. ACS Applied Materials & Interfaces. 2016; 8; 7037-7046, DOI: 10.1021/acsami.5b12423.

[23] Jiang Yiqing, Qian Xing, Niu Yudi, Shao Li, Zhu Changli, Hou Linxi, Cobalt iron selenide/sulfide porous nanocubes as high-performance electrocatalysts for efficient dye-sensitized solar cells. Journal of Power Sources. 2017; 369; 35-41, DOI: 10.1016/j.jpowsour.2017.09.080.

[24] Jian Siou-Ling, Huang Yi-June, Yeh Min-Hsin, Ho Kuo-Chuan, A zeolitic imidazolate framework-derived ZnSe/N-doped carbon cube hybrid electrocatalyst as the counter electrode for dye-sensitized solar cells. Journal of Materials Chemistry A. 2018; 6; 5107-5118, DOI: 10.1039/c8ta00968f.

[25] Li Chun-Ting, Chang Hung-Yu, Li Yu-Yan, Huang Yi-June, Tsai Yu-Lin, Vittal R., Sheng Yu-Jane, Ho Kuo-Chuan, Electrocatalytic zinc composites as the efficient counter electrodes of dye-sensitized solar cells: Study on the electrochemical performances and density functional theory calculations. ACS Applied Materials & Interfaces. 2015; 7; 28254-28263, DOI: 10.1021/acsami.5b07724.

[26] Kung Chung-Wei, Chen Hsin-Wei, Lin Chia-Yu, Huang Kuan-Chieh, Vitta R., Ho Kuo-Chuan, CoS acicular nanorod arrays for the counter electrode of an efficient dye-sensitized solar cell. ACS NANO. 2012; 6; 7016-7025, DOI: 10.1021/nn302063s.

[27] Song J., Li G. R., Xi Kai, Lei B., Gao X. P., Kumar R. Vasant, Enhancement of diffusion kinetics in porous MoN nanorods-based counter electrode in a dye-sensitized solar cell. Journal of Materials Chemistry A. 2014; 2; 10041-10047, DOI: 10.1039/c4ta01342e.

[28] Zhou Huawei, Shi Yantao, Dong Qingshun, Wang Liang, Zhang Hong, Ma Tingli, High electrocatalytic activity of $W_{18}O_{49}$ nanowires for cobalt complex and ferrocenium redox mediators. RSC Advances. 2014; 4; 42190-42196, DOI: 10.1039/c4ra07906j.

[29] Yang Xiao, Zhou Lei, Feng Ali, Tang Huaibao, Zhang Haijun, Ding Zongling, Ma Yongqing, Wu Mingzai, Jin Shaowei, Li Guang, Synthesis of nickel sulfides of different phases for counter electrodes in dye-sensitized solar cells by a solvothermal method with different solvents. Journal of Materials Research. 2014; 29; 935-941, DOI: 10.1557/jmr.2014.74.

[30] Sun Hong, Zhang Lu, Wang Zhong-Sheng, Single-crystal $CoSe_2$ nanorods as an efficient electrocatalyst for dye-sensitized solar cells. Journal of Materials Chemistry A. 2014; 2; 16023-16029, DOI: 10.1039/c4ta02238f.

[31] Yuan Hong, Jiao Qingze, Liu Jia, Liu Xiufeng, Li Yongjian, Shi Daxin, Wu Qin, Zhao Yun, Li Hansheng, Facile synthesis of $Co_{0.85}Se$ nanotubes/reduced graphene oxide nanocomposite as Pt-free counter electrode with enhanced electrocatalytic performance in dye-sensitized solar cells. Carbon. 2017; 122; 381-388, DOI: 10.1016/j.carbon.2017.06.095.

[32] Huang Yi-June, Lee Chuan-Pei, Pang Hao-Wei, Li Chun-Ting, Fan Miao-Syuan, Vittal R., Ho Kuo-Chuan, Microemulsion-controlled synthesis of $CoSe_2$/$CoSeO_3$ composite crystals for electrocatalysis in dye-sensitized solar cells. Materials Today Energy. 2017; 6; 189-197, DOI: 10.1016/j.mtener.2017.10.004.

[33] Huang Shoushuang, Ma Dui, Hu ZhangJun, He Qingquan, Zai Jiantao, Chen Dayong, Sun Huai, Chen Zhiwen, Qiao Qiquan, Wu Minghong, Qian Xuefeng, Synergistically enhanced electrochemical performance of Ni_3S_4-PtX (X = Fe, Ni) heteronanorods as heterogeneous catalysts in dye-sensitized solar cells. ACS Applied Materials & Interfaces. 2017; 9; 27607-27617, DOI: 10.1021/acsami.7b05418.

[34] Jin Zhitong, Zhang Meirong, Wang Min, Feng Chuanqi, Wang Zhong-Sheng, Cobalt selenide hollow nanorods array with exceptionally high electrocatalytic activity for high-efficiency quasi-solid-state dye-sensitized solar cells. Journal of Power Sources. 2018; 378; 475-482, DOI: 10.1016/j.jpowsour.2017.12.064.

[35] Geim A. K., Grigorieva I. V., Van der Waals heterostructures. Nature. 2013; 499; 419-425, DOI: 10.1038/nature12385.

[36] Hou Wenjing, Xiao Yaoming, Han Gaoyi, The dye-sensitized solar cells

based on the interconnected ternary cobalt diindium sulfide nanosheet array counter electrode. Materials Research Bulletin. 2018; 107; 204-212, DOI: https://doi.org/10.1016/j.materresbull.2018.07.040.

[37] Chen Haijie, Xie Yian, Cui Houlei, Zhao Wei, Zhu Xiaolong, Wang Yaoming, Lu Xujie, Huang Fuqiang, In situ growth of a $MoSe_2$/Mo counter electrode for high efficiency dye-sensitized solar cells. Chemical Communications. 2014; 50; 4475-4477, DOI: 10.1039/c3cc49600g.

[38] Lemme Max C., Li Lain-Jong, Palacios Tomás, Schwierz Frank, Two-dimensional materials for electronic applications. MRS Bulletin. 2014; 39; 711-718, DOI: 10.1557/mrs.2014.138.

[39] Saadi Fadl H., Carim Azhar I., Velazquez Jesus M., Baricuatro Jack H., McCrory Charles C. L., Soriaga Manuel P., Lewis Nathan S., Operando synthesis of macroporous molybdenum diselenide films for electrocatalysis of the hydrogen-evolution reaction. ACS Catalysis. 2014; 4; 2866-2873, DOI: 10.1021/cs500412u.

[40] Fan Miao-Syuan, Lee Chuan-Pei, Li Chun-Ting, Huang Yi-June, Vittal R., Ho Kuo-Chuan, Nitrogen-doped graphene/molybdenum disulfide composite as the electrocatalytic film for dye-sensitized solar cells. Electrochimica Acta. 2016; 211; 164-172, DOI: 10.1016/j.electacta.2016.06.047.

[41] Ibrahem Mohammed Aziz, Huang Wei-Chih, Lan Tian-wey, Boopathi Karunakara Moorthy, Hsiao Yu-Chen, Chen Chih-Han, Budiawan Widhya, Chen Yang-Yuan, Chang Chia-Seng, Li Lain-Jong, Tsai Chih-Hung, Chu Chih Wei, Controlled mechanical cleavage of bulk niobium diselenide to nanoscaled sheet, rod, and particle structures for Pt-free dye-sensitized solar cells. Journal of Materials Chemistry A. 2014; 2; 11382-11390, DOI: 10.1039/c4ta01881h.

[42] Huang Yi-June, Fan Miao-Syuan, Li Chun-Ting, Lee Chuan-Pei, Chen Tai-Ying, Vittal R., Ho Kuo-Chuan, $MoSe_2$ nanosheet/poly(3,4-ethylenedioxythiophene): poly(styrenesulfonate) composite film as a Pt-free counter electrode for dye-sensitized solar cells. Electrochimica Acta. 2016; 211; 794-803, DOI: 10.1016/j.electacta.2016.06.086.

[43] Mohammadnezhad Mahyar, Liu Mimi, Selopal Gurpreet Singh, Navarro-Pardo Fabiola, Wang Zhming M., Stansfield Barry, Zhao Haiguang, Lai Cheng-Yu, Radu Daniela R., Rosei Federico, Synthesis of highly efficient $Cu_2ZnSnS_xSe_{4-x}$ (CZTSSe) nanosheet electrocatalyst for dye-sensitized solar cells. Electrochimica Acta. 2020 340; 135954-135964, DOI: 10.1016/j.electacta.2020.135954.

[44] Antonelou Aspasia, Syrrokostas George, Sygellou Lamprini, Leftheriotis George, Dracopoulos Vassileios, Yannopoulos Spyros N, Facile, substrate-scale growth of mono and few-layer homogeneous MoS_2 films on Mo foils with enhanced catalytic activity as counter electrodes in DSSCs. Nanotechnology. 2016; 27; 045404-045414, DOI: 10.1088/0957-4484/27/4/045404.

[45] Chiu I. Ting, Li Chun-Ting, Lee Chuan-Pei, Chen Pei-Yu, Tseng Yu-Hao, Vittal R., Ho Kuo-Chuan, Nanoclimbing-wall-like $CoSe_2$/carbon composite film for the counter electrode of a highly efficient dye-sensitized solar cell: A study on the morphology control. Nano Energy. 2016; 22; 594-606, DOI: 10.1016/j.nanoen.2016.02.060.

[46] S Infant Raj, Xu Xiuwen, Yang Wang, Yang Fan, Hou Liqiang, Li Yongfeng, Highly active and reflective MoS_2 counter electrode for enhancement of photovoltaic efficiency of dye sensitized solar cells. Electrochimica Acta. 2016; 212; 614-620, DOI: 10.1016/j.electacta.2016.07.059.

[47] Chiu Jian-Ming, Chen E. Ming, Lee Chuan-Pei, Shown Indrajit, Tunuguntla Venkatesh, Chou Jui-Sheng, Chen Li-Chyong, Chen Kuei-Hsien, Tai Yian, Geogrid-inspired nanostructure to reinforce a $Cu_xZn_ySn_zS$ nanowall electrode for high-stability electrochemical energy conversion devices. Advanced Energy Materials. 2017; 7; 1602210-1602219, DOI: 10.1002/aenm.201602210.

[48] Patil Supriya A., Hussain Sajjad, Shrestha Nabeen K., Mengal Naveed, Jalalah Mohammed, Jung Jongwan, Park Jea-Gun, Choi Hyosung, Kim Hak-Sung, Noh Yong-Young, Facile synthesis of cobaltenickel sulfide thin film as a promising counter electrode for triiodide reduction in dye-sensitized solar cells. Energy. 2020; 202; 117730-117737, DOI: 10.1016/j.energy.2020.117730.

[49] Lee Chi-Ta, Peng Jia-De, Li Chun-Ting, Tsai Yu-Lin, Vittal R., Ho Kuo-Chuan, Ni_3Se_4 hollow architectures as catalytic materials for the counter electrodes of dye-sensitized solar cells. Nano Energy. 2014; 10; 201-211, DOI: 10.1016/j.nanoen.2014.09.017.

[50] Li Chun-Ting, Lee Chuan-Pei, Chiu I. Ting, Vittal R., Huang Yi-June, Chen Tai-Ying, Pang Hao-Wei, Lin Jiann T., Ho Kuo-Chuan, Hierarchical $TiO_{1.1}Se_{0.9}$-wrapped carbon cloth as the TCO-free and Pt-free counter electrode for iodide-based and cobalt-based dye-sensitized solar cells. Journal of Materials Chemistry A. 2017; 5; 14079-14091, DOI: 10.1039/c7ta02474f.

[51] Jiang Xiancai, Li Hongmei, Li Shuolin, Huang Shaowei, Zhu Changli, Hou Linxi, Metal-organic framework-derived Ni–Co alloy@carbon microspheres as high-performance counter electrode catalysts for dye-sensitized solar cells. Chemical Engineering Journal. 2018; 334; 419-431, DOI: 10.1016/j.cej.2017.10.043.

[52] Jiang Yiqing, Qian Xing, Zhu Changli, Liu Hongyu, Hou Linxi, Nickel cobalt sulfide double-shelled hollow nanospheres as superior bifunctional electrocatalysts for photovoltaics and alkaline hydrogen evolution. ACS Applied Materials & Interfaces. 2018; 10; 9379-9389, DOI: 10.1021/acsami.7b18439.

[53] Su Lijun, Xiao Yaoming, Han Gaoyi, Lin Jeng-Yu, One-step hydrothermal synthesis of feather duster-like NiS@ MoS_2 with hierarchical array structure for the Pt-free dye-sensitized solar cell. Journal of Nanoparticle Research. 2018; 20; 115-125, DOI: 10.1007/s11051-018-4223-5.

[54] Huang Yi-June, Chen Han-Ting, Ann Shiuan-Bai, Li Chun-Ting, Lin Jiann T., Lee Chuan-Pei, Ho Kuo-Chuan, Hierarchical urchin-like $CoSe_2/CoSeO_3$ electrocatalysts for dye-sensitized solar cells: up to 19% PCE under dim light illumination. Journal of Materials Chemistry A. 2019 7; 26089-26097, DOI: 10.1039/C9TA09166A.

[55] Liao Wei, Gao Yuan, Wang Wen, Zuo Xueqin, Yang Qun, Lin Yunxiang, Tang Huaibao, Jin Shaowei, Li Guang, Boosted reactivity of low-cost solar cells over a CuO/Co_3O_4 interfacial structure integrated with graphene oxide. ACS Sustainable Chemistry & Engineering. 2020; 8; 7308-7315, DOI: 10.1021/acssuschemeng.0c00282.

[56] Huang Jie, Qian Xing, Yang Jiahui, Niu Yudi, Xu Chong, Hou Linxi, Construction of Pt-free electrocatalysts based on hierarchical CoS_2/N-doped C@Co-WS_2 yolk-shell nano-polyhedrons for dye-sensitized solar cells. Electrochimica Acta. 2020; 340; 135949-135959, DOI: 10.1016/j.electacta.2020.135949.

Chapter 18

Excited-State Dynamics of Organic Dyes in Solar Cells

Ahmed M. El-Zohry

Abstract

Organic dyes are promising candidates for wide applications in solar cells, due to their controlled environmental impact, and low-cost. However, their performances in several solar cell architectures are not high enough to compete with the traditional semiconductor based solar cells. Therefore, several efforts should be gathered to improve the efficiency of these organic dyes. Herein, we discuss several deactivation processes recently found in several organic dyes using optical spectroscopic techniques. These processes are believed to be mostly detrimental for the performance of organic dyes in solar cells. These processes include deactivation phenomena such as isomerization, twisting, and chemical interactions with redox couple. Thus, based on similar studies, more optimized synthetic procedures for organic dyes could be implemented in the near future for high efficient solar cells based on organic dyes.

Keywords: charge dynamics, deactivation pathways, chemical interactions, spectroscopic tools

1. Introduction

1.1 DSSCs design

Dye Sensitized Solar Cells (DSSCs) have drawn the attention of renewable energy scientists, since the proof of concept done by O'Regan and Grätzel in 1991 [1]. In that concept, it was shown that an adsorbed photosensitizer on a low-cost low bandgap semiconductor can generate electricity with a reasonable efficiency from the incident sun light. Such a process was a breakthrough at the time, despite the low efficiency of the utilized sensitizer, as only highly crystalline semiconductor, such as Si, was believed to be the only way to capture the sunlight and convert it into electricity. In a typical Si solar cell, the light is absorbed by the crystalline Si atoms and the energetic charges are generated within the bandgap of the semiconductor, which later can be extracted by the external circuit [2]. However, in the DSSC, the light is absorbed by the photosensitizer "adsorbed dye" and then transfers its energetic charge to the low-cost semiconductor that is responsible for transferring the charge to the external circuit. The first utilized photosensitizer was based on metal complex, a Ru-complex, thus, many metal-based complexes were tested later on the best performances in DSSCs [3]. The main excited state charge dynamics for metal-based complexes for DSSCs are based on a MLCT (metal to ligand charge transfer process) state, in which the incident light moves an electron from the metal core to the surrounding ligands in the complex, then the charge hops from the ligand to the CB (conduction band of the semiconductor) via a triplet state. Thus, heavy metal ions with low oxidation potentials were utilized such as Ru atoms [4]. Later on, plenty of attempts have been done to replace these costly metal photosensitizers by the metal-free photosensitizers, organic dyes, to further reduce the cost of the working cell [5].

2. Organic dyes and strategic designs

Several synthetic strategies have been implemented for optimizing the metal-free, pure organic photosensitizers for working conditions in DSSCs [6]. One of the successful approaches for building organic photosensitizers is based on **D-L-A** (Donor-Linker-Acceptor) approach [7]. In this approach, the D unit is an electron-rich moiety, the L unit is typically a single or several consecutive π-bonds, then the A unit is an electron-deficient moiety that is connected by an anchoring group, such as COOH (carboxylic acid), which binds to the low-band gap semiconductor, see **Figure 1** for a graphical illustration of the organic dye. The ultimate dye should absorb most of the incident solar spectrum especially in the visible and the infrared regions, with high oscillator strength [3]. One of the most successful organic photosensitizers in DSSCs is the indoline family, which is based on Indoline moiety as a D unit [9–11]. While other acceptor groups (A) have been utilized such as rhodanine and cyanoacrylic moieties [9–11].

3. DSSCs working mechanism

Different than Ru-complexes, the organic dyes in DSSCs inject the energetic electrons from the singlet states as the triplet state population has mostly a very low quantum yield [12]. As the spin state for both the excited and the ground state of the organic dyes is the same, various deactivation mechanisms can occur for the adsorbed dyes on semiconductor surfaces. These deactivation processes include large scale motions such as isomerization [13], twisting [14], and local chemical interactions such as interactions with electrolyte components surface species [15–17].

Figure 2 summarizes the main processes for exciting an adsorbed dye on low-band gap semiconductor such as TiO_2. There processes are such as follow:

1. **Excitation:** The adsorbed dye absorbs part of the incident solar spectrum and an electron is transferred from the ground state to the excited state instantaneously.

2. **Decay:** The populated electron recombines back again to the ground state due to various excited state processes within the dye.

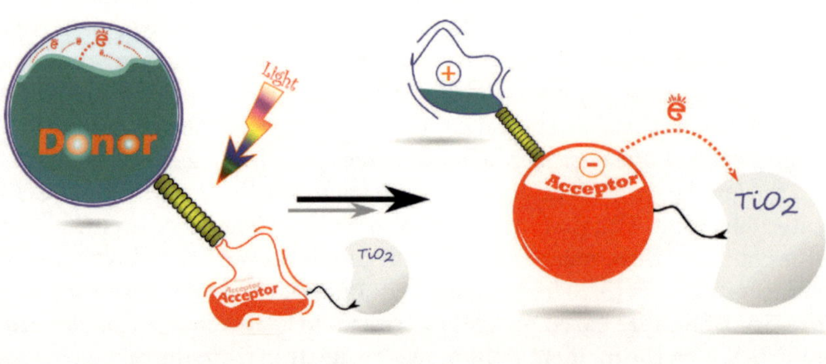

Figure 1.
Schematic representation for the successful design of organic photosensitizers for utilization in DSSCs based on D-L-A strategy, readapted from reference [8].

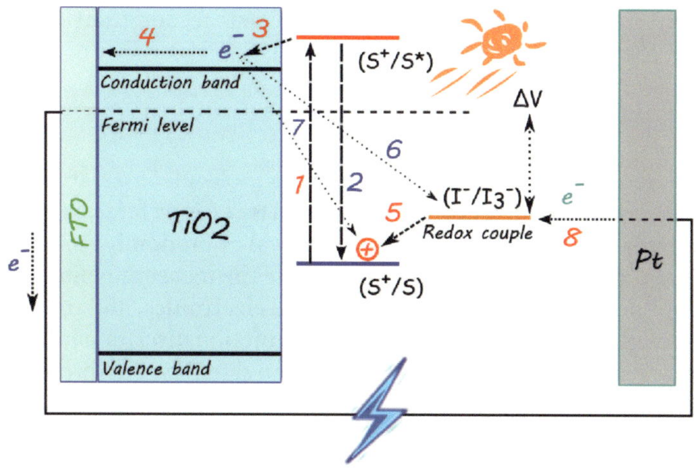

Figure 2.
Schematic representation for electron dynamics in DSSCs. Each process has its number that is mentioned in the main text. Red numbers are for deactivating processes and blue numbers are for favorable processes, readapted from reference [8].

3. **Electron Injection:** The excited electron is transferred from the dye to the CB of the semiconductor, leaving an oxidized adsorbed dye behind.

4. **Electron Diffusion:** The injected electron is diffused through the mesoporous semiconductor area reaching the conducting glass such as FTO (Fluorine doped Tin Oxide).

5. **Electron Regeneration:** The oxidized dye recovers its electron from the utilized electrolyte in the DSSCs, which has a low oxidation potential, such as (I^-/I_3^-) redox couple.

6. **Electron Recombination to Redox:** The injected electron in the CB diffuses backward to the adsorbed species on the semiconductor surfaces such as, the oxidized redox couple.

7. **Electron Recombination to dye:** The injected electron in the CB diffuses backward to the adsorbed species on the semiconductor surfaces such as, the oxidized dye.

All these processes contribute both positively and negatively to the overall performance of the DSSC. These processes are marked in different colors in **Figure 2**, depending on their role. However, due to the sake of this chapter, I will mainly be focusing on the exited state dynamics of organic dyes that improve or reduce the total performance of the DSSC. However, before presenting these dynamics, I will illustrate in the following section the main optical tools utilized for investigating these processes in DSSCs.

4. Optical spectroscopic tools

Several optical spectroscopic tools have been utilized to follow the charge dynamics for organic dyes in DSSCs. These common tools include TCSPC

(Time-correlated single photon counting), fs-TA (Femtosecond transient absorption), and fs-TE (Femtosecond transient Emission).

4.1 TCSPC

TCSPC helps to measure the emission decay of a molecule in a fast and an accurate way, due to the high repetition rate of the laser (ps or fs lasers). The accuracy of the measurements depends on the arrival of randomly emitted photons to the detector at different time channels. To initiate the measurements, a reference signal from the laser source is registered at the electronics, and the arrival time of the laser signal is measured by a constant function discriminator (CFD). Then, a linear increase in the voltage starts when the signal passes through time to amplitude converter (TAC). In the meanwhile, an electrical signal is registered from the emitted photon at the CFD, and another signal is sent to the TAC to stop the voltage increase. The time difference between the start and stop corresponds to the time delay after examining signal by the rest of the electronics. Repeating these measurements many times gives the histogram plot at the end. **Figure 3** presents the components of TCSPC [8, 12, 13].

4.2 fs-TA

As many ongoing processes of DSSCs are relatively fast ones, one needs a technique with high time-resolution to follow such processes in DSSCs. One of the most utilized techniques to follow such processes is the fs-TA setup [18–25]. Simply, in fs-TA, one needs a laser source of short pulses in the range of 100 fs per pulse, and by overlapping two laser pulses at the measuring sample (one to start the reaction 'pump', and another to probe it), the resulted spectrum at the detector provide exceptional information about both the ground state and the excited state of the reaction, **Figure 4** shows a simple scheme for utilizing fs-TA setup. The pump pulse is usually in the visible range to promote a charge transfer, and the probe pulse can be usually in the visible or in the infrared range [14, 26, 27]. The main advantages of fs-TA are the ability to detect dark states that are not observed by other time-resolved emission techniques such as charge transfer, energy transfer, intersystem crossing, and charge recombination [28–30].

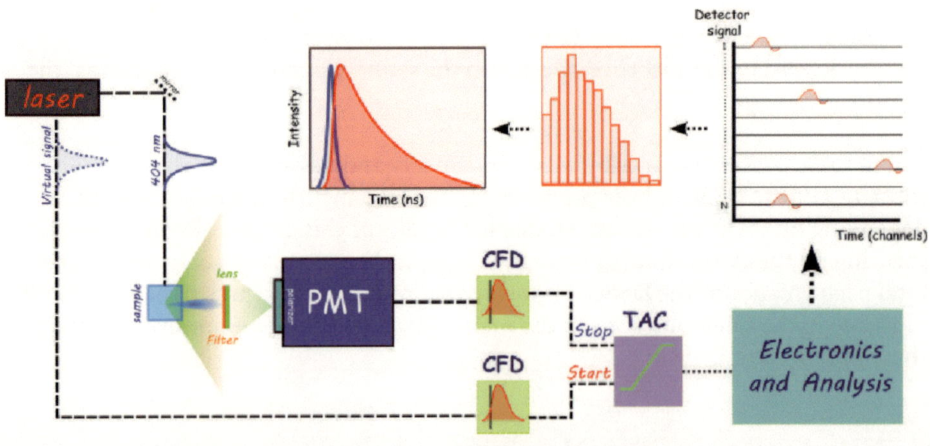

Figure 3.
Schematic representation for a typical TCSPC setup, readapted from reference [8].

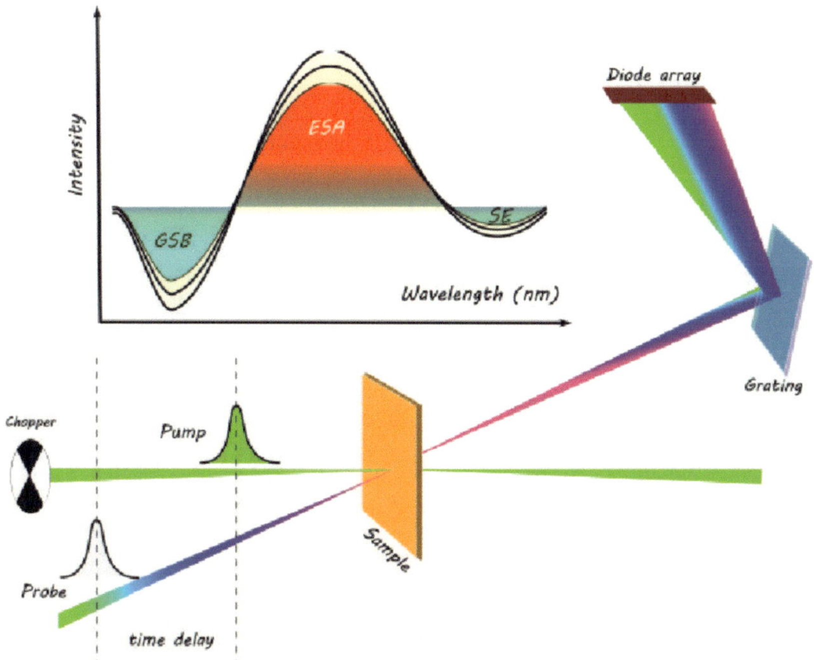

Figure 4.
Illustration for the generation of TA signal by fs-laser pulses.

4.3 fs-TE

Time-resolved transient emission techniques are more versatile to follow the charge dynamics in general for the charge dynamics for dyes in DSSCs. To be able following the emission spectral information along with the emission lifetimes of the studied dyes, one commonly uses time-resolved emission streak camera, **Figure 5** shows the basic components for measuring emission using streak camera. The main advantage of using streak camera is the ease of utilizing it in comparison with other techniques such as fs-TA. Using emission streak camera, one needs only one laser source to excite the sample, then the emitted photons are collected and directed inside the streak camera, in which the photons can be spatially and temporally separated, resulting of a 2D-image that contain information about the time and energy of the emitted photons [12, 13].

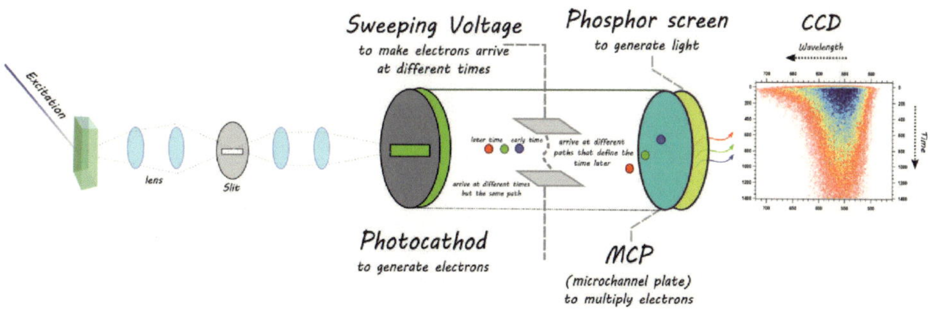

Figure 5.
Typical design for a streak camera. Readapted from reference [8].

5. Excited state dynamics

5.1 Electron injection

Electron injection process is the transfer of a charge such as an electron from the excited dye to the CB of the semiconductor after light absorption, and it is considered the first beneficial process for high performance in DSSCs. The electron injection rate depends on the coupling strength between the adsorbed dye and the semiconductor, which includes the energy alignments of both the excited state of the dye and the fermi level of the semiconductor. For a long time, the electron injection time scale was trusted to be only in the range of 100 fs, however, this is not the case for all organic dyes as shown later by showing slower electron injections lifetimes [26, 27, 31]. The detection of slow electron injection in the picosecond time scale was mainly achieved by utilizing the IR (infrared) probe light in the fs-TA instead of the visible probe light [27, 29, 32, 33]. The advantage of using the IR versus the visible probe was mainly attributed the sole sensitivity of the IR to the vibrations of the electrons in the CB of the semiconductor, while the visible probe interacts with several species at the semiconductor surface such as the oxidized dye and the redox couple [8, 33, 34]. Famous organic indoline dyes were measured on TiO_2 mesoporous surfaces using fs-TA in the IR region centered at 5000 nm, and multi-exponential injection rates were detected including fast lifetimes of 100 fs and slow ones in the range of tens of ps [27, 33]. **Figure 6** shows the captured data for various indoline dyes, in which the D131 dye shows a fast injection lifetime of 100 fs, while other dyes (D102, D149, and D205) show additional slow injection lifetimes that can reach to 30 ps as in the case of D149 dye. These slow injection rates are connected to large scale motions on mesoporous surfaces as shown later on, such as isomerization. The presence of slow injection rates is thought to be beneficial to the overall efficiency of the DSSC, due to the expected minimized charge recombination afterwards [26, 31].

5.2 Deactivation processes

- Aggregation

Aggregation is a common problem for adsorption of organic dyes on mesoporous surfaces, in which the dyes are stacking in various ways very close to each other

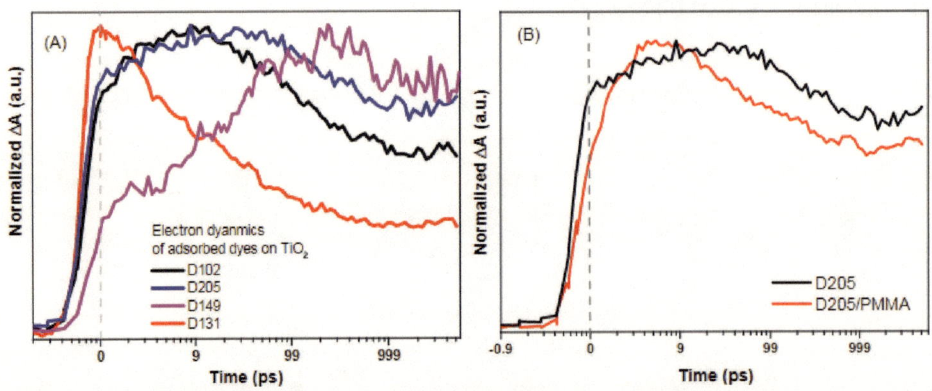

Figure 6.
Time resolved transient absorption for the electron injection process of indoline dyes adsorbed on TiO_2. (A) Comparison between various dyes indicated in the legend. (B) Comparison between D205 on TiO_2 versus impeding the dye in PMMA, readapted from reference [27].

due to the high concentration utilized during the adsorption process, resulting of side deactivation pathways that hindered the charge transfer processes in DSSCs [13, 35, 36]. Reducing the dye aggregations can happen by utilizing co-adsorbent agents such as CDCA (cheno-deoxycholic acid) [37], or by impeding organic dyes in the MOF-ZIF8 structures, which increases the dye's emission lifetime by putting the dyes at far distances from each other [38]. **Figure 7** presents the appearance of fast emission lifetime components for the D149 dye upon the presence of aggregation. However, upon using low concenrtation of the D149 dye, the short lifetimes disappears due to the absence of aggregation. In DSSCs, the presence of aggregation reduces the amount of charges transferred to the CB of the semiconductor, and thus, reduces the overall efficiency of the cell.

- Isomerization

The local movement of adsorbed organic dyes was overlooked for a long time due to the expected well-packed order of adsorbed dyes, and many argued that isomerization is not a competing process with the electron injection as the latter is very fast. However, as electron injection process can be slow as well, the isomerization and the change of local arrangements of molecules on surfaces can reduce the DSSC efficiency due to uncontrolled deactivation processes [13, 39, 40]. **Figure 8** shows the absorption spectra changes of L0Br organic dye labeled by heavy bromine atom on the mesoporous ZrO_2 surfaces under photo-irradiation [40]. The changes in absorption spectra along the NMR measurements revealed the formation of *cis-trans* equilibrium on the mesoporous surfaces [13, 40].

- Twisting

Isomerization of organic dyes in DSSCs is not always spectroscopically detectable especially when the resulted isomers such as *cis* and *trans* isomers are chemically identical [14]. However, this is not the case for many organic molecules containing subunits such as phenyl groups that can rotate or twist without spectroscopic signatures. For instance, the parent molecule of the organic dye D149 dye has a diphenyl groups attached to a double bond. In solution, the lifetime of the parent molecule is very short of ca. 20 ps, and upon impeding this parent molecule in polymer matrix PMMA (Poly methyl methacrylate), the lifetime is extended to 2.5 ns, the time-resolved data for the parent molecule is shown in **Figure 9**. This ultrafast deactivation process is present in the derived dyes utilized in solar cells and it can

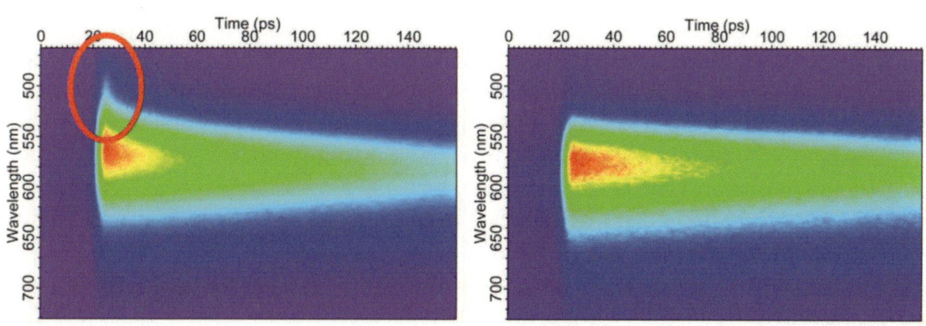

Figure 7.
Time-resolved emission for D149 organic dye inside PMMA matrix showing the effect of concentration and the aggregation formation on the appearance of fast emission lifetime components (to the left), readapted from reference [13].

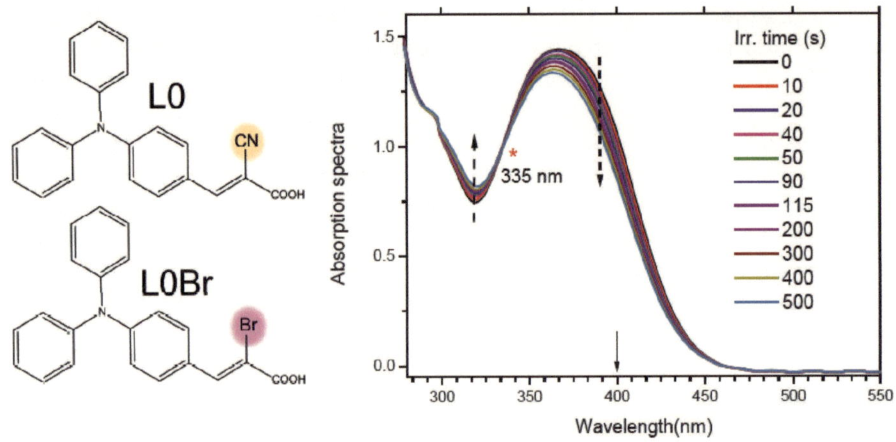

Figure 8.
Two organic dyes, Lo, and LoBr were utilized to investigate the isomerization process on ZrO_2 surfaces under 400 nm photo-irradiation, readapted from reference [40].

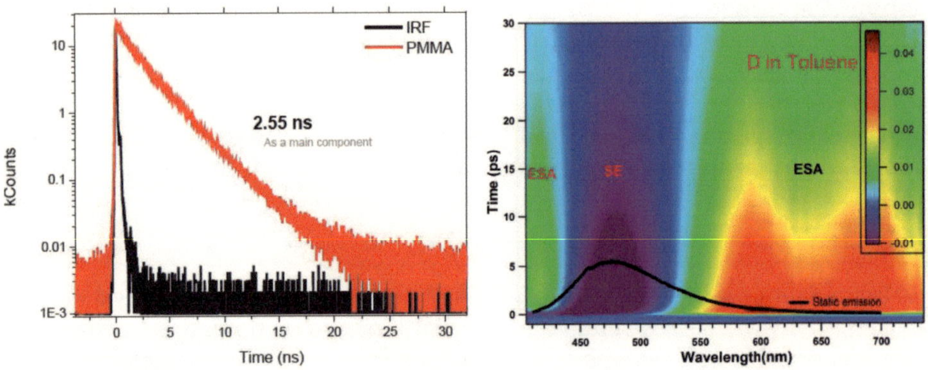

Figure 9.
The time-resolved emission data for the parent molecule of D149 in solution and in PMMA matrix (left). fs-TA data for the parent molecule in toluene (right), readapted from reference [14].

potentially compete with the electron injection process, minimizing the amount of charges extracted through the DSSC.

- **TICT (Twisted Intramolecular Charge Transfer)**

Although the previous large scale motions of organic dyes apparently compete with the electron injection process, the TICT process of some studied organic dyes seems to help boosting the DSSC efficiency through an indirect pathway [31, 41]. Upon comparing organic dyes with the twisting ability on mesoporous semiconductors surfaces with the corresponding ones that do not show such a process, both the electron dynamics and the DSSC efficiency have been correlated [26, 31]. An organic dye named L1 dye shows the TICT process in solution as depicted in **Figure 10**. This dye shows a high performance in DSSCs of ca. 5.5% [26]. While the modified dye L1Fc that do not show any TICT state, instead shows a LCT (local charge transfer state), its efficiency in DSSCs was lower L1 of ca. 1.1% [26, 31].

Using fs-TA in the infrared region to investigate the electron dynamics in the CB of TiO_2 revealed that the presence of TICT state allows for slower electron injection from the L1 dye to the TiO_2, and due the structural rearrangements of the L1 dye on the mesoporous surfaces, the back electron recombination is hindered allowing for

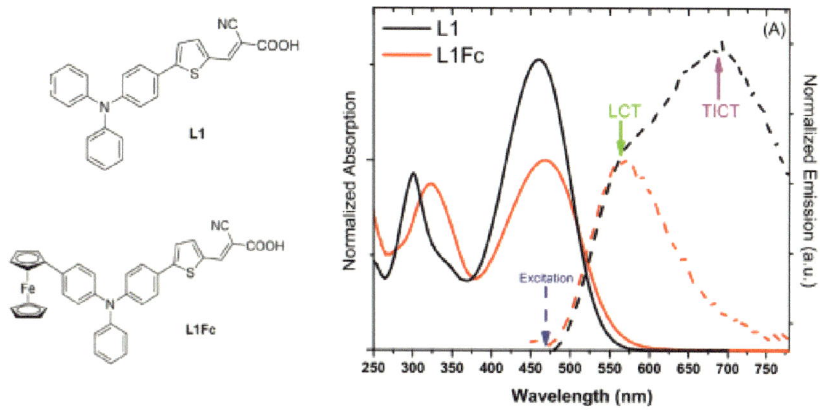

Figure 10.
Chemical structures of L1 and L1Fc dyes along with their absorption and emission data in acetonitrile, readapted from reference [31].

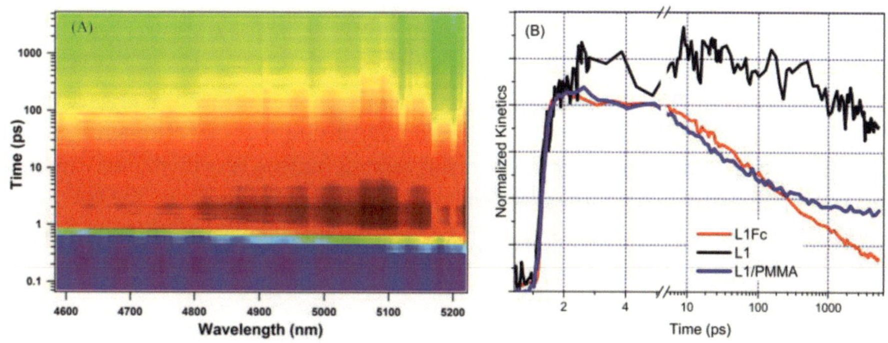

Figure 11.
(A) False 2D plot for the electron injection of the L1 dye to the CB of TiO_2 in the infrared. (B) Normalized kinetic traces for L1, L1Fc, and L1/PMMA on TiO_2, readapted from reference [31].

high performance in DSSCs. However, for the L1Fc case along with the L1/PMMA case, the TICT state is blocked and thus the electron injection was faster from the LCT state, but the electron recombination was order of magnitudes faster than in the L1 dye case, resulting of poor efficiency in DSSCs. **Figure 11** shows the time resolved data for electron injection for the discussed three cases. Thus, although the presence of TICT process can consume some energy to populate the TICT state, the benefit of reducing the charge recombination process is much larger on the DSSC efficiency.

- Chemical Interactions with the Redox Couple

Traditionally, the utilized electrolyte in DSSCs is solely assumed to regenerate the adsorbed oxidized dye on the mesoporous surface after the electron injection. This regeneration process is typically in the pico- to nano- second time scale [42–44]. However, just recently, it has been shown that the utilized electrolyte can form ground state interactions with the adsorbed dye on the surface that both affect the electron injection and recombination processes [15]. These effects will have detrimental effects on the performance of organic dyes in DSSCs. The formation of ground state complexes have been confirmed by using steady state absorption and emission measurements. **Figure 12** shows the kinetic traces for the electron dynamics of adsorbed organic dye D149 on TiO_2 in contact with different components of the

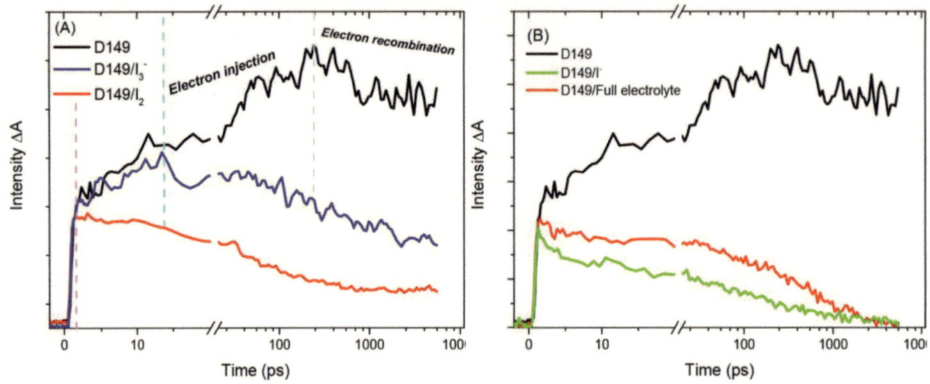

Figure 12.
Effect of chemical interactions between the D149 organic dye and the redox couple electrolyte (Iodide, iodine, tri-iodide) on the electron dynamics of D149 dye on mesoporous TiO_2, the rise of the signal is due to electron injection, while the signal decay is due to the electron recombination, readapted from reference [15]. (A) Comparison between D149 and complexes of D149 with tri-iodide, and iodine. (B) Comparison between D149 and complexes of D149 with iodide, and full electrolyte.

traditional iodide electrolyte used in various DSSC sets [45]. For the case of D149/TiO_2, slower electron injection and recombination processes have been observed. However, upon adding I_3^-, I^-, or I_2, the electron injection process was much faster of ca. 100 fs, and more importantly the electron recombination was increased dramatically, due to the adsorbed complexes species on the surface [15]. Thus, the chemical interactions between the chemical substances should be considered upon optimizing the DSSC efficiency.

6. Conclusion

Although the DSSC shows promising results with respect to low-cost and moderate efficiency in comparison with inorganic semiconductor solar cells, the ongoing processes in DSSC are quite complex and lots of studies are required to increase the output efficiency. In this chapter, we highlighted the fact that organic dyes have many excited state processes that have been overlooked in the past. Most of these processes showed detrimental effects on the overall performance of the DSSC. However, other exited state processes, such as the formation of TICT state, illustrated that high efficiency can also be attained through the excited state dynamics of the adsorbed dye. Understanding the dye's excited state processes will allow for fine tuning of such processes, via the chemical synthesis of organic dyes, correlating with the output efficiency of the DSSC.

Author notes

On leave from Chemistry Department, Assiut University

Author details

Ahmed M. El-Zohry
Department of Physics, Stockholm University Stockholm, Sweden

*Address all correspondence to: ahmed.elzohry@fysik.su.se; amfzohry@yahoo.com

References

[1] O'Regan, B.; Grätzel, M. A Low-Cost, High-Efficiency Solar Cell Based on Dye-Sensitized Colloidal TiO2 Films. Nature 1991. DOI: 10.1038/353737a0

[2] Polman, A.; Knight, M.; Garnett, E. C.; Ehrler, B.; Sinke, W. C. Photovoltaic Materials: Present Efficiencies and Future Challenges. Science (80-.). 2016, 352 (6283), 307-3017.

[3] Hagfeldt, A.; Boschloo, G.; Sun, L. C.; Kloo, L.; Pettersson, H. Dye-Sensitized Solar Cells. Chem. Rev. 2010, 110 (11), 6595-6663. DOI: 10.1021/Cr900356p.

[4] Polo, A. S.; Itokazu, M. K.; Iha, N. Y. M. Metal Complex Sensitizers in Dye-Sensitized Solar Cells. Coord. Chem. Rev. 2004, 248 (13-14), 1343-1361. doi:10.1016/j.ccr.2004.04.013.

[5] Jung, M. R.; Jo, H. J.; Yang, H. S.; Kim, H.; Kang, J. K.; Kim, D. H.; Ahn, K. S.; Kim, J. H. Molecular Design and Photovoltaic Performances of Organic Dyes Containing Triphenylamine for Dye-Sensitized Solar Cell. Mol. Cryst. Liq. Cryst. 2011, 538, 278-284. DOI: 10.1080/15421406.2011.564102.

[6] Kim, B. G.; Chung, K.; Kim, J. Molecular Design Principle of All-Organic Dyes for Dye-Sensitized Solar Cells. Chem. Eur. J. 2013, 19 (17), 5220-5230. DOI: 10.1002/chem.201204343.

[7] Baheti, A.; Thomas, K. R. J.; Li, C. T.; Lee, C. P.; Ho, K. C. Fluorene-Based Sensitizers with a Phenothiazine Donor: Effect of Mode of Donor Tethering on the Performance of Dye-Sensitized Solar Cells. ACS Appl. Mater. Interfaces 2015, 7 (4), 2249-2262. https://doi.org/10.1021/am506149q.

[8] El-Zohry, A. M. Exploring Organic Dyes for Grätzel Cells Using Time-Resolved Spectroscopy, Uppsala University, 2015, Vol. Doctoral D. https://doi.org/diva2:857014.

[9] Higashijima, S.; Miura, H.; Fujita, T.; Kubota, Y.; Funabiki, K.; Yoshida, T.; Matsui, M. Highly Efficient New Indoline Dye Having Strong Electron-Withdrawing Group for Zinc Oxide Dye-Sensitized Solar Cell (Vol 67, Pg 6289, 2011). Tetrahedron 2011, 67 (43), 6289-6293. DOI: 10.1016/j.tet.2011.08.092.

[10] Horiuchi, T.; Miura, H.; Uchida, S. Highly-Efficient Metal-Free Organic Dyes For Dye-Sensitized Solar Cells. Chem. Commun. 2003, No. 24, 3036-3037. DOI: 10.1039/B307819a.

[11] Kim, J. Y.; Kim, Y. H.; Kim, Y. S. Indoline Dyes with Various Acceptors for Dye-Sensitized Solar Cells. Curr. Appl. Phys. 2011, 11 (1), S117–S121. DOI: 10.1016/j.cap.2010.11.098.

[12] Lakowicz, J. R. Principles of Fluorescence Spectroscopy; Springer, 2007.

[13] El-Zohry, A.; Orthaber, A.; Zietz, B. Isomerization and Aggregation of the Solar Cell Dye D149. J. Phys. Chem. C 2012, 116 (50). https://doi.org/10.1021/jp306636w.

[14] El-Zohry, A. M.; Roca-Sanjuán, D.; Zietz, B. Ultrafast Twisting of the Indoline Donor Unit Utilized in Solar Cell Dyes: Experimental and Theoretical Studies. J. Phys. Chem. C 2015, 119 (5). https://doi.org/10.1021/jp505649s.

[15] El-Zohry, A. M.; Zietz, B.; M. El-Zohry, A.; Zietz, B.; El-Zohry, A. M.; Zietz, B. Electron Dynamics in Dye-Sensitized Solar Cells Influenced by Dye–Electrolyte Complexation. J. Phys. Chem. C 2020, 124 (30), 16300-16307. https://doi.org/10.1021/acs.jpcc.0c03436.

[16] M. El-Zohry, A.; Agrawal, S.; De Angelis, F.; Pastore, M.; Zietz, B. Critical Role of Protons for Emission

Quenching of Indoline Dyes in Solution and on Semiconductor Surfaces. J. Phys. Chem. C 2020, 0 (ja). https://doi.org/10.1021/acs.jpcc.0c07099.

[17] El-Zohry, A. M.; Zietz, B. Concentration and Solvent Effects on the Excited State Dynamics of the Solar Cell Dye D149: The Special Role of Protons. J. Phys. Chem. C 2013, 117 (13). DOI: 10.1021/jp400782g.

[18] Rosspeintner, A.; Lang, B.; Vauthey, E. Ultrafast Photochemistry in Liquids. Annu. Rev. Phys. Chem. 2013, 64, 247-271. DOI: 10.1146/annurev-physchem-040412-110146.

[19] Furube, A.; Katoh, R.; Hara, K. Electron Injection Dynamics in Dye-Sensitized Semiconductor Nanocrystalline Films. Surf. Sci. Rep. 2014, 69 (4), 389-441. DOI: 10.1016/j.surfrep.2014.09.003.

[20] Klimov, V. I.; McBranch, D. W.; Leatherdale, C. A.; Bawendi, M. G. Electron and Hole Relaxation Pathways in Semiconductor Quantum Dots. Phys. Rev. B 1999, 60 (19), 13740-13749. DOI: 10.1103/PhysRevB.60.13740.

[21] Maiuri, M.; Garavelli, M.; Cerullo, G. Ultrafast Spectroscopy: State of the Art and Open Challenges. J. Am. Chem. Soc. 2020. https://doi.org/10.1021/jacs.9b10533.

[22] Knorr, F. J.; McHale, J. L.; Clark, A. E.; Marchioro, A.; Moser, J. E. Dynamics of Interfacial Electron Transfer from Betanin to Nanocrystalline TiO2: The Pursuit of Two-Electron Injection. J. Phys. Chem. C 2015, 119 (33), 19030-19041. https://doi.org/10.1021/acs.jpcc.5b05896.

[23] Anderson, A. Y.; Barnes, P. R. F.; Durrant, J. R.; O'Regan, B. C. Quantifying Regeneration in Dye-Sensitized Solar Cells. J. Phys. Chem. C 2011, 115 (5), 2439-2447. DOI: 10.1021/Jp1101048.

[24] Rohwer, E.; Richter, C.; Heming, N.; Strauch, K.; Litwinski, C.; Nyokong, T.; Schlettwein, D.; Schwoerer, H. Ultrafast Photodynamics of the Indoline Dye D149 Adsorbed to Porous ZnO in Dye-Sensitized Solar Cells. Chemphyschem 2013, 14 (1), 132-139. https://doi.org/10.1002/cphc.201200715.

[25] Debnath, T.; Maity, P.; Lobo, H.; Singh, B.; Shankarling, G. S.; Ghosh, H. N. Extensive Reduction in Back Electron Transfer in Twisted Intramolecular Charge-Transfer (TICT) Coumarin-Dye-Sensitized TiO2 Nanoparticles/Film: A Femtosecond Transient Absorption Study. Chem. Eur. J. 2014, 20 (12), 3510-3519. DOI: 10.1002/chem.201303903.

[26] El-Zohry, A. M.; Cong, J.; Karlsson, M.; Kloo, L.; Zietz, B. Ferrocene as a Rapid Charge Regenerator in Dye-Sensitized Solar Cells. Dye. Pigment. 2016, 132. https://doi.org/10.1016/j.dyepig.2016.05.021.

[27] El-Zohry, A. M. The Origin of Slow Electron Injection Rates for Indoline Dyes Used in Dye-Sensitized Solar Cells. Dye. Pigment. 2019, 160. https://doi.org/10.1016/j.dyepig.2018.09.002.

[28] Liu, D.; El-Zohry, A. M.; Taddei, M.; Matt, C.; Bussotti, L.; Wang, Z.; Zhao, J.; Mohammed, O. F.; Di Donato, M.; Weber, S. Long-Lived Charge-Transfer State Induced by Spin-Orbit Charge Transfer Intersystem Crossing (SOCT-ISC) in a Compact Spiro Electron Donor/Acceptor Dyad. Angew. Chemie - Int. Ed. 2020, 59 (28). https://doi.org/10.1002/anie.202003560.

[29] Abdellah, M.; El-Zohry, A. M.; Antila, L. J.; Windle, C. D.; Reisner, E.; Hammarström, L. Time-Resolved IR Spectroscopy Reveals a Mechanism with TiO2 as a Reversible Electron Acceptor in a TiO2-Re Catalyst System for CO2 Photoreduction. J. Am. Chem. Soc. 2017, 139 (3). https://doi.org/10.1021/jacs.6b11306.

[30] Hussain, M.; El-Zohry, A. M.; Gobeze, H. B.; Zhao, J.; D'Souza, F.; Mohammed, O. F. Intramolecular Energy and Electron Transfers in Bodipy Naphthalenediimide Triads. J. Phys. Chem. A 2018, 122 (29). https://doi.org/10.1021/acs.jpca.8b03884.

[31] El-Zohry, A. M.; Karlsson, M. Gigantic Relevance of Twisted Intramolecular Charge Transfer for Organic Dyes Used in Solar Cells. J. Phys. Chem. C 2018, 122 (42). https://doi.org/10.1021/acs.jpcc.8b08326.

[32] Gao, J.; El-Zohry, A. M.; Trilaksana, H.; Gabrielsson, E.; Leandri, V.; Ellis, H.; D'Amario, L.; Safdari, M.; Gardner, J. M.; Andersson, G.; Kloo, L. Light-Induced Interfacial Dynamics Dramatically Improve the Photocurrent in Dye-Sensitized Solar Cells: An Electrolyte Effect. ACS Appl. Mater. Interfaces 2018, 10 (31). https://doi.org/10.1021/acsami.8b06897.

[33] Juozapavicius, M.; Kaucikas, M.; van Thor, J. J.; O'Regan, B. C. Observation of Multiexponential Pico- to Subnanosecond Electron Injection in Optimized Dye-Sensitized Solar Cells with Visible-Pump Mid-Infrared-Probe Transient Absorption Spectroscopy. J. Phys. Chem. C 2012, 117 (1), 116-123. https://doi.org/10.1021/jp309732z.

[34] Antila, L. J.; Santomauro, F. G.; Hammarström, L.; Fernandes, D. L. A.; Sá, J. Hunting for the Elusive Shallow Traps in TiO2 Anatase. Chem. Commun. 2015, 51 (54), 10914-10916.

[35] Geiger, T.; Kuster, S.; Yum, J. H.; Moon, S. J.; Nazeeruddin, M. K.; Gratzel, M.; Nuesch, F. Molecular Design of Unsymmetrical Squaraine Dyes for High Efficiency Conversion of Low Energy Photons into Electrons Using TiO(2) Nanocrystalline Films. Adv. Funct. Mater. 2009, 19 (17), 2720-2727. DOI: 10.1002/adfm.200900231.

[36] Horiuchi, T.; Miura, H.; Uchida, S. Highly Efficient Metal-Free Organic Dyes for Dye-Sensitized Solar Cells. J. Photochem. Photobiol. a-Chemistry 2004, 164 (1-3), 29-32. DOI: 10.1016/j.jphotochem.2003.12.018.

[37] Ito, S.; Miura, H.; Uchida, S.; Takata, M.; Sumioka, K.; Liska, P.; Comte, P.; Pechy, P.; Grätzel, M. High-Conversion-Efficiency Organic Dye-Sensitized Solar Cells With A Novel Indoline Dye. Chem. Commun. 2008, No. 41, 5194-5196. DOI: 10.1039/B809093a.

[38] Abdelhamid, H. N.; El-Zohry, A. M.; Cong, J.; Thersleff, T.; Karlsson, M.; Kloo, L.; Zou, X. Towards Implementing Hierarchical Porous Zeolitic Imidazolate Frameworks in Dye-Sensitized Solar Cells. R. Soc. Open Sci. 2019, 6 (7). https://doi.org/10.1098/rsos.190723.

[39] Zhang, L.; Cole, J. M. TiO2-Assisted Photoisomerization of Azo Dyes Using Self-Assembled Monolayers: Case Study on Para-Methyl Red Towards Solar-Cell Applications. ACS Appl. Mater. Interfaces 2014, 6 (5), 3742-3749. https://doi.org/10.1021/am500308d.

[40] Zietz, B.; Gabrielsson, E.; Johansson, V.; El-Zohry, A. M.; Sun, L.; Kloo, L. Photoisomerization of the Cyanoacrylic Acid Acceptor Group-a Potential Problem for Organic Dyes in Solar Cells. Phys. Chem. Chem. Phys. 2014, 16 (6). https://doi.org/10.1039/c3cp54048k.

[41] Grabowski, Z. R.; Rotkiewicz, K.; Rettig, W. Structural Changes Accompanying Intramolecular Electron Transfer: Focus on Twisted Intramolecular Charge-Transfer States and Structures. Chem. Rev. 2003, 103 (10), 3899-4031. DOI: 10.1021/Cr940745l.

[42] Friedrich, D.; Valldecabres, L.; Kunst, M.; Moehl, T.; Zakeeruddin,

S. M.; Grätzel, M. Dye Regeneration Dynamics by Electron Donors on Mesoscopic TiO2 Films. J. Phys. Chem. C 2014.

[43] Hamann, T. W.; Ondersma, J. W. Dye-Sensitized Solar Cell Redox Shuttles. Energy Environ. Sci. 2011, 4 (2), 370-381. DOI: 10.1039/C0ee00251h.

[44] Antila, L. J.; Myllyperkiö, P. M.; Mustalahti, S. A.; Lehtivuori, H.; Korppi-Tommola, J. E. I. Injection and Ultrafast Regeneration in Dye–Sensitized Solar Cells. J. Phys. Chem. C 2014.

[45] Bella, F.; Gerbaldi, C.; Barolo, C.; Gratzel, M. Aqueous Dye-Sensitized Solar Cells. Chem. Soc. Rev. 2015, 44 (11), 3431-3473. https://doi.org/10.1039/c4cs00456f.

Chapter 19

Improvement of Efficiency of Dye Sensitized Solar Cells by Incorporating Carbon Nanotubes

Md. Mosharraf Hossain Bhuiyan, Fahmid Kabir, Md. Serajum Manir, Md. Saifur Rahaman, Prosenjit Barua, Bikrom Ghosh, Fumiaki Mitsugi and Tomoaki Ikegami

Abstract

Dye-sensitized solar cells (DSSCs) have aroused intense attention over the past three decades owing to their low cost, inexpensive raw materials, simple fabrication process, and employment of eco-friendly materials. Recently, to take advantage of their lower electrical resistance, excellent electrocatalytic operation, mechanical integrity, low cost, and flexibility, carbon nanotubes CNTs have been incorporated into DSSCs with a view to improve the efficiency further. CNT can be used in the anode, electrolyte, and counter electrode. The incorporation of CNTs into the anode's semiconductor material decreases the host material's resistance and increases thermal conductivity, electrical conductivity, mechanical strength, and durability. CNTs in ionic liquids have been investigated as a potential alternative for traditional liquid electrolytes for DSSC application because of low viscosity, low vapor pressure, high diffusion coefficient, high electrochemical, and thermal stability. CNT based counter electrode has attracted considerable interest because of its fast electron transfer kinetics and large surface area. This book chapter provides an insight into the fabrication of DSSCs by incorporating CNT and its effects on cell conversion efficiencies.

Keywords: dye sensitized solar cell, carbon nanotubes, efficiency

1. Introduction

In recent years, world energy consumption has increased significantly due to rapid global population growth, improved quality of life, and new technologies. United States Department of Energy (USDoE) anticipates that global energy demand will double by 2050 and triple by 2100 [1]. About 87% of energy demands are supplied by burning fossil fuels, such as coal, oil, and natural gases [2]. However, it has become a major concern about the environmental damage caused by the combustion method. Excessive use of fossil fuel energy sources urges researchers to think about the alternatives to eternal clean energy sources. The use of renewable energy is of great significance due to the rising cost of fossil fuels in tandem with a decrease in carbon dioxide emissions that avoids global warming. Significant improvements have been made in this regard, including energy technologies focused on wind power, biofuels, solar panels, and fuel cells [3, 4].

Solar energy will effectively meet some of the energy needs of future generations. The $3x10^{24}$ Joule/year supply of energy from the sun to the earth is 10 000 times more than the global energy requirement. This suggests that the use of 10% efficient photovoltaic cells will cover just 0.1% of the earth's surface area that could supply our current electricity needs [5]. In addition to photovoltaic device technologies, protect the global environment, and ensure economic growth with sustainable resources [6]. A solar cell or photovoltaic device is a solid-state device that is generally used for converting solar energy into useable electricity. The main advantage of the solar cells is that it does not require fossil fuel burning and does not produce any harmful emission [2]. The development of solar cells can be divided into three generations: First generation, second generation, and third generation solar cell. First generation or crystalline-silicon based solar cell has high cell efficiency(~26.7% against theoretical limit ~29%), which dominates the global solar cell market (~90%) [7]. However, the rigid cell structure and high cost related to manufacturing silicon wafers limit the use of first generationsilicon-based solar cells. Second generation or thin-film solar cell is fabricated by depositing multiple layers of photovoltaic material on plastic, glass, or metal substrate. Thin-film solar cell thickness can vary from few nanometers to 10 microns. However, Indium and Tellurium's scarcity makes it somewhat difficult to commercializeand large-scale production of the thin-film solar cell. Also, cadmium is a highly toxic material, which poses both severe health and environmental hazards. Third-generation solar cells are targeted to achieve both high efficiency and low cost. The theoretical cell conversion efficiency of third-generation solar cells varies from 31 to 41%, and it is expected that this limit can be easily overcome by using semiconductor nanoparticles [8]. Perovskite solar cell (PSC), copper zinc tin sulfide (CZTS) solar cell, quantum dot solar cell (QDSC), organic solar cell (OSC), and dye-sensitized solar cell (DSSC) are the example of the third-generation solar cells. Most of the third-generation solar cells are still in the research stage.

DSSC is a third-generation solar cells co-invented by Michael Gratzel and Brian O'Regan in 1988 at UC Berkeley, USA and further developed by Michael GratzelÉcolePolytechniqueFédérale de Lausanne, Switzerland. The DSSCs are regarded as successful applicants for the substitution of high-cost conventional solar cells. Throughout the last twenty years, Gratzel and hisits colleagues have made considerable efforts to develop further. The different factors involved in the DSSCs system are still under development in several ways to make it ready for commercialization. To date, the maximum 14.7% power conversion efficiency (PCE) has been recorded for the DSSCs [9]. The theoretical PCE limit of the single junction DSSC under standard test conditions (STC) is 32%, according to Professor Michael Graetzel, and a two-level tandem DSSC structure could reach PCE of 46% [10]. Since there is a big difference between theoretical and experimental PCE of DSSC, there is aroom for further improvement. Different researchers have suggested and worked on adifferent methods to improve the PCE of DSSC, such as anode modification by doping material, anode modification by carbon nanotube, surface modification by $TiCl_4$, dye modification, electrolyte modification, and cathode modification by different carbon variant [11–13]. Advancements in DSSCs technology are occurring at an ever-increasing rate, as the development of novel carbon-based materials, such as carbon nanotubes (CNTs). The CNT structure consists of enrolled graphite sheets, in a word, and can be classified as either single-walled (SWCNT) ormulti-walled (MWCNT) depending on its preparation method [14]. Cell performance can be increased by adding SWCNTs/MWCNTs. The incorporation of CNTs into the semiconductor material (i.e., TiO_2, ZnO, SnO_2, ctc.) decreases the host material's resistance and increases thermal conductivity, electrical conductivity, mechanical strength, and durability. The main aim of using CNTs in the DSSC is to accelerate the flow of electrons from the semiconductor material to the transparent conductive oxide (TCO) glass substrate through the CNTs without any resistance inside the grain boundary [15].

In this book chapter, the next five sections are focused on the basics of DSSCs, CNTs, the effect of CNTs in the cell performance of DSSC, and the degradation study of CNT based DSSC, respectively. In the sixth section, the development of CNT based DSSC is summarized with an outlook. In addition, it is also focused on the improvement of efficiency of DSSC by incorporating carbon nanotubes, which is of great significance for enhancing the performance of DSSC.

2. Dye sensitized solar cell

DSSC is different from other conventional solar cells in terms of both cell architecture and the physical process behind its operation. DSSC combines both solid and liquid phase in contrast to typical crystalline silicon or thin-film solar cell technology based on solid-state semiconductor materials. A typical DSSC consists of a transparent conducting oxide (TCO) glass substrate (i.e., ITO, FTO, AZO, etc.) as an anode, a wide band-gap semiconductor (a nanocrystalline semiconductor material, such as TiO_2, ZnO, SnO_2, $SrTiO_3$, Zn_2SnO_4,and Nb_2O_5, etc. deposited on the TCO), dye sensitizer (anchored on to the surface of nanocrystalline semiconductor material), a volatile electrolyte (I^-/I_3^-, Br^-/Br_3^-, $SCN^-/(SCN)_3^-$, and $SeCN^-/(SeCN)_3^-$, etc. redox couple), and a platinum (or carbon) coated TCO glass substrate as a counter electrode [16–24]. A typical DSSC structure is shown in **Figure 1**.

Figure 2 illustrates the basic operating principle of DSSC. Nanocrystalline semiconductor material (i.e., TiO_2) is deposited on the TCO and provides the indispensable surface area for dye photosensitizer absorption. Photon energy from the sunlight is absorbed (or collected) by the dye photosensitizer layer and produce excited electron (D+) from the highest occupied molecular orbital (HOMO) to the lowest unoccupied molecular orbital (LUMO) (Eq. (1)). The excited dye injects an electron to the conduction band (C.B.) of the semiconductor material and the dye molecule oxidized and losses an electron (Eq. (2)). The injected electron travels

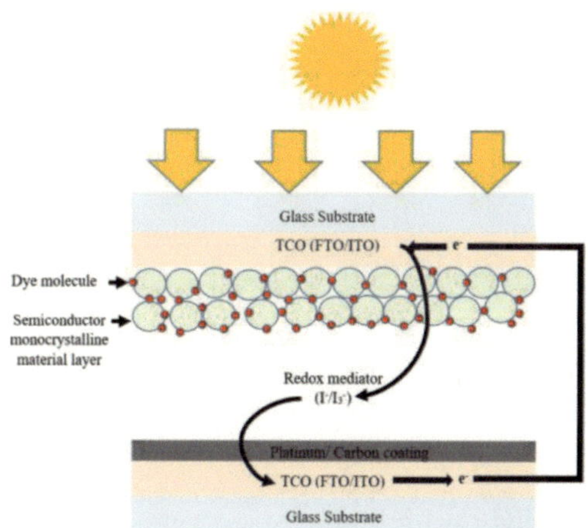

Figure 1.
Basic cell structure of DSSC.

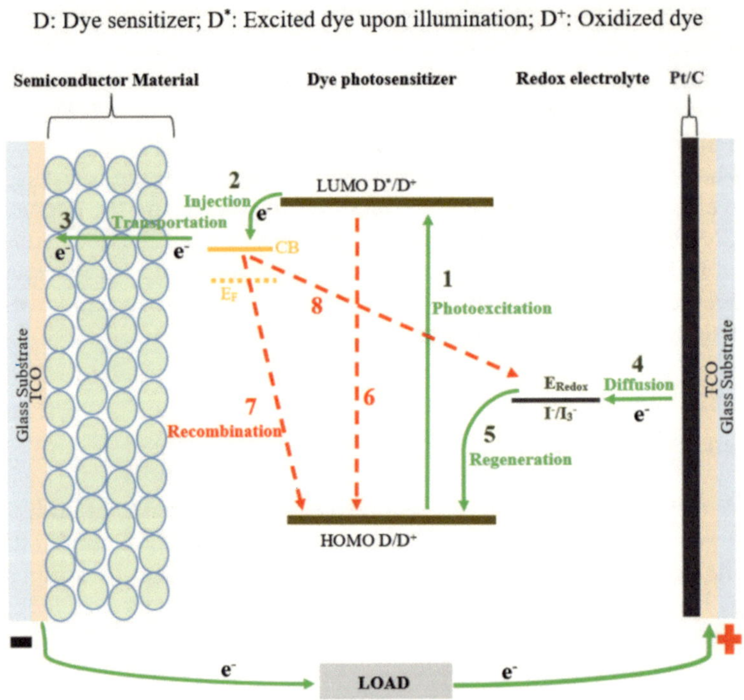

Figure 2.
Basic operating principles of DSSC.

through the semiconductor material toward the anode, and electrical energy is delivered to the external load (Eq. (3)). Then the electron further travels to complete the circuit and reaches the counter electrode (C.E.). The electron is transferred from the C.E. to the electrolyte. Dye regenerates when the dye accepts an electron from the I^- and I^- gets oxidized to I_3^- (Eq. (4)). I_3^- ion float around, and they receive ion from the C.E. (Eq. (5)) [25]. However, some unwanted reaction occurs, such as the recombination of dye (Eq. (6)), dye recombination to the ground state (Eq. (7)), and recombination of electrolyte (Eq. (8)) that reduces the overall cell's electron circulation performance.

Excitation of dye upon illumination

$$D + h\nu \rightarrow D^* \quad (1)$$

Oxidation of dye due to injection of electrons in TiO_2 photoanode

$$D^* \rightarrow D^+ + e^- \left(TiO_2\right) \quad (2)$$

Energy generation

$$e^- \left(TiO_2\right) + C.E. \rightarrow TiO_2 + e^-_{(C.E.)} + \left(\text{electrical energy}\right) \quad (3)$$

Regeneration of dye

$$D^+ + \frac{3}{2} I^- \rightarrow D + \frac{1}{2} I_3 \quad (4)$$

Restoration of electrolyte at the counter electrode

$$\frac{1}{2} I_3^- + e^-_{(C.E.)} \rightarrow \frac{3}{2} I^- + C.E. \quad (5)$$

Recombination of dye

$$TiO_2|D^* \rightarrow TiO_2|D \quad (6)$$

Dye recombination to ground state

$$TiO_2|D^+ + e^-_{(C.B.)} \rightarrow TiO_2|D \quad (7)$$

Recombination of electrolyte

$$I_3^- + 2e^- \ (TiO_2) \rightarrow 3I^- \quad (8)$$

Green arrows (path (1)–(5)) represent the electron transfer and movement of electrons within the solar cell, while red arrows (path (6)–(8)) represent potential recombination losses within the solar cell.

3. Carbon nanotubes (CNTs)

Carbon nanotubes (CNTs) are hollow cylinders consisting of single or multiple concentric layers of carbon atoms in a honeycomb lattice structure [26]. The CNT structure consists of enrolled graphite sheets, in a word, and can be classified as either or multi-walled (MWCNT) (**Figure 3(a)**) or single-walled CNT (SWNT) (**Figure 3(b)**) depending on its preparation method. In transmission electron microscopy (TEM) studies, MWCNTs were first observed by Iijima in 1991, while SWCNTs were independently developed by Iijima and Bethune in 1993 [26]. CNT has a sp^n hybridization (where n = 2) state of carbon material. However, because of the curved surface of CNT, it does not have a genuine sp^2 hybridization. CNT has a $sp^{2 + ?}$hybridization, which is in between n = 2 and 3. It is understood that CNT is a material lying between fullerenes and graphite as a new member of carbon allotropes [27]. Carbon nanotubes (CNTs) show very excellent adsorption characteristics because they have a high specific surface area and a nanoscale formation that constitutes many sites. It also has high electrical conductivity, mechanical strength, and a high aspect ratio [28].

3.1 CNT based photoanode in DSSC

CNT incorporated semiconductor material on the conducting electrode surface, offers efficient charge collection and transportation of charge carriers. The electrons

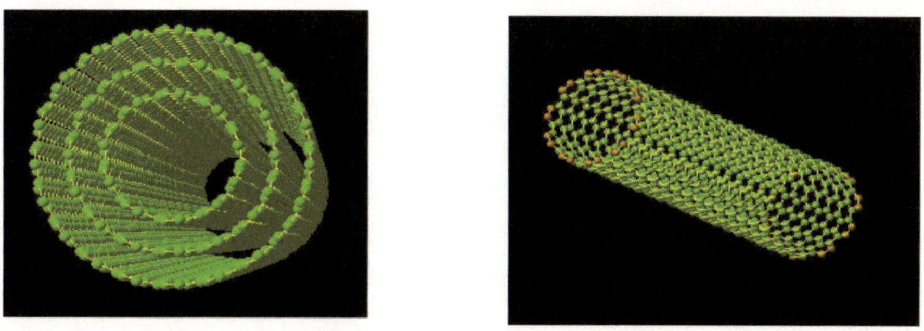

(a) (b)

Figure 3.
(a) MWCNTs, (b) SWCNTs.

injected from the excited dye into the semiconductor materials are then transferred through a CNT scaffold to generate photocurrent. Such 1-D nanostructures have been successfully exploited to improve the performance of DSSC [29]. **Figure 4** illustrates the role of CNT incorporated semiconductor material (i.e., TiO_2) for efficient electron transportation in DSSC.

Studies by Brown et al. has shown that the presence of CNT (i.e., SWCNT) does not directly affect the primary charge injection process in the D^*/TiO_2 system. SWCNT incorporated TiO_2 (TiO_2/SWCNT/D^*) films collect photoinduced electrons more effectively by charge separation than TiO_2/D^* films [29]. Studies have also shown that SWCNT accepts and stores electrons when in contact with photo-irradiated TiO_2 semiconductor materials. The fermi equilibrium with photo-irradiated TiO_2 and SWCNT can store to 1 electron per 32 carbon atoms. When the dyes are linked to the TiO_2-SWCNT suspension, the stored electrons are ready to discharge on demand [30]. SWCNT incorporated TiO_2 showed ~30% higher photoinduced current compared to without SWCNT incorporated TiO_2. Though the SWCNT incorporated TiO_2 based DSSC showed increases in the photoinduced current, the open-circuit voltage degrades in the SWCNT incorporated TiO_2 based DSSC. This phenomenon can be explained by the electron capture properties of SWCNT. At equilibrium condition, a positive shift (~20–30 mV) of the SWCNT causes loweropen-circuit voltage (which is directly related to the difference between fermi level of photoanode and redox electrolyte), as shown in **Figure 5**.

The photoinduced electrons are transferred to the SWCNT network, which minimizes the possibility of charge recombination at grain boundaries [29].

CNTs (i.e., SWCNTs) can be pristine (p-SWCNTs), metallic (m-SWCNTs) and semiconducting (s-SWCNTs) depending on their delocalized electrons occupying a 1-D density of states, chiral angles and diameters. The efficiency of SWCNT/TiO_2 based DSSC depends on different parameters, such as eventual charge separation, charge transfer, charge transport, and recombination rates. Studies by Guai et al. showed that m-SWCNT incorporated TiO_2 based photoanode does not significantly improve the recombination; however, it still enhances cell conversation efficiency of plain TiO_2 based DSSC. This indicates that plain TiO_2 has poor conductivity, which reduces the efficient charge transportation. On the other hand, s-SWCNT incorporated TiO_2 based photoanode showed superior conductivity. s-SWCNT incorporated TiO_2 based photoanode can also suppress the election recombination; thus, significant cell conversion efficiency is observed [31].

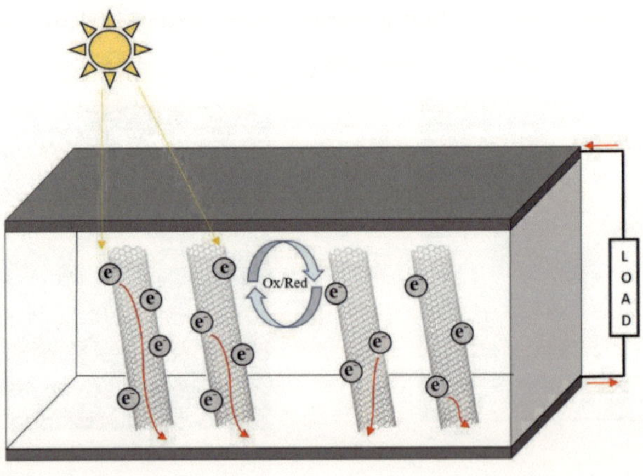

Figure 4.
CNT incorporated photoanode for DSSC.

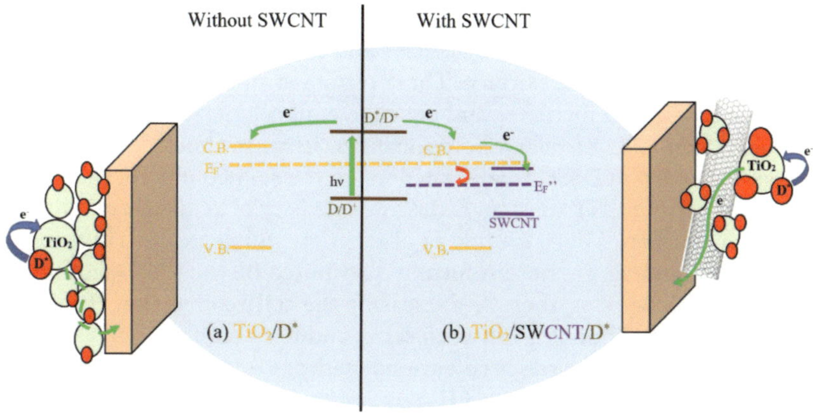

Figure 5.
Energy diagram illustrating D/TiO_2 and transportation of photoinduced electrons without (a) and with (b) SWCNT network.*

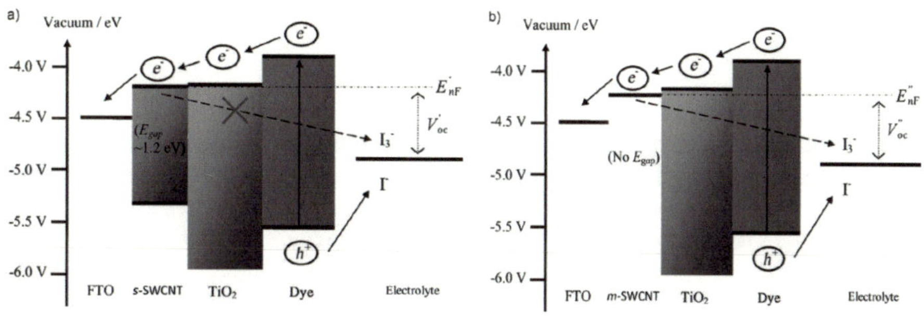

Figure 6.
Energy-band diagrams of DSSCs with incorporated (a) s-SWCNTs and (b) m-SWCNTs [31].

Figure 6 illustrates the energy band diagrams of s-SWCNT and m-SWCNT incorporated TiO_2 based DSSC. **Figure 6a** shows the photoinduced electron transfers from the photosensitizer's excited state to the TiO_2 and quickly moves from s-SWCNT to FTO. Since there is less possibility of electrons transported back (or recombination) to the liquid electrolyte to cause I_3^- reduction, more photoinduced electrons are effectively transported and collected by the FTO. This results in enhanced photocurrent generation than p- and m-SWCNT based DSSCs. Also, the effective collections of the photoinduced electron at the anode results in a positive shift in E_{nf}, increasing the open-circuit voltage of the SWCNT/TiO_2 based DSSC. On the contrary, though m-SWCNT has better electron mobility than s-SWCNT, it has a higher disruption in charge carrier transportation, leading to increased back reaction, as shown in **Figure 6b**. As a result, fewer electrons are collected at the FTO, which leads to lower photocurrent and cell conversion efficiency [31]. Hence the relative cell conversion efficiencies can be estimated as follows:

$$TiO_2 > p - SWCNT/TiO_2 > m - SWCNT/TiO_2 > s - SWCNT/TiO_2.$$

However, for the development of the high performance of SWCNT based DSSC, the combination of both m-SWCNT and s-SWCNT is used. Numerous researchers have been working on the combination of m-SWCNT and s-SWCNT. And the w/w % of s-SWCNT varies between 88 and 97% in the mix [32].

The introduction of CNTs into the semiconductor materials causes better dispersion of semiconductor materials particles and smaller crystalline size, structure with high porosity and coarse surface. These results in anincreasein the total surface area thus dye absorption increases and hence overall cell performance. However, the photosensitizer (i.e., metallic photosensitizer, organic photosensitizer, natural photosensitizer) used in DSSCusually anchors on the semiconductor materials (i.e., TiO_2) surface, not the CNT surface. If the mass density of semiconductor materials decreases, the number of dyes loading or absorption will decrease. Thus, if the semiconductor materials are not uniformly distributed on the CNT's surface, total dye absorption will be poor, thereby decreasing the cell's conversion efficiency. To solve this problem, the bonding between semiconductor material and CNTs should be increased. Different research used various methods to solve the problem. CNTs treated with concentrated HNO_3 and H_2SO_4 results in the introduction of -COOH anchoring group on the CNTs, and provides further improved bonding between semiconductor material and CNTs. The acid-treated CNT (i.e., SWCNT) can be used in two ways; either incorporated into the semiconductor material to improve charge transfer or introduced in semiconductor material/electrolyte interface as light scattering centers. In the first case, when acid-treated SWCNTs were incorporated into the semiconductor material (i.e., TiO_2) films, the fabricated DSSC showed a 25% more enhancement in photocurrent than the untreated CNTs. In the second case, when acid-treated SWCNTs were introduced at the TiO_2/electrolyte interface, the value of open-circuit voltage increases, whereas the value of photogenerated current remains constant. The improvement in open-circuit voltagegenerally impliesdecreased dark current and the negative shift of the conduction band of semiconductor material [33, 34]. **Figure** 7 illustrates the TiO_2 films with untreated SWCNTs and with acid treated SWCNT.

Similar to SWCNT, MWCNT also improves cell performance. However, the cell performance can be further improved by employing different treatment techniques. Numerous researchers are working on the topics and discovered different methods. Employing these methods can further improve the cell efficiency of DSSC. For example, Zhang et al. introduced RF induced oxygen plasma treatment for the

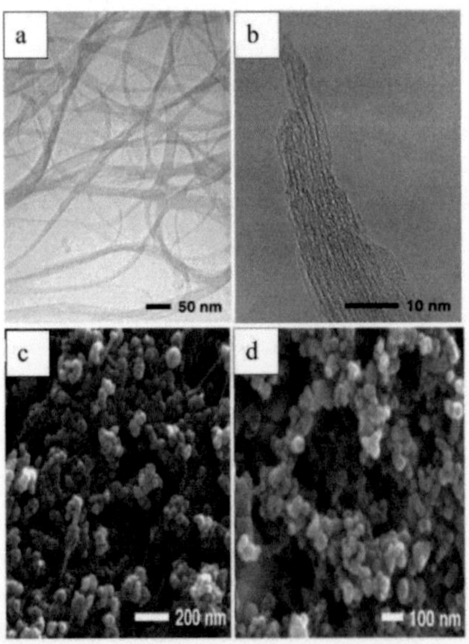

Figure 7.
(a) TEM of untreated bundles of SWCNTs, (b) HRTEM of a single fragmented bundle of acid treaded SWCNTs (c) SEM of TiO_2 films with untreated SWCNTs and (d) SEM of TiO_2 films with acid treated SWCNTs [33].

oxygen containing groups on the surface of the MWCNTs. The oxygen plasma treated MWCNT makes the surface more hydrophilic and improves the dispersion in TiO_2, which leads to high surface area and enhanced dye absorption and hence the cell performance increased. According the Zhang et al. study, the plasma treated MWCNT can improve around 75% cell performance than untreated MWCNT/TiO_2 photoanode based DSSC [34, 35].

Figure 8 illustrates the surface morphology of the coated films of pure TiO_2, MWCNT-TiO_2, and plasma-treated MWCNT-TiO_2 photoanodes using scanning electron microscopic (SEM). The introduction of MWCNT into the TiO_2 semiconductor material causes immobilized uniformly. The FESEM images also indicate thatincorporating MWCNT into the TiO_2 metal oxide material causes a highly porous and coarse surface (**Figure 8b**), which leads to a higher surface area for dye absorption. However, MWCNT-TiO_2 photoanodes has irregular pore sizes and a non-uniform porous structure, which affect the total surface area improvement. On the other hand, plasma treated MWCNT has a higher dispersion in TiO_2 metal oxide material leading to a more uniform porous surface structure (**Figure 8c**), which successfully increases the total surface area for better dye absorption [35].

Researchers have also explored other methods for employing CNTs for high surface area and hierarchical nanoporous structurefor higher cell conversion efficiency. Yun et al. employed TiO_2 hollow sphere/CNT by direct mixing and showed 4.71% with 0.1 wt.% [36]. Muduli et al. sensitized TiO_2/MWCNT composite by hydrothermal method and achieved the crystalline phase, which showed 50% more cell conversion efficiency than the one without the MW-CNTs [37]. Patrick et al. submerged TiO_2 colloid in the optically transparent electrode and electrophoreticallydeposited SWCNTs for working photoanode. The modified photoanode showed better charge separation and prevented back reaction/recombination, showing 45% improvement in photocurrent [29]. Subha et al. fabricated single-crystalline 1D rutile TiO_2 nanorods/MWCNT composite template-free synthesis method and reported 60% improvement in cell performance. Due to the single

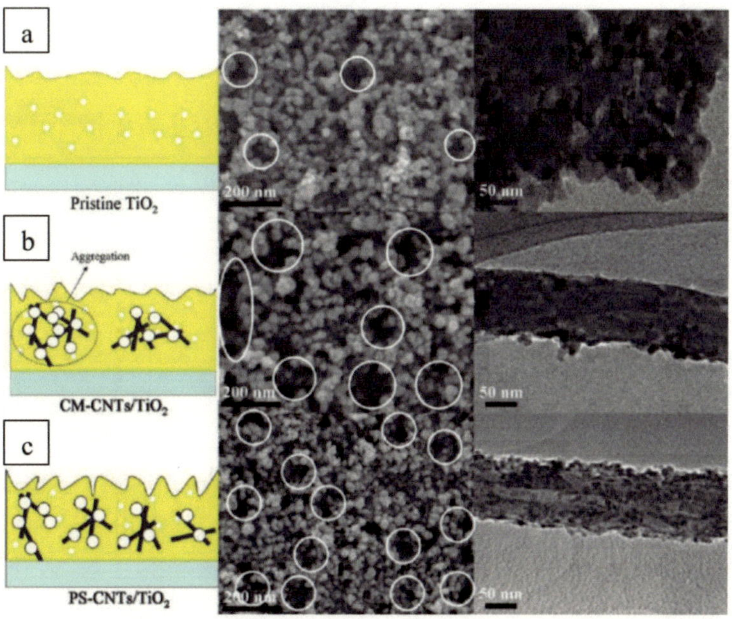

Figure 8.
FESEM and TEM images of the (a) pristine/pure TiO_2, (b) MWCNTs/TiO_2, and (c) plasma treated-MWCNTs/TiO_2photoanode [35].

crystalline structure, there were no grain boundaries, which provides a smooth surface for electron transportation [38]. Zhu et al. sensitized rice grain-shaped TiO_2/MWCNT composite by electrospinning process. Due to the single crystalline structure and high surface area of the rice grain-shaped TiO_2/MWCNT composite, DSSC showed a 32% improvement in cell performance with 0.2 wt.% MWCNT [39].

3.2 CNT based electrolyte in DSSC

In recent years, carbon-based materials in ionic liquids have been investigated as a potential alternative for traditional liquid electrolytes for DSSC application. An effective electrolyte has low viscosity, low vapor pressure, high diffusion coefficient, high electrochemical, and thermal stability. Conventional liquid redox electrolyte has low viscosity and high diffusion coefficient. However, liquid redox electrolyte uses a volatile solvent, which causes a problem in the commercialization of DSSC, such as cell leakage of electrolyte, performance degradation, high-temperature instability, and pressure build-up after in the fabricated cell due to the volatile solvent. Also, liquid electrolyte creates obstacles for the flexible structure and large-scale solar cell. Quasi-solid-state electrolytes based on ionic liquids can easily solve the drawbacks of liquid electrolyte efficiently. Organic hole conducting materials, p-type inorganic semiconductors, or different nano-components (i.e., graphene, CNTs) were diffused into ionic liquids to the sensitized quasi-solid-state electrolyte for DSSC application. **Figure 9** shows a basic schematic diagram of a CNT based electrolyte for DSSC.

3.3 CNT based cathode in DSSC

In recent years, carbon-based materials, such as graphite, carbon black, and carbon nanotubes, have been studied to replace traditional platinum (Pt) counter

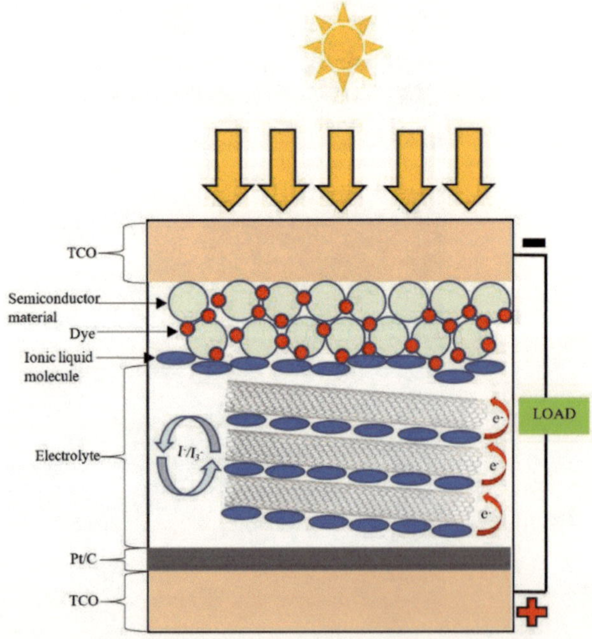

Figure 9.
CNT based electrolyte for DSSC application.

electrodes for low-cost DSSC. Carbon materials are not only abundant but also highly resistant to corrosion. Carbon-based material, especially CNT, has attracted considerable interest because of its fast electron transfer kinetics and large surface area. **Figure 10** shows a basic schematic of a CNT based counter electrode for DSSC.

Different methods have been explored for CNT based counter electrode. Nam et al. used paste printing and CVD growing methods for the counter electrode. The paste printing MWCNT based counter electrode based DSSC has lower cell efficiency (8.03%) than the Pt. counter electrode based DSSC's cell efficiency (8.80%). However, the CVD has grown MWCNT had higher cell conversion efficiency (10.04%) than the Pt. counter electrode based DSSC's cell efficiency (8.80%) [40]. Ramasamy et al. fabricated spray-coated MWCNT counter electrode and showed the effect of spray time/coating thickness [41]. Widodo also fabricated spray-coated CNT on FTO glass substrate for counter electrode for DSSC [42]. Prasetio et al. used different weight (0.01, 0.02 and 0.04 gram) of CNT and observed the cell performance of DSSC [43]. **Figure 11** illustrates the SEM of (a) the surface and (b) the cross-section CNT based counter electrode [43].

4. Effect of CNT on the cell performance of DSSC

4.1 Effect of CNT based photoanode on the cell performance of DSSC

As mentioned earlier (in Section 3.1.), adding CNT nanoparticles into mesoporous structure provides a strong light-harvesting capability and a large surface area for high-efficiency DSSC. Mesoporous semiconductor materials anchor on the long tubular CNT's outer surface and this assembly ensure efficient electron transport through CNTs. CNT improves the electron transport and increases the coating's thickness; thus, dye building on the anode material increase. CNT results in gains in the photocurrent without compromising the electron injection to the electrode.

From the previous discussion (Section 3.1.), CNT can be used either CNT/semiconductor material composite photoanode or counter electrode. For CNT-based photoanode for different types of dyes, such as metal complex dye sensitizer, natural dye sensitizer has been explored for DSSC operation.

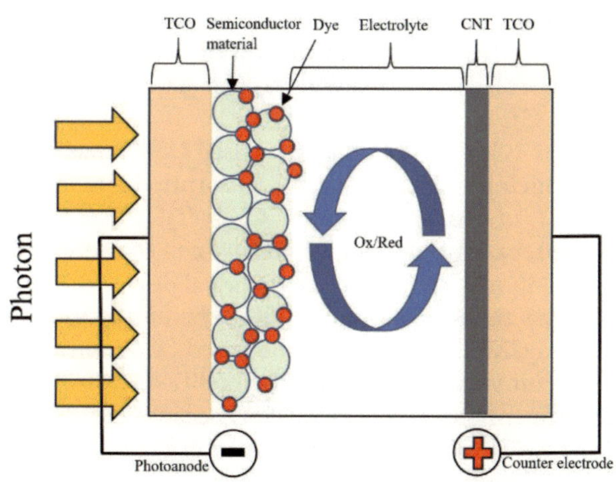

Figure 10.
CNT based counter electrode for DSSC.

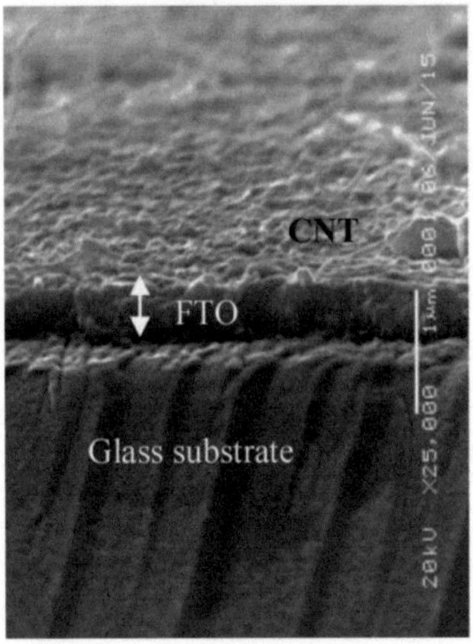

Figure 11.
SEM of the cross-section CNT based counter electrode [43].

Photoanode	J_{sc} (mA/cm^{-2})	V_{oc} (V)	Fill factor (FF)	Efficiency (%η)	Improvement (%)
PristineTiO_2	6.48	0.84	0.65	3.63	
Chemically modified CNT/TiO_2	8.56	0.83	0.66	4.66	28
O_2 plasma treated CNT/TiO_2	11.04	0.85	0.68	6.34	75

Table 1.
Different CNT/TiO_2 photoanode based DSSC [35].

Zhang et al. used N719 dye on the CNT photoanode-based DSSC. They prepared two different types of CNT based photoanode: pristine, chemically modified, and O_2 plasma-treated CNT. For chemically modified CNT, 500 mg CNT was mixed with 100 ml H_2SO_4/HNO_3 solution (1:1) and kept at 140°C for 6 hours. Afterward, the solution was filtered, cleaned with distilled water, and vacuum dried. For O_2 plasma-treated CNT, CNT was treated with O_2 plasma treated for 40 minutes at 0.26 Torr (at 50 W) [35].

O_2 plasma-treated CNTs/TiO_2 photoanode based DSSCs has more uniform holes and rough surface, which provides higher dye adsorption and less charge recombination than either TiO_2 or chemical modified CNTs/TiO_2. The O_2 plasma-treated CNTs/TiO_2 photoanodebased DSSC showed 6.34% cell efficiency, which is ~75% higher than the conventional TiO_2 photoanode based devices (**Table 1**). **Figure 12** illustrates the I-V characteristics of pristine TiO_2, chemically modified CNT/ TiO_2, and O_2 plasma-treated CNT/ TiO_2 photoanode based DSSC with N719 dye. The value of the open-circuit voltage of chemically modified CNT/ TiO_2 based DSSC was lower than the TiO_2 photoanode based DSSC because of poor adhesion between chemically modified CNT/TiO_2; however, the overall cell conversion efficiency was 28% higher than the traditional TiO_2 based DSSC [35].

Dembele et al. sensitized an ethanolic suspension of MWCNTs (0.006 g of MWCNTs in 15 mL of ethanol) and then mixed with a known weight of TiO_2 paste

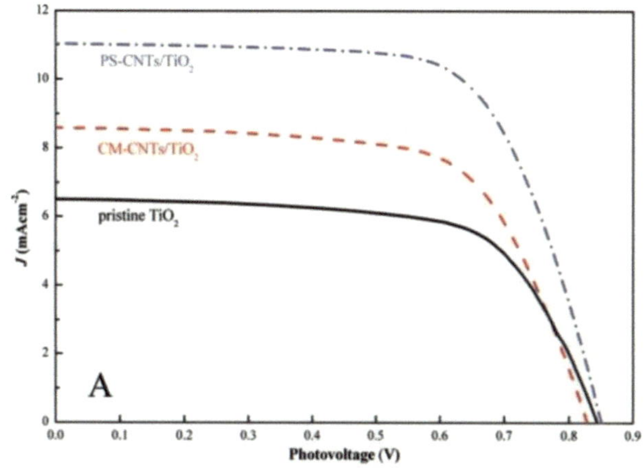

Figure 12.
I-V characteristics of pristine TiO_2, chemically modified CNT/ TiO_2 and O2 plasma treated CNT/TiO_2 photoanode based DSSC with N719 dye [35].

to obtain a fixed percentage of MWCNT/TiO_2 composite. The MWCNT/TiO_2 composites were tape cast on FTO glass substrate. The concentration of MWCNTs was varied in the range of 0–0.250 wt. %. The fabricated photoanodes were soaked in a 0.5 mM ethanolic solution of N719 sensitizer. They were increasing the percentages of MWCNTs in the MWCNT/TiO_2 composite, which results in a slight improvement in dye loading (from 0.010–0.020%). In comparison, the average number of dye mole per volume unit improved almost 3.5 times for the 0.25 wt.% MWCNT/ TiO_2film than the pure TiO_2film. However, this chemisorption of dye multilayers are detrimental for overall cell efficiency: for optimum cell efficiency, the dye should be chemisorbed in a closely packed monolayer. **Figure 13** illustrates the SEM image of the bare TiO_2 and MWCNT (0.25 wt.%)/TiO_2 film [44].

From **Table 2**, optimized concentration (0.010–0.020 wt.%) of MWCNTs based TiO_2 photoanode significantly increases the overall cell conversion efficiency. Introducing a reflection layer into the cell increases the electron lifetime and decreases recombination, leading to improved overall cell performance and photoconversion efficiency (9.0%) [44].

Not only DSSC sensitized with metallic dye, but also DSSC sensitized with natural dye shows similar characteristics to the concentration of CNT (i.e., MWCNT). Kabir et al. used natural yellow dye sensitizer extracted from the turmeric (*Curcuma longa*). They have fabricated DSSC with natural yellow dye as a sensitizing source for TiO_2 photoanode with different MWCNT concentrations. The concentration of MWCNT ranges from 0.005 wt. % to 0.050 wt. %. **Figure 14** shows surface morphology of (a) bare TiO_2, (b) MWCNTs, (c) MWCNTs/TiO_2, and (d) TiCl4 treated MWCNTs/TiO_2 film [45].

The integration of MWCNTs improved the cell efficiency of the DSSC by developing a special charge carrier transport channel that is distributed uniformly throughout the TiO_2 semiconductor. Cell efficiency of all concentrations of MWCNT incorporated by TiO_2 is higher than that of the without MWCNT incorporated by TiO_2. The concentration of the MWCNT from 0.000 wt. % to 0.015 wt. %, cell efficiency has improved dramatically (From **Table 3**). The optimum concentration is 0.015 wt. % MWCNTs have a maximum cell efficiency of 1.653%, whereas for without MWCNT integrated TiO_2, cell efficiency is 0.921%. After achieving the best possible combination, further increasing the concentration of MWCNT leads to a negative effect on cell parameters such as I_{sc}, V_{oc}, and FF.

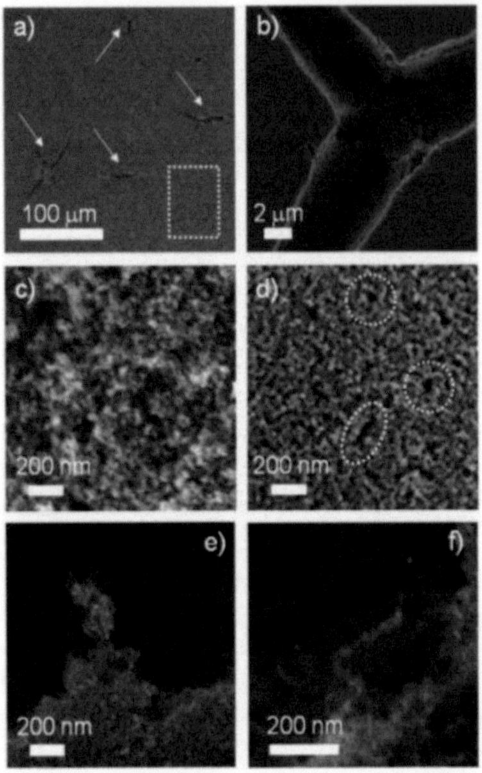

Figure 13.
SEM images of (a, b, d, e, f) the 0.25 wt.% CNT photoanode at various magnifications, and (c) bare TiO_2 photoanode [44].

Higher concentrations of MWCNTs result in a loss of transparency and low absorption of light, which decreases the photogenerated current [45].

4.2 Effect of CNT based electrolyte on the cell performance of DSSC

As mention earlier (Section 3.2.), CNT based electrolyte not only enhance cell performance of DSSC but also provide improvement in cell structure. Ahmad et al. sensitized a new type of quasi-solid-state electrolyte by dispersing graphene and CNT into the 1-methyl 3-propyl imidazolium iodide (PMII) ionic liquid. They also varied the CNT (i.e., SWCNT) content from 1 wt.% to 16 wt.% in the PMII ionic liquid. Maximum cell efficiency of 1.43% was observed for 7% SWCNT +93% PMII ionic liquid electrolyte (**Table 4**). They have also combined graphene with SWCNT and observed the effect of SWCNT in the quasi-solid-state electrolyte in DSSC application. Combining SWCNTand graphene with PMII enhances cell performance significantly, which is higher than both SWCNT + PMII combination and PMII (**Table 5**) [46].

Lee et al. sensitized MWCNT–polymethyl methacrylate (PMMA) composite electrolyte by thermal polymerization for solid state DSSC. The MWCNT-PMMA composite has made a homogenous solution of 0.26 g MWCNT, 5 g of methyl methacrylate (MMA), and 2-hydroxy-2-methyl-propylphenone (initiator). The solution was vacuum dried and cleaned thoroughly (with dichloroethane (DCE) to remove initiator residue). The MWCNT-PMMA composite was mixed with iodide couples (0.1 M of LiI, 0.015 M of I2, and 0.2 M of t-butyl pyridine) in acetonitrile solvent and stirred for 20 hours for MWCNT-PMMA composite electrolyte.

Anode structure	CNTs (wt %)	$T_{annealing}$ (°C)	Thickness (μm)	PCE (%)	FF (%)	V_{oc} (mV)	J_{sc} (mA cm^{-2})	Dye loading (mol $mm^{-3} \times 10^7$)
Transparent layer	0	450	12.6	6.5	71.0	745	12.7	1.05
Transparent layer	0.003	450	12.5	5.6	68.0	700	11.9	
Transparent layer	0.007	450	11.5	6.4	67.1	695	13.8	
Transparent layer	0.010	450	12.8	8.1	71.0	724	15.6	1.05
Transparent layer	0.015	450	9.9	7.9	71.0	734	15.3	1.65
Transparent layer	0.020	450	11.3	7.4	71.0	704	14.8	1.25
Transparent layer	0.045	450	8.8	6.7	71.0	698	13.6	1.35
Transparent layer	0.075	450	10.1	5.9	73.0	707	11.6	1.35
Transparent layer	0.250	450	16.4	1.1	62.0	789	2.2	3.70
Transparent layer + reflecting layers	0	500	15.9	7.0	69.0	755	13.6	
Transparent layer + reflecting layers	0.010	500	16.4	9.0	74.0	758	16.0	

Table 2.
Effect of CNT (MWCNT) concentration in the cell performance of N719 bye based DSSC [44].

Figure 15 shows the FESEM image of MWCNT and MWCNT-PMMA composite, and **Figure 16** illustrates the TEM and HR-TEM images MWCNT and MWCNT-PMMA composite [47].

Table 6 lists the photovoltaic performance of DSSC fabricated with MWCNT-PMMA composite. Lee et al. Lee et al. achieved higher cell efficiency of 2.9% (than the reference cell's efficiency of 1.9%), which was possible because of enhanced amorphous structure and ionic conductivity [47].

4.3 Effect of CNT based counter electrode on the cell performance of DSSC

To achieve high cell performance in a large application area, platinum is used for its excellent electrochemical activity. However, using Pt as a counter electrode increases overall cell cost. Many efforts have been made for large surface area and fast electron transportation to apply CNTs to the cathode. CNTs prices are lower than Pt, but it has high electrochemical activity, which makes DSSCs commercially viable. Prasetio et al. sensitized CNT-based cathode by doctor blade method and observed CNT concentration's effect by varying the weight of CNT (0.01, 0.02, and 0.04 g). A slurry was prepared by mixing a fixed mass of (0.01 or 0.02 or 0.04 g) CNT, 0.2 g ethylcellulose, 2 ml ethanol, and 0.8 g terpineol. The slurry was doctor bladed on FTO substrate and dried in the air, followed by annealing at 450°C for 60 minutes. **Figure 17** illustrates SEM images of prepared CNT cathode fabricated with different CNT masses (0.01, 0.02, and 0.04 g) [43].[Use the similar format for gram, either g or gr or gram throughout the chapter].

Table 7 lists the photovoltaic performance of DSSC fabricated with different masses of CNT as a cathode. Increasing the mass of CNT in the cathode increases the photogenerated current; however, CNT does not increase other cell parameters, such as open-circuit voltage and fill factor. An increase in short circuit current resulted in higher cell conversion efficiency.

Ramasamy et al. spray-coated MWCNT on FTO and used it as a cathode. Dispersed MWCNT was sprayed onto FTO glass substrate with a spray gun, which was

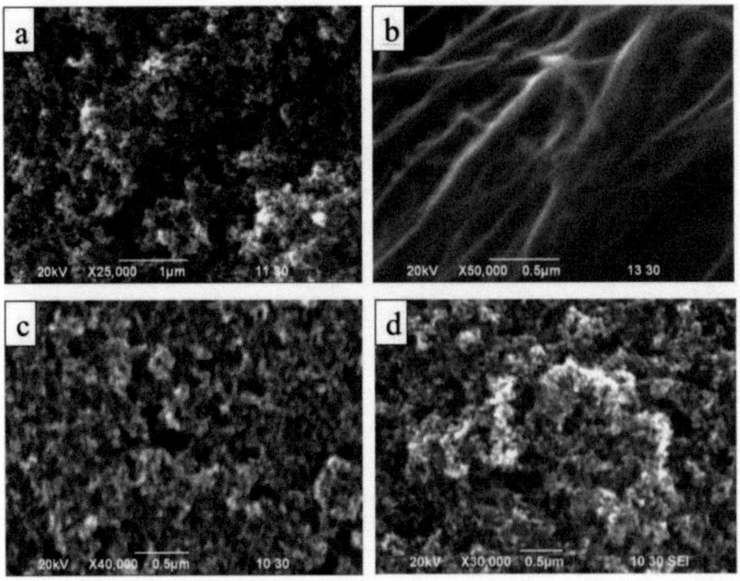

Figure 14.
SEM image of (a) bare TiO_2, (b) MWCNTs, (c) MWCNTs/TiO_2, and (d) $TiCl_4$ treated MWCNTs/TiO_2 film [45].

MWCNT concentration	V_{OC} (V)	I_{sc} (mA)	FF	η%	Dye loading (mol mm^{-3} × 10^7)
0 wt.%	0.515 ± 0.010	3.792 ± 0.024	0.472 ± 0.007	0.921 ± 0.037	1.13
0.005 wt.%	0.512 ± 0.008	4.973 ± 0.017	0.535 ± 0.003	1.362 ± 0.038	1.15
0.010 wt.%	0.513 ± 0.001	5.346 ± 0.011	0.564 ± 0.001	1.546 ± 0.008	1.22
0.015 wt.%	0.502 ± 0.006	5.995 ± 0.028	0.553 ± 0.009	1.653 ± 0.054	1.37
0.020 wt.%	0.499 ± 0.005	5.119 ± 0.023	0.549 ± 0.010	1.404 ± 0.046	1.34
0.025 wt.%	0.498 ± 0.001	4.873 ± 0.020	0.522 ± 0.002	1.267 ± 0.017	1.31
0.030 wt.%	0.489 ± 0.004	4.543 ± 0.018	0.504 ± 0.006	1.119 ± 0.027	1.28
0.040 wt.%	0.486 ± 0.011	4.543 ± 0.023	0.499 ± 0.008	1.080 ± 0.047	1.28
0.050 wt.%	0.481 ± 0.013	4.361 ± 0.028	0.494 ± 0.003	1.036 ± 0.041	1.25

Table 3.
I-V performance of different concentrations of the MWCNT incorporated TiO_2 based DSSC fabricated with natural yellow dye [45].

SWCNT content (wt%)	J_{sc} (mA/cm^2)	V_{oc} (V)	FF	Efficiency (%)
0% (only PMII)	0.370	0.575	0.64	0.16 ± 0.01
1%	0.524	0.573	0.70	0.25 ± 0.01
7%	5.19	0.540	0.41	1.43 ± 0.13
10%	2.15	0.616	0.36	0.56 ± 0.02
13%	1.64	0.614	0.41	0.46 ± 0.02
16%	2.09	0.541	0.32	0.40 ± 0.02

Table 4.
I-V performance of DSSCs with quasi-solid-state electrolytes containing different wt.% of SWCNTs in PMII [46].

Content (wt%)	J_{sc} (mA/cm^2)	V_{oc} (V)	FF	Efficiency (%)
100% PMII	0.370	0.575	0.64	0.16 ± 0.01
85% PMII +3% SWCNT +12% graphene	7.32	0.594	0.44	2.50 ± 0.10
85% PMII +12% SWCNT +3% graphene	4.66	0.561	0.43	1.39 ± 0.10

Table 5.
I-V performance of DSSCs with quasi-solid-state electrolytes containing different wt.% of SWCNT and graphene in PMII [46].

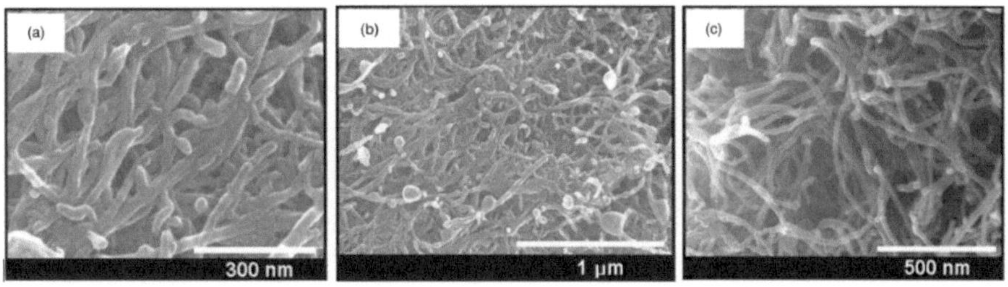

Figure 15.
SEM images of (a) MWCNT, and (b and c) MWCNT–PMMA composite film [47].

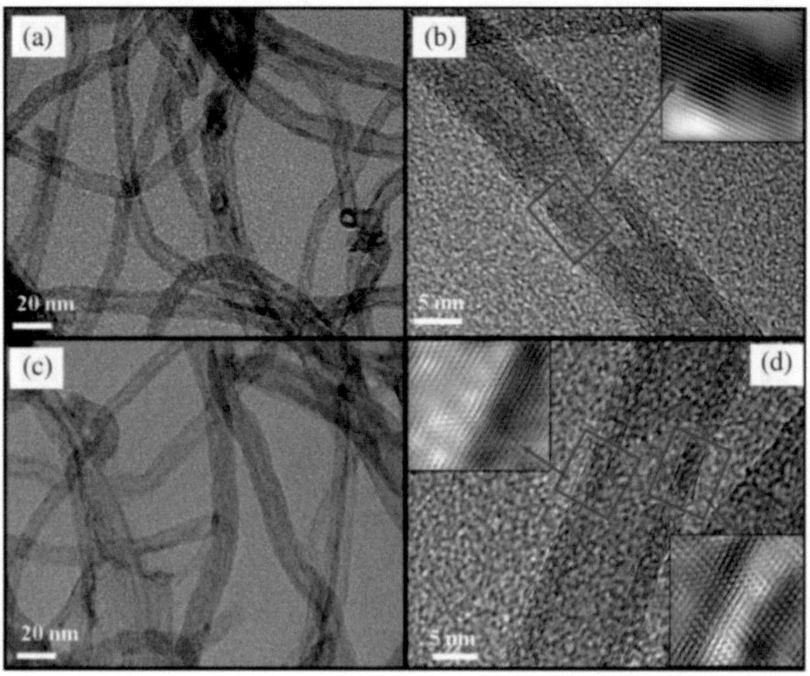

Figure 16.
TEM and HR-TEM images of (a, c) MWCNT and (b, d) MWCNT–PMMA composites [47].

connected to an air compressor. Ramasamy et al. varied the spray time and observed the effect of spraying time on the cell performance of DSSC. Since the spray time is directly related to the thickness of the MWCNT layer, in other words, they have observed the effect of MWCNT coating thickness on the cell performance of DSSC [41].

Sample	Differential scanning calorimetry (DSC) data		Conductivity (mS/cm)	Photovoltaic performance			
	Melting temperature T_m (°C)	Heat of melting ΔH (J/g)		J_{sc} (mA/ cm^2)	V_{oc} (V)	FF (%)	Efficiency (%)
PEO	74.3	119..5	1.2	5.7	0.61	53.4	1.9
MWCNT-PMMA	81.2	97.8	2.3	8.9	0.57	61.8	2.9

Table 6.
I-V performance of DSSC fabricated with MWCNT-PMMA composite [47].

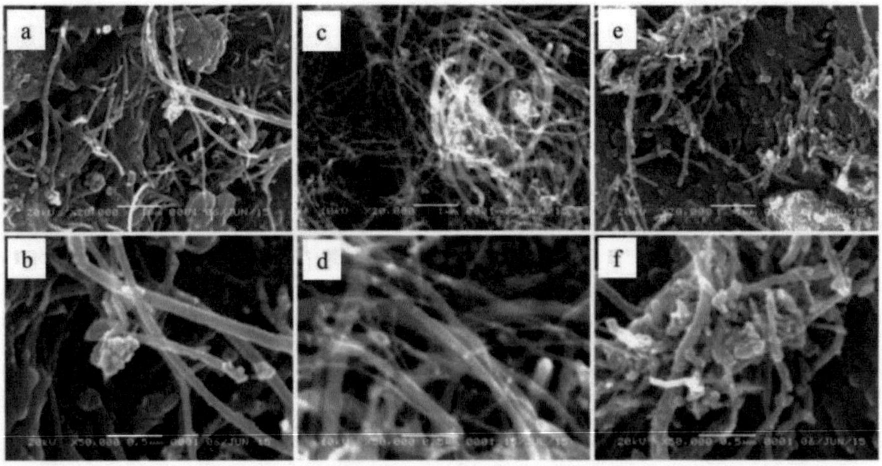

Figure 17.
SEM images of prepared CNT cathode fabricated with different masses of the CNT (a, b) 0.01 gram (c, d) 0.02 gram, and (e, f) 0.04 gram [43].

Mass of CNT in cathode (in gram)	J_{sc} (mA/cm^2)	V_{oc} (V)	FF (%)	Efficiency (%)
0.01	2.093	0.49	31	0.32
0.02	4.829	0.48	32	0.74
0.04	6.413	0.45	32	0.91

Table 7.
I-V performance of DSSC fabricated with different mass of CNT as cathode [43].

Table 8 lists the I-V parameters of various spraying times of the MWCNT counter electrode. The value of open-circuit voltage is more or less independent of the spray time; however, both short circuit current and fill factor showed a strong dependence on the spraying time of MWCNTs. The charge transfer resistance of the MWCNT cathode in the iodide/tri-iodide electrolyte solution was decreased by increasing the spraying time, which results in a significant improvement in the cell performance of the MWCNT counter electrode based DSSC [41].

Nam et al. used two different methods for CNT counter electrode-based DSSC: screen printing and chemical vapor deposition. Screen printed MWCNT cathode based DSSC showed lower cell performance (8.03%) than the reference Pt cathode based DSSC's cell performance (8.80%) because of unfavorable contact resistance between the MWCNT and FTO. On the contrary, chemical vapor deposited

Spraying time (s)	V_{oc} (V)	J_{sc} (mA/cm^2)	FF (%)	Efficiency (%)
Bare FTO	0.428	1.48	0.07	0.04
5	0.772	8.03	0.11	0.68
10	0.784	12.81	0.15	1.51
30	0.773	15.67	0.28	3.39
60	0.778	15.92	0.47	5.82
100	0.778	15.86	0.57	7.03
200	0.783	15.64	0.62	7.59

Table 8.
I-V performance of DSSC fabricated with MWCNT based cathode [41].

Cathode	J_{sc} (mA/cm^2)	V_{oc} (V)	FF (%)	Efficiency (%)
Reference Pt coated cathode	17.68	746.27	0.65	8.80
Paste printed MWCNT cathode	15.27	738.43	0.69	8.03
CVD grown MWCNT cathode	17.62	755.89	0.73	10.04

Table 9.
I-V performance of DSSC with different MWCNT based cathode (different deposition method for MWCNT) [40].

$H_2PtCl_6.6H_2O$ (%)	SWCNT (%)	Light transmittance (%)	J_{sc} (mA/cm^2)	V_{oc} (V)	FF (%)	Efficiency (%)
0.24	0	87	7.29	0.74	0.56	3.00
0.48	0	83	9.42	0.74	0.68	4.71
0.72	0	72	5.86	0.75	0.67	2.91
0.48	0.03	81	9.61	0.75	0.68	4.88
0.48	0.06	80	11.20	0.75	0.71	5.96
0.48	0.012	74	8.53	0.75	0.68	4.33

Table 10.
I-V performance of DSSC fabricated with different $H_2PtCl_6.6H_2O$ and SWCNT content [48].

MWCNT cathode based DSSC showed higher cell performance (10.04%) due to aligned MWCNT and improved charge carrier conduction path (**Table 9**) [40].

Xiao et al. used Pt/SWCNT film to spray onto the ITO coated polyethylene naphthalate substrate using a vacuum thermal decomposition method at 120°C. The fabricated cathode showed higher light transmittance, higher chemical stability, higher electrocatalytic activity for redox electrolyte, and lower charge carrier transfer resistance [48]. **Table 10** lists the photovoltaic properties of DSSC fabricated with different $H_2PtCl_6.6H_2O$ and SWCNT content [48].

5. Prospects of CNT in DSSC application

DSSCs have attracted considerable attention due to their simple fabrication process, inexpensive raw materials, and employment of eco-friendly materials. Recently, to take advantage of their lower electrical resistance, excellent electrocatalytic operation, mechanical integrity, low cost, and flexibility, CNTs have been incorporated into DSSCs. Kongkaland et al. used a different approach using carbon fiber electrodes (CFE) than conventional transparent conduction oxide, such as ITO, FTO, etc. They have deposited TiO_2 semiconductor material on the CFE and CFE-SWCNT film. The incorporation of SWCNT in the CFE increased cell performance from 7.36–16%. This two times improvement in the cell

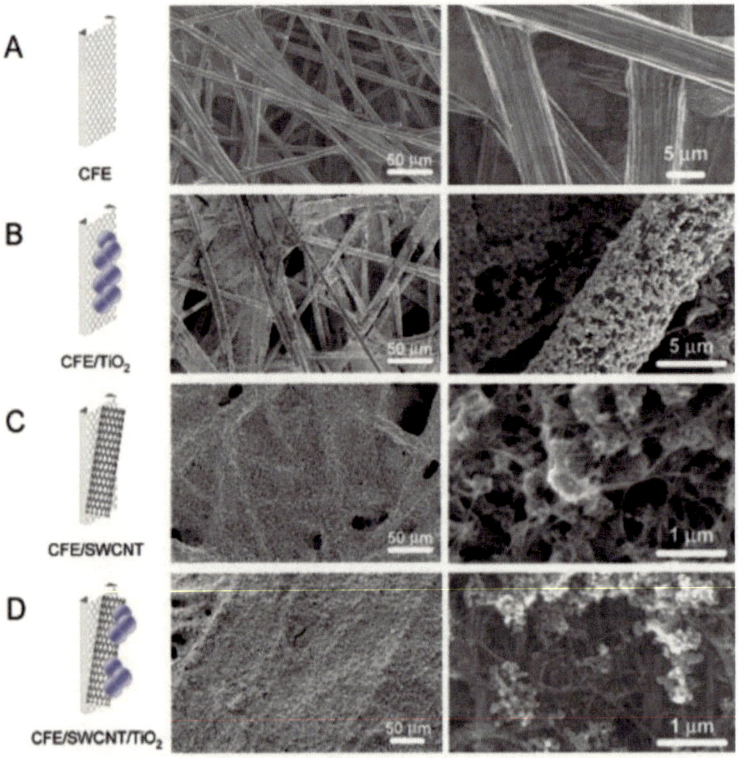

Figure 18.
SEM of a (A) CFE before surface modification; (B) deposition of TiO_2 on CFE film; (C) electrophoretic deposition of SWCNT on CFE; (D) after deposition of deposition of TiO_2 on CFE-SWCNT film [49].

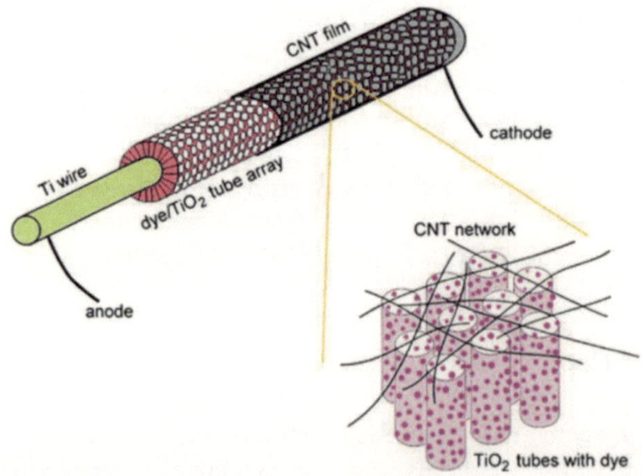

Figure 19.
Coaxial single-wire structure DSSC [50].

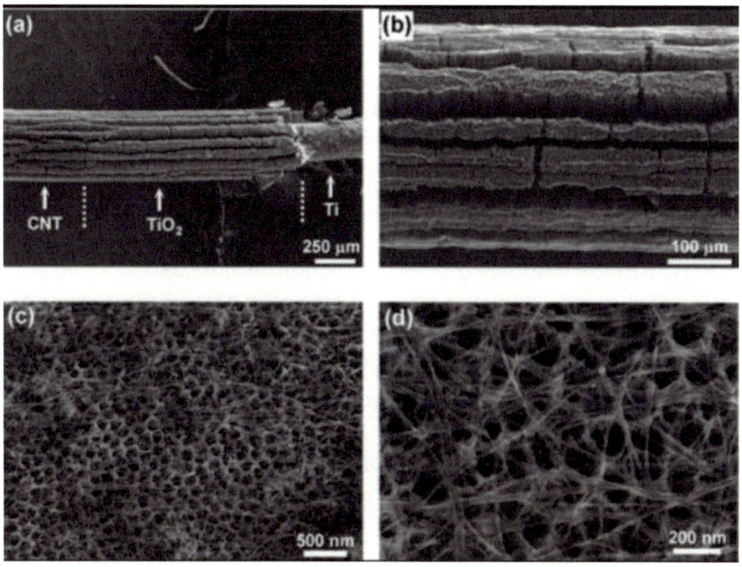

Figure 20.
SEM image of (a) different parts of the wire, (b) a wire section uniformly wrapped by CNT film. (c) CNT network deposited on porous TiO_2, and (d) CNT-TiO_2 nanotubes [50].

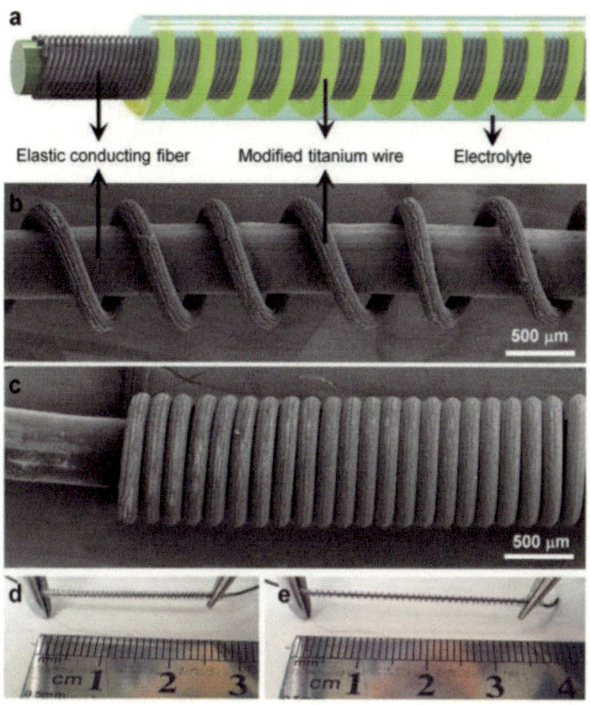

Figure 21.
(a) Basic structure of a textile DSC (b) SEM images of a textile DSC with 560μmpitch distances (c) SEM images of a textile DSC with 164 μm pitch distances, (d) A textile DSC before and after stretch by 30%, and (e) A textile DSC after 30%stretch [51].

performance indicates that CNTs can serve a valuable role in facilitating charge collection in DSSC application. **Figure 18** illustrates the SEM image of CFE and CFE-SWCNT base anode for DSSC application [49].

Powering next-generation wearable/implantable biomedical devices, or smart textile, have gained extensive attention in recent years. Among the developed energy harvesting devices, DSSCs cell structures have become ideal candidates for developing practical self-powered biomedical devices or smart textile due to their lightweight, flexibility, high power-per-weight ratios, and superior mechanical stability/robustness. The conventional planar-shaped DSSCs with sandwich-like configuration includes five primary parts: TCO, an anode (semiconductor material), dye, redox electrolyte, and cathode (Pt/C). Based on the conventional planar structure, DSSCs can also be made into flexible configurations. Zhang et al. fabricated fiber-shaped DSSC and made a double-wire structure (shown in **Figure 19**). They have manufactured a flexible DSSC structure on a single wire (Ti-TiO_2) and wrap the CNT around the tube array. CNT provides full contact with the active layer, unlike Pt, and provides uniform light absorption throughout the entire circumference ofDSSC [50]. **Figure 20** illustrates the fabricated fiber shaped DSSC [50].

Yang et al. further developed a wearable DSSC textiles method based on electrically conducting fibers. They have prepared fiber electrodes by aligning winding multiwalled carbon nanotube (MWCNT) sheets on rubber fibers. The working fiber electrode was prepared by incorporating modified Ti onto the MWCNT fiber electrode (**Figure 21c**). The wire-shaped DSSCs could weave into wearable photovoltaic textile solar cells. The maximum cell efficiency of the wire-shaped DSSC reached 7.13% [51].

6. Conclusion

Incorporating CNT in the DSSC increases the interaction between electrodes and electrolyte, enhancing the cell performance of DSSC. In addition, incorporating CNT in the semiconductor material decreases resistance to the grain boundaries. It provides a unique charge carrier transport channel distributed uniformly in the host semiconductor to absorb the charge carrier from the collector. Cell performance of CNT based photoanode can be improved by optimizing the CNT concentration and deposition method. For CNT based electrolyte, the ionic electrolyte can be an alternative for traditional redox electrolyte. Also, CNT based ionic electrolyte provides better cell performance with enhanced durability. Finally, the CNT-based cathode can offer a large surface area and fast electron transportation, which reduces the chances of recombination.

Lastly, incorporation of CNTs into DSSC will serve a significant role in producing solar cells that produce energy at an affordable rate relative to the existing energy generation approaches. New methods are frequently being published and will present opportunities for innovation in both research and industrial growth.

Author details

Md. Mosharraf Hossain Bhuiyan[1,4*†], Fahmid Kabir[2†], Md. Serajum Manir[3], Md. Saifur Rahaman[1], Prosenjit Barua[2], Bikrom Ghosh[2], Fumiaki Mitsugi[5] and Tomoaki Ikegami[5]

1 Institute of Nuclear Science and Technology, Atomic Energy Research Establishment, Bangladesh Atomic Energy Commission, Dhaka, Bangladesh

2 Institute of Energy, University of Dhaka, Dhaka, Bangladesh

3 Institute of Radiation and Polymer Technology, Atomic Energy Research Establishment, Bangladesh Atomic Energy Commission, Dhaka, Bangladesh

4 Department of Computer Science and Engineering, Central University of Science and Technology, Mirpur, Dhaka, Bangladesh

5 Faculty of Advanced Science and Technology, Kumamoto University, Kumamoto, Japan

*Address all correspondence to: mosharraf22003@yahoo.com; mosharraf22003@baec.gov.bd

† These authors contributed equally.

References

[1] M. I. Hoffert, K. Caldeira, A. K. Jain, E. F. Haites, L. D. Harvey, S. D. Potter, M. E. Schlesinger, S. H. Schneider, R. G. Watts, T. M. Wigley and D. J. Wuebbles, "Energy implications of future stabilization of atmospheric CO 2 content," Nature, vol. 395, no. 6705, pp. 881–884, 1998.

[2] F. Kabir, M. M.H. Bhuiyan and M. S. Manir, "Effect of combination of natural dyes and theblocking layer on the performance of DSSC," in Solar cells, IntechOpen, 2021.

[3] M. M.H. Bhuiyan, M. Hussain and M. A. Asgar, "Evaluation of a Solar Photovoltaic Lantern in Bangladesh Perspective," Journal of Energy & Environment, vol. 4, p. 75–82, 2005.

[4] M. M. H. Bhuiyan, M. A. Asgar, R. K. Mazumder and M. Hussain, "Economic evaluation of a stand-alone residential photovoltaic power system in Bangladesh," Renewable energy, vol. 21, no. 3–4, pp. 403–410, 2000.

[5] J. Wu, Z. Lan, S. Hao, P. Li, J. Lin, M. Huang, L. Fang and Y. Huang, "Progress on the electrolytes for dye-sensitized solar cells," Pure and Applied Chemistry, vol. 81, no. 11, pp. 2241–2258, 2008.

[6] M. M. H. Bhuiyan and M. A. Asgar, "Sizing of a stand-alone photovoltaic power system at Dhaka," Renewable Energy, vol. 28, no. 6, pp. 929–938, 2003.

[7] L. C. Andreani, A. Bozzola, P. Kowalczewski, M. Liscidini and L. Redorici, "Silicon solar cells: toward the efficiency limits," Advances in Physics: X, vol. 4, no. 1, p. 1548305.

[8] S. Ananthakumar, J. R. Kumar and S. M. Babu, "Third-Generation Solar Cells: Concept, Materials and Performance - An Overview," in Emerging Nanostructured Materials for Energy and Environmental Science, 2019, p. 305–339.

[9] K. Kakiage, Y. Aoyama, T. Yano, K. Oya, J. I. Fujisawa and M. Hanaya, "Highly-efficient dye-sensitized solar cells with collaborative sensitization by silyl-anchor and carboxy-anchor dyes," vol. 51, no. 88, pp. 15894–15897, 2018.

[10] Dye Sensitized Solar Cells [Online]. Available from: https://gcell.com/dye-sensitized-solar-cells [Accessed 08 01 2021].

[11] D. Sengupta, P. Das, B. Mondal and K. Mukherjee, "Effects of doping, morphology and film-thickness of photo-anode," Renewable and Sustainable Energy Reviews, vol. 60, pp. 356–376, 2016.

[12] F. Kabir, S. N. Sakib, S. T. E. E. S. Uddin and M. T. F. Himel, "Enhance cell performance of DSSC by dye mixture, carbon nanotube and post TiCl4 treatment along with degradation study," Sustainable Energy Technologies and Assessments, vol. 35, pp. 298–307, 2019.

[13] J. Jaksik, H. J. Moore, T. Trad, O. I. Okoli and M. J. Uddin, "Nanostructured functional materials for advanced three-dimensional (3D) solar cells," Solar Energy Materials and Solar Cells, vol. 167, pp. 121–132, 2017.

[14] M. M.H. Bhuiyan, F. Mitsugi, T. Ueda and T. Ikegami, "Nox sensing characteristics of single wall carbon nanotube gas sensor prepared by pulsed laser ablation," International Journal of Nanomanufacturing, vol. 4, no. 1–4, pp. 186–196, 2009.

[15] F. Kabir, B. M. M. H. M. M. S. M. S. Rahaman, M. A. Khan and T. Ikegami, "Development of dye-sensitized solar

cell based on combination of natural dyes extracted from Malabar spinach and red spinach," Results in Physics, vol. 14, p. 102474, 2019.

[16] F. Kabir, M. Bhuiyan, M. R. Hossain, M. H. Bashar, M. S. Rahaman, M. S. Manir, S. M. Ullah, S. S. Uddin, M. Z.I. Mollah, R. A. Khan and S. Huque, "Improvement of efficiency of Dye Sensitized Solar Cells by optimizing the combination ratio of Natural Red and Yellow dyes," Optik, vol. 179, pp. 252–258, 2019.

[17] D. Sengupta, P. Das, U. Kasinadhuni, B. Mondal and K. Mukherjee, "Morphology induced light scattering by zinc oxide polydisperse particles: promising for dye sensitized solar cell application," Journal of Renewable and Sustainable Energy, vol. 6, no. 6, p. 063114, 2014.

[18] K. Manseki, T. Sugiura and T. Yoshida, "Microwave synthesis of size-controllable SnO2 nanocrystals for dye-sensitized solar cells," New Journal of Chemistry, vol. 38, no. 2, pp. 598–603, 2014.

[19] E. Guo and L. Yin, "Tailored SrTiO3/TiO2 heterostructures for dye-sensitized solar cells with enhanced photoelectric conversion performance," Journal of Materials Chemistry A, vol. 3, no. 25, pp. 13390–13401, 2015.

[20] Y. F. Wang, K. N. Li, Y. F. Xu, H. S. Rao, C. Y. Su and D. B. Kuang, "Hydrothermal fabrication of hierarchically macroporous Zn2SnO4 for highly efficient dye-sensitized solar cells," Nanoscale, vol. 5, no. 13, pp. 5940–5948, 2013.

[21] X. Jin, C. Liu, J. Xu, Q. Wang and D. Chen, "Size-controlled synthesis of mesoporous Nb2O5 microspheres for dye sensitized solar cells," RSC Advances, vol. 4, no. 67, pp. 35546–35553, 2014.

[22] Z. S. Wang, K. Sayama and H. Sugihara, "Efficient Eosin Y Dye-Sensitized Solar Cell Containing Br-/Br3- Electrolyte," The Journal of Physical Chemistry B, vol. 109, no. 47, pp. 22449–22455, 2005.

[23] O. Byrne, A. Coughlan, P. K. Surolia and K. R. Thampi, "Succinonitrile-based solid-state electrolytes for dyesensitised solar cells," Progress in Photovoltaics: Research and Applications, vol. 23, no. 4, pp. 417–427, 2015.

[24] P. Wang, S. M. Zakeeruddin, J. E. Moser, R. Humphry-Baker and M. Grätzel, "A Solvent-Free, SeCN−/(SeCN)3- Based Ionic Liquid Electrolyte for High-Efficiency Dye-Sensitized Nanocrystalline Solar Cells," Journal of the American Chemical Society, vol. 126, no. 23, pp. 7164–7165, 2004.

[25] N. A. Ludin, A. A.A. Mahmoud, B. M. A. A. A. H. Kadhum, and N. S.A. Karim, "Review on the development of natural dye photosensitizer for dye-sensitized solar cells," Renewable and Sustainable Energy Reviews, vol. 31, pp. 386–396, 2014.

[26] P. Avouris, J. Appenzeller, R. Martel and S. J. Wind, "Carbon Nanotube Electronics," Proceedings of the IEEE, vol. 91, no. 11, pp. 1772–1784, 2003.

[27] K. Tanaka, "Classification of Carbon," in Carbon Nanotubes and Graphene, Elsevier, 2014, pp. 1–5.

[28] T. Ueda, M. M. H. Bhuiyan, H. Norimatsu, S. Katsuki, T. Ikegami and F. Mitsugi, "Development of carbon nanotube-based gas sensors for NOx gas detection working at low temperature," Physica E: Low-dimensional Systems and Nanostructures, vol. 40, no. 7, pp. 2272–2277, 2008.

[29] P. Brown, K. Takechi and P. V. Kamat, "Single-Walled Carbon Nanotube Scaffolds for Dye-Sensitized Solar Cells," The Journal of Physical

Chemistry C, vol. 112, no. 12, pp. 4776–4782, 2008.

[30] A. Kongkanand and P. V. Kamat, "Electron storage in single wall carbon nanotubes. Fermi level equilibration in semiconductor–SWCNT suspensions," ACS nano, vol. 1, no. 1, pp. 13–21, 2007.

[31] G. H. Guai, Y. Li, C. M. Ng, C. M. Li and M. B. Chan-Park, "TiO2 Composing with Pristine, Metallic or Semiconducting Single-Walled Carbon Nanotubes: Which Gives the Best Performance for a Dye-Sensitized Solar Cell," ChemPhysChem, vol. 13, no. 10, pp. 2566–2572, 2012.

[32] C. Liu and H. M. Cheng, "Controlled growth of semiconducting and metallic single-wall carbon nanotubes," Journal of the American Chemical Society, vol. 138, no. 21, pp. 6690–6698, 2016.

[33] S. R. Jang, R. Vittal and K. J. Kim, "Incorporation of Functionalized Single-Wall Carbon Nanotubes in Dye-Sensitized TiO2 Solar Cells," Langmuir, vol. 20, no. 22, pp. 9807–9810, 2004.

[34] M. Janani, P. Srikrishnarka, S. V. Nair and A. S. Nair, "An in-depth review on the role of carbon nanostructures in dye-sensitized solar cells," Journal of Materials Chemistry A, vol. 3, no. 35, pp. 17914–17938, 2015.

[35] S. Zhang, H. Niu, Y. Lan, C. Cheng, J. Xu and X. Wang, "Synthesis of TiO2 nanoparticles on plasma-treated carbon nanotubes and its application in photoanodes of dye-sensitized solar cells," The Journal of Physical Chemistry C, vol. 115, no. 44, pp. 22025–22034, 2011.

[36] J. Yu, J. Fan and B. Cheng, "Dye-sensitized solar cells based on anatase TiO2 hollow spheres/carbon nanotube composite films," Journal of Power Sources, vol. 196, no. 18, pp. 7891–7898, 2011.

[37] S. Muduli, W. Lee, V. Dhas, S. Mujawar, M. Dubey, K. Vijayamohanan, S. H. Han and S. Ogale, "Enhanced conversion efficiency in dye-sensitized solar cells based on hydrothermally synthesized TiO2– MWCNT nanocomposites," ACS applied materials & interfaces,, vol. 1, no. 9, pp. 2030–2035, 2009.

[38] S. Sadhu and P. Poddar, "Template-Free Fabrication of Highly-Oriented Single-Crystalline 1D-Rutile TiO2-MWCNT Composite for Enhanced Photoelectrochemical Activity," The Journal of Physical Chemistry C, vol. 118, no. 33, pp. 19363–19373, 2014.

[39] Z. Peining, A. S. Nair, Y. Shengyuan, P. Shengjie, N. K. Elumalai and S. Ramakrishna, "Rice grain-shaped TiO2–CNT composite—A functional material with a novel morphology for dye-sensitized solar cells," Journal of Photochemistry and Photobiology A: Chemistr, vol. 31, no. 1, pp. 9–18, 2012.

[40] J. G. Nam, Y. J. Park, B. S. Kim and J. S. Lee, "Enhancement of the efficiency of dye-sensitized solar cell by utilizing carbon nanotube counter electrode," Scripta Materialia, vol. 62, no. 3, pp. 148–150, 2010.

[41] E. Ramasamy, W. J. Lee, D. Y. Lee and J. S. Song, "Spray coated multi-wall carbon nanotube counter electrode for tri-iodide (I3-) reduction in dye-sensitized solar cells," Electrochemistry Communications, vol. 10, no. 7, pp. 1087–1089, 2008.

[42] S. Widodo, G. Wiranto and M. N. Hidayat, "Fabrication of Dye Sensitized Solar Cells with Spray Coated Carbon Nano Tube (CNT) Based Counter Electrodes," Energy Procedia, vol. 68, no. 37–44, 2015.

[43] A. Prasetio, A. Subagio, A. Purwanto and H. Widiyandari, "Dye-sensitized solar cell based carbon

nanotube as counter electrode," in AIP Conference Proceedings, 2016.

[44] K. T. Dembele, G. S. Selopal, C. Soldano, R. Nechache, J. C. Rimada, I. Concina, G. Sberveglieri, F. Rosei and A. Vomiero, "Hybrid carbon nanotubes–TiO2 photoanodes for high efficiency dye-sensitized solar cells," The Journal of Physical Chemistry C, vol. 117, no. 28, pp. 14510–14517, 2013.

[45] F. Kabir, S. N. Sakib and S. S. Uddin, "Effect of MWCNT's concentration in TiO2 based DSSC and degradation study of the cell," Journal of Renewable and Sustainable Energy, vol. 11, no. 2, p. 023502, 2019.

[46] I. Ahmad, U. Khan and K. Gun'ko, "Graphene, carbon nanotube and ionic liquid mixtures: towards new quasi-solid state electrolytes for dye sensitised solar cells," Journal of Materials Chemistry, vol. 21, no. 42, pp. 16990–16996, 2011.

[47] H. C. Lee, M. S. Akhtar, J. G. Park, K. J. Kim, S. K. Lee and O. Yang, "Carbon Nanotube (CNT)–Polymethyl Methacrylate (PMMA) Composite Electrolyte for Solid-State Dye Sensitized Solar Cells," Journal of nanoscience and nanotechnology, vol. 10, no. 5, pp. 3502–3507, 2010.

[48] Y. Xiao, J. Wu, G. Yue, J. Lin, M. Huang and Z. Lan, "Low temperature preparation of a high performance Pt/SWCNT counter electrode for flexible dye-sensitized solar cells," Electrochimica Acta, vol. 56, no. 24, pp. 8545–8550, 2011.

[49] A. Kongkanand, R. Martínez Domínguez and P. V. Kamat, "Single wall carbon nanotube scaffolds for photoelectrochemical solar cells. Capture and transport of photogenerated electrons," Nano letters, vol. 7, no. 3, pp. 676–680, 2007.

[50] S. Zhang, C. Ji, Z. Bian, R. Liu, X. Xia, D. Yun, L. Zhang, C. Huang and A. Cao, "Single-Wire Dye-Sensitized Solar Cells Wrapped by Carbon Nanotube Film Electrodes," Nano letters, vol. 11, no. 8, pp. 3383–3387, 2011.

[51] Z. Yang, J. Deng, X. Sun, H. Li and H. Peng, "Stretchable, wearable dye-sensitized solar cells," Advanced Materials, vol. 26, no. 17, pp. 2643–2647, 2014.

Chapter 20

Mechanism for Flexible Solar Cells

Ghaida Salman Muhammed

Abstract

Flexible photovoltaics are covering the way to low-cost electricity. The build-up of organic, inorganic and organic–inorganic solar cells on flexible substrates by printing technologies is to provide lightweight and economic solar modules that can be incorporated in various surfaces. Progress of flexible and lightweight solar cell is interesting for many terrestrial and space applications that require a very high specific power. Thin-film solar cells on polymer films can produce more than 2KW\Kg specific power. Flexible solar cells are proposed to open up a numerous of possibilities for enabling new applications in consumer electronics and space satellites. Recent research in thin-film electronics has been concentrated on the replacement of the traditional rigid glass plate substrate with plastic or metallic foils. Organic materials bear the potential to develop a long-term technology that is economically viable for large-scale power generation based on environmentally safe materials with unlimited availability. Organic and organic-inorganic photovoltaics (PVs) (third generation solar cells) continue to attract great attention from the PV community, due to their promising features such as low organic–inorganic cost, flexibility and light weight. In this chapter, many of the possible materials for manufacturing of flexible solar cells are discussed.

Keywords: photovoltaics, organic–inorganic, fabrication, properties, flexible solar cell

1. Introduction

Flexible photovoltaics are including the way to cheap electricity. The preparation of organic, inorganic and organic–inorganic solar cells over flexible substrates by often roll-to-roll printing technologies is to provide lightweight and economic solar modules that can be incorporated in various surfaces. Under standard conditions the current conversion efficiencies are in the 3–15% range for potentially flexible, thin film devices. Meanwhile heavy, stiff, and fragile inorganic materials can exceed 30% efficiency [1, 2], but in real applications the overall productivity is high. In 2012, Luke F. Lester et al. [1] discussed the preliminary studies on the integration of GaAs-based InAs/InGaAs quantum dots-in-a-well (DWELL) solar cells onto flexible surfaces such as Kapton and nanopaper including weight considerations for all the integrated materials. The 2 billion people who lack access to the grid as well as to energy-eager companies and families in the industrialized world facing the increasing costs of electricity generated using fossil fuel resources can get cheap and clean electricity by these new photovoltaic technologies [2].

Progress of flexible and lightweight solar cell is fascinating for many terrestrial and space applications that require a very high specific power. More than 2 kW\kg specific power can be produced by thin-film solar cells on polymer films. They are

Figure 1.
Picture of a solar cell fabricated on a flexible substrate [4].

important for the development of novel products, such as lightweight and portable sources of power for emergencies and recreational use, photovoltaics incorporated buildings (roof and facades), consumer electronics (smart cards, data and telecommunication products), and solar boats and cars, etc. [3]. **Figure 1** shows a solar cell fabricated on a flexible substrate [4].

Flexible solar cells are proposed to accelerate a numerous of possibilities for providing new applications in consumer electronics and space satellites. Organic and amorphous semiconductors are very important materials to achieve flexible and light-weight solar cells, essentially due to their strong light absorption properties, process temperature compatibility with flexible substrates and potentially cheap processing cost. Due to the highly disordered and defective crystalline structure in these materials, the poor minority carrier lifetime prevent their use for making high efficiency and reliable solar cells [5].

The replacement of the traditional rigid glass plate substrate with plastic or metallic foils has been concentrated by the recent research in thin-film electronics. Metallic materials, stainless steel and molybdenum foils have been used as substrates in the fabrication of thin-film transistors. A number of plastic materials (organic polymers) also have been verified successfully in a variety of thin-film applications [6]. The glass substrate may contribute to more than 90% of the total weight of the solar cells. The glass substrate should be substituted with a lightweight and flexible thin substrate, such as metal or polymer foils to maximize the high specific power. This gives flexibility to the solar panels to change to any kind of shape for incorporation in buildings, and for application in a variety of products. Flexible solar modules can help low cost and easily deployable power generators in space. Solar cells with AM1.5 efficiency of 11.4% on foils (highest efficiency recorded for flexible CdTe cell) have been developed. A comparison of the cells prepared on different polyimides is presented by A. Romeo et al. [7]. Plastic substrates solar cells can also well used in the solar car because of those characteristic [4]. Inexpensive solar cells would help game reserve the environment. Coating existing roofing materials with its plastic photovoltaic cells which are inexpensive enough to cover a home's entire roof with solar cells, then enough energy could be captured to power almost the entire house. Then, the dependence on the electric grid (fossil fuels) would decrease and help to reduce pollution [8].

2. Organic semiconductors

The toleration of the ability to develop a long-term technology that is economically active for large-scale power generation based on environmentally safe materials with unlimited availability is caused by the organic materials. The organic semiconductors are less expensive materials than the inorganic semiconductors like Si; they have high optical absorption coefficients which offer the opportunity for the production of very thin solar cells. Also, thin flexible devices can be fabricated using high throughput and low temperature approaches that employ well established printing techniques in a roll-to-roll process. The electronic structure of all organic semiconductors is based on conjugated π-alternation between single and double carbon-carbon bonds. Single bonds are known as σ-bonds and are associated with localized electrons, and double bonds contain a σ-bond and a π-bond. The π-electrons are much more mobile than the σ-electrons; they can jump from site to site between carbon atoms thanks to the mutual overlap of π orbital's along the conjugation path, which causes the wave functions to delocalize over the conjugated backbone. The π-bands are either empty (called the Lowest Unoccupied Molecular Orbital (LUMO)) or filled with electrons (called the Highest Occupied Molecular Orbital (HOMO)). The band gap of these materials ranges from 1 to 4 eV. This π-electron system has all the essential electronic features of organic materials: light absorption and emission, charge generation and transport [9]. Also, molecular orbitals which form σ and π bonds represent the energy levels for organic semiconductor materials [10].

The denoted bonding molecular orbitals (σ and π) form the highest molecular orbital (highest energy levels) where the denoted anti-bonding molecular orbitals (σ* and π*) form the lowest molecular orbitals (lowest energy levels). These molecular orbitals are similar to energy bands levels in inorganic materials. **Figure 2** shows the method of creating energy gap levels in organic semiconductor [11].

The anti-bonding π* molecular orbitals (conduction band) joined the π bonding molecular orbitals (valance band) to create the Lowest Unoccupied Molecular Orbital (LUMO) and Highest Occupied Molecular Orbital (HOMO). The gap between the (LUMO) and (HOMO) is the energy gap where the conductivity in organic semiconductor depends on. Thus, from **Figure 2** it is clear that σ bonds are extremely filled with electrons where π bonds are empty. On the other hand, if the energy band gap becomes as small as possible the tolerance of electrons to move from the (HOMO) to (LUMO) increases. Some organic semiconductors have a very small band gap of <2 eV, which mean that it is good materials compared to some inorganic semiconductors, which haves a large energy band gap [10].

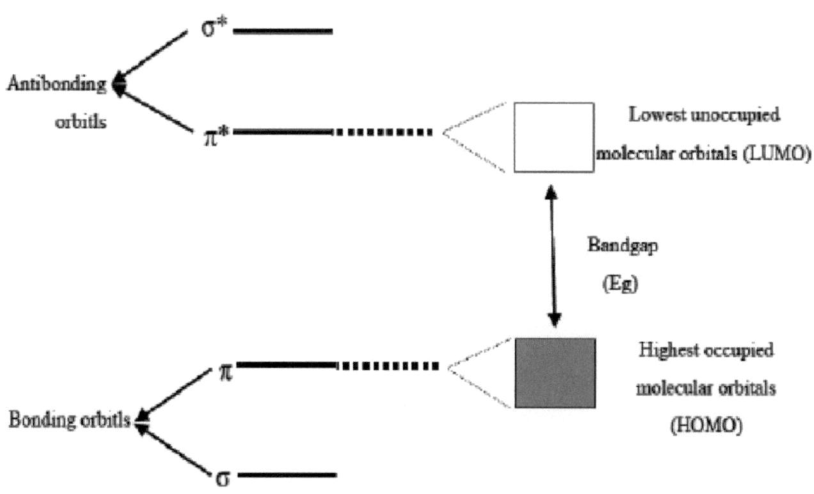

Figure 2.
The energy levels in organic semiconductors [11].

In organics the small radius excitons generated as a result of photon absorption is the source of photocarriers. Excitons in organics have a binding energy in the range 200–400 meV, which is significantly higher than the binding exciton energy for semiconductor materials ~2–40 meV. The exciton dissociation should occur at the interface between the donor and acceptor materials with suitable (HOMO) and (LUMO) energy levels because that thermal energy at room temperature is not sufficient (~25 meV) for exciton dissociation to hole and electron in the bulk medium. The mechanism of exciton dissociation is not completely known, however, the charge transfer process between the donor and acceptor components is the major factor controlling the charge separation at the interface. After charge separation, holes and electrons move to the opposite electrodes because of drift and diffusion [12].

The working principle of a polymer bulk heterojunction (BHJ) solar cell as shown in **Figure 3** is summarized by the creation of an exciton in the active layer, due to light absorption, and then this exciton will separate into two charge carriers at the interfaces between the species that constitute the active layer (typically, a binary blend of a polymer and a fullerene or two polymers, which act as the donor phase and the acceptor phase), with subsequent collection by the electrodes. To get efficient steps, all of these steps must follow very strict limitations. At first, the generated excitons must hop between the molecules reaching an interface between the two phases before recombining (radiatively or non-radiatively). This means that the two phases should be mixed in an optimal structure, with phase domains usually in the order of (10–30) nm (the average exciton diffusion length in polymers). Then, the position of the energy levels at the interface must be favorable for a fast exciton dissociation followed by charge separation (*i.e.*, the electron in the acceptor phase and the hole in the donor phase without successive recombination). After that, the charges must travel inside the respective phases, reaching the collecting electrodes again without a charge recombination: at this point, the energetic level structure at the electrode interfaces plays an essential role, ideally the interface being an ohmic contact [13].

The conjugated polymer-PCBM bulk heterojunction is currently the best conjugated polymer-based PV cell. One significant improvement to this device structure was made recently by many researchers, who found that the morphology of the blend could be optimized by casting the polymer and PCBM from a solvent that prevents long-range phase separation and enhances the polymer chain packing. This

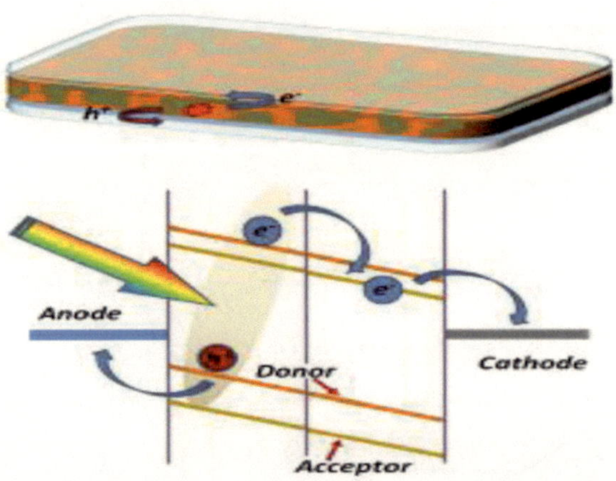

Figure 3.
A schematic diagram for the working principle of a polymeric bulk heterojunction solar cell [13].

avoids the formation of isolated regions of polymer and PCBM in the film and gives the polymer increased hole mobility. This fabricates a device with more than double the EQE of the previous best device and with 2.5% power efficiency under AM1.5G conditions [14].

For organic solar cells, the magnitude of J_{SC}, V_{OC}, and FF depends on parameters such as: light intensity, temperature, composition of the components, thickness of the active layer, the choice of electrodes used, and the solid state morphology of the film. A clear understanding of the device operation and photocurrent J_{ph} generation and its limitations in these devices are required for their optimization and maximization. In order to allow for further design of new materials that can improve the efficiency of this type of solar cells, the relation between the experimental J_{ph} and material parameters (charge-carrier mobility, band gap, molecular energy levels, or relative dielectric constant) needs to be understood and controlled [9].

Two intrinsic issues can be resolved by the bulk heterojunction (BHJ) structure, charge separation and charge transport, in organic layers. These are representative structures; though, they have a fundamental limitation in terms of open circuit voltage (V_{OC}) that is basically determined by the offset energy between the highest occupied molecular orbital (HOMO) of p-type organic semiconductors and the lowest unoccupied molecular orbital (LUMO) of n-type organic semiconductors, even though the work functions of electrodes often affect V_{OC}. A tandem structure can be applied to maximize the power conversion efficiency in organic solar cells by increasing V_{OC}, because the overall voltage becomes the sum of individual V_{OC} values in each sub cell (front and back cells). In addition, the selecting complementary BHJ layers enhances the overall short circuit current density (J_{SC}) of tandem cells, because these layers have different absorption ranges for maximizing solar light harvesting. An adverse effect is also present owing to marginally increased electrical resistances by the presence of additional interfaces and active layers in series connection. As illustrated previously, both normal- type and inverted-type structures are possible by placing suitable electrodes with appropriate work functions on each side. To date, the most popular bulk heterojunction structure is made with the composites of conjugated polymers (p-type) and fullerene derivatives (n-type), leading to polymer:fullerene solar cells. Hence, P3HT polymers have been introduced because they can absorb visible light up to 650 nm and their glass transition temperature approaches ~110C°. The external quantum efficiency (EQE) of

Material	Molecular structure	HOMO (eV)	LUMO (eV)	Carrier mobility (cm^2/Vs)*
P3HT	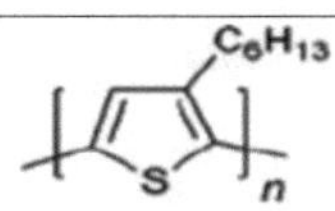	5.2	3.2	$\mu_h = 2 \times 10^{-4}$
PCPDTBT	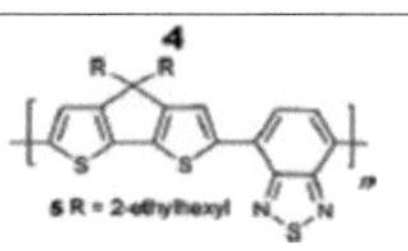	4.9	3.5	$\mu_h = 2 \times 10^{-2}$
PCBM	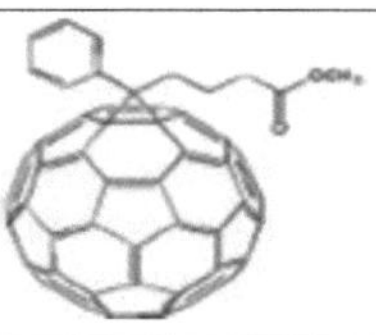	6	4.2	$\mu_e = 3 \times 10^{-3}$

Table 1.
Organic molecular structure [18].

devices at a maximum wavelength reached more than 70% for P3HT:PC_{61}BM solar cells, compared to ~50% for MDMOPPV: PC_{61}BM solar cells [15]. Because of the semicrystalline nature of P3HT spin-cast films, it has among the highest reported field-effect transistor mobilities for a conjugated polymer. The morphology and the mobility of pure rr-P3HT and blended rr-P3HT: PCBM films are highly dependent on casting conditions [16, 17].

The flexible side chain of the P3HT molecule introduces a good solubility in organic solvents in spite of its kind as stiff polymers. The long and narrow fibrils which are produced by P3HT crystalizes are from a network that is able to create a good percolation paths for the charge carriers' transportation. This leads to high carrier mobility. Differing from the P3HT, PCBM is a derivative of fullerene; it has a high electron affinity which makes it a qualified electron acceptor material in the organic thin film solar cells as shown in **Table 1** [18].

3. Organic-inorganic photovoltaic solar cells (hybrid)

Organic and organic-inorganic photovoltaics (PVs) (third generation solar cells) follows the second generation (thin film inorganics such as amorphous silicon (a-Si), cadmium telluride (CdTe), and copper indium gallium selenide (CIGS)) and first generation (semiconducting, crystalline) PVs. Third generation solar cells continues to attract great attention from the PV community, due to their promising features such as low fabrication cost, flexibility and light weight. The organic PVs include devices with flat and bulk heterojunction between the various types of conjugated polymers, small molecules, fullerene derivatives, and carbon nanotubes. Organic PVs are still unable to overcome the high 6–7% barrier of conversion efficiency, despite considerable progress in solar cell architecture, design and rational choice of the donor-acceptor materials, [19].

The heterojunction in organic-inorganic hybrid solar cells is formed between inorganic semiconductors and organic compounds (small molecules, oligomers, polymers, carbon nanotubes). The hybrid PVs has higher carrier mobility of the semiconductor and the light absorption at longer wavelengths than for organic compounds. Whereas, the organic component allows hybrid solar cells to be superior over conventional semiconducting PVs in terms of cost efficiency, scalable wet processing, the variety of organic materials (mismatch between inorganic components can be minimized or prevented), light weight, and flexibility. The progress in advanced semiconducting nanostructures in combination with organic nanomaterials (fullerenes and carbon nanotubes) opens the way to overcome the 10% barrier of conversion efficiency for hybrid solar cells. Although the band engineering of the hybrid solar cell is not as facile as for semiconducting PVs, but it is a useful instrument in the design of the hybrid solar cell architecture. For example, the chemical functionalizing of the organic component effects on the band gap energy and position of the Fermi level for conducting polymers and small molecules. **Figure 4** illustrates the types of hybrid PVs depending on the nature of organic and inorganic component and the morphology of the devices [20].

One potential alternative to crystalline silicon PV cells is cells made from thin films (<1 μm) of conjugated (semiconducting) polymers, which can easily be cast onto flexible substrates over a large area using wet-processing techniques. Polymer or hybrid solar cells often utilize a nanostructured interpenetrating network of electron-donor and electron-acceptor materials. The hybrid polymer solar cells using blends of the conjugated polymer and inorganic materials to convert sunlight into electricity. These devices will combine the advantages of two materials, high electron mobility and photosensitivity of inorganic semiconductors, and high hole

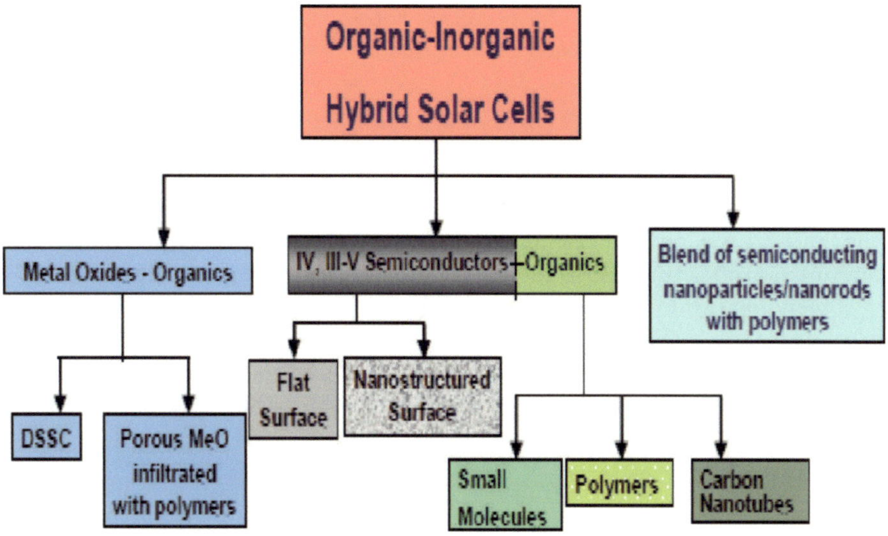

Figure 4.
Classification of hybrid solar cells [20].

mobility of conjugated polymers. Due to the poor interfacial junction between the organic and inorganic materials, the power conversion efficiency of the hybrid photovoltaic devices is still very low. Improving the heterojunction between two materials is concentrated by many researchers. Therefore, an important method is to use two materials having complementary operation of the p-type and n-type electronic properties for the hybrid polymer solar cell [21].

In bulk heterojunction, electron accepting nanoparticles are mixed with the electron donating polymer and the exciton created in the polymer material diffuse to the donor-accepter interface for charge separation. Bulk heterojunction solar cell are preferred than multilayer or heterojunction polymer solar cells, because of the binding energy of the polymeric excitons, which is in the range of 0.2 eV-0.4 eV and that is considerably higher than the binding energy for inorganic semiconductor materials. Also, the life time of the exciton in the conjugated polymer is about sub-nanoseconds and the small diffusion range which is about (5–10) nm. After absorption of light and for efficient charge generation, each exciton has to find a donor – acceptor interface within femtoseconds within few nano-meter, otherwise it will be lost without charge generation. Because of these properties, a poor efficiency results by the heterojunction or multilayer organic photovoltaic devices. To solve this problem the semiconducting nanoparticles are incorporated into the polymer matrices since polymer materials phase separate on a nanometer dimension. The Junctions throughout the bulk of the material are created due to the mixing of the p and n type materials that ensure quantitative dissociation of photo generated excitons irrespective of the thickness. In such hybrid materials, organic polymer acts as electron donor and the inorganic nanoparticles as an electron acceptor. So that the positive charges move by hole hopping, and the negative charges by electron hopping via charge transfer between molecules. It is obvious from this mechanism; that both the organic and inorganic materials contribute to the photocurrent in hybrid bulk heterojunction cells [22].

Organic-inorganic hybrid solar cells are typically thin film devices consisting out of photoactive layer(s) between two electrodes of different work functions. The anode which is often with high work function is a conductive and transparent indium tin oxide (ITO) on a flexible plastic or glass substrate. The photoactive light absorbing thin film consists out of a conjugated polymer as organic part and an inorganic part out of e.g. semiconducting nanocrystals [23]. A large number of various semiconductors have been investigated for organic–inorganic hybrid solar cells. For example, titanium dioxide (TiO_2) nanoparticles have been extensively employed due to their high power conversion efficiency. Also, Zinc sulfide (ZnS) is another promising semiconductor, but it has been less studied. A top metal electrode (e.g. Al, LiF/Al, Ca/Al) is vacuum deposited onto the photoactive layer finally. **Figure 5** is a schematic illustration of a typical device structure. For photoactive layers there are two different structure types, the bilayer structure and the bulk heterojunction structure. The bulk heterojunction structure is usually realized by just blending the donor and acceptor materials and depositing the blend on a substrate [24].

The charge separation process at the donor- acceptor interface in a hybrid solar cell is shown in **Figure 6**. The major photovoltaic steps include: photo-excitation

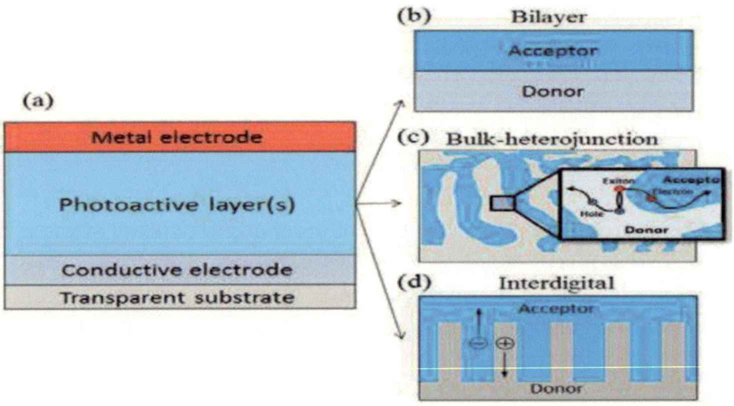

Figure 5.
A schematic illustration of typical device structures for hybrid solar cells [24].

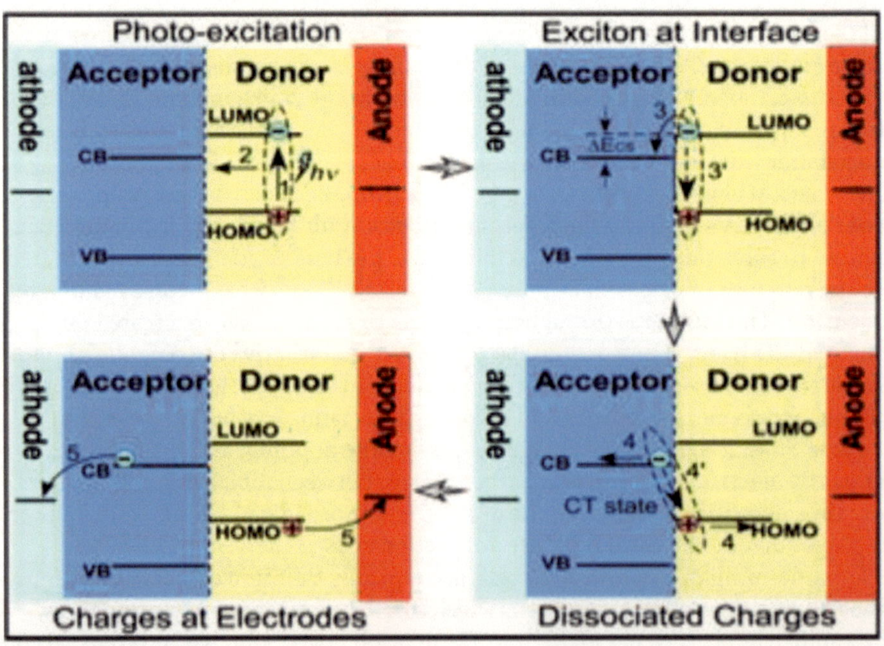

Figure 6.
A schematic diagram of the charge separation process at the donor-acceptor interface in a hybrid solar cell [25].

into excitons (1), excitons migration to interfaces (2), charge transfer from the donor to the acceptor at the interface (3), charge migration to electrodes (4) and charge injections into electrodes. The recombination of the excitons in the donor and separated charges at the interfaces are the processes which resulting in the reduction of the photovoltaic conversion efficiency [25].

4. Conclusions

Nano-science is defined as the study of small dimensions materials that exhibit remarkable properties, functionality and phenomena.

Nanotechnology controls current progress in chemistry, physics, material science, biotechnology and electronics to create novel materials that have unique properties because their structures are determined on the nanometer scale [26]. The world needs to curb CO_2 emissions soon and reduce our dependence on expensive hydrocarbons; thus, a renewable materials and solar energy on a massive scale are required. Therefore, flexible photovoltaic modules will be among the main tools used to get rid of this dependency.

This chapter has highlighted the advancements that have been made for the fabrication of flexible solar cell and the progress in this field, aimed at facilitating diffusion of these technologies [2].

Author details

Ghaida Salman Muhammed
Department of Physics, College of Science, University of Baghdad, Baghdad, Iraq

*Address all correspondence to: ghaidasalman7@gmail.com

References

[1] Luke F. Lester, Kai Yang, Mohamed A. El-Emawy, Therese Saiz, Ryan A. Clark, Andrew Hollowell, Olga Lavrova: Flexible Solar Cells for Micro-Autonomous Systems Technology. Proc. of SPIE. 2012; 7679: 1-8. DOI: 10.1117/12.852598.

[2] Mario Pagliaro, Rosaria Ciriminna, Giovanni Palmisano: Flexible Solar Cells. Chem SusChem. 2008; 1: 880-891. DOI: 10.1002/cssc.200800127.

[3] A.N. Tiwari, A. Romeo, D. Baetzner, H. Zogg: Flexible CdTe Solar on Polymer Films. Prog. photovolt: Res. Appl. 2001; 9: 211-215.

[4] Qingfeng Su1, Jianming Lai1, Genfa Zhang, Shijun Feng, Weimin Shi: Flexible Thin Film Solar Cells Using in the Car. World Electric Vehicle Journal. 2010; 4: 793-797. ISSN 2032-6653.

[5] Davood Shahrjerdi, Stephen W. Bedell , Can Bayram, Cristina C. Lubguban , Keith Fogel, Paul Lauro, John A. Ott, Marinus Hopstaken, Michael Gayness, Devendra Sadana: Ultralight High-Efficiency Flexible InGaP/(In)GaAs Tandem Solar Cells on Plastic. Adv. Energy Mater. 2013; 3: 566-571. DOI: 10.1002/aenm.201200827.

[6] Helena Gleskova, I-Chun Cheng, Sigurd Wagner, James C. Sturm, Zhigang Suo: Mechanics of thin-film transistors and solar cells on flexible substrates. Solar Energy. 2006; 80:687-693. DOI:10.1016/j.solener.2005.10.010.

[7] A. Romeo, G. Khrypunov, F. Kurdesau, D.L. Bätzner, H. Zogg, A.N. Tiwari: High Efficiency Flexible CdTe Solar Cells on Polymer Substrates. Technical Digest of the International. 2004; 14: 715-716.

[8] Dr. V.K. Sethi, Dr. Mukesh Pandey, Ms. Priti Shukla: Use of Nanotechnology in Solar PV Cell. International Journal of Chemical Engineering and Applications. 2011; 2: 77-80. DOI: 10.1155/8293.

[9] Askari. Mohammad Bagher: Introduction to Organic Solar Cells. Sustainable Energy. 2014; 2: 85-90. DOI: 10.12691/rse-2-3-2.

[10] A. Alias, Z. Zabidi, A. Ali, M. Harun: Optical characterization and properties of polymeric materials for optoelectronic and photonic applications. International journal of applied science and technology. 2013; 3: 11-38.DOI: 10.30845/ijast

[11] Thamraa Mohammed Alshahrani, "Fabrication and Charactrasation of PAMAM-Based Organic Photovoltics [thesis]. UK: Bangor University; 2016.

[12] Pang-Leen Ong 1 and Igor A. Levitsky: Organic/IV, III-V Semiconductor Hybrid Solar Cells. Energies. 2010; 3: 313-334. DOI: 10.3390/en3030313.

[13] Sandro Lattante: Electron and Hole Transport Layers: Their Use in Inverted Bulk Heterojunction Polymer Solar Cells. Electronics. 2014; 3: 132-164. DOI: 10.3390/electronics3010132.

[14] Kevin M. Kevin M. Coakley and Michael D. McGehee: conjugated polymer photovoltaic cells. American Chemical Society, mmcgehee@stanford.edu.

[15] Hwajeong Kim, Sungho Nam, Jaehoon Jeong, Sooyong Lee, Jooyeok Seo, Hyemi Han, Youngkyoo Kim: Organic solar cells based on conjugated polymers: History and recent advances. Korean J. Chem. Eng. 2014; 31: 1095-1104. DOI: 10.1007/s11814-014-0154-8.

[16] Alex C. Mayer, Shawn R. Scully, Brian E. Hardin, Michael W. Rowell,

Michael D. McGehee: Polymer-based solar cell. Materials today. 2007; 10: 28-33. ISSN: 1369 7021.

[17] Michael D. McGehee," Nanostructured Organic–Inorganic Hybrid Solar Cells", MRS BULLETIN, Vol. 34, p.p.95-100, 2009. www.mrs.org/bulletin.

[18] Xiong Gong, Ming-Hong Tong, Sung Heum Park, Michelle Liu, Alex Jen, Alan J. Heeger: Semiconducting Polymer Photodetectors with Electron and Hole Blocking Layers: High Detectivity in the Near- Infrared. Sensors. 2010; 10: 6488-6496. DOI: 10.3390/s100706488.

[19] Pang-Leen Ong, Igor A. Levitsky: Organic/IV, III-V Semiconductor Hybrid Solar Cells. Energies. 2010; 3: 313-334. DOI: 10.3390/en3030313.

[20] Gledhill. S.E., Scott. B., Gregg. B.A.: Organic and nano-structured composite photovoltaics: An overview. J. Mater. Res. 2005; 20: 3167-3179.

[21] Kin-Tak Lam, Yu-Jen Hsiao, Liang-Wen Ji, Te-Hua Fang, Wei-Shun Shih, Jun-Nan Lin: Characteristics of Polymer-Fullerene Solar Cells with ZnS Nanoparticles. Int. J. Electrochem. Sci. 2015; 10: 3914-3922. (http://creativecommons.org/licenses/by/4.0/).

[22] T. Abdul Kareem, A. Anu Kaliani: Fabrication and Characterization of ZnS Cubic: P3HT, ZnS Hexa: P3HT and ZnS Hexa: P3HT: PVA-Ag Bulk Heterojunction Solar Cells. journal of nano and electronic physics. 2015; 7: 1-6.

[23] Yunfei Zhou, Michael Eck, Michael Krüger. Organic-Inorganic Hybrid Solar Cells: Prof. Leonid A. Kosyachenko, editors. Solar Cells - New Aspects and Solutions; 2011. 512 pages. ISBN 978.

[24] M. Sookhakian , Y.M. Amin, S. Baradaran, M.T. Tajabadi, A. Moradi Golsheikh, W.J. Basirun: A layer-by-layer assembled graphene/zinc sulfide/polypyrrole thin-film electrode via electrophoretic deposition for solar cells. Thin Solid Films. 2014; 552: 204-211.

[25] Ruchuan Liu: Hybrid Organic/Inorganic Nanocomposites for Photovoltaic Cells. Materials. 2014; 7: 2747-2771.DOI:10.3390/ma7042747.

[26] T. Satyanarayana1, S. Sudhakar Reddy: A Review on Chemical and Physical Synthesis Methods of Nanomaterials. International Journal for Research in Applied Science & Engineering Technology. 2018; 6: 2885-2889. DOI:10.22214/ijraset.2018.1396.

Section 4

Solar Energy Technologies

Chapter 21

Advanced Laser Processing towards Solar Cells Fabrication

Jhantu Kumar Saha and Animesh Dutta

Abstract

The ultra-short pulse laser has the potential in selective nano-structuring of thin-films layers by adjusting the wavelength of laser radiation depending on optical properties of the thin- film and the substrate that will solve its efficiency and stability issues in a one-step process, which is a promising methodology for thin-film solar cell fabrication that are fabricated through a sequence of vapor deposition and scribing processes. The review is performed to further understand the structure of the laser modified surface and the nature of dopants and defects in the crystalline grains. Using low temperature studies, the electronic levels of the dopant and its configuration with the lattice could be probed. The review is also explored the concept of using thin films of silicon as the laser irradiation substrate and for enhanced the visible and infrared absorption of films of silicon with thicknesses of few micrometer. Although the review is made good progress studying the properties of new material and incorporation into device but there are many unanswered questions and exciting avenues of research are also explored with femtosecond laser irradiated silicon.

Keywords: femtosecond laser, photovoltaics, silicon, thin-film, intermediate band solar cells

1. Introduction

Photovoltaics (PV), the conversion of sunlight to electricity, is a promising technology that could allow for the generation of electrical power on a very large scale and contribute considerably to solving the energy problem that the next generation must face. The factors motivating the solar cell research are not only to reduce cost of the manufacturing cost of solar cell technology but also increase to the efficiency of solar cell.

Research on solar cells falls into two general categories, both aimed at reducing the cost per kilowatt-hour. The first category (eg. single crystalline Si and GaAs solar cells) involves using expensive materials and advanced processing techniques to obtain the highest possible efficiency. The increased efficiency will hopefully offset the extra cost. The second category (Poly and thin-film Si, CdTe, and $CuInSe_2$ solar cell) involves using cheaper materials and cheaper processes [1–5]. The lower quality material sacrifices efficiency, but this is hopefully offset by its low cost.

Crystalline silicon solar cells are transparent to wavelengths of light longer than 1.12 μm, due to their electronic band gap of 1.07 eV means they are transparent to 23% of solar energy. Whereas thin film amorphous silicon solar cells have a larger band gap of 1.75 eV and are transparent to light longer than 0.71 μm means they loss 53% of solar energy.

From a photovoltaic standpoint, the most attractive property of femtosecond laser irradiated silicon is that it absorbs nearly all light that is emitted by the sun. This offers an opportunity to tap into that lost energy and, therefore, appears to be an attractive option for solar cells. The ultra-short pulse laser has the potential to improve the optical properties of different layer of solar cell. There are many defects in the laser modified surface, thus, it is unlikely that femtosecond laser irradiated silicon will be able to improve upon the already high efficiency of single crystal silicon solar cells (25%) or even improve upon the lower efficiency of polycrystalline silicon solar cells (14%).

Thin-film solar cells have stimulated enormous research interest as a cheap alternative to bulk crystalline silicon solar cells [6–9]. The limitation of all thin-film solar cells, made from a variety of semiconductors, is that the absorbance of the near band gap is small, especially for the indirect band gap semiconductor silicon. Therefore, structuring the thin-film solar cell so that light is trapped inside to increase the absorbance is very important. On the other hand, femtosecond laser irradiated silicon can be used as a photovoltaic device; it can convert wavelengths of light that are not normally absorbed by silicon into an electrical signal. Over the past several decades ultra-short, pulsed laser irradiation of silicon surfaces has been an active area of materials science research [10–17]. The ultra-short duration of the laser pulses leads to extremely high energy densities in the material. The real advantage of femtosecond laser irradiated silicon is that it not only absorbs nearly all the wavelengths of light but does so in a laser modified surface film that is less than 500 nm thick. This makes it ideal for incorporation with thin film silicon. However, since thin film silicon already contains a large number of defects and exhibits a much lower efficiency (typically 10%), it would seem to be a good candidate for use with our femtosecond laser irradiation process [12]. There are few high-efficiency PV concepts, photon management for photovoltaics as well as several ways to increase the performance in solar cells such as isotropic acidic texturing, mechanical grooving, reactive ion etching, anisotropic silicon etching, rapid crystallization of amorphous silicon for thin-film silicon solar cell and laser processing for photovoltaics are reviewed in the following section.

2. High-efficiency PV concepts

The current state of the art of high efficiency single-junction monocrystalline silicon (c-Si) based solar cells are the PERL (Passivated Emitter, Rear Locally diffused) cell, the SunPower A300 cell, and the Sanyo HIT cell [18–20]. There are also few high-efficiency PV concepts as reviewed in the following section.

The intermediate band (IB) solar cell concept can be used to increase the efficiency of current of solar cells, ideally is above the limit established by Shockley and Queisser in 1961 but fabricating IB is difficult and no current design has demonstrated the theoretically predicted efficiency improvements [21–24].

Quantum dots (QDs) are nanocrystals of a material immersed in a matrix material usually with a higher bandgap. Quantum dots are also used to make the cells but the efficiencies that have been achieved so far are not yet satisfactory [24]. QDs can also be used in IB solar cell [25].

3. Photon Management for Photovoltaics

Photon management for photovoltaics (PV) is focused for increasing the solar absorbance of solar irradiance through some modification of the photovoltaic material. Very small sized (micrometer-sized) photovoltaic material (eg. silicon) spikes decrease geometric light trapping for reflection. The extent of geometric light trapping enhances optical path length. The light trapping depends on the height, spacing, and subtended angle although structures with large, graded density and the spike heights will not play for visible and near-IR light. Multiple reflection on the surface is the dominant effect of light trapping [26–30]. Treating the case of spikes subtending a cone angle of 42°, observed on a silicon surface after irradiation in pulsed-laser hyper doping and surface texturing as shown in **Figure 1a**. Incident light undergoes four reflections before escaping as illustrated in **Figure 1b** of the optical path of light incident on laser-textured silicon surfaces, with cones subtending 42°.

There are also several ways to increase the performance in solar cells as reviewed in the following section.

3.1 Isotropic acidic texturing

The novel technique of isotropic texturing of the multi-crystalline surface basically use acidic solution (such as HNO3–HF–CH3COOH) followed by a simple chemical treatment to make the surface more uniform in terms of roughness [31]. Due to high reflectivity of acid textured surface helps to improve the open circuit voltage but gives lower short circuit current of the multi-crystalline silicon (mc-Si) solar cell [32].

3.2 Mechanical grooving

The mechanically texturized structures created by anisotropic etching of mono-crystalline silicon, resulted the first silicon solar cell to exceed 20% energy conversion efficiency at 1-sun illumination because of positive trend of improvement in the electronic quality of the c-Si substrates [33].

3.3 Reactive ion etching (RIE)

Texturing surface have been developed using the reactive ion etching (RIE) method, which is expected to form a low reflectance surface on various crystalline orientations of grains [34, 35].

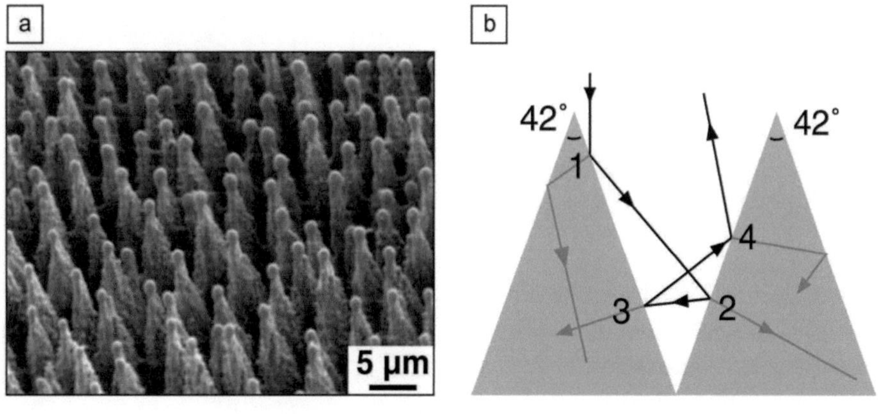

Figure 1.
Pulsed-laser hyper doping and surface texturing, Cambridge (2001) (ref. R. Younkin, PhD dissertation, Harvard University), (a) scanning electron micrographs, (b) illustration of the optical path (adapted from [31] with permission).

3.4 Anisotropic silicon etching

Textured silicon surface with 100% pyramid density can be obtained on the surface texturization of monocrystalline wafers with solutions containing sodium-hydroxide and isopropanol [36].

3.5 Rapid crystallization of amorphous silicon for thin-film silicon solar cell

The novel crystallization technique for synthesizing crystalline Si film from a-Si film utilizing a VHF thermal micro-plasma jet is used as shown in **Figure 2** [37]. **Figure 3** shows the photocurrent–voltage characteristics and collection efficiency for p–i–n Si thin-film solar cells with crystallized Si films as an intrinsic layer [37].

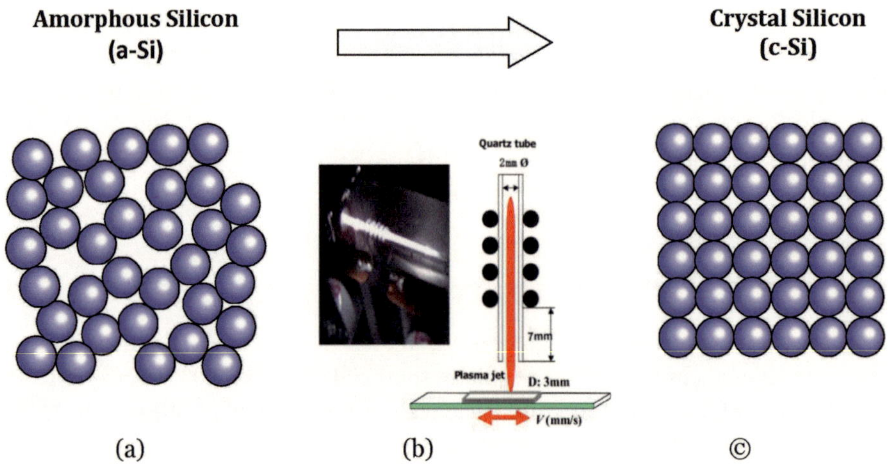

Figure 2.
Rapid crystallization of amorphous silicon utilizing a very-high-frequency micro-plasma jet.

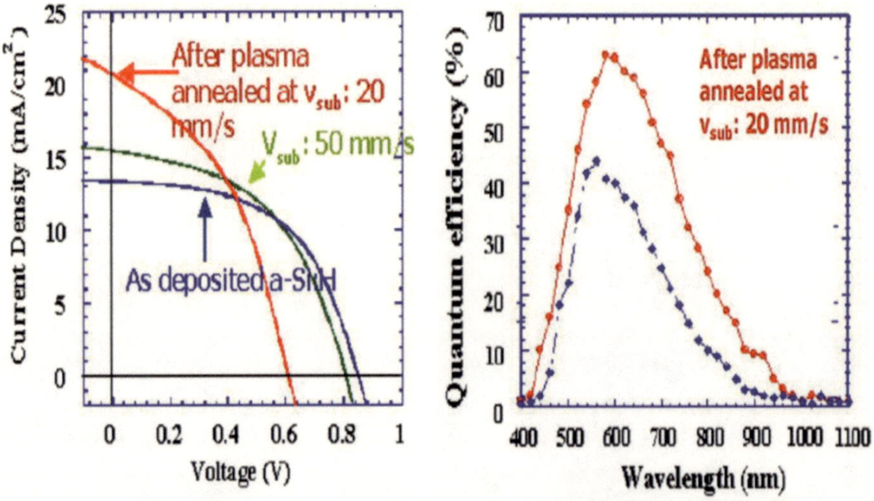

Figure 3.
(a) Photocurrent–voltage characteristics, (b) collection efficiencies spectra of solar cells [adapted from [37] with permission].

4. Laser processing for photovoltaics

The laser modified surface and the nature of dopants and defects in the crystalline grains are crucial to improve the performance of solar cells. Pulsed-laser hyper doping & surface texturing for photovoltaics, laser processing for thin-film (TF) photovoltaic, Light trapping for thin silicon solar cells by Femtosecond Laser Texturing, Patterning of Transparent Conducting Oxide (TCO) layers by Femtosecond Laser as well Solar cells based on laser-modified materials are discussed in the following section.

4.1 Pulsed-laser hyper doping and surface texturing for photovoltaics

The two different approaches eg. pulsed-laser hyper-doping and surface texturing are used to enhance photon absorption enhancement from the pulsed-laser processing of semiconductors with nanosecond, picosecond, or femtosecond laser pulses. The absorptance A is obtained from the expression $A = 1–R–T$, where R and T are reflectance and transmittance, respectively, measured with an integrating sphere to collect both specular and diffuse light. **Figure 4a** shows the untreated crystalline silicon (c-Si) which has negligible absorption of light with a wavelength longer than 1.1 μm due to its energy bandgap. **Figure 4b** shows the pulsed-laser hyper-doping with sulfur enables absorption of sub-bandgap light. **Figure 4c** shows the pulsed laser texturing that enhances above-bandgap light absorption with geometric light trapping. **Figure 4d** shows the broadband near-unity absorption is achieved with both pulsed-laser hyper-doping and surface texturing.

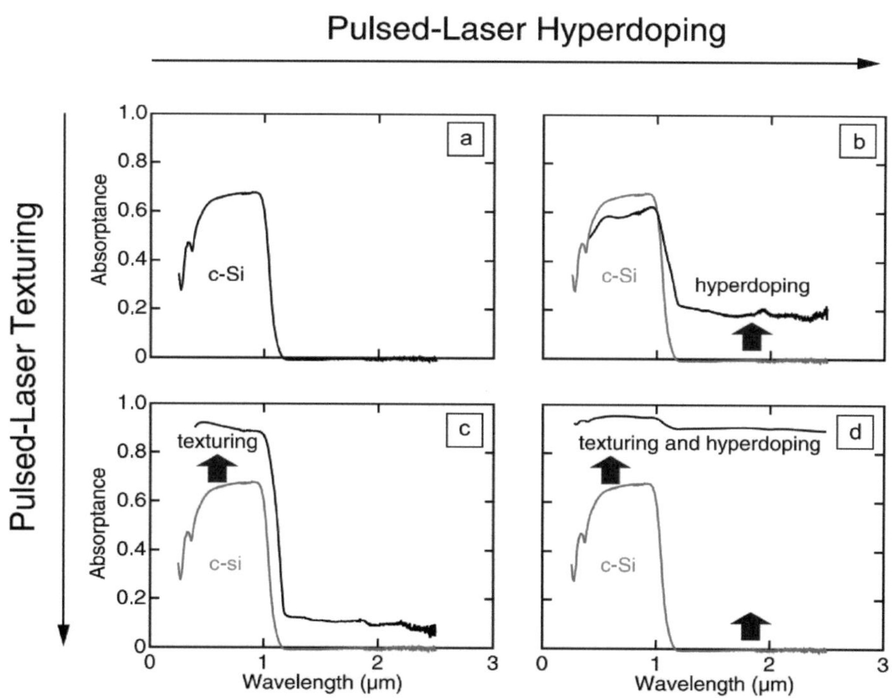

Figure 4.
(a) Untreated crystalline silicon (c-Si), (b) pulsed-laser hyper-doping with sulfur, (c) pulsed laser texturing, (d) both pulsed-laser hyper-doping and surface texturing (adapted from [31] with permission).

Table 1 compares various texturing methods and reflectivity values reported for c-Si and mc-Si wafers. While anisotropic chemical etching method using KOH or NaOH with IPA is applicable for c-Si materials, it could not be applied to multi-crystalline materials due to the anisotropic nature of the chemical etchant. Isotropic chemical texturing uses acidic mixture of HF and HNO3 and organic additives for multi-crystalline silicon (mc-Si). On the other hand, lasers are unique energy sources and laser ablation is an isotropic process. Lasers could texture surfaces by selectively removing materials by ablation process. Texturing could be achieved irrespective of the crystallographic orientation of material surface. Under shorter pulse regimes (few nanoseconds to femtoseconds), a very different types of self-assembled micro/nano structures are formed [32].

Total reflection of as-laser-treated samples is very low and increases by a few more percentages after post-chemical cleaning as shown in reference [42].

4.2 Laser processing for thin-film (TF) photovoltaic

The serial monolithic interconnection of thin-film solar cell can be achieved by laser scribing three patterns during fabrication. The layers are numbered in the order in which they are deposited. They are micromachined by laser ablation, an established material removal process [35].

Nayak et al. [37] reported nanocrystalline Si material following femtosecond-laser-induced crystallization of a-Si:H. Despite the number of structural defects, which for the time being prohibits PV applications, the process produces remarkable light-trapping microstructures at the surface.

Author	Year	Technique	Substrate used	Applicable to	Approx. R (%) @550 nm (as textured)
Gangopadhay et al.	2017	Isotropic acidic texturing	mc-Si	mc-Si	15
Zechner et al.	1997	Mechanical grooving	mc-Si	c-Si & mc-Si	>15
Inomata et al.	1997	RIE	mc-Si	c-Si & mc-Si	<2
Vazsonyi et al.	1999	NaOH+IPA	c-Si	c-Si	10
Nashimoto et al.	2000	Na_2O_3	c-Si	c-Si	10
Abbott et al.	2006	Laser texture	c-Si	c-Si & mc-Si	<5
Nishioka et al.	2009	Ag nanoparticle	c-Si	c-Si & mc-Si	<5
Branz et al.	2009	Au nanoparticle	c-Si	c-Si & mc-Si	<2
Younkin et al.	2003	Femtosecond laser-induced microstructure	c-Si	c-Si & mc-Si	<3
Nayak et al.	2010	Ultrafast laser micro/nano structure	c-Si	c-Si, mc-Si, and thin a-Si	<3

Table 1.
Various Texturing Methods and Reflectivity values reported for mono-crystalline (c-Si) and multi-crystalline (mc-Si) wafers (Adapted/modified from references [32, 36–42]).

4.3 Light trapping for thin silicon solar cells by femtosecond laser texturing

A variety of surface morphologies can be obtained from fs laser treatments, depending on laser parameters and the ambient gas environment. The efficacy fs laser texturing of solar cell devices is also demonstrated in ref. 26 and the problem of laser-induced damage may have significantly improved cell performance in the future by increased absorption (vs. un-textured cells) for infrared photons, due to enhanced light-trapping.

4.4 Patterning of transparent conducting oxide (TCO) layers by femtosecond laser

A method for patterning crystalline indium tin oxide (c-ITO) patterns on amorphous indium tin oxide (a-ITO) thin films is proposed by femtosecond laser irradiation at 80 MHz repetition rate. The laser patterning technique provides a versatile and highly precise means of fabricating the transparent electrode structures required in a wide range of modern optoelectronic devices. High repetition rate femtosecond (80 MHz) laser-induced crystallization and proposed laser patterning technique provides a versatile and highly precise means of fabricating TCO structures [42].

4.5 Femtosecond laser induced crystallization & simultaneous formation of light-trapping nanostructures

Femtosecond laser induced crystallization & simultaneous formation of light-trapping nanostructures is a one-step laser process, which could lead to fabricate the highly efficient solar cells [42–51]. Mmicrostructures and small spikes have been spontaneously formed upon laser treatment. Interestingly the a-Si:H films turned completely dark from an original shiny reddish gray color. A similar effect has been extensively studied by Mazur's group and others in crystalline bulk silicon wafers [42–52].

5. Solar cells based on laser-modified materials

Solar cells based on laser-modified materials focus on three major thrusts that will lead to more efficient and economic thin-film solar cell fabrication by (i) combining femtosecond laser irradiation processing of a-Si:H surface and simultaneous crystallization occurs in a one step process [16]. Optical absorption will be enhanced by light trapping via multiple reflections through the surface geometry changes, and the formation of mixture of μc-Si:H and a-Si:H after crystalline suggests that the overall stability will be potentially increased; (ii) Laser with a shorter, femtosecond pulse duration will be applied for nano-structuring of TCO deposited on glass as a plasmonic nanostructure for efficient light trapping; (iii) For scribing thin-film solar cells with femtosecond laser will be applied for electrical isolation, hermetic sealing of the module, glass cutting, the complete removal of all layers from the edges of fully processed thin-film solar cells on glass substrates [52]. **Figure 5** is a schematic presentation of how femtosecond laser irradiated silicon can be incorporated into the p-i-n configuration of a thin-film silicon solar cell.

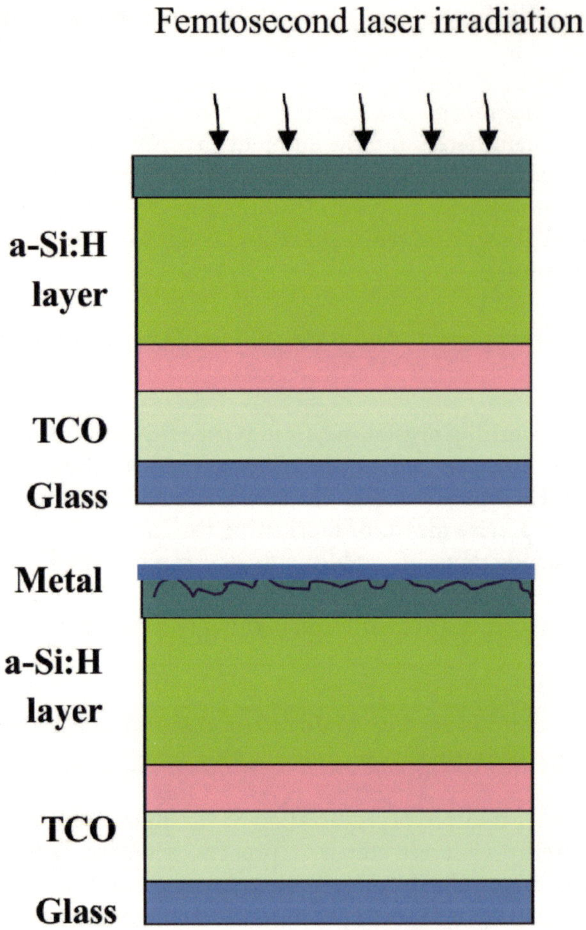

Figure 5.
Schematic cross-section of the fabrication process of a femtosecond laser irradiated thin film silicon solar cell.

6. Conclusions and outlook

Pulsed laser processing provides two routes for effective photon management: Surface texture and hyper doping process are distinct and independently achievable. Surface texture using intense pulsed-laser light to create quasi-periodic surface features reduces reflection and increases path length through the material.

The creation of femtosecond laser irradiated thin film silicon solar cells is an exciting area of research. Thin-film solar cell requires a highly efficient light-trapping design to absorb a significant fraction of the incident sunlight and material property changes to increase stability. Laser based treatment of thin-film is required for resolve efficiency and stability issues in a one-step process, which is a promising methodology for thin-film solar cell fabrication. Laser with a shorter, femtosecond pulse duration will be applied for nano-structuring of TCO deposited on glass as a plasmonic-nanostructure for efficient light trapping. The ultra-short pulse of femtosecond laser at small fluences will also overcome the parasitic losses and decreased electrical power output of the solar module resulting from the active material remaining over the active layer scribing with picosecond laser of thin-film solar cells which is critical in forming the series interconnects between cells.

Thin-film materials eg. Si, CdTe, $CuInSe_2$ are becoming more and more attractive based on their potential for low-cost solar modules, possibly to create tandem junctions and large-scale manufacturing. They can reduce the cost of the material at the expense of efficiency. The a-Si:H is the most popular material for use in thin film form due to its low energy economy (watt/cost). But due to their instability and low efficiency, a thin-film a-Si:H solar cell requires a highly efficient light-trapping design to absorb a significant fraction of the incident sunlight and material property changes to increase stability. On the other hand, microcrystalline silicon is one of the promising materials for thin-film solar cells of achieving high conversion efficiency. In addition, microcrystalline silicon films show enhanced carrier mobility, excellent stability against light-induced degradation and improved longer wavelength response. But deposition rate of microcrystalline silicon thin-film fabricated by conventional PECVD is lower compared with the a-Si:H. Therefore, laser-based treatment of a-Si:H is required to resolve its efficiency and stability issues in a one-step process, which is a promising methodology for thin-film solar cell fabrication.

Author details

Jhantu Kumar Saha* and Animesh Dutta
School of Engineering, University of Guelph, Canada

*Address all correspondence to: jsaha@uoguelph.ca

References

[1] Green MA, Dunlop ED, Ebinger JH, Yoshita M, Kopidakis N, and Ho-Baillie AW, Solar cell efficiency table (version 55), Progress in Photovoltaics: Research and Applications, 28, 3 (2019). DOI: https://doi-org.subzero.lib.uoguelph.ca/10.1002/pip.1021

[2] Guha S, Yang J, Nath P, and Hack M, Enhancement of open circuit voltage in high efficiency amorphous silicon alloy solar cells, Applied Physics Letters, 49, 218 (1986). DOI: https://doi.org/10.1063/1.97176

[3] I. Gordon, L. Carnel, D. Van Gestel, G. Beaucarne and J. Poortmans, 8% Efficient thin-film polycrystalline-silicon solar cells based on aluminum-induced crystallization and thermal CVD, Progress in Photovoltaics: Research and Applications, 15, 575 (2007). DOI: 10.1002/pip.765.

[4] Repins, M. Contreras, *M. Romero*, Y. Yan, W. Metzger, J. Li, S. Johnston, B. Egaas, C. DeHart, J.Scharf, B. McCandless, R. Noufi, Characterization of 19.9%-efficient CIGS absorbers, *Photovoltaic Specialists Conference*, 2008. *PVSC '08. 33rd IEEE*, 11-16 May 2008, San Diego-CA, P. 1-6

[5] Wu X, Keane JC, Dhere RC. DeHart C, Duda A, Gesert TA, S. Asher, DH Levi, P. Sheldon, Time-resoloved photoluminescence studies of CdTe solar cells, Proceedings of 17th European Photovoltaics conference, Munich, October 2001, p. 995-1000. https://doi.org/10.1063/1.1597974

[6] Saha JK, Haruta K, Yeo M and Shirai H, "Rapid crystallization of amorphous silicon utilizing very-high-frequency plasma jet for Si thin-film solar cells", Solar Energy Materials & Solar Cells 93 (2009) 1154-1157. DOI: 10.1016/j.solmat.2009.03.001

[7] Chen X, Jia B, Saha JK, Cai B, Stokes N, Qiao Q, Wang Y, Shi Z and Gu M, Broadband enhancement in thin-film amorphous silicon solar cells enabled by nucleated silver nanoparticles, Nano Lett. 12 (5) (2012). doi: 10.1021/nl203463z

[8] Kanemitsu P, Nakada I, and Kuroda H, Picosecond laser induced anomalous crystallization in amorphous silicon, Appl. Phys. Lett. **47**, 939 (1985). DOI:10.1063/1.95934

[9] Saha, JK, Ohse N, Kazu HK, Tomohiro K, and Hajime Shirai, "Fast deposition of microcrystalline Silicon films from SiH_2Cl_2 using a high-density microwave plasma source for Si thin-film solar cells", Solar Energy Materials & Solar Cells 94 (2010) 524-530. doi.org/10.1016/j.solmat.2009.11.017

[10] Bartlome R, Strahm B, Sinquin Y, Feltrin A and Ballif C, Laser applications in thin-film photovoltaics, Appl Phys B 100, 427-436 (2010). DOI:10.1007/s00340-009-3890-4

[11] Carey J E, Femtosecond-laser Microstructuring of Silicon for Novel Optoelectronic Devices, PhD thesis, Harvard University, 2004

[12] Tull B R, Femtosecond Laser Ablation of Silicon: Nanoparticles, Doping and Photovoltaics, Harvard University, 2007

[13] Wang X C, Zheng H Y, Tan C W, Wang F, Yu H Y and Pey KL, Femtosecond laser induced surface nanostructuring and simultaneous crystallization of amorphous thin silicon film, Optical Express, 18 (18), 19379-19385 (2010). DOI:10.1364/OE.18.019379

[14] Herman J, Benfarah M, Bruneau S, Axente E, Coustillier G, Itina T, Guillemoles J F, and Alloncole P, Comparative investigation of solar cell thin film processing using nanosecond

and femtosecond laser, J. Phys. D: Appl. Phys. 39, 453-460 (2006). DOI:10.1088/0022-3727/39/3/005

[15] Wang H, Kongsuwan P, Satoh G, Lawrence Y, Femtosecond laser-induced surface texturing and crystallization of a-Si:H thin film, Int J Adv Manuf Technol 65,1691-1703 (2012). DOI:10.1007/s00170-012-4291-0

[16] Zhao J, Wang A, and Green MA, "High-efficiency PERL and PERT silicon solar cells on FZ and MCZ substrates," *Solar Energy Materials & Solar Cells,* vol. 65, 2001, pp. 429-435. https://doi.org/10.1016/S0927-0248(00)00123-9

[17] SunPower, "SunPower Announces World-Record Solar Cell Efficiency," May 2008. [Online]. Available: http://investors.sunpowercorp.com/releasedetail.cfm?ReleaseID=309613

[18] SANYO, "SANYO Develops HIT Solar Cells with World's Highest Energy Conversion Efficiency of 23.0%, May 2009. [Online]. Available: http://sanyo.com/news/2009/05/22-1.html

[19] Shockley W, Queisser H J, Detailed balance limit of efficiency of p-n junction solar cells, J. Appl. Phys. 32, 510 (1961). DOI:10.1063/1.1736034.

[20] Araujo G L, Marti A, Absolute limiting energy efficiencies for photovoltaic energy conversion, Solar Energy Mater. Solar Cells 33, 213 (1994). DOI:10.1016/0927-0248(94)90209-7

[21] Luque A, Marti A, Increasing the efficiency of ideal solar cells by photon induced transitions at intermediate levels, Phys. Rev. Lett. 78, 5014 (1997). DOI: 10.1103/PhysRevLett.78.5014.

[22] Luque A, Marti A, The Intermediate Band Solar Cell: Progress Toward the Realization of an Attractive Concept, Advanced Materials, 22, 160-174, (2010), DOI: 10.1002/adma.200902388.

[23] Marti A, Antolin E, Cánovas E, Lopez N, Linares P G, Luque A, Stanley R, Farmer C D, Elements of the design and analysis of quantum-dot intermediate band solar cells, Thin Solid Films 2008, 576, 6716. DOI: 10.1016/j.tsf.2007.12.064.

[24] Luque A, Martı´, Stanley AC, Lo'pez N, Cuadra L, Zhou D, Mc-Kee A, General equivalent circuit for intermediate band devices: Potentials, currents and electroluminescence, Journal of Applied Physics 96, 903 (2004). DOI:10.1063/1.1760836.

[25] Marti A, Lopez N, Antolin E, Canovas E, Luque A, Stanley CR, Farmer CD, Diaz P, Emitter degradation in quantum dot intermediate band solar cells, Appl. Phys. Lett. 2007, 90, 233510. DOI: 10.1063/1.2747195.

[26] Luque A, Martı A, Lo'pez. Antolın NE, Canovas E, Stanley CR, Farmer C, Dı'az P, Operation of the intermediate band solar cell under nonideal space charge region, J. Appl. Phys. 2006, 99, 094503. DOI: 10.1063/1.2193063.

[27] Marti A, Antolin E, Canovas E, Lopez N, Linares PG, A. Luque A, Stanley CR, Farmer CD, Elements of the design and analysis of quantum-dot intermediate band solar cells, Thin Solid Films 2008, 516, 6716. DOI: 10.1016/j.tsf.2007.12.064.

[28] Tull BR, Carey JE, Mazur E, McDonald JP, Yalisove SM, Silicon surface morphologies after femtosecond laser irradiation, MRS Bull. 31, 626 (2006), DOI: 10.1557/mrs2006.160.

[29] Sher M, M. Winkler MT, and Mazur E, Pulsed laser hyperdoping and surface texturing for photovoltaics, MRS Bull. 36, 626 (2011). DOI:10.1557/mrs.2011.111.

[30] Gangopadhyay U, Dhungel S K, Basu P K, Dutta S K, Saha H, Yi J,

Comparative study of different approaches of multicrystalline silicon texturing for solar cell fabrication. DOI: 10.1016/j.solmat.2006.08.011.

[31] Blakers AW, Green MA, "20% Efficiency Silicon Solar Cells", Appl. Phys. Lett. 48, 1986, pp.215-217. DOI:10.1063/1.96799.

[32] Inomata Y, Fukui K, Shirasawa K, Surface texturing of large area multicrystalline silicon solar cells using reactive ion etching method, Solar Energy Materials and Solar Cells 48 (1997) 237-242. DOI: 10.1016/S0927-0248(97)00106-2.

[33] Yoo J ,Yu G, Yi J, Large area multicrystalline silicon solar cell fabrication using reactive ion etching (RIE), Solar Energy Materials & Solar Cells 95 (2011) 2-6. DOI: 10.1016/j.solmat.2010.03.029.

[34] Vazsonyi E, Clercq K D, Einhaus R, Kerschaver E V, Said K, Poortmans J, Szlufcik J, Nijs J, Improved anisotropic etching process for industrial texturing of silicon solar cells. Solar Energy Materials & Solar Cells 57 (1999) 179—188. DOI:10.1016/S0927-0248(98)00180-9.

[35] Saha J K, Shirai H, Rapid crystallization of amorphous silicon utilizing a very-high-frequency micro-plasma jet for silicon thin-film solar cells, Vol. 93, issues 6-7, 1154 (2009). DOI: 10.1016/j.solmat.2009.03.001.

[36] Gangopadhay U, Dhungel S K, Basu P K, Dutta S K, Saha, H, Yi J, Comparative study of different approaches of multicrystalline silicon texturing for solar cell fabrication, Solar Energy Materials and Solar Cells, 91(4), 285-289 (2007). DOI: 10.1016/j.solmat.2006.08.011

[37] Zechner C, Hahn G, Jooss W, Wibral M, Bitnar B, Keller S, Spiegel M, Fath P, Willeke G, Bucher E., Systematic study towards high efficiency multicrystalline silicon solar cells with mechanical surface texturization, Conference Record of the 26th IEEE Photovoltaic Specialists Conference 1997, 243-246. DOI:10.1109/PVSC.1997.654074

[38] Vázsonyi É, Dücső Cs, Pekker Á, Characterization of the anisotropic etching of silicon in two-component alkaline solution, J. Micromech. Microeng. 17 (2007) 1916-1922. DOI:10.1088/0960-1317/17/9/02

[39] Nishimoto Y, Namba K, Investigation of texturization for crystalline silicon solar cells with sodium carbonate solutions, Solar Energy Materials & Solar Cells 61 (2000) 393-402. DOI: 10.1016/S0927-0248(99)00162-2

[40] Branz H M, Yost V E, Ward S, Jnes K M, To B, Stradins P, Nanostructured black silicon and the optical reflectance of graded-density surfaces, Appl. Phys Lett. 94, (2009), 231121. DOI: 10.1063/1.3152244

[41] Younkun R, Carey J E, Mazur E, Levinson J A, Friend C M, Infrared absorption by conical silicon microstructures made in a variety of background gases using femtosecond-laser pulses, Journal of Applied Physics 93 (5), 2626-2629. DOI: 10.1063/1.1545159

[42] Nayak B R, Ivyengar V V, Gupta M C, Nayak B R, Ivyengar V V, Gupta M C, Efficient light trapping in silicon solar cells by ultrafast-laser-induced self-assembled micro/nano structures Progress in Photovoltaics: research and applications, 19 (2011), DOI: 10.1002/pip.1067.

[43] Crouch C H, Carey J E, Shen M, Mazur E, Genin F Y, Infrared absorption by sulfur-doped silicon formed by femtosecond laser irradiation, Appl. Phys. A 79, 1635 (2004), DOI: 10.1007/s00339-004-2676-0.

[44] Bartlome R, Strahm B, Sinquin Y, Feltrin A, Ballif C, Laser applications for thin-film photovoltaics, Applied Physics B: Lasers and Optics, 2010, Volume 100, Number 2, Pages 427-436, DOI: 10.1007/s00340-009-3890-4.

[45] Lee B G, Lin Y T, Sher M J, Mazur E, Branz H M, Light Trapping for Thin Silicon Solar Cells by Femtosecond Laser Texturing, presented at the 2012 IEEE Photovoltaic Specialists Conference, Austin, Texas, USA, June 3-8, 2012, pages 001606-001608.

[46] Cheng C W , Lin C Y, Shen W C, Lee Y J, Chen J S, "Patterning crystalline indium tin oxide by high repetition rate femtosecond laser-induced crystallization", Thin Solid Films 518 (2010)7138-7142. DOI: 10.1016/j.tsf.2010.07.025.

[47] Nayak B K, Gupta M C, "Femtosecond-laser-induced-crystallization and simultaneous formation of light trapping microstructures in thin a-Si:H films", Appl. Phys. A 89, 663-666 (2007). DOI: 10.1007/s00339-007-4268-2.

[48] Sher M J, Hammond K, Christakis L, Mazur E, "The photovoltaic potential of femtosecond-laser textured amorphous silicon", Proc. of SPIE Vol. 8608, 2013.DOI: 10.1117/12.2005451.

[49] Her T H, Finlay R J, Wu C, Mazur E, Femtosecond laser-induced formation of spikes on silicon, Appl. Phys. A 70, 383 (2000). DOI: 10.1007/s003390051052.

[50] Tull B R, Carey J E, Mazur E, McDonald J P and Yalisove S M, Silicon surface morphologies after femtosecond laser irradiation, Materials Research Society (Ultrafast Lasers in Materials Research), Volume 31, Issue 8, P626-633, 2006. DOI: 10.1557/mrs2006.160.

[51] Barmina E V, and Shafeev G A, Solar Cells Based on Laser-Modified Silicon, Physics of Wave Phenomena, 2018, Vol. 26, No. 2, pp. 93-98. DOI:10.3103/S1541308X18020036.

[52] Renewable Energy: Laser Assisted Processing of Composite Materials and Solar Cells, Advanced Manufacturing Laboratory, Available from http://www.aml.engineering.columbia.edu/ [Accessed:2020:10.15]

Chapter 22

Application of Solar Energy in Medical Instruments (Microscope)

Badria Ibrahim Eisa Idris, Ahmed Mohamed, Ayah Salah, Osman Abdalrahman Almahdi Alamin, Fatehia Garma and Alnazier Osman

Abstract

Investments in solar PV capacities are now rapidly growing in both grid connected and off grid mode. Solar generation has been a reliable source for supplying electricity in regions without access to the grid for long. Development of renewable energy sources, therefore, has a vast potential in Sudan. Solar energy is a radiant energy which produces by the sun as result of nuclear fusion reaction. Medical services cannot be reached to people in rural areas and war's zones which remotely isolated because of poor road links with the urban centers, and remoteness from the national electrical transmission grid. So, to make the medical services available, a PV encapsulation and manufacturing solar system is used to generate an electric supply which used to supply them, and the microscope's circuit is changed to achieve the required results.

Keywords: Solar energy, Renewable energy, PV system, Microscope

1. Introduction

Solar Energy is radiant energy produced in the sun as a result of nuclear fusion reactions. It is transmitted to the earth through space by electromagnetic radiation [1].

Global installed capacity for solar-powered electricity has seen an exponential growth, reaching around 227 GWe at the end of 2015. It produced 1% of all electricity used globally. Major solar installation has been in regions with relatively less solar resources (Europe and China) while potential in high resource regions (Africa and Middle East) remains untapped [2].

Residents of rural and remote communities continue to show poorer health outcomes than residents in metropolitan centers, while the health of Indigenous communities remains unacceptable. While residents face increasing difficulties in accessing appropriate care in situations where integration and continuity of care are woefully inadequate [3].

In today's climate of growing energy needs and increasing environmental concern, alternatives to the use of non-renewable and polluting fossil fuels have to be investigated, one such alternative is solar energy.

Using a solar energy to operate medical devices one of an important applications to an alternative energy. Development of renewable energy sources, therefore, has a vast potential in Sudan as it known because of great sun peak during the year, in Sudan medical services cannot be reached to people in rural areas and war's zones which remotely isolated because of poor road links with the urban centers, and remoteness from the national electrical transmission grid. So, to make the medical services available, a PV encapsulation and manufacturing solar system is used to generate an electric supply which used in health centers.

Operating a microscope (60 watts) using a solar energy depends on: Time of sun peak in Sudan which is about 9 hours/day generate 6.3 KW, The time of turning on the device which is 6 hours/day and solar modules.

By two ways had been tested:

1. AC method: by converting the DC solar output into AC using an inverter at this process.

2. DC method: by modulating the microscope power supply from AC to DC by canceling the transformer with his protection circuit, and changing the AC 6 volt lamb into DC 12 volt lamb using a Zener diode 12 volt as a voltage regulator.

2. Theoretical backgrounds

2.1 Background of the study

Since 1960's the use of the energy becomes as a necessary part of life. The concept of solar cells takes a place on 1970 when researchers looking for another kind of energy sources which is clean and environment friendly. Many countries use this technology until this moment, but the solar cells produce low amount of energy that makes it use as narrow as seen. The researchers made many research on this field to increase the production of energy from the solar cells, one of these researches is to enhance the silicon capability by using the material science and technologies [4].

2.2 Renewable energy

Renewable energy is energy which comes from natural resources, it flows involve natural phenomena such as sunlight, wind, tides, plant growth, and geothermal heat [4]. It provides 19% of electricity generation worldwide [5]. **Figure 1** shows the different type of renewable energy.

2.3 Solar energy

Solar Energy is radiant light and heat from the sun that is harnessed using a range of ever-evolving technologies, It is an important source of renewable energy and its technologies.

Solar Radiation reaches the Earth's upper atmosphere at a rate of 1.37 kwatts per square meter (kW/m2) in Sudan [7].

Advantages of solar energy:

Here are the main advantages of solar energy [7]:

1. One of the cleanest forms of energy.

2. Harmonious with nature.

3. Safe to handle.

4. Easy to install, operate and maintain.

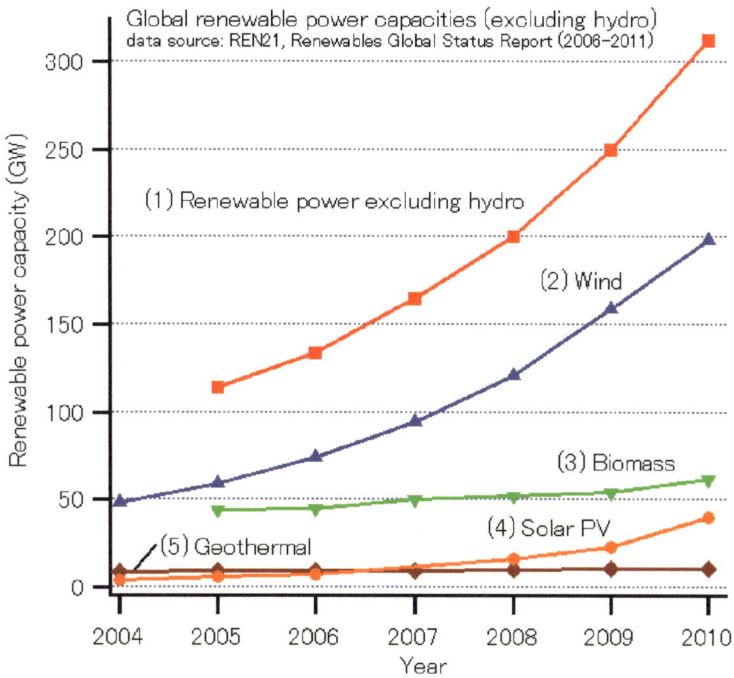

Figure 1.
Renewable energy progress [6].

5. Long life. Solar panels can last up to 20 years or more.

6. Ideal for remote areas, where electricity is not reliable.

2.4 Photovoltaic cells (PV)

Solar cells (also called photovoltaic cells or photoelectric cells) are the most basic components of solar units that convert light energy into electricity. The cells are wired in series, sealed between sheets of glass or plastic, and supported inside a metal frame. These frames are called solar modules or panels [7].

The main parameters that characterize a PV panel are:

1. Short circuit current (I_{SC}): the maximum current provided by the panel when the connectors are short circuited.

2. Open circuit voltage (V_{OC}): the maximum voltage that the panel provides when the terminals are not connected to any load (an open circuit).

3. Maximum power point (Pmax): the point where the power supplied by the panel is at maximum, where P max = Imax*Vmax.

4. Fill factor (FF): the relation between the maximum power that the panel can actually provide and the product ISC. VOC.

5. Efficiency (h): the ratio between the maximum electrical power that the panel can give to the load and the power of the solar radiation (PL) incident on the panel [8].

Types:

There are two primary types of PV technologies available commercially are crystalline silicon and thin film. **Table 1** shows two different type of solar module.

2.5 Typical solar system components

2.5.1 PV module

A number of solar cells electrically connected to each other and mounted in a support structure or frame is called a photovoltaic module.

2.5.2 The battery

Battery serves two important purposes in a photovoltaic system: to provide electrical energy to the system when energy is not supplied by the array of solar panels, and to store excess energy generated by the panels whenever that energy exceeds the load [8].

2.5.3 The regulator

It used to ensure that the battery is working in appropriate conditions, and to avoid overcharging or over discharging the battery, both of which are very detrimental to the life of the battery [8].

2.5.4 Convertor

There are two main types of convertor's:

	Crystalline silicon		**Thin film**
	Monocrystalline silicon (c-Si):	**Multicrystalline silicon (mc-Si):**	**Amorphous silicon**
Efficiency Cell	20–30%	18–22%	13.5%
Production cell	12–15%	11–14%	2–6%
Advantage	• Well established and test technology. • Stable. • Can be made in square cells better power density.	• Well established and tested technology • Stable efficiency • less expensive than single crystal silicon. • Square cells allow efficient packing density.	• Low material use. • Preparation of thin films is possible. • Higher Absorption properties. • Potential for low cost. • Less affected by shading.
Disadvantages	Uses expensive material. • Waste in slicing wafers.	• Uses expensive material. • Waste in slicing wafers.	• Pronounced degradation in power output. • Low efficiency. • It is difficult to maintain uniformity of the film over large areas.

Table 1.
Shows types of solar modules [9].

2.5.4.1 DC/DC converters

DC/DC converters transform a continuous voltage to another continuous voltage of a different value.

2.5.4.2 DC/AC converter (or inverter)

Inverters are used when your equipment requires AC power. Inverters chop and invert the DC current to generate a square wave that is later filtered to approximate a sine wave and eliminate undesired harmonics. Be aware that not all the equipment will accept a modified sine wave as voltage input [10].

2.5.5 Equipment or load

The load (microscope) is the equipment that consumes the power generated by your energy system. The load may include medical equipments, routers, workstations, lamps, TV sets, etc. In this type of system it is absolutely necessary to use efficient and low power equipment to avoidwasting energy. It should be obvious that as power requirements increase, the expense of the photovoltaic system also increases.

2.6 Microscope

Is an instrument used to see objects that are too small to be seen by the naked eye [11], optical instrument that have a magnifying lens for inspecting objects which are too small to be seen by human's eyes. Microscopes are used to observe the shape of bacteria, fungi, parasites and host cells in various stained and unstained preparations [12].

2.6.1 Types of microscope

1. Light microscope.
2. A compound light microscope.
3. Electron microscope.

2.6.2 Parts of the microscope

The main parts of microscope will show bellow in **Figure 2**.

2.6.3 Functioning of the microscope

There are three main optical pieces in the compound light microscope. All three are essential for a sharp and clear image. These are:

1. Condenser

   It illuminates the object by converging a parallel beam of light on it from a built-in or natural source.

2. Objectives

   It forms a magnified inverted (upside down) image of the object.

3. Eye-pieces

It magnifies the image formed by the objective. This image is formed below the plane of the slide. **Figure 3** shows an electrical circuit of microscope:

2.7 Collecting all the components together

As seen in **Figure 4** all the component of photovoltic were collecting together, solar panels generate power when solar energy is available, the regulator to prevent batteries' damage, the battery for storing collected energy to be used later,

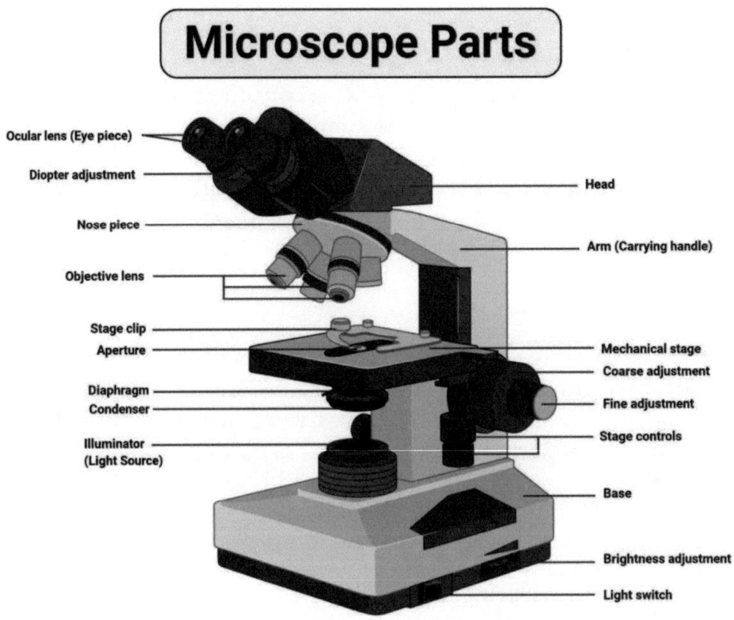

Figure 2.
Parts of microscope [13].

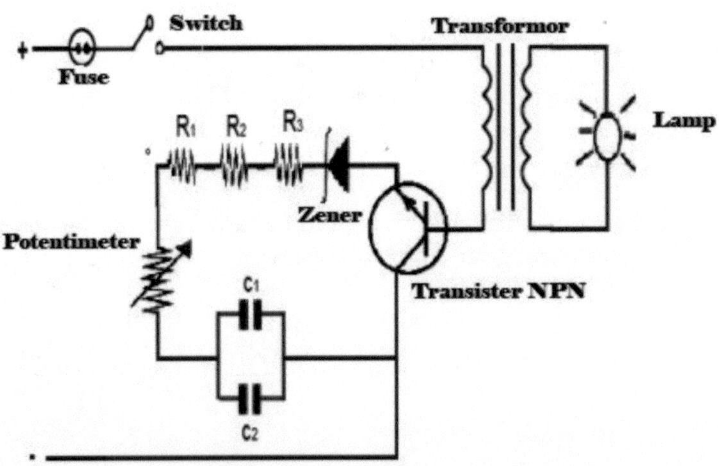

Figure 3.
Microscope' circuit diagram.

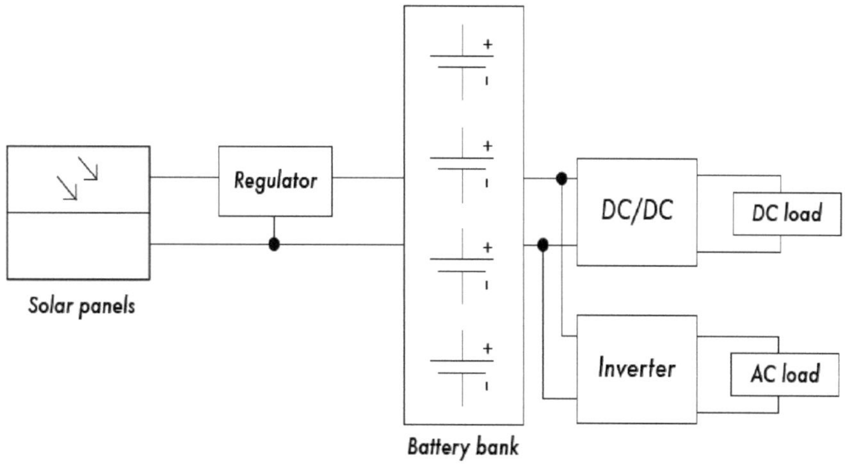

Figure 4.
A solar installation with DC and AC loads.

converters and inverters for adapting the stored energy, and finally, the load (microscope) consumes the stored energy to do work. When all of the components are in balance and are properly maintained, the system will support itself for years [9].

3. Research methodology

3.1 Introduction

This chapter contains experimental procedure and measurement which has been recorded during the testing period medical devices operating solar (For example: microscope). In addition it presents operational problems.

3.2 Design circuit

The circuit of the microscope was modulated because:

1. The device work with AC but solar cell generate DC.

2. To avoid the use of the inverter because it is expensive. Transformer has been canceled, two potentiometers were connected in series (one static and variable), adding Zener with a value of 12 volt (to maintain the value of effort fixed), it was used for halogen bulb 12v, switch and fuse. And finally connect all components of the circuit taking into account the polarity. **Figure 5** shows that.

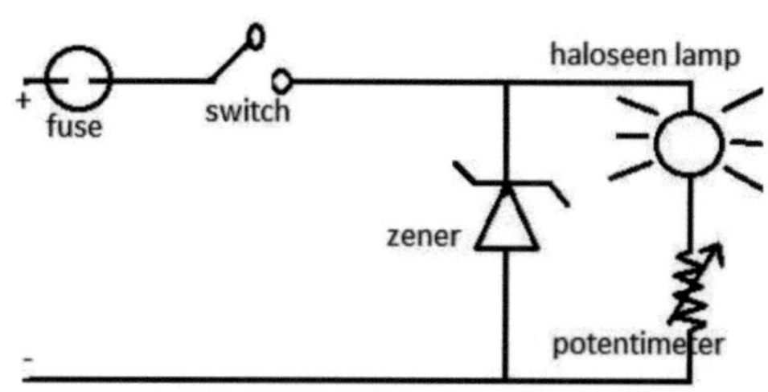

Figure 5.
Modulated circuit of microscope.

Fuse 1.6 A, Switch (on\off), Zener 12 voltage, DC haloseen lamp 12 v Potentiometer 1MΩ. A new microscope circuit was designed on experimental board and expert in university lab using a power supply unit, it was adjusted on DC current 3A and 12 V, the board connected to the supply and turned on the lamp was illuminated.

3.3 Sizing photovoltaic calculations

System sizing calculations are important, because unless the system components are balanced, energy (and ultimately, money) is wasted. A microscope needs 20w income to run any device operates on–off actuation 6 hours.

3.3.1 Calculate the number of module

To calculate the number of panels required to cover a given load, you just need to know the current and voltage at the point of maximum power: IPmax and Vpmax (Ipmax and Vpmax are provided by the manufacturer).

3.3.2 About The Efficiency

1. Battery efficiency = 0.90

2. Charge controller efficiency = 0.90

3. Inverter efficiency = 0.85

4. Conversion factor for DC load = 0.81

5. Conversion factor for AC load = 0.72

3.3.3 Load calculations

The Power = voltage*current
P = 6*3.3 = 20 W
Total power (Energy consumption) = the power*number of devices
Total power = 20 W *1 = 20 W
Power (watts/hour) = total power*required time loads
P (W/h) = 20*6 = 120 W/h
The required power of cells = energy consumption/system factor
The required power of cells = 120 W/0.81 = 148 Wh
Battery capacity = the required power/system voltage
Battery capacity = 148/12 = 12.31AH
The cell Production = 18
The No. of Modules = battery capacity/cell production
The No. of Modules = 12.31/18 = 0.68 ≈ 1modules

3.3.4 Battery calculations

The required battery capacity = battery capacity/depth of discharge
The required battery capacity = 12.31/0.4 = 30
The available capacity 70AH

The No. of batteries = capacity batteries/the available capacity
Number of batteries = 30.78 / 70 = 0.44 ≈ 1 battery

3.3.5 Regulator calculations

The charge controller load current = maximum current cell*the number of modulus
Isc = 3.5A
Appropriate regulator = 3.5*1 = 3.5A

3.3.6 Inverter calculation

Inverter = overall capacity/inverter efficiency
Inverter = 21.05 Watts

3.3.7 System components

Solar cell 50 W
Battery 60 Ah
Regulator 20A
Inverter 21.05 W

3.4 Testing instruments

3.4.1 Load

Microscope device is selected because it is a simple device does not consume high power compared with other devices, Electrical specifications for the Microscope:
In the case (AC):
Input 220 V; output 6 V, 3.3A, 20.4 W.
In the case (DC):
Input 12 V, 1.6 A, 19.2 W.

3.4.2 PV module

The **Table 2** shows specifications of solar panel.

3.4.3 Electrical parameters of a Solar cell (by the manufacturer)

Maximum power: 50 ± 3% Wp
Short circuit current: 3.09A
Open circuit voltage: 21.6 V
Max. Power current: 2.89A
Max. Power voltage: 17.3 V
Standard test condition at: 1,000 w\m2 solar irradiance. Cell temperature 25°C, AM 1.5.

Type of module Wp	Isc	Ipm
50	3.01	2.76

Table 2.
Solar modules electrical specifications.

3.4.4 Inverter

Specification:

1. DC input 10-15 V

2. AC output 220-240 V, 50 Hz/60 Hz.

3. Constant output power 300Watts.

3.5 Experiments

At the first experiment was wired between the charge controller and the load but it was not work because output current from it was less than 3A to operate the load.

3.5.1 DC way

After modifying the load it can work immediately with DC current. This solar photovoltaic lighting system consists of a 20Watt single crystalline module, a self-contained photovoltaic controller with (regulator), a sealed lead-acid battery and a single 100 Watt fluorescent light. The system will provide daily lighting throughout the year. **Figure 6** shows block diagram of DC way.

3.5.2 Inverter way

The four primary components for producing electricity using solar power, which provides common 220volt AC power for daily use, are: Solar panels, charge controller, battery and inverter. Solar panels charge the battery, and the charge regulator as it will be shown in **Figure 7** ensures proper charging of the battery. The battery provides DC voltage to the inverter, and the inverter converts the DC voltage to normal AC voltage.

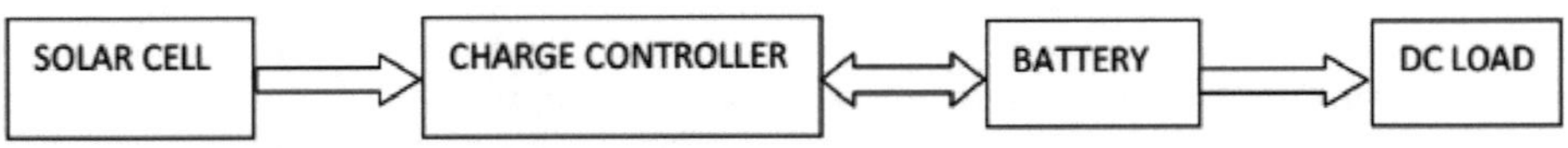

Figure 6.
Black diagram of solar system without inverter.

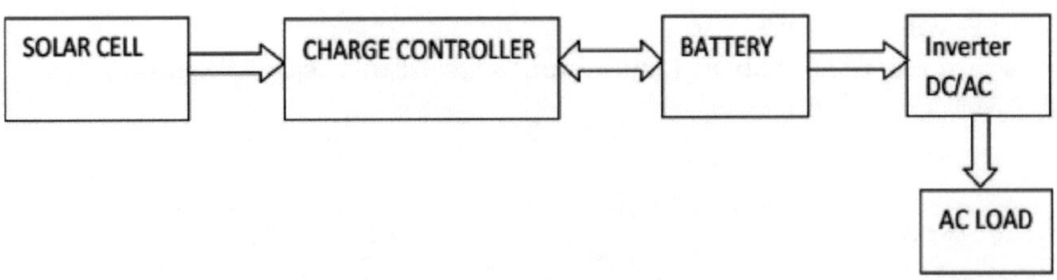

Figure 7.
Black diagram of solar system with inverter.

4. Results

The design of modulated circuit shown below in **Figure 8** discuss that solar energy (direct current) can be used to turn on the load for continues six hours to solve problem as it mentioned earlier.

4.1 This device was tested with solar (AC way)

The load can work in another way .by converting the output direct current from the batteries to AC current using a DC/AC inverter .this current can operate immediately without any modulating in circuit as shows bellow in **Figure 9**.

Figure 8.
Microscope's modulated circuit with direct current.

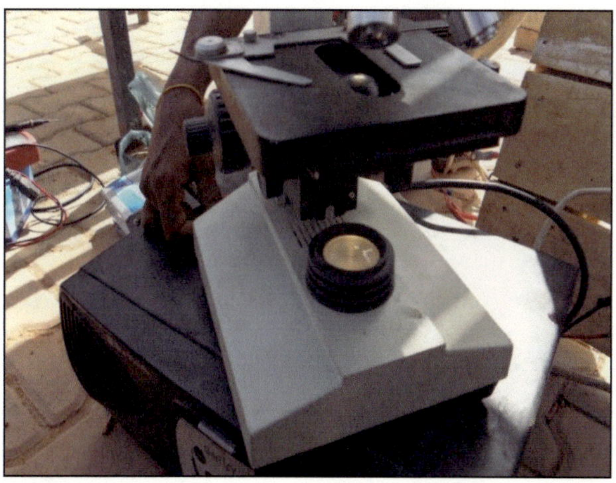

Figure 9.
AC way.

4.2 Solar was tested with solar (DC way)

After the circuit was modulated in order to work with solar current, it should be tested using solar to perform project, **Figure 10** come next shows the result of problem which mentioned previously.

The circuit was operated with DC of solar, finally after that the sensitivity of this light should be measure to avoid eyes' damage.

4.3 Photo sensitivity of light

Photo sensitivity (PS) is ratio of radiant energy expressed in watt(W) incident on the device, to the resulting photo current expressed in amperes(A) it may be represented as either an absolute sensitivity(A\W) or as relative sensitivity normalized, usually expressed in percentage (%), PS should be more than 5%,**Table 3** shows the PS of light from electrical power and from solar power, PS was calculated for both, the PS of them in acceptance percentage.

When solar power was used; PS of it = 8.3% it's > 5%, so, it is in range of acceptance and then the doctors could see with this light without any effective to their eyes.

Figure 10.
DC way.

	Solar energy		Electrical supply	
	DC	AC	DC	AC
Ampere	1.6	3.3	1.6	3.4
Voltage	12	6	11.9	5.8
Photosensitivity	0.083	0.167	0.084	0.172

Table 3.
PS of electrical and solar energy.

Tool	The cost(SDG)
Panels (50 W)	600
Charge control (10A)	300
Batteries (12 V,65 Ah)	750
Inverter (300 W)	450
Wiring	50
∑equipment	2150

Table 4.
The cost of solar system.

4.4 Economic results

The cost of inverter way is 2150 SD pounds and it can work for 25 years without any additional cost. But DC way is cheaper and cost 1700 SD pounds without any additional cost. **Table 4** shows calculated the cost of solar system.

The annual cost of the electric power grid is calculated for a period of 25 years by Sudanese pound (SDP).

The cost of electricity power =

Total power (Wh)*0.76 SDP/kW*365.25 day *25 years = 833 SDP

The electricity power is the cheaper than PV system; but the solar energy can be used in rural areas where there is no electricity power and offer medical services to fulfill the project's objective.

5. Recommendations

1. Solar PV system can be alternative technology for remote rural areas where grid electric power **is not** available. Grid electric power is more cost effective power supply than Solar system. But this system can be alternative for some of medical devises.

2. Then and finally as if the microscope was modulated and operated with solar energy, we recommend that to operate small healthcare units in rural areas.

6. Conclusions

The microscope illuminated by solar cell. Some medical devices operated by using solar energy, which this energy produces a DC current. These devices modulated by canceling step down transformer and rectifier, and offered appropriate DC current by sizing photovoltaic cell and a charging battery or used inverter to convert DC current to AC current. Solar cell is more expensive than electricity power, but it can use in areas where there are no electricity power.

In comparison to other energy sources such as coal, natural gas, nuclear, wind, and hydro-electric, solar energy is one of the least cost effective. For the most part, the exorbitant cost of solar energy is due to the high cost of manufacturing high quality solar panel material that has high energy conversion efficiency. Today, most cheaply made solar cells have an energy efficiency of about 14% while the higher quality materials achieve efficiencies of about 20–25% solar energy is still relatively expensive to a consumer.

Acknowledgements

At the beginning we would like to thank our project supervisor, Dr. Alnazeir Osman Hamza for the valuable guidance and advices, also the ministry of science and technology Renewable Energy Institute (the Photovoltaic Encapsulation & Manufacturing Unit (Khartoum-south)) for providing us with a good environment and facilities to complete this project.

Also we will take this opportunity to thank Dr. Mohamed Altaib Alfadni the Executive Manager of Sharg alneel specialized hospital & Dr. Alfatih Abdelgader Alfadni who support us with a microscope device for testing.

Special declarations

This project could not be completed without our parents' prayers so heartily thanking for them besides special thanking to Dr. Altahir Mohammed, Fatehia Garma.

Author details

Badria Ibrahim Eisa Idris*, Ahmed Mohamed, Ayah Salah,
Osman Abdalrahman Almahdi Alamin, Fatehia Garma and Alnazier Osman
Biomedical Engineering Department, Sudan University of Science and Technology, Khartoum, Sudan

*Address all correspondence to: bdriaibrahim@yahoo.com

References

[1] Definition of solar energy available from: https://en.wikipedia.org/wiki/Solar_energy [Accessed: 2018-02-02]

[2] World energy resources. 2016. Available from: https://www.worldenergy.org/wpcontent/uploads/2017/03/WEResources_Solar_2016.pdf Accessed: 2018-02-02]

[3] Nwafor S. Chimela*, Ogbonna K. Nkem, Onyemuwa Nnaemeka, Olaniyi Abiodun http://escijournals.net/index.php/IJAE/article/download/1923/916 [Accessed: 2018-02-07]

[4] International Energy Agency," renewable energy party", 2002.

[5] Renewable Energy N21, "renewable 2010 global report", 2010.

[6] renewable energy progress, available from: http://upload.wikimedia.org/wikipedia/commons/thumb/6/65/GlobalREPowerCapacity-exHydro-Eng.png/290px-GlobalREPowerCapacity-exHydro-Eng.png [accessed: 2018-02-07]

[7] Tamil Nadu," best practices in solar water pumping", Auroville's Centre for Scientific Research, India, 2002.

[8] Rob Flickenger, Corinna Aichele, Carlo Fonda. Wireless Networking in the Developing World, Canada's International Development Research Centre, Canada, December 2007.

[9] Green, Martin. A., Solar Cells: Operating Principles, Technology and System.

[10] Jenny Nelson, "The Physics Of Solar Cells", Imperial College, UK,2008.

[11] definition of microscope [Internet]. Available from: https://en.wikipedia.org/wiki/Microscope [Accessed: 2018-02-01]

[12] Mr K.K. Khanna, "THE MICRO-SCOPE", New Delhi, India, 1999.

[13] parts of microscope Available from: https://i.ytimg.com/vi/4rerHeaXSio/maxresdefault.jpg Accessed: [2018-02-07].

Chapter 23

Solar Energy in Industrial Processes

Guillermo Martínez-Rodríguez and Amanda L. Fuentes-Silva

Abstract

A design methodology to integrate solar heat into industrial process is showed in this chapter, attending restrictions like availability for area of installation, economic, environmental, and operating conditions. The evaluation of each of the restrictions allows responding to real situations that arise in the industrial sector and thereby determining the scenario that best suits the industry. To achieve this objective, the evaluation of two real scenarios was carried out; in the first one there are no installation area limitations, while in the second, only the 50% of required installation area is available. The results obtained when evaluating the scenarios exhibit a direct relationship between the available space, the capital of the investment and the CO_2 emissions, but this is not reflected in the same proportion in the operation of the process. In scenario one, the payback of the integrated system is 5.99 years with zero emissions to the atmosphere. For scenario two, the reduction of CO_2 emissions is 80.62% with a recovery time of the investment of the integrated system of 2.61 years. In this context, Chemical Vapour Deposition is proposing as a innovative technology to improve the solar devices efficiency.

Keywords: solar collector network, spatial restriction, payback time, thermal solar heat integration, real scenarios to integrate solar heat

1. Introduction

Industrial sector demands large amounts of energy, two thirds of it is in the form of heat [1]. Of the heat demand in industry, almost half, 48%, is required for high temperature heat (more than 400 °C) mainly in the intensive industries of iron and steel and other minerals, which reach temperatures of up to 1450 °C; chemical and petrochemical, 900 °C; among others, 22% is used for medium-temperature heat (150 to 400 °C) and the remaining 30% is used for low-temperature heat (less than 150 °C) [2]. Currently, only 9% of the heat demand of industrial processes is supplied from renewable energy sources, 79% of solar thermal installations use flat plate solar collectors and evacuated tubes, 11% correspond to the use of parabolic-cylinder collectors, and the remaining percentage of other technologies [3]. Chemical Vapour Deposition, CVD, plays an important role in the development and improvement of solar technology. CVD technologies have been innovating for at least 50 years to increase the efficiency of solar energy collection, both from solar cells and, more recently, from evacuated tubes [4].

When a protocol to solar heat for process industrial integration (SHIP integration) is being designed, there are some parameters that determine the solar heat integration potential: a) inherent to the process: energy demand [5], hourly heat demand profile, seasonal heat demand profile [6], temperature intervals, continuous, semicontinuous or batch processes, different kinds of solar heat (steam, drying, hot water) [7]; b) regarding to the facilities: location [5], surface area availability; c) depending on the expected objectives: solar fraction, outlet temperatures, payback time, lower emissions of greenhouse gases (GHG), saving costs [8]. However, the picture is not complete if the limitations or restrictions that represent serious challenges to overcome to achieve efficient use of solar energy are not given equal importance: a) inherent to the process: higher process heat demand than solar heat produced; b) regarding to the facilities: limited flexibility of the systems, use of outdated or non-optimal technology for process conditions, higher costs of solar heating systems than fossil fuel conventional systems [6]. And all these considerations must in turn take into account energy policy and the associated investment, which can also limit or restrict the optimal use of solar energy: lack of economic support to research and innovation to tuning and updating of technology, lack of standard procedures for the implementation and evaluation of technological systems, difficulty promoting attractive investment and business models for the deployment and integration of renewable energies, and few market incentives [9], prohibitions to produce and distribute renewable heat [10], among other.

With the technologies that currently exist for the use of solar energy for heat production, the processes that demand low-temperature heat are the most convenient for integrating solar heat [11], besides, when solar storage is introduced, solar fraction increases markedly compared to a process without storage [5].

The evaluation of some of the restrictions or limitations such as: available installation space for the collector network, availability of capital or the low prices of the fossil fuels used and the supply time of the collector network, allows to evaluate the real impact of each scenario when compared with the one where there are no restrictions for the installation of the solar thermal device and thus seek the profitability of the device in whatever the scenario. Next, the proposed methodology for the integration of solar thermal energy with some real restrictions is described.

The objective of this chapter is to make a general approach that includes some real scenarios that arise when solar thermal energy is integrated into industrial processes, whether they operate continuously or in batches.

2. Methodology for the integration of solar thermal energy

The integration of solar thermal energy must be economically attractive to compete with fossil fuels and it must be flexible in such a way that it can be applied to various real scenarios where there are spatial, economic, operational, and environmental limitations, limiting the amount of solar energy that can be supplied and with it the solar fraction. The irradiance levels of the site constitute a variable that can limit the maximum output temperatures that a network of solar collectors can reach in a period to guarantee the temperature level and the supply of the total or partial thermal load. Any of the previously stated limitations can define the integration of solar thermal energy.

Whatever the application, the integration of solar thermal energy must be cost-effective and environmentally friendly by reducing the generation of greenhouse gases. Although any reduction in the generation of greenhouse gases is an important point for any industrial sector, the design objective should always be the total elimination of the use of fossil fuels. However, the limitations to achieve a solar fraction of unity are real and it is important to evaluate how they can define the installation of the solar thermal device. Each of the restrictions raised is a challenge that must be addressed and resolved to respond to the real needs that arise in the industrial sector when the transition towards the use of renewable energies is sought to replace fossil fuels.

The methodology developed is general and can be used or coupled to medium temperature solar collectors or mixed systems (low and medium temperature solar collector technologies) for a specific application. The range of applications can be expanded by defining some other objectives of industrial interest.

The approach contemplates the integration of solar thermal energy into a real case, using low-temperature solar collectors, specifically, flat plate solar collectors, for the selection, design, and operation of the collector network. On the other hand, the design of the collector network is based on the most critical conditions of the year, which correspond to the winter period and guarantee the supply of the thermal load throughout the year. It is important to mention that the rest of the year there will be surplus energy that can be used in other applications. In the selected case study, two scenarios will be evaluated. In the first scenario, the total supply of the thermal load is considered at the temperature level required by the process, with a solar fraction of one and zero greenhouse gas emissions to the environment. In this scenario there are no space limitations for the installation of the solar collector network. In the second scenario, it is proposed that only 50% of the area required for the total supply of the thermal load is available, reducing the fraction as the generation of greenhouse gases. In this last scenario, it is analysed how the restrictions impact and what implications it has with the rest of the variables such as:

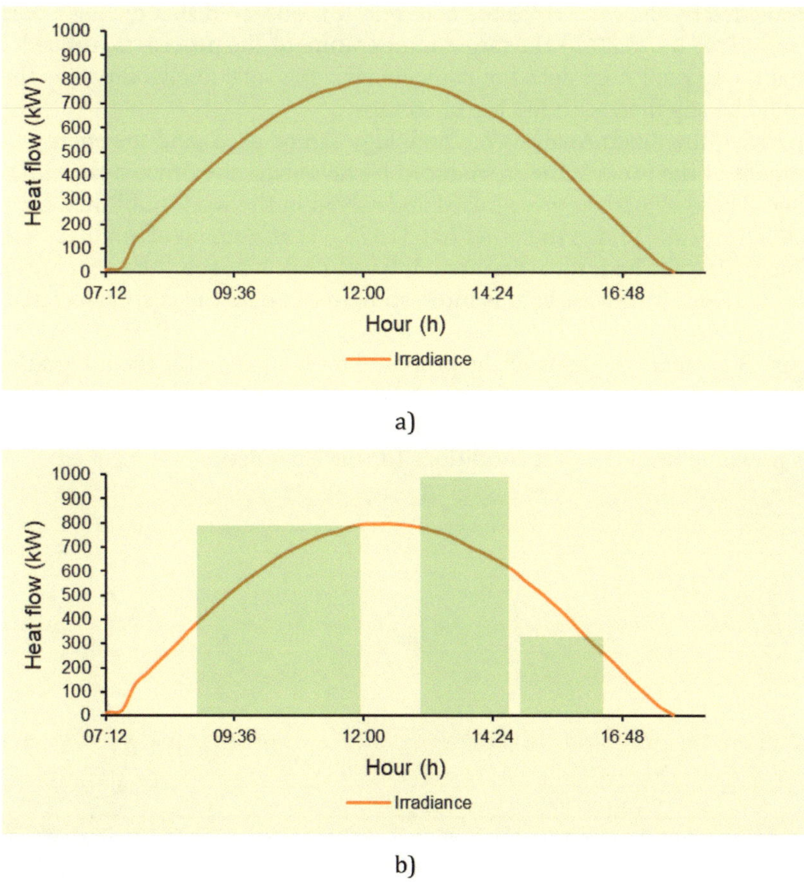

Figure 1.
Main stages that require thermal energy in an industrial process (green colour): Continuous process (a), batch process (b).

reduction in emissions, time in the recovery of investment (payback time) and time of direct supply of the collector network.

To carry out the integration of solar thermal energy, it is important to quantify the available solar resource to determine the maximum outlet temperature of the solar collector network, determining the supply time at target temperature and the size of the network. For a batch process, it would be sought that the direct supply time from the network is equal to the time required by the process, otherwise it is necessary to match the availability of the solar resource with the energy requirement of the process at the temperature required by it.

In **Figure 1**, reference is made to the energy requirements of an industrial process and to solar availability, to match these requirements and to be able to integrate solar thermal energy. For the solar energy supply or production curve, it is considered that there is an available area of 1000 m^2. Section a) represents a continuous process and the energy demand of the process is above the amount that is captured with the available surface (1000 m^2), to supply the thermal load, the capture area would be increased. In part b) a batch process is shown where it is observed that the energy requirement can be provided entirely by solar energy.

In this approach it is possible to supply the thermal energy of the process through solar thermal energy using a thermal storage system to match the demand and energy production. The temperature level required for each process operation is not considered in the diagram.

Figure 2 displays the temperature level required by the process and the temperature provided by the solar collector network. It is observed that during a period of 5 h 45 min (9:45 h - 15:30 h) the target temperature of the process is reached. In the time that the requirement does not coincide with the solar availability, the thermal load would be supplied with the use of storage.

In parallel, the Pinch Analysis methodology can be used, and the energy requirement of the process can be reduced by favouring the process-process heat exchange. These objectives were raised and solved in the work published by Fuentes-Silva et al. [8]. On the other hand, ΔT_{min} is a function of auxiliary services, when increasing the heat recovery area, it is reduced, however, the area of the collector network increases, so it is interesting to evaluate the costs associated with this relationship.

Figure 3 presents the logic of the proposed methodology for the integration of solar thermal energy into industrial processes, it is represented by a block diagram. The proposed methodology is carried out in some stages simultaneously and feedback is provided until the best conditions for the final design are reached.

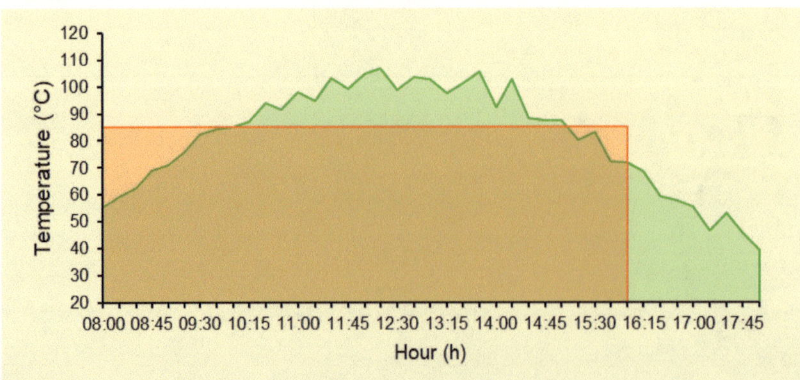

Figure 2.
Temperature level required by the process (85 °C) and that supplied by the solar collector network (variable).

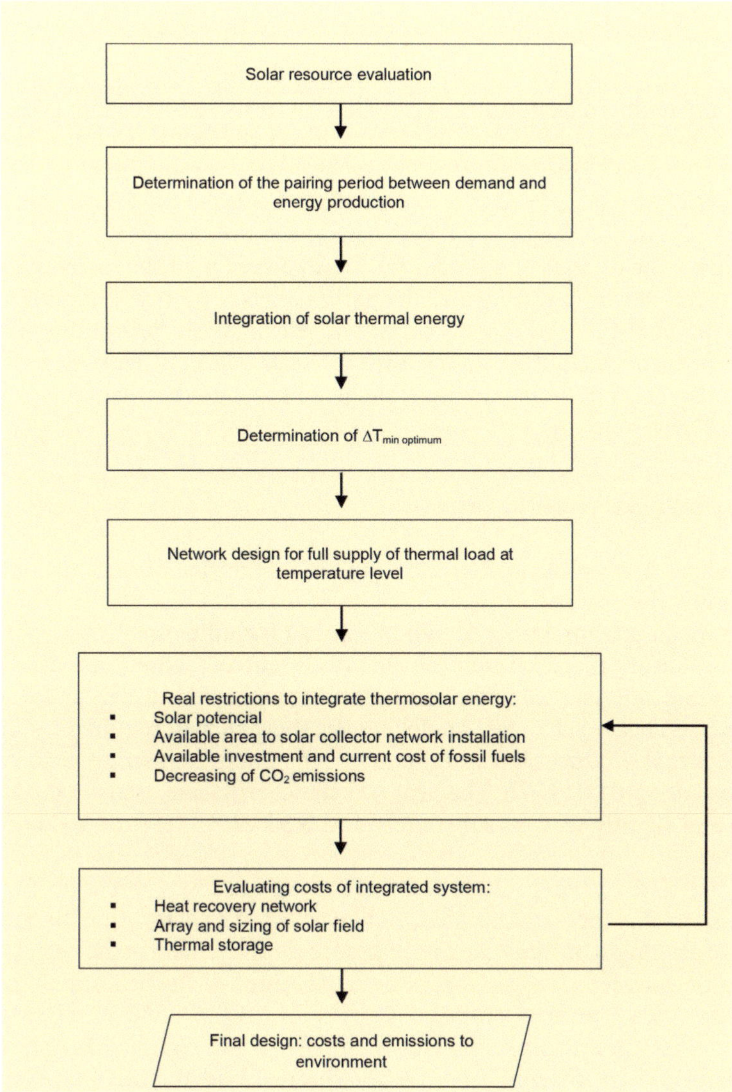

Figure 3.
Design algorithm to integrate solar thermal energy.

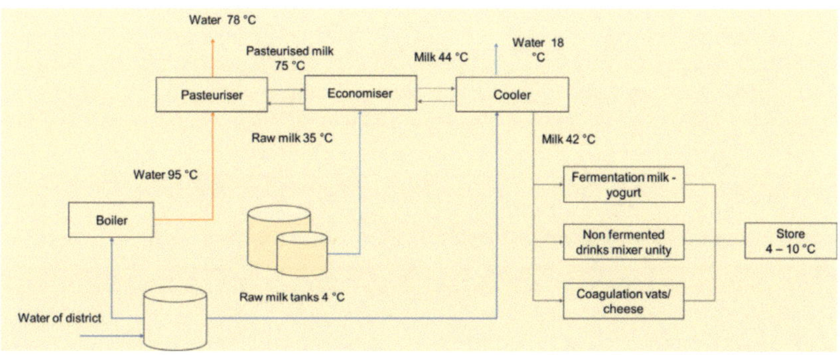

Figure 4.
Flow diagram of dairy process.

2.1 Case study

The case study is a dairy process described in the literature [12]. It operates on batch and the required heat utility is from 8:00 h to 13:00 h (five hours), enough to carry out pasteurisation of milk at 85 °C. Hot water is needful to curd the milk from 8:30 h to 13:00 h (4.5 hours) at 40 °C; more hot water is required at 62 °C to clean the equipment during a period of 2 hours (16:00 h – 18:00 h). **Figure 4** represents the diverse production stages in a dairy plant at required temperatures.

In **Figure 5** the periods of demand for thermal energy of the main operations of the process are shown (squares) and the period of solar energy production is also displayed (parabolic line). **Figure 5** shows the pairing periods, and this allows to determine if the heat supply is direct from the solar collector network or if requires storage, it also helps to define the possible heat exchanges between the process streams to recover energy.

2.2 Energy integration of the process

Next, Pinch Analysis is used to carry out the integration of solar thermal energy and determine the optimal ΔT_{min}.

Dairy process stream data is shown in **Table 1** including inlet and outlet temperatures and heat capacity. From the data of the currents, the Composite Curves are constructed and determine the minimum hot and cold utilities, and thus the process-process heat exchanges. To deliver the thermal load required by the process, a boiler that produces hot water at 95 °C for 5 hours is used, providing a thermal load of 4,401.01 kWh. The heat transfer coefficients used are 0.8 kW/m^2 °C for water and slightly viscous substances and 0.3 kW/m^2 °C for viscous substances.

The Composite Curves provide a scheme of pure countercurrent heat transfer of the process streams for a chosen ΔT_{min}. They are used to determine the minimum heating and cooling requirements of the process and this allows the calculation of the minimum transfer area prior to the detailed design of the heat transfer network.

In **Figure 6** we have the Composite Curves for a ΔT_{min} of 10 °C where the minimum utilities are determined, which are 294.78 and 260.68 kW of heating and cooling, respectively. It is observed that the thermal load and the temperature level of the hot utility can be supplied by solar thermal energy.

Determination of the minimum area assumes that there is vertical heat exchange between the Composite Curves throughout the entire enthalpy range. In the graph of the Composite Curves, we can read the amount of heat that will be transferred

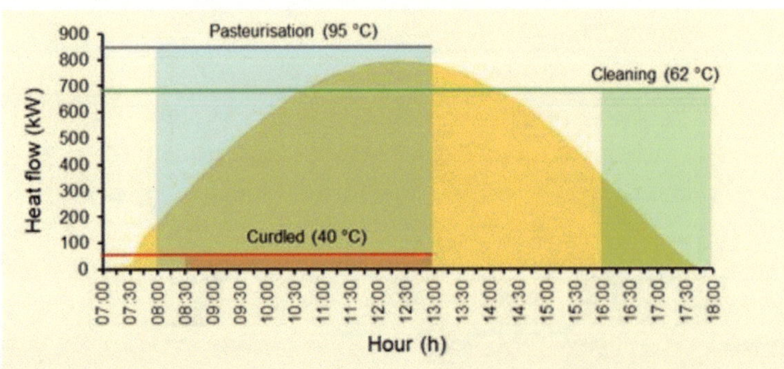

Figure 5.
Pairing between major process operations and available irradiance in a day.

Description	T_{inlet} (°C)	T_{outlet} (°C)	CP (kW/°C)
Mains water	12.20	38.00	2.64
Milk- rennets vats	34.00	35.00	3.89
Pasteurised milk- effluent	75.00	44.00	4.38
Water for boiler	95.00	78.00	10.63
Raw milk- input	4.00	35.00	4.38
Milk- input	35.00	75.00	4.38
Mains water- for cooler	12.20	18.00	6.35
Input milk	44.00	36.00	5.84
Water – of storage tank	73.30	40.00	1.94
Water – rennets vats	40.00	38.00	3.89
Water – of storage tank	62.40	25.00	9.30
Water – of storage tank, surplus	62.40	25.00	12.90
Main water -for boiler	12.20	95.00	10.63

Table 1.
Stream data for dairy process

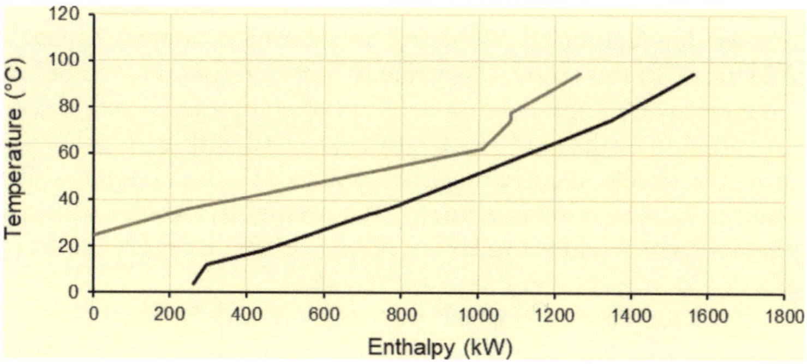

Figure 6.
Composite curves for a ΔTmin of 10 °C.

between the streams, as well as the values of the outlet and inlet temperatures (at that interval) of the hot streams, like so the values of the outlet temperatures and entry of cold currents. So if the curves, both for hot and cold currents, correspond to only one of them, as happens between an exchange of process current - auxiliary service, we can estimate the necessary area of heat exchange given by Eq. (1) known as the Bath equation, which is based on the countercurrent heat exchange assumption implicit in the compound curves and this leads to the minimum area, only if the global heat transfer coefficients are equal for all exchanges [13].

$$A_{HRN} = \sum_{k}^{k\ intervals} \frac{1}{\Delta T_{ML}} \left(\sum_{i}^{hot\ i.} \frac{q_i}{h_i} + \sum_{j}^{cold\ j} \frac{q_j}{h_j} \right) \tag{1}$$

The estimate of the total A_{HRN} area is obtained by adding the area of each interval (m^2), q_i and q_j are the thermal exchange loads obtained from the enthalpies in the Composite Curves (kW). h_i and hj are the individual transfer coefficients (kW/m^2 °C). The subscripts k, i and j.

The calculated minimum area is used to determine the cost of the heat recovery network, C_{HRN} (USD), of the process based on Eq. (2).

$$C_{HRN} = N_e \left[26600 + 6500\left(\frac{A_{HRN}}{N_e}\right)^{0.9}\right] \tag{2}$$

Where N_e is the number of equipment, A_{HRN} is the area of the heat recovery network (m^2) this value corresponds to the exchange area [14]. To determine the cost of auxiliary services, C_{AS}, Eq. (3), reported by [15].

$$C_{AS} = Q_h\, C_{ST} + Q_c C_{CO} \tag{3}$$

In Eq. (3) Q_h and Q_c are the minimum requirements for heating and cooling the process (kW), while C_{ST} and C_{CO} are the costs associated with steam and cooling water, respectively (USD/kW year). In this study, the cost of heating services is proposed to be 150 USD/kW year and cooling services 35 USD/kW year.

The total cost of the annualised heat recovery network, $C_{TA\ HRN}$ (USD/year), is obtained by adding the estimated cost of capital (C_{HRN}) and the annualised service costs (C_{AS}), given by Eq. (4)

$$C_{TA\ HRN} = f_A C_{HRN} + f_A C_{AS} \tag{4}$$

The annualisation factor f_A is calculated considering an annual interest of 5% for a period of 25 years. In this study, this value is considered as the number of years of useful life of a piece of equipment or a network of equipment. To find the optimal ΔT_{min} it is required to evaluate the total cost associated with each chosen ΔT_{min}. **Table 2** shows the results to determine the optimal ΔT_{min} in an interval from 1 to 30 °C, for each ΔT_{min} the minimum energy requirements, the heat exchange area and the annualised costs of: heat recovery (C_{HRNA}), auxiliary services (C_{ASA}) and the total cost of the heat recovery network ($C_{TA\ HRN}$).

In **Figure 7** the total cost of the heat recovery network is shown, it can be seen the optimal ΔT_{min} is located at 13 °C and the minimum energy requirements are: 339.81 kW and 305.71 kW, corresponding to the minimum services heating and cooling, respectively.

The design of the solar collector network will be carried out considering the optimal ΔT_{min} and its corresponding minimum hot utility (thermal load).

2.3 Solar resource available

The irradiance data and other environmental variables (ambient temperature and wind speed) were taken from a meteorological station of the Solar Energy Laboratory of the University of Guanajuato, Guanajuato city, Mexico. The geographical location of the meteorological station is latitude of 21°01′0” N, longitude 101°15′24” W, at an average sea level elevation of +2 000 meters, central time zone, UTC − 6 and in summer UTC − 5. **Figure 8** shows the radiation data throughout the year under clear sky.

2.4 Design of the solar collector network

The integration of solar energy in an industrial process presents a challenge for existing process integration techniques, due to the non-continuous nature of supply, available irradiance levels associated with direct supply time from the solar collector network to the process. The efficiency of solar technologies must be

ΔTmin (°C)	Q_h (kW)	Q_c (kW)	N_e	A_{HRN} (m^2)	C_{HRN} (USD)	C_{HRNA} (USD/y)	C_{ASA} (USD/y)	$C_{TA\ HRN}$ (USD/y)
1	159.69	125.65	17	531.63	3,199,987.73	227,046.99	28,351.41	255,398.40
2	174.70	140.66	17	428.05	2,721,302.35	193,083.09	31,128.26	224,211.35
3	189.71	155.67	17	369.75	2,446,943.96	173,616.69	33,905.11	207,521.80
4	204.72	170.68	17	329.89	2,256,861.40	160,129.86	36,681.96	196,811.82
5	219.73	185.63	17	301.89	2,121,959.05	150,558.21	39,456.78	190,014.99
6	234.74	200.64	17	278.22	2,006,954.08	142,398.32	42,233.63	184,631.95
7	249.75	215.65	17	258.87	1,912,212.04	135,676.14	45,010.48	180,686.62
8	264.76	230.66	17	242.64	1,832,205.01	129,999.45	47,787.33	177,786.78
9	279.77	245.67	17	228.77	1,763,388.35	125,116.74	50,564.18	175,680.92
10	294.78	260.68	17	216.70	1,703,160.69	120,843.44	53,341.03	174,184.47
11	309.79	275.69	17	206.12	1,650,130.10	117,080.79	56,117.88	173,198.67
12	324.80	290.70	17	196.77	1,603,030.89	113,738.98	58,894.73	172,633.71
13	339.81	305.71	17	188.45	1,560,908.19	110,750.27	61,671.58	172,421.85
14	354.82	320.72	17	180.99	1,522,985.26	108,059.55	64,448.43	172,507.98
15	369.83	335.73	17	174.26	1,488,643.65	105,622.92	67,225.28	172,848.20
16	384.84	350.74	17	168.16	1,457,391.84	103,405.53	70,002.13	173,407.66
17	399.85	365.75	17	162.60	1,428,847.98	101,380.28	72,778.98	174,159.26
18	414.86	380.76	17	157.53	1,402,675.38	99,523.27	75,555.83	175,079.10
19	429.87	395.77	17	152.88	1,378,625.23	97,816.85	78,332.68	176,149.53
20	444.88	410.78	17	148.61	1,356,470.80	96,244.94	81,109.53	177,354.47
21	459.89	425.79	17	144.67	1,336,008.06	94,793.05	83,886.38	178,679.43
22	474.90	440.79	17	141.04	1,317,077.19	93,449.86	86,662.92	180,112.78
23	489.91	455.81	17	137.69	1,299,526.66	92,204.61	89,440.08	181,644.69
24	504.92	470.82	17	134.58	1,283,244.98	91,049.38	92,216.93	183,266.31
25	521.52	487.41	18	130.09	1,293,338.46	91,765.54	95,286.82	187,052.36
26	539.17	505.12	18	125.69	1,270,040.65	90,112.50	98,554.10	188,666.60
27	556.82	322.05	19	121.57	1,281,326.20	90,913.24	94,794.15	185,707.39
28	576.80	542.75	18	117.34	1,225,502.94	86,952.45	105,516.39	192,468.84
29	596.01	561.96	18	113.47	1,204,793.70	85,483.07	109,069.50	194,552.57
30	613.66	579.61	18	110.03	1,186,270.01	84,168.77	112,334.75	196,503.52

Table 2.
Results of pinch analysis to determine optimal ΔTmin.

determined since it is influenced by the environmental conditions of the place, this must be evaluated to guarantee the supply of the hot utility. This methodology combines this information to lead to the design of the solar thermal installation to reach the target temperature required by the process and satisfy the thermal requirements of the process with the lowest cost and taking care of the restrictions that the process presents. This must be attractive to compete with fossil fuels.

The design of the solar collector network is based on the methodology proposed by Martínez-Rodríguez et al., [16] to supply the thermal load at the required process temperature level. The design variables of the low temperature solar collector

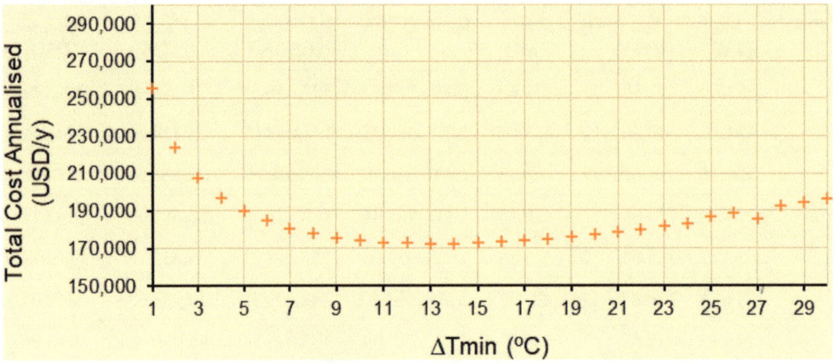

Figure 7.
Annualised total cost from heat recovery network against ΔTmin.

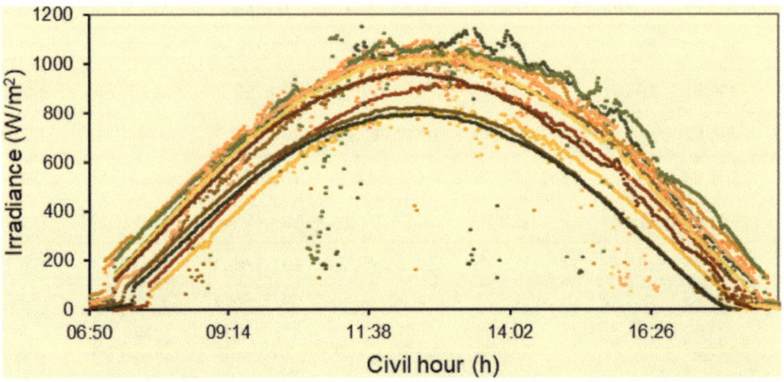

Figure 8.
Irradiance throughout a year for clear sky days.

network are the operating conditions of the process (temperature and required thermal load), the environmental parameters (irradiance and ambient temperature), geometric dimensions and characteristics of the flate-plate solar materials, the properties of the working fluid, and the network operating conditions (flow and feed temperature).

The minimum number of collectors connected in series can be calculated considering that the minimum difference between the outlet temperature of the fluid from the collector, T_O (°C), and the temperature of the fluid at the entrance to the collector, T_i (°C), is equal to or greater than 1 °C. Generalising this difference for any collector or series have Eq. (5)

$$\Delta T = T_o^n - T_o^{n-1} \tag{5}$$

Where n refers to outlet temperature of the *n-th* element and *n-1* refers to the outlet temperature of one minus to the *n-th* element (°C) (**Figure 9**).

The number of branches or lines in parallel is calculated by Eq. (6)

$$N_p = \frac{Qi}{Q} \tag{6}$$

Q_i is the thermal load provided by a series of n collectors connected in series (kW) and Q is the total thermal load (kW) required by the process, whether to cover partially or totally.

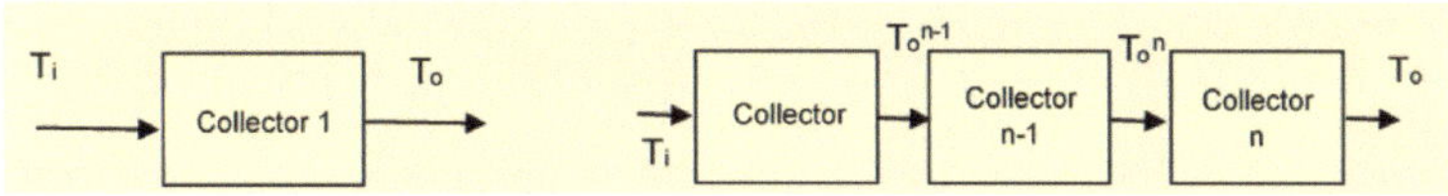

Figure 9.
Series of n *collectors.*

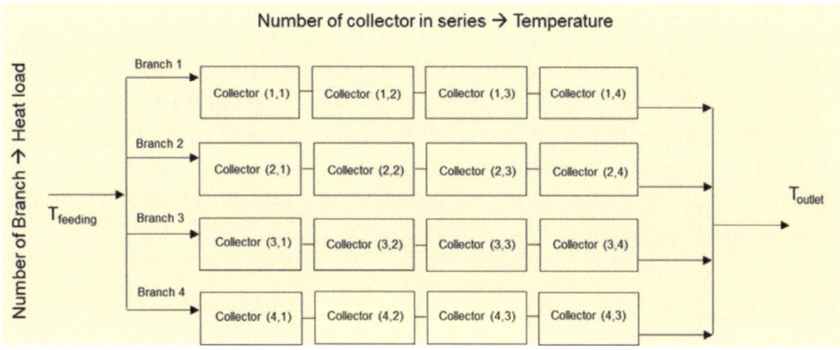

Figure 10.
Structure of a solar collector network with 16 units.

Several solar collectors, N_c, make up the structure of the solar collector network as show Eq. (7)

$$N_c = N_p N_s \tag{7}$$

N_s is the number of series collectors and N_p is the number of parallel branches.

In **Figure 10** the arrangement of a network of 16 collectors in a 4 x 4 arrangement (parallel series) is displayed. Generalising the arrangement for any collector network it can be denoted as m x n, where the lines in parallel to be placed (m) and the number of collectors connected in series (n) are shown. In this way, it is possible to meet the temperature level and the thermal load required for the process, of these, the first is achieved by connecting n collectors in series and the second is achieved through the determination of branches given by Eq. (6).

The absorber surface of the solar collector network is calculated from Eq. (8)

$$A_{SCN} = LWN_c \tag{8}$$

Where A_{SCN} is the area of the solar collector network (m^2), N_c is the number of collectors, L is the length (m), and W is the width of the solar collector (m).

Then we proceed to determine the cost of the collector network, C_{SCN} (USD), using Eq. (9) reported in [17]:

$$C_{SCN} = N_c\left[\gamma_0 + \frac{A_t N_t}{\pi}\left(\gamma_1 d + \gamma_2 + \frac{\gamma_3}{d}\right) + WL\gamma_4 + \gamma_{10}\frac{\dot{m}L\mu}{\pi\rho d^4}\right] + \gamma_5\left(\frac{\dot{m}H_b}{e_{ff}}\right) \tag{9}$$

Where N_c is the number of collectors, A_t is the lateral area of the tube, N_t is the number of tubes, d is the internal diameter of a tube, W and L are the width and length of a solar collector, H_b and *eff* are the load and pump efficiency, respectively. The γ_i terms are as follows: 6,768.82 (USD); 202,822.47 (USD/m^3); –1,576.96 (USD/m^2); 32,576 (USD/m); 994.1 (USD/m^2); 3.52 (USD h m^{-1} kg^{-1}); 0.14 (h^2/m^5); 0.45 (h/m^2); 1 dimensionless, 0.54 (m) and 261.61 (USD m h^2/kg), respectively.

For the estimation of the costs of the solar collector network, they can also be annualised, a 25-year amortisation period of the investment is considered with fixed interest of 5%, as shown in Eq. (10)

$$C_{TA\ SCN} = f_A C_{SCN} \tag{10}$$

Table 3 shows the arrangement area, and costs of the flat solar collector network where two scenarios are analysed. In scenario 1, it is considered that solar thermal energy supplies the entire thermal load required for the process at a temperature of 101 °C. In scenario two, a limitation of 50% is considered in the area available for the installation of the solar collector network in the plant. The first scenario would be the most desirable where it seeks to substitute the use of fossil fuels in a profitable way. In this scenario, it is also considered that there are no restrictions on available area or capital investment. Scenario 2 has a lower investment, however, there are still emissions to the environment.

2.5 Thermal storage system

The cost of the thermal storage system represents 30% of the cost of the solar thermal installation [17] and its operation is essential to store heat, also to dampen fluctuations in environmental variability, and increase the solar fraction of the process, when there is a gap between energy production and demand.

To determine the size of the thermal storage system, V_{TS} (m^3), we have Eq. (11) [18]:

$$V_{TS} = \frac{3600\, Q_{TS}\, t}{C_p \Delta T_{TS}\, \rho\, eff_{TS}} \tag{11}$$

In Eq. (11), Q_{TS} is the total heat load to be stored in the day (kW), t is the storage time of the system (h), C_p is the heat capacity of the working fluid (kJ/kg °C) and ρ is the density of the thermal fluid (kg/m^3), in this case water, ΔT_{TS} and eff_{TS} are the temperature variation of the thermal storage system (°C) and its efficiency (dimensionless).

The size, storage time and cost of the storage system is conditioned by the sizing of the collector network and the operation of the solar thermal device. Next, Eq. (12) used to determine the cost of storage, C_{TS} (USD) [18].

$$C_{TS} = 5800 + 1600\, {V_{TS}}^{0.7} \tag{12}$$

The cost of the thermal storage system is also annualised with an interest of 5% in a period of 25 years, this function is by Eq. (13) where the cost of the thermal storage system is multiplied by the annualisation factor, f_A.

$$C_{A\ TS} = f_A\, C_{TS} \tag{13}$$

Heat load to supply (kW)	Temperature level (°C)	Solar collector network array	A_{SCN} (m^2)	f	Supply period (h)	C_{SCN} (USD)	$C_{TA\ SCN}$ (USD/y)
339.81	101.20	23x29	1293.98	1	5	367,412.97	26,068.85
169.91	101.20	12x29	675.12	0.5	5	204,448.15	14,506.10

Table 3.
Results of the design of the solar collector network for different operating conditions.

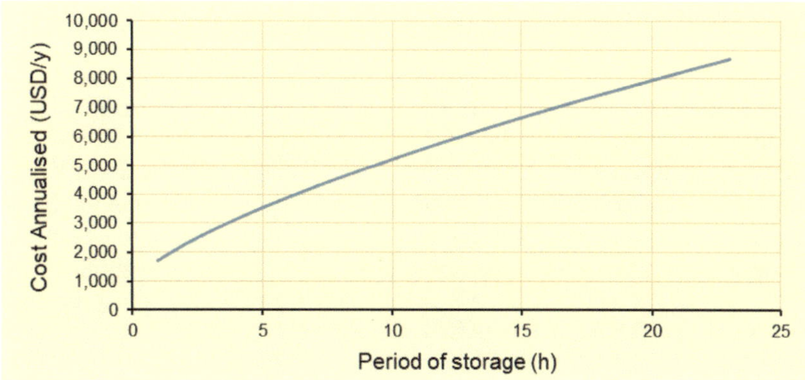

Figure 11.
Annualised cost of the thermal storage system as a function of accumulation time.

In **Figure 11** the annualised cost of the thermal storage system is shown as a function of the accumulation time, costs increase the longer the storage time. **Figure 11** shows the annualised cost of the thermal storage system for a load of 339,812 kW, which corresponds to the total heating service. If we assume that this thermal load is stored for 23 hours, the cost would be 8669.73 USD/year, compared to the cost of the solar collector network, this represents 32%. To define the operation and size of the storage system, the process conditions and the costs associated with them must be considered.

2.6 Total cost of the solar thermal installation

The total cost of the solar thermal installation (USD) includes: the total cost of the solar collector network and the total cost of the thermal storage system that represents around 80% of the total, the rest corresponds to the control and pumping system. The costs are also annualised at a fixed interest rate (5%) for a repayment period of 25 years and the costs are obtained per year for that period.

2.6.1 Cost evaluation

To evaluate the total cost of the integrated system, the costs of each of the components must first be updated: the heat recovery network, the solar field, and the thermal storage system. In this case the values of the equations are calculated for the year 2010. To obtain the updated cost values, the costs obtained must be multiplied by the value of the ratio that exists between the cost index for 2019 (I_C 2019) and the 2010 cost index (I_C 2010), which are 607.5 and 550.8, respectively [15] as seen in Eq. (14).

$$I_A = \frac{I_{C\ 2019}}{I_{C\ 2010}} \tag{14}$$

One method of comparing the magnitude of a dollar equity investment with a future stream of income is to convert the cost of capital into a future annual charge. The calculation is done with Eq. (15) [15], where i is the annual interest and n is the number of years of useful life of a piece of equipment or of a network of equipment or process in general. According to IRENA, solar plants have a useful life of 25 years.

$$f_A = C_T * \left[\frac{i * (1+i)^n}{(1+i)^n - 1}\right] \tag{15}$$

2.6.2 Environmental impact assessment

Currently, solar thermal technology is competitive against those that operate with fossil fuels, however, the negative environmental impact generated by these energy sources is global and tangible. One of the most well-known parameters is the quantification of CO_2 emissions as greenhouse gas, however, they are not the only gases emitted, in addition, there are other no less important parameters that are not considered. In this study, only the tons of CO_2 that are ceased to be emitted into the atmosphere are quantified when integrating solar thermal energy into the process. The factor of 0.203 kg CO_2/kWh is considered for fixed combustion equipment that works with natural gas as fuel, this data is reported for the European Union [19].

3. Analysis of results

The original energy consumption of the process is 880.202 kW for 5 h/day with a generation of CO_2 emissions of 312.7 tons/year (893.40 kg/day). The average price of natural gas reported worldwide for 2020 was 0.047 USD/kWh [20]. While the reported price of the same fuel in Mexico, for first month of 2019, was 0.3589 USD/kWh [21]. As can be seen, they present a significant difference that will directly impact the investment recovery time. It should be noted that the price of natural gas in Mexico has varied significantly from January 2019 (0.3589 USD/kWh) to December 2020 (0.024 USD/kWh).

When analysing the match between the energy requirement of the process and the availability of solar energy, it was determined that it is 2 h. However, there is availability of solar thermal energy at the target temperature for 1 h more that can be used as most convenient in the operation of the solar thermal device. To supply the entire thermal load and cover the remaining three hours, a storage tank of 32.58 m^3 is required.

In scenario 1 it is considered that there are no restrictions due to limitation of available area and capital investment. When carrying out the integration of energy, the requirement of the hot utility is reduced by 61.4%, the rest can be provided with solar thermal energy and the generation of greenhouse gases is zero. The total cost of the integrated system (heat recovery network and solar thermal installation) is 1,995,266.49 USD. If the investment is made for 25 years, the cost is 145,488.85 USD/year. The payback for the integrated system is 27.56 years (according to international data) and 3.6 years (Mexican data). When considering only the solar thermal installation, the payback is 5.99 years (according to the international estimate) and 0.78 years (according to the national evaluation).

In scenario 2 there is an available area restriction of 50% (650 m^2). The resulting network, depending on the available space, presents an arrangement of 11x29 with a thermal load of 181.45 kW that can supply 2 h of the thermal requirement of the process. The supply of the thermal load for those 2 h can be given directly from 11:00 h - 13:00 without storage or, it can be supplied from 8:00 h - 10:00 h with storage. Based on the economic aspect for scenario 2, the supply of the thermal load will be for 2 h without storage. The cost of the integrated system is 1,750,541.36 USD with a reduction of 12% compared to scenario 1. CO_2 emissions into the atmosphere are 60.36 tons/year. The payback of the integrated system is 24.17 years (according to the international price of gas) and 3.166 years (according to the national price of gas). The payback of the solar collector network is 2.61 years (according to the international gas price report) and 0.34 years (for the national gas price reported).

4. Chemical vapour deposition as a route to improve solar technology

Other solar technology that is heading the solar integration to industry in a big part of world, are the photovoltaic solar panels. Photovoltaics (PV) implies the direct conversion of sunlight into electricity by mean of semiconducting materials with a photovoltaic effect. Solar panels are widely used because its property of magnify the inlet micro-power, by the relatively constant production of electricity and by the possibility to use the stored electrical energy even in the absence of sunlight. Each solar panel is made up of a multitude of solar cells which manufacturing is, in this moment, in a high-tech period (third generation) of research (i. e. Dye-sensitised solar cells, DSSC, Perovskite solar cells, PVSC, Polymer hetero-junction solar cells, PSC, among others) [4]. Function, materials, characteristics, power-conversion-efficiency of solar cells are widely described in meticulous reviews and papers [4, 22–23]. The aim in this paragraph is to display the benefits that use of Chemical Vapour Deposition, CVD, has implied to improve solar cells performance and to present the novelties in evacuated solar tubes.

Among the most used thin film deposition processes to manufacture solar cells, are: evaporation, sputtering technique and chemical vapour deposition (CVD), with some variants in each technique. Briefly, they can be described as follows [24]:

Evaporation. The source material is evaporated in vacuum, this lets vapour particles to travel until the substrate, then, they condense to a solid state. Unfortunately, could occur that the different components of an alloy vaporise at different speeds, which will cause the composition of the deposited layer to be different from the original composition.

Sputtering deposition or Physical Vapour Deposition. The source materials are sputtered by the hitting of high energy ions in an oxidising atmosphere and deposited on a heated substrate, following the growth of thin films.

Chemical Vapour Deposition (CVD). The CVD technique consists of the reaction of a gas mixture inside a vacuum chamber followed by diffusion of reactants to a heated substrate to produce a material in the form of a thin layer. A useful variant is the reaction of metal–organic precursors (MO-CVD) because these ones improve the efficiency of solar cells.

Using CVD and PVD techniques give added value to solar cells with not too high costs, thanks to these techniques the efficiency of solar cells has increased from 10% in the 70ies to 20% today, since the different thin films that can be deposited perform various functions such as: antireflection, passivation layers, thickening of the absorbent layer, among others that have not yet been explored [24].

It is important to mention that exists an innovative report in literature about the use of CVD technique to deposite selective coating in evacuated solar tubes. The novel absorber layers have a base of carbon nanotube sheets that have showed their capability to converting solar radiation into electricity and heat [25], this is a promising result in increasing the efficiency of evacuated tubes.

5. Conclusions

The proposed methodology allows integrating solar thermal energy, in a profitable way, and replace, totally, the use of fossil fuels (scenario 1).

In the event of any restriction such as an available space of 50% respect scenario 1 (scenario 2), cost savings can be up to 12% during the two hours that the thermal load can be supplied directly, with a payback time of the solar device of 2.61 years, eliminating completely the use of thermal storage. The reduction of the requirement of hot utility was 80.62% being 19.40% by integration from solar energy.

Whenever solar energy is integrated, there is a reduction in greenhouse gas emissions, but when the first objective is to reduce GHG, there should be no limitation in the economic aspect to achieve the objective. Because current international prices of conventional fuels constitute a determining restriction to integrate solar heat to industrial processes. Therefore, economic speculation of conventional fuel prices constitutes a relevant challenge to be considered in the proposal and implementation of energy policies that really intend to encourage business models with renewables, and more specifically, with solar thermal energy.

Acknowledgements

The authors are very grateful to the University of Guanajuato and Trade Union Association, ASPAAUG, for the partial support granted for the publication costs of this chapter.

Conflict of interest

The authors declare no conflict of interest.

Nomenclature

A_{HRN}	Area of the heat recovery network, m^2.
A_{SCN}	Area of the solar collector network, m^2.
A_t	Lateral area of the tube, m^2.
C_{HRN}	Cost of the heat recovery network, USD.
C_{AS}	Cost of auxiliary services, USD/year.
C_{ATS}	Annualised cost of the thermal storage system, USD/year.
C_{CO}	Costs associated with cooling water, USD/kW year.
C_{SCN}	Cost of the collector network, USD.
C_{ST}	Costs associated with steam, USD/kW year.
$C_{TA\ HRN}$	Cost of the annualised heat recovery network, USD/year.
$C_{TA\ SCN}$	Annualised costs of the solar collector network, USD/year.
C_{TS}	Cost of the system storage thermal, USD.
Cp	Heat capacity of the working fluid, kJ/kg° C.
CP	Mass flow multiplied by the specific heat of the stream, kW/ °C.
d	Internal diameter of a tube of the solar collector, m.
eff	Pump efficiency, dimensionless.
eff_{TS}	Efficiency, dimensionless.
f_A	Annualisation factor, dimensionless.
h_i	Heat transfer coefficient of i-th stream, kW/m^{2o} C.
h_j	Heat transfer coefficient of j-th stream, in kW/m^{2o} C.
H_b	Load pump, m.
i	Annual interest, %.
$I_{C\ 2010}$	Cost index for 2010, dimensionless.
$I_{C\ 2019}$	Cost index for 2019, dimensionless.
L	Length of the solar collector, m.
$\dot{m}$	Mass flow rate, kg/s.
n	Number of years, y.
N_c	Number of solar collectors, dimensionless.
N_e	Number of equipment, dimensionless.

N_p	Number of branches or lines in parallel, dimensionless.
N_s	Number of series collectors, dimensionless.
N_t	Number of tubes, dimensionless.
q_i	Enthalpy change of i-th stream, kW.
q_j	Enthalpy change of j-th stream, kW.
Q	Total thermal load required by the process, kW.
Q_c	Minimum requirements for cooling the process, kW.
Q_h	Minimum requirements for heating, kW.
Q_i	Thermal load provided by a series of n collectors connected in series, kW.
Q_{TS}	Total heat load to be stored, kW.
t	Storage time of the system, h.
T_i	Temperature of the fluid at the entrance to the collector, °C.
T_{inlet}	Stream inlet temperature, °C.
T_o	Outlet temperature of the fluid from the collector, °C.
$T_o^{\ n}$	Outlet temperature of the n-th element, °C.
$T_o^{\ n-1}$	Outlet temperature of one minus to the n-th element, °C.
T_{outlet}	Stream outlet temperature, °C.
V_{TS}	Volume of the thermal storage system, m^3.
W	Width of the solar collector, m.
Greek symbols	
ΔT	Temperature difference, °C.
ΔT_{min}	Delta temperature minimum, °C.
ΔT_{ML}	Logarithmic mean temperature difference, °C.
ΔT_{TS}	Delta temperature variation of the thermal storage system, °C.
γ_0	Materials adjustment parameter, 6768.82 USD.
γ_1	Raisers adjustment parameter, 202,822.47 USD/m^3.
γ_2	Raisers diameter adjustment parameter, 1576.96 USD/m^2.
γ_3	Raisers diameter adjustment parameter, 32.576 USD/m.
γ_4	FPSC area adjustment parameter, 994.1 USD/m^2.
γ_5	Pumping costs adjustment parameter, 3.52 USD h/m kg.
γ_6	Pumping costs adjustment parameter, 0.14 h^2/m^5.
γ_7	Pumping costs adjustment parameter, 0.45 h/m^2.
γ_8	Pumping costs adjustment parameter, 1.00 dimensionless.
γ_9	Pumping costs adjustment parameter, 0.54 m.
γ_{10}	Pumping costs adjustment parameter, 261.61 USD m h^2/kg.
ρ	Density of the thermal fluid, kg/m^3.

Author details

Guillermo Martínez-Rodríguez* and Amanda L. Fuentes-Silva
Department of Chemical Engineering, University of Guanajuato, Guanajuato, Mexico

*Address all correspondence to: guimarod@ugto.mx

References

[1] World Energy Outlook 2016, OECD/IEA [Internet]. 2016. Available from: https://www.iea.org/reports/world-energy-outlook-2016 [Accessed: 2020-12-01]

[2] Schoeneberger C, McMillan CA, Kurup P, Akar S, Margolis R, Masanet E. Solar for industrial process heat: A review of technologies, analysis approaches, and potential applications in the United States. Energy. 2020; 206:118083. DOI: 10.1016/j.energy.2020.118083

[3] International Renewable Energy Agency (IRENA), calculations made by Deger Saygin based on the source International Energy Agency (IEA), World Energy Statistics 2018 [Internet]. 2018. Available from: https://www.irena.org/-/media/Files/IRENA/Agency/Publication/2018/Jul/IRENA_Renewable_Energy_Statistics_2018.pdf [Accessed: 2020-12-01]

[4] Babar F, Mehmood U, Asghar H, Mehdi M H, Khan A U H, Khalid H, Huda N U, Fatima Z. Nanostructured photoanode materials and their deposition methods for efficient and economical third generation dye-sensitized solar cells: A comprehensive review. Renewable and Sustainable Energy Reviews. 2020; 129: 109919. DOI: 10.1016/j.rser.2020.109919

[5] Baniassadi A, Momen M, Amidpour M, Pourali O. Modeling, and design of solar heat integration in process industries with heat storage. Journal of Cleaner Production. 2018; 170: 522–534. DOI: 10.1016/j.jclepro.2017.09.183

[6] Oosthuizen D, Jurgens Goosen N, Hess S. Solar thermal process heat in fishmeal production: prospects for two South African fishmeal factories. Journal of Cleaner Production. 2020; 253: 119818. DOI: 10.1016/j.jclepro.2019.119818

[7] Jia T, Huan J, Li R, He P, Dai Y. Status, and prospect of solar heat for industrial processes in China. Renewable and Sustainable Energy Reviews. 2018; 90: 475–489. DOI: / 10.1016/j.rser.2018.03.077

[8] Fuentes-Silva AL, Velázquez-Torres D, Picón-Núñez M, Martínez-Rodríguez G. Solar Thermal Integration With and Without Energy Storage: The Cases of Bioethanol and a Dairy Plant. Chemical Engineering Transactions. 2020; 81: 493–498. DOI:10.3303/CET2081083.

[9] Denholm P, Brinkman G, Mai T. How low can you go? The importance of quantifying minimum levels for renewable integration generation. Energy Policy. 2018; 115: 249–257. DOI: 10.1016/j.enpol.2018.01.023

[10] Eiholzer T, Olsen D, Hoffman D. Integration of a solar thermal system in a medium-sized brewery using pinch analysis: methodology and case study. Applied Thermal Engineering. 2017; 113: 1558–1568. http://dx.doi.org/10.1016/j.applthermaleng.2016.09.124.

[11] Suresh N, Rao BS. Solar energy for process heating: a case study of select Indian industries. Journal of Cleaner Production. 2017. 151: 439–451. DOI: 10.1016/j.jclepro.2017.02.190

[12] Quijera JA, González-Alriols M, Labidi J. Integration of a solar thermal system in a dairy process. Renewable Energy. 2011; 36: 1843–1853. DOI: 10.1016/j.renene.2010.11.029.

[13] Linnhoff, B. Linnhoff March: Introduction to Pinch Technology [Internet]. 1998. Available from: https://www.ou.edu/class/che-design/a-design/Introduction%20to%20Pinch%20Technology-LinhoffMarch.pdf [Accessed: 2020-12-05

[14] Smith R. Chemical Process Design and Integration. In John Wiley & Sons, Ltd. 2005.

[15] Towler G, Sinnott R. Chemical Engineering Design. 2nd ed. Oxford: Butterworth-Heinemann/Elsevier; 2013. 1320 p. DOI: 10.1016/C2009-0-61216-2

[16] Martínez-Rodríguez G, Fuentes-Silva AL, Lizárraga-Morazán JR, Picón-Núñez M. Incorporating the concept of flexible operation in the design of solar collector fields for industrial applications. Energies. 2019; 12: 570: 1–20. DOI: 10.3390/en12030570.

[17] Karagiorgias M, Botzios A, Tsoutsos T. Industrial solar thermal applications in Greece economic evaluation, quality requirements and case studies. Renewable and Sustainable Energy Reviews 2001; 5: 157–173. DOI: 10.1016/S1364-0321(00)00012-5.

[18] Yang P, Liu LL, Du J, Li JL, Meng QW. Heat exchanger network synthesis for batch processes by involving heat storages with cost targets. Applied Thermal Engineering. 2014; 70, 1276–1128. DOI: 10.1016/j.applthermaleng.2014.05.041

[19] Emission factors carbon footprint registration, offsetting, and carbon dioxide absorption projects version 12nd, Spanish Office for Climate Change (OECC) [Internet]. 2019. Available from: https://www.miteco.gob.es/es/cambio-climatico/temas/mitigacion-politicas-y-medidas/factores_emision_tcm30-479095.pdf [Accessed: 2020-12-01]

[20] Global Petrol Prices [Internet]. 2020. Available from: https://es.globalpetrolprices.com/Mexico/natural_gas_prices/ [Accessed: 2021-02-12]

[21] Comisión Reguladora de Energía (CRE) [Internet]. 2021. Available from: https://www.cre.gob.mx//IPGN/index.html [Accessed: 2021-02-12]

[22] Tao F, Green M, Valenzuela-Garcia A, Xiao T, Van-Tran A T, Zhang Y∗, Yin Y, Chen X. Recent progress of nanostructured interfacial solar vapor generators. Applied Materials Today. 2019; 17: 45–84. DOI: 10.1016/j.apmt.2019.07.0112352-9407

[23] Dapkus P D,*, Chi C Y, Choi S J, Chu H J, Dreiske M, Li R, Lin Y, Nakajima Y, Ren D, Stevenson R, Yao M, Yeh T W, Zhao H. Selective area epitaxy by metalorganic chemical vapor deposition–a tool for photonic and novel nanostructure integration. Progress in Quantum Electronics. DOI: 10.1016/j.pquantelec.2020.100304

[24] Hoffmann W, Pellkofer T. Thin films in photovoltaics: Technologies and perspectives. Thin Solid Films. 2012; 520: 4094–4100. DOI: 10.1016/j.tsf.2011.04.146

[25] Sobhansarbandi S, Martinez P M, Papadimitratos A, Zakhidov A, Hassanipour F. Evacuated tube solar collector with multifunctional absorber layers. Solar Energy. 2017;146: 342–350. DOI: 10.1016/j.solener.2017.02.038